S0-AVX-627

CARTOGRAPHIC SCIENCE

A Compendium of Map Projections, with Derivations

Donald Fenna

CRC Press
Taylor & Francis Group
Boca Raton London New York

CRC Press is an imprint of the
Taylor & Francis Group, an informa business

CRC Press
Taylor & Francis Group
6000 Broken Sound Parkway NW, Suite 300
Boca Raton, FL 33487-2742

Printed in the United States of America on acid-free paper
10 9 8 7 6 5 4 3 2 1

International Standard Book Number-10: 0-8493-8169-X (Hardcover)
International Standard Book Number-13: 978-0-8493-8169-0 (Hardcover)

Library of Congress Cataloging-in-Publication Data

Fenna, Donald.
Cartographic science : a compendium of map projections, with derivations / Donald Fenna.
p. cm.
Includes bibliographical references and index.
ISBN 0-8493-8169-X (alk. paper)
1. Map projection. 2. Cartography. I. Title.

GA110.F46 2006
526'.8--dc22 2006047546

Visit the Taylor & Francis Web site at
http://www.taylorandfrancis.com

and the CRC Press Web site at
http://www.crcpress.com

Dedication

To my late parents, George and Nellie Fenna, who, despite their very modest circumstances, always placed learning second only to survival,

and

to John P. Snyder, whose books provided a key base for this work, and for which this work is hopefully an acceptable explanatory companion.

Donald Fenna

Dr. Donald Fenna, Professor Emeritus, Faculty of Medicine, University of Alberta, is the author of numerous journal papers covering mathematics, computing science, and medicine and of

Elsevier's Encyclopedic Dictionary of Measures [1998]

and

Oxford Dictionary of Weights, Measures, and Units [2002].

WRE DIGITAL COMPUTER

PROGRAMME SHEET

TITLE : Sine and cosine of two angles. REFERENCE L.4.1.8.
DATE 8.1.58

Uses β, δ; 3 odd and 7.

Preset Parameters : Y ÷ IF Ø
B ÷ CH b α, first of two locations holding angles.
D = CH d α, first of four locations holding results.
G and H such that CHG α CHH α is the pseudo order for 360°.

NO.	ORDERS EVEN	ORDERS ODD	NOTES
0	CO 15 Aα	CO 3 α	
1	CA 16 Aα	CE 3 α	←15 (δ)=-4, -3, -1.
2	SM 26 Aα	SO 3 α	
3	NE 5 Aα	CA - Bα	If $(3)_0=1$, put $(3)_0=0$
4	SW 1 Bα	CL - Bα	If $(3)_0=0$, swap (B) and (1B), and put $(3)_0=1$
3→ 5	AE 16 Aα	CO 3 α	
6	LD 1 Bα	SL 2 α	
7	SO 3 α	SR 1 α	Form(7)=y= $\frac{x}{\pi}-\frac{1}{2}$ for sin x
8	CL 7 α	SM 7 α	= $\frac{x}{\pi}$ for cos x
9	MA 7 α	CL 7 α	
10	SM 7 α	SB 504 α	
12→11	CL 7 α	MA 7 α	
12	AD 25 Aβ	JE 11 Aα	
13	SL 2 α	CL 4 Dδ	(Acc)= cos y CLD, CL1D,... CL3D.
14	AE 3 α	AE 5 α	
15	NO 1 Aα	(CH α)	LINK
16	LD α	SE 508 α	
17	CH 501 Y	CH α	a_8
18	NO 427 α	SE 511 Yδ	a_7
19	CO 292 α	CH 1 Yδ	a_6
20	CL 12 Y	SE 485 Y	a_5
21	CE 237 Y	CH 240 Yδ	a_4
22	CA 453 Y	SO 168 YY	a_3
23	SE 16 α	LD 60 α	a_2
24	PE 108 Y	CT 66 δ	a_1
25	CH β	PO α	a_0
26	CH Gα	CH Hα	2π in user's scale.

Frontispiece: A sample from the author's program that tracked Sputnik in southern latitudes. The program was written in 1957 for the Elliott Brothers' 403 computer operating as WREDAC at the Weapons Research Establishment, Salisbury, South Australia. The sample is from a general trigonometrical sub-routine used in several programs. It is reproduced from the pertinent program, published in 1958 as the Department of Supply, Weapons Research Establishment, Tech. Memo. TRD 26. WREDAC Programme T 1.5 General Kinetheodolite Calculation by D. Fenna Ph.D. See page 441 for reference.

Preface

Cartography is the process of depicting a curved world on a flat surface. It is inherently a frustrating task, since it can be achieved only at the expense of discarding some desirable attributes; for instance correct shape and correct area cannot co-exist. The beginning of the word *cartography*, along with the word *chart* as used for maritime maps, appears to come to us through Latin from the Ancient Egyptian word for *cut* adverting to chopping up the leaf of papyrus in the process of paper making [1931a][†]. The latter part of *cartography*, like many other word parts in the field, derives from Greek[&] while the word *map* is again from Latin, *mappa* meaning a cloth sheet of any form.

Both art and science are used in the production of a map. For the simple sketch map that guides us down a few streets and through a few turns, little science is required. For the map that guides a missile to its target, little art is required. The typical map, however, has a substantial component of each – the science to give it accuracy and the art to give it readability and aesthetic enhancement. The use of colour alone is important to the readability of most maps, while the hatching of slopes represents an effort through art to avoid the 2-dimensional limitations of a flat piece of paper. This text deals purely with the science of map making – essentially with the transformation of a curved surface onto a planar piece of paper, in black and white. It covers transformations that can be seen as a literal projecting through to numerous others that can be expressed only in complicated mathematical form.

Cladius Ptolemy of Alexandria wrote the first known book on cartography, and his *Geographia* of circa 150C.E. was an amazing compilation of locational data as well as maps. Most of the data, as well as many of the maps, were undoubtedly of older provenance. His near-contemporary Marinus of Tyre was the admitted author of some of those maps. Ptolemy, however, pioneered the use of sophisticated map projections and established the practice of basing the content of maps routinely on longitude and latitude [2000a]. Some projections in use today appear in Ptolemy's work, but most have been developed since and new ones keep emerging. The latest serves the special needs

---

[†] Four digits followed by a lower-case letter is a reference to the chronological bibliography in Appendix E.
[&] See Appendix A for elaboration of the impact of Greek words in cartography.

of seeing Earth from artificial satellites and the challenges of mapping solar bodies far less regular than our own planet. All this is elegantly surveyed by John P. Snyder in his essentially literary *Flattening the Earth* [1993a]. In the separate, austere publication *An Album of Map Projections* [1994a], Snyder and Philip M. Voxland also show a large selection of projections along with the mathematical plotting formulae for many, but without an intervening explanation of the development of those formulae or any discussion of cartography. It is hoped that this text will fill the gap between those two notable works, thereby a companion to the *Album* and a bridge to there from Snyder's literary work.

There are many existing texts that do explain the development of such formulae and of the various approaches to projecting, but characteristically they cover only the more familiar projections. A few range more widely, and beyond simple trigonometry more deeply into mathematics, many aspects of which are involved in the overall repertoire of projections. Works by Frederick Pearson [1994a] and by Lev Bugayevskiy and Snyder [1995a] are notable examples, but both are challenging in their presentation to the typical cartographer and neither covers a large number of projections.

This text endeavours to provide explanation of the mathematical development for a large range of projections within a framework of different cartographic methodologies, all with a gentle and progressive immersion in the mathematics involved. To help in this cause, the book typically exploits any one facet of mathematics as much as possible before introducing another, and supports such steps with intermittent tutorials. The mathematical development may sometimes appear laboured to a mathematician, but the needs of readers with less mathematical background are given priority. Even the writing of equations has been addressed with concern for the reader less adept with them, by merely increasing their spaciousness, for instance. This matter is discussed further at the close of Chapter 1.

Mathematics is a highly symbolic language, and its extensive and deep involvement in this cartographic work brings a demand for a great variety of symbols. Typically the usage fits with standard practice but that applies essentially only to general mathematical practice, however, the notations in cartography being extremely variable. Trying to make the mathematics as clear and easy to follow as possible, various standards have been adopted for consistent use throughout the work, as indicated below. The deepening intricacies of the cartographical mathematics, however, demand much adornment of the letters used to represent variables. The adopted standards are defined as needed; all are collected and shown fully in Appendix B.

One practice entrenched without exception is that letters representing angles imply radians when in lower case, degrees when in upper case. All such letters are

Greek. Examples that occur only in lower case include α, β, γ, δ, η, ν, ξ, υ, χ, ψ and ω. In addition,

θ is the standard for angle in 2-dimensional polar co-ordinates,

λ and Λ always refer to longitude, and

ϕ and Φ always refer to latitude.

The λ sometimes has an underline to stress when it is not the usual Greenwich-based angle. Because longitude is inherently arbitrarily based, and because it appears in so many equations differenced from a central meridian, λ is often used unadorned to represent longitude relative to that central meridian; that fact is intermittently stressed in words. The letter ψ is always used for a function dependent solely on latitude, as is the symbol $\overline{\varphi}$, which denotes specifically isometric latitude (see Chapter 12).

Other Greek letters given standard meaning as angular variables are:

δ for arc angle, hence great-circle distance on a sphere of unit radius

ζ for azimuth angle

χ for rotational angle

The letter γ is used mostly for the angle between a central line and a general projection line at the centre of the globe. The letter ξ is used mostly for the same at any other point.

Other standard usages of Greek letters for variables are:

ε for the eccentricity of an ellipse

ρ for the radial distance in polar co-ordinates

σ for the factor that rescales a map to have true scale repositioned

Variables are typically written in italics, but upper-case Greek letters are among the exceptions. Other exceptions are vectors and matrices, both of which are always upright bold Roman letters.

The Greek letter π is used in upright form for the number so familiar with circles and radians. The Roman letter e is also used in upright form to mean the special number 2.7182~ used among other places for natural logarithms (leaving the italic version of the letter available to be a variable). (The tilde symbol as just used indicates that the number is incomplete i.e., truncated. The digits shown for such numbers are rounded.) The Roman letter i is used upright for the imaginary entity $\sqrt{-1}$.

The lower-case duo x, y is used in the standard way for the two co-ordinates of 2-dimensional space. They are used unadorned for mathematics of a general nature but have dotted central superscripts (*overscripts*) when used for map variables, for which they are the routine final-stage plotting variables. As such they are then denoted as $\dot{x}, \dot{y}$, but if such a map is an intermediate step before a further transformation to a revised map, they are $\ddot{x}, \ddot{y}$ and, where necessary, $\dddot{x}, \dddot{y}$ for maps successively derived mathematically from the first. Associated intermediate variables are likewise adorned.

The upper-case trio X, Y, Z is used for 3-dimensional space, oriented in the usual mathematical manner. When relating to its use in generating the 3-dimensional ellipsoid, the 2-dimensional ellipse is plotted with Y and Z axes as they occur 3-dimensionally.

Consistent with common mathematical practice, the symbol $\boldsymbol{z}$ is used for complex numbers, with real numbers x and y as its constituent parts, viz $\boldsymbol{z} = x + \mathrm{i}y$. The equivalent involving longitude and latitude uses a Greek symbol in the form $\boldsymbol{\mu} = \lambda + \mathrm{i}\phi$. In this book complex numbers are shown in bold italic style. Derivatives in calculus terms are written usually in full form but sometimes with a bracketed right-superscript containing an Indo-Arabic numeral (e.g. $f^{(1)}, f^{(2)}, f^{(3)}$).

Since most projections are used for small-scale maps of the whole world and of large regions, most notably within atlases, for which the detailed shape of Earth is sheer pedantry, most of the book deals with mapping a sphere. The ellipsoidal† shape is only taken into account in the last few chapters. While this involves minor repetition for those projections used in the more precise way, it avoids unnecessary complication through much of the initial explanation.

This book is divided into four main parts:

A The Curved World – two chapters of introductory material
B A Spherical World – the main content spanning fourteen chapters
C An Ellipsoidal World – two chapters
D The Real World – the final chapter.

Except for some brief footnotes and a few places where extended mathematical development is removed to the end of a chapter as a lengthy endnote, reference material occurs in four forms:

- bibliographic material as already mentioned
- equations
- tutorials
- diagrams and maps

The first has already been described as using a four-digit number (being the year of publication) followed by a lower-case letter. Equations are labelled and referred to via the page number of 1 to 3 digits followed by a lower-case sequencing letter; references to them use the styles "Equation 876e", "Equations 876e & 724b" and "Equations 765p" where the label applies to a pair of equations. Maps and other illustrations have the generic title "Exhibit" and are numbered sequentially within each chapter, e.g.

† The mathematically unambiguous word *ellipsoid* is preferred to the perhaps more frequent *spheroid* for the figure obtained by revolving an ellipse. More precisely the figure used is in *ellipsoid of revolution* and, being revolved about its minor axis, an *oblate* ellipsoid.

Exhibit 10–7, but reference to them is usually by page. Tabulations of numbers are not distinguished by any rubric; they usually sit continuous with associated text, and referenced by page number when otherwise. The tutorials interspersed through the work are sequentially numbered, but sometimes referenced also by page. Most are of two-column format, so readily distinguished; all have boxed headers to break the reader's attention. The overall arrangements are designed for maximal efficiency in accessing referenced material.

While sometimes adjacent to referring text, most exhibits and tutorials are placed prior to their need, on a left-hand page to facilitate cross-reference attention. Many of each type are placed at points that fit with pagination preferences.

The published work was created on a Macintosh computer and produced in camera-ready form on a Lexmark E322 printer. It was written directly into Microsoft Word using related software including Equation Editor. Illustrations were prepared and embedded using Adobe Freehand 8 and maps were created using Geocart [™ Daniel R. Strebe). The Palatino typeface is used for general text, the Times typeface for mathematical expressions, including mathematical elements within ordinary sentences.

Among English usages that may prove unusual to some readers is the strict use of capitals on proper nouns (e.g., Earth, North Pole) and the distinctive *else* in place of the ambiguous *or* for exclusive circumstances (leaving *or* to relate synonyms).

The completion of this text was assisted by many librarians in the University of Alberta Library and others in the municipal library of the Town of Canmore. The author wishes to put on record his appreciation of all, particularly David Jones of the former and the inter-library loan staff at the latter, who happily and successfully wandered far from their routine paths. The quality of the finished work has been greatly enhanced by the diligent attention of my brother Dr. Douglas Fenna of England to selected sections and the overall review of Dr. Douglas Hube of Edmonton. Others that have provided assistance include Dr. Waldo Tobler of California, Dr. Tom Overton of Victoria (British Columbia), Daniel 'daan' Strebe of Seattle, Dr. Ivar Ekeland of Vancouver, and Dr. Robert Buck, Robert Morse, MeiMei Chung and Leni Honsaker of Edmonton. The author also appreciates the permission readily granted by the Commonwealth of Australia for the illustration used as frontispiece and by the American Geographical Society for the use of a reproduction from the *Geographical Review* as Exhibit 15–11. The three maps in Chapter 19 were derived, with appreciation, from illustrations in the U.S. government publication *An Album of Map Projections* mentioned earlier. Finally, the author wishes to renew the appreciation due to his wife Sydney for her wide support of his writing endeavours.

Readers are welcome to contact the author via email to don.fenna@ualberta.ca.

Contents

Tutorials

PART A

The Curved World

The ship vanishing below the horizon has demonstrated an apparent curvature to Earth for millennia. That Earth is round appears to have been realized thousands of years ago too, perhaps by the most ancient of known peoples; reasonably accurate measurement of its circumference was made at least two millennia ago. Yet at various times and places, not only presumably in the earliest of human days but even into relatively recent centuries, many believed that Earth was flat. That Earth was curved and indeed roughly spherical was proved by the completion of the famous Magellan voyage of the 1520s. The surface of Earth is undulating — indeed precipitously so in places — but, for the purposes of map projecting, we need a smooth surface devoid of mountains, etc., (without precluding those mountains appearing, with their elevations, on the final maps). Traditionally this has been seen as Mean Sea Level (MSL) — the conceptual surface were the seas to cover the whole planet. Map projecting involves the transformation of points from the 3-dimensional curved world to flat paper in some unambiguous way. In the two chapters of this opening part we first enquire into the shape and size of Earth and into schemes for uniquely identifying points thereon. We then investigate the means for relating the curved surface to a flat surface, developing plotting equations for two simple map projections, and address the matter of scaling.

TUTORIAL 1 2-DIMENSIONAL CO-ORDINATES

The familiar graph on paper uses a 'horizontal' axis termed the x axis, with values increasing to the right, and a 'vertical' (i.e., within the 2-dimensional context) axis termed the y axis, with values increasing upwards. The two lines usually intersect at their zero points, known as the *origin* and commonly labelled O.

Any point in the plane of the paper can be identified uniquely by its x and y values — the two conventionally being written in that order, within parentheses, and separated by a comma. The generic point, written (x, y), is usually labelled P.

The value of x is referred to as the x co-ordinate, the value of y as the y co-ordinate. Together these are called *Cartesian co-ordinates,* after their creator, Rene Descartes of France (1596 – 1650).

If the axes are *orthogonal* (i.e., at right angles to each other), the system is said to be rectangular or *rectilinear,* which is usual and assumed unless stated to be otherwise.

The two orthogonal axes divide the whole plane into four quadrants, usually accorded ordinal numbering applied counter-clockwise, with the first quadrant being that with both x and y positive; the second that with x negative, y positive; the third having both co-ordinates negative; the fourth that with x positive, y negative.

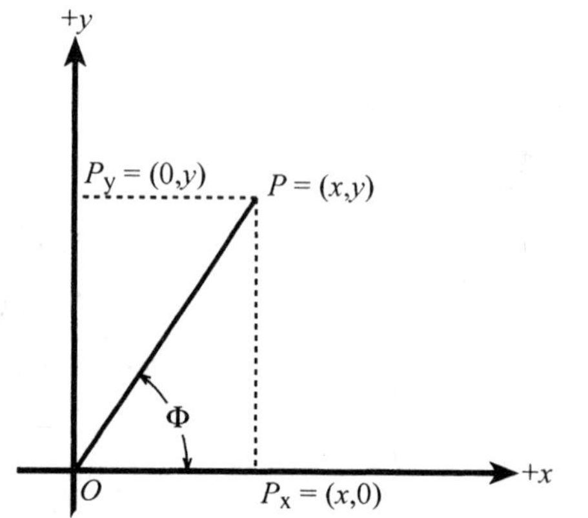

Seeing Θ as the counter-clockwise angle from the x axis in the above illustration, we define the trigonometric functions sine (sin) and cosine (cos) as

$$\sin\Theta = \frac{y}{\rho}, \quad \cos\Theta = \frac{x}{\rho}$$

both being positive and ≤ 1. Set in the first quadrant, these definitions of sin and cos are restricted to angles between 0° and 90°, but they are readily extendable. With x negative in the second and third quadrants and y negative in the third and fourth, both functions change sign accordingly. Whichever the quadrant, we have

$$P = (\rho\cos\Theta, \rho\sin\Theta)$$

The variables ρ and Θ are referred to as the *polar co-ordinates.*

If $r > 0$, the set of points such that

$$\left(\frac{\rho}{r}\right)^2 = \left(\frac{x}{r}\right)^2 + \left(\frac{y}{r}\right)^2 = 1$$

forms the circle centred on O of radius r.

CHAPTER 1

The Task:
Curved Earth to flat map

Preamble

Mankind has long endeavoured to depict the world, in miniature, on a flat surface. With the development of satellites and electronics during the closing stages of the 20th Century, our knowledge of the shape and dimensions of our planet has risen enormously. Such knowledge has only compounded the intricate variations from ellipsoidal of the perceived surface. Were Earth perfectly spherical, however, projection onto flat paper would present intrinsic challenges; its elaboration merely compounds the mathematics. In this chapter we examine the shape and size of Earth and devise schemes for identifying location on Earth and maps before developing the simplest of projections from curved Earth to flat map.

Mathematically, elementary geometry and trigonometry, including 3-dimensional co-ordinates, suffice for this chapter (though more advanced tutorials are interleaved in it).

The two maps in this chapter are both from the most-elementary projection.

TUTORIAL 2 TRIGONOMETRIC FUNCTIONS

The two basic trigonometric functions sin and cos were defined in Tutorial 1. The derived functions tangent (tan), cosecant (csc), secant (sec) and cotangent (cot) complete the basic set. In full, using the illustration of the first quadrant and extending beyond as discussed in the Tutorial 1, they are, for angles expressed in degrees,

$$\sin\Theta = \frac{y}{\rho} = \frac{1}{\csc\Theta} = \cos(90-\Theta)$$

$$\cos\Theta = \frac{x}{\rho} = \frac{1}{\sec\Theta} = \sin(90-\Theta)$$

$$\tan\Theta = \frac{y}{x} = \frac{\sin\Theta}{\cos\Theta} = \cot(90-\Theta)$$

$$\csc\Theta = \frac{\rho}{y} = \frac{1}{\sin\Theta} = \sec(90-\Theta)$$

$$\sec\Theta = \frac{\rho}{x} = \frac{1}{\cos\Theta} = \csc(90-\Theta)$$

$$\cot\Theta = \frac{x}{y} = \frac{\cos\Theta}{\sin\Theta} = \tan(90-\Theta)$$

In the first quadrant all functions have positive values but all other than sin and cos are unbounded in size. The functions sin, tan, and sec increase continually as the angle increases from zero to 90°, with sin going from 0 to 1, tan from 0 to ∞, and sec from 1 to ∞. The functions cos, cot, and csc decrease continually as the angle increases from zero to 90°, with cos going from 1 to 0, cot from ∞ to 0, and csc from ∞ to 1. The range and signage for all four quadrants are shown in the accompanying table.

Writing $\sin^2\Theta$ for $(\sin\Theta)^2$ etc., Pythagoras shows that for any angle Θ

$$\sin^2\Theta + \cos^2\Theta = 1$$

$$1 + \cot^2\Theta = \csc^2\Theta$$

$$\tan^2\Theta + 1 = \sec^2\Theta$$

While degrees are the most familiar unit for measuring angles — and widely used in cartography — the alternative unit radian (rad) is significantly more convenient for developing the related mathematics. Defined as *the angle subtended by a section of circumference equal in length to the radius,* the radian is thus such that

$$2\pi\,\text{rad} = 360°, \pi\,\text{rad} = 180°, \tfrac{\pi}{2}\,\text{rad} = 90°$$

$$1\,\text{rad} = \frac{180°}{\pi} = 57.295\,78_{\sim}° = 57°17'44.8_{\sim}''$$

We routinely use lower-case letters for angles expressed in radians.

quadrant:	*1st*	*2nd*	*3rd*	*4th*
sin	0 to +1	+1 to 0	0 to −1	−1 to 0
cos	+1 to 0	0 to −1	−1 to 0	0 to +1
tan	0 to +∞	−∞ to 0	0 to +∞	−∞ to 0
csc	+∞ to +1	+1 to +∞	−∞ to −1	−1 to −∞
sec	+1 to +∞	−∞ to −1	−1 to −∞	+∞ to +1
cot	+∞ to 0	0 to −∞	+∞ to 0	0 to −∞

BASIC SHAPE AND SIZE OF EARTH

A round world

Earth is nearly spherical, with a circumference of about 40 000 km[†] (nearly 25 000 miles[‡]) and an average radius of about 6371 km (nearly 4000 miles). It rotates continuously, one rotation taking close to 23 hours and 56 minutes (on the familiar clock) relative to the fixed stars — the *sidereal day* (the 24 hours of the *solar day* accommodate the changing orientation to the Sun as Earth travels its orbit about that key body).

The surface end points of the axis of rotation are referred to as the *poles* (specifically as the *geographic poles*), the two distinguished as the *North Pole* and the *South Pole*. The line of surface points equidistant from the Poles is the *Equator*. As a relic of an earlier epoch when Earth was largely molten from centre to surface, the sphere is distorted by the centrifugal effect of its daily rotation (spin) — the diameter along the axis of rotation is over 40 km (26 miles) less than any transverse diameter. The Moon and various other planetary bodies are very much more distorted from spherical than Earth. In contrast, the endless "heavens" are depicted on star maps as though on a precise sphere — termed the *celestial sphere*.

Earth and other astronomical bodies have irregular surfaces, with mountains, valleys and lakes, giving the land a detailed relief. In addition, Earth (at least) has seas and oceans, then submarine relief. Maps are used to show both surface and submarine relief, and are used to depict even underlying geology. From the point of view of this text, however, a nominal smooth surface is assumed. For Earth, this is traditionally *Mean Sea Level* (*MSL*) — nominally the surface level of the open oceans, extended horizontally through the land areas; because of the spin, this is essentially ellipsoidal rather than spherical. Originally, and still essentially, MSL is established at a coastal place by very frequent observations (e.g. hourly) from a stable platform over many years — enough to mask the effects of random waves and regular tides, variations in saltiness of the water and atmospheric air pressure, and even multi-year phenomena like El Niño. Currents and protracted winds, however, can move water against some coasts and away from others persistently; the routine discrepancy of local mean level around North America is about 50 cm — a phenomenon directly measurable by the string of water surfaces of the Panama Canal. The level used over any region is therefore defined by averaging a variety of such evaluations in a sophisticated way — hence rarely identical to the actual mean level observed at any one coastal station.

[†] The metre was original defined as one ten-millionth the distance from the Equator to the North Pole [2002a].

[‡] The mile originated in the geographical or nautical form, making the circumference of Earth about 360 x 60 = 21 600 miles, but Roman then British machinations changed its common meaning [1998a].

Relating the observed levels at different sites is a matter of *geodesy* (< Greek *Earth* + *divide*), which term covers the determination of Earth's size and shape along with the land surveying. MSL is extrapolated through the land area as the *level surface* — *level* meaning that everywhere the surface is orthogonal to the plumb line, which points along a line perpendicular to the ellipsoidal surface rather than towards Earth's centre. (Being not tied to Earth, a plumb bob has its own minor spin-induced deviation from that line.)

A level water surface is implicitly a surface of uniform gravity (i.e., of gravitational attraction less the centrifugal effect of rotation), but MSL is defined mathematically on the basis of gravity potential rather than gravity itself, as the equipotential surface with the value applying at the agreed sea level. Because of vagaries in Earth's crust, the equipotential surface varies from being precisely ellipsoidal. Called more correctly the *geoid*, its details are discussed in Chapter 19.

The geoid is the datum from which elevations (including depth or negative elevation) and altitudes are normally measured. However, for map projecting a surface amenable to mathematical operations is necessary. A sphere provides adequately for the maps of atlases and for many others; for more accurate mapping an ellipsoidal surface is markedly preferable. As described in the Preface, this book develops most map projections on the basis of a true sphere before stepping later (in Chapter 17) into the ellipsoidal. But, whichever the choice of shape, elevations on any final map are relative to the geoid. (Civil engineering maps are often based on a plane surface rather than a curved level surface. Even over a span of 100 miles or 160 km chordal and surface distances differ by less than 0.005%. Until recently, U.S. authorities used maps built on a mosaic of 123 *State Planes*, fitted within state boundaries defined by geodesic surveying.)

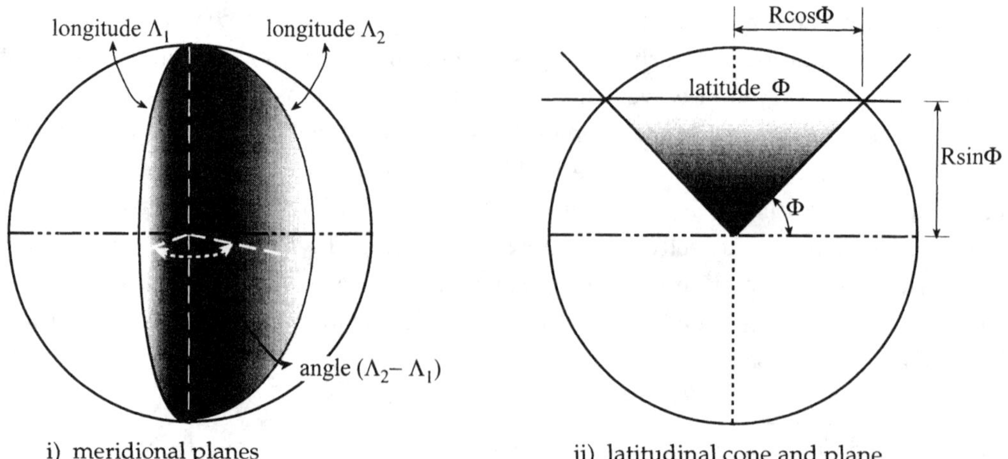

i) meridional planes

ii) latitudinal cone and plane

Exhibit 1–1: Earth – longitude and latitude.

LONGITUDE AND LATITUDE

Concept

The Equator is one example of a *great circle* — the intersection with Earth's surface of any plane that passes also through the centre of Earth (any circle that is not a great circle is termed a *lesser circle*). The Equatorial plane divides Earth into two hemispheres called the Northern Hemisphere and the Southern Hemisphere — terms usually implicitly meaning the surface rather than the solid.

The standard means for expressing the position of any point on Earth's nominal surface is by a pair of angles called longitude and latitude. The former represents the distance around Earth from some given datum line and the latter represents the distance from the Equator towards a Pole.

Longitude

If we take two great circles that each pass through the Poles (clearly, if a great circle passes through one Pole it must pass through the other), their planes cut Earth into pieces that could be described as wedges of a spherical pie. The curved edges of each wedge are halves of great circles reaching from one Pole to the other; these are called *meridians*. The two defining planes of any wedge intersect along the axis of Earth, at a consistent angle when measured between lines in a plane at right angles to the axis (e.g., the angle in the Equatorial Plane. This angle, technically a dihedral angle, is called the *meridional angle* in geography. Exhibit 1–1 illustrates the geometry and this basic geography.

Longitude refers to the angular distance to the meridian of a point from an adopted reference meridian, usually expressed as the lesser of the two angles between the two planes and qualified appropriately with E for East else W for West.

For centuries, each significant nation used its own reference meridian — commonly that of the capital city of a European country. French maps, therefore, differed in their notation from British maps — not by a lot but more than enough to have you miss an island, maybe hit a reef. Following the 1884 International Prime Meridian Conference in Washington, the British reference meridian passing through a specific telescope at Greenwich (just east of London) was adopted as the standard international reference meridian. The time scheme set to register 12:00 when the Sun is nominally[†] in the plane of that meridian became the world standard, as Greenwich Mean Time (GMT). GMT was

---

[†] Because of inherent variation in the length of the precise solar day as Earth travels its elliptic orbit, the Sun can be discrepant relative to a clock by up to 15 minutes on a given date; see Equation of Time [2002a].

renamed Universal Time (UT)[†] in 1972. That 1884 Conference also adopted the standard scheme of 24 1-hour time zones girdling Earth in 15° steps of longitude.

The two furthest meridians in each direction from Greenwich, namely 180°E and 180°W, are one (nominally the International Date Line, though that line steps aside from the 180° meridian over extended distances). In general cartography, East is regarded as positive, West as negative, and the 180° meridian is regarded as positive. Using Λ to represent longitude, we thus get $-180 < \Lambda \leq +180$.

The length of meridians is uniform. Neglecting the ellipticity, all are half the length of the Equator. Allowing for it, they are about 0.2% shorter than that.

Latitude

Latitude refers to the angular distance from the Equator towards one or other Pole, measured at the centre of the planet relative to the Equatorial plane. It therefore goes from zero to a maximum of 90° in either direction, with any latitude having to be qualified as North else South for uniqueness. The points with a given value of latitude form the intercept with Earth's surface of a cone with its apex at Earth's centre. Those on Earth's smooth nominal surface form a ring of points equidistant from the Equator (and from each Pole). That ring is identically the intersection with Earth's surface of a plane parallel to the Equatorial plane, resulting in the ring being called a *parallel*. It is obviously a lesser circle for non-zero values of latitude. The planes of any two parallels are a uniform distance apart. Obviously the parallels are shorter as the distance from the Equator increases, to the limit of zero length at each Pole. Neglecting the oblateness, the length of a parallel relative to that of the Equator is the cosine of the latitude. It decreases progressively with latitudinal angle, rapidly so in higher latitudes, being half that of the Equator at 60° and barely a quarter at 75° (accommodating the ellipticity displaces the figures by less than 0.35%). By the same geometry, the distance of any parallel plane from the Equator is proportional to the sine of the latitude (see Exhibit 1–1, also Tutorial 2 for trigonometric functions). Mathematically, North is regarded as positive and South as negative. Using Φ to represent latitude, we thus get

$$-90 \leq \Phi \leq +90$$

with the extreme values occurring uniquely at the two Poles (with all values of longitude).

Subtracting any latitude from that of a Pole gives its *co-latitude*. Relative to the North Pole, the usual context, this is subtracting from +90°, yielding a value from 0° at that Pole through +90° at the Equator to +180° for the other Pole. Relative to the South

[†] There are several forms of UT, progressively accommodating detailed peculiarities of Earth's rotation. UT_0 is equivalent to GMT. [2002a].

Pole, the subtraction is from –90°, yielding a value from 0° at the South Pole through –90° at the Equator to –180° for the other Pole. Although used predominantly within the adjacent hemisphere, the term is routinely applied across the Equator, in which circumstance the reference Pole should be made clear. (Co-latitude is often denoted by an appended prime mark to the symbol for latitude, but that practice is not used in this text.)

The modulus of co-latitude — needed widely for projections — is called the *polar distance* (being the linear surface distance on a sphere of unit radius). Hence

relative to the North Pole, polar distance $= |90° - \Phi| \qquad\qquad = 90° - \Phi$

relative to the South Pole, polar distance $= |-90° - \Phi| = |90° + \Phi| = 90° + \Phi$

Each is zero at its own Pole and strictly positive otherwise.

Measuring longitude

Because Earth rotates steadily, the meridional angle between any two points (regardless of their respective latitudes) can be obtained from the difference in time at which some celestial body crosses over the respective meridians (i.e., passes through what are called the *celestial meridians* — the intersection lines of the celestial sphere and extended planes that define the meridians). It is better to use a star than our Sun, because of the complications from Earth's orbit. Accuracy depends overwhelmingly on the clock used to measure that time. It was only late in the 18th Century that a suitably accurate and transportable *chronometer* became available, created by James Harrison [1995b].

Measuring latitude

Latitude is usually ascertained by observing and measuring the elevation in the sky of some celestial object and taking the difference between the measured value and the value that applies at the Equator for the point on the same meridian (the latter value can be obtained from a reference book, but must accommodate the seasonal variation). Such measurements, routinely effected with a sextant, measure elevation relative to the tangent plane to Earth at the observation point. Since Earth is oblate rather than truly spherical, this plane is not orthogonal to a line through the precise centre of Earth, except at the Poles and the Equator. (A plumb line or other gravitational vertical does not point precisely at the centre of Earth, except at the Poles and the Equator — it hangs perpendicularly to the local tangent plane.) Latitude so ascertained — the *geodesic latitude* — is generally discrepant from the *geocentric latitude* inherently used in map projecting. In mid latitudes it is discrepant by approximately half a degree. This is of no consequence in the many following chapters that are based on a spherical world. It is discussed further, after turning to an ellipsoidal world, in Chapter 17.

TUTORIAL 3 3-DIMENSIONAL CO-ORDINATES

The 2-dimensional systems for co-ordinates can readily be extended to a 3rd dimension, for which we choose to use upper-case letters. We accord with the standard convention, illustrated in the accompanying diagram, of having the Z axis vertical with positive upwards, the others placed to meet the right-hand-screw rule. Hence, we have the $+Y$ half-axis on the paper pointing to the right and the $+X$ half-axis coming forward from the paper. To avoid the last appearing as a dot, an oblique drawing is used, as shown.

Let O be the zero point of all three axes and let $P = (X, Y, Z)$ be the generic point. The value of each co-ordinate then equals the distance, along a line parallel to the pertinent axis, from P to the plane containing the other two axes. For example, the value Z is the distance to the XY plane measured along a line parallel to the Z axis (i.e., the distance from P to $P_{XY} = (X, Y, 0)$). If ρ is again the length of the line from O to P and ϕ denotes the angle at O to that line from the line from O to P_{XY}, then

$$\text{length of } OP_{XY} = \rho\cos\phi,$$

$$\text{length of } PP_{XY} = \rho\sin\phi$$

the latter being Z. The values for X and Y can be evaluated within the XY plane by drawing orthogonal lines to the respective axes (i.e., to the points P_X and P_Y). If λ denotes the angle at O from the $+X$ half-axis to the line from O to P_{XY}, then

$$X = (\text{length of } OP_{XY})\cos\lambda$$

$$Y = (\text{length of } OP_{XY})\sin\lambda$$

Hence the three Cartesian co-ordinates are:

$$X = \rho\cos\phi\cos\lambda$$

$$Y = \rho\cos\phi\sin\lambda$$

$$Z = \rho\sin\phi$$

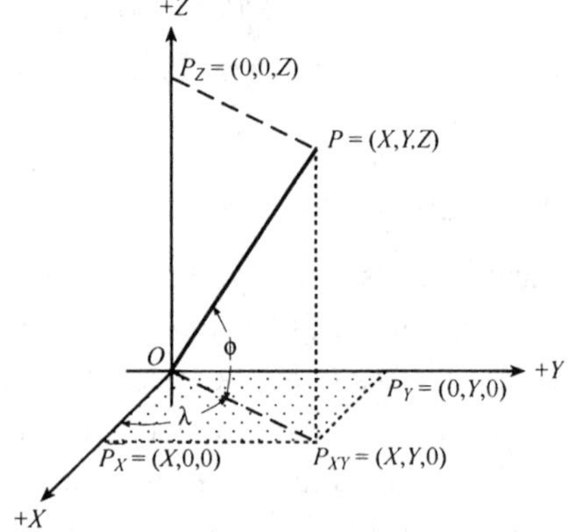

For ρ we have

$$\rho = \text{length of } OP = \sqrt{X^2 + Y^2 + Z^2}$$

the positive-valued root being implied.

The set of all points such that

$$\rho^2 = X^2 + Y^2 + Z^2 = 1$$

forms the spherical surface centred on O of radius 1, the set of all points such that

$$\left(\frac{\rho}{2}\right)^2 = \left(\frac{X}{2}\right)^2 + \left(\frac{Y}{2}\right)^2 + \left(\frac{Z}{2}\right)^2 = 1$$

forms the spherical surface of radius 2, etc.

Point P can be represented directly by the variables ρ, λ and ϕ referred to again as the *polar co-ordinates*, now 3-dimensional.

EXPRESSION OF ANGLE

Degrees versus radians

In discussing longitude and latitude the degree was used as the unit of measurement. It is the most familiar unit for angles, and the usual major unit for expressing longitude and latitude, with 360° equalling one revolution, each divided into 60 minutes, each of those into 60 seconds. (These are all of Babylonian provenance, the $360 = 6 \times 6$ being an adaptation to their 60-based numbering scheme from what was originally $365\frac{1}{4}$ degrees — each such degree being the amount by which the night sky visibly moves each day through the year.) More clearly called *arcminutes* and *arcseconds* (abbreviated as arcmin and arcsec), these are usually symbolized by the single and double prime marks ′ and ″, the latter also called the second symbol. On maps and in related work, it is usual to express longitude and latitude in such degrees, minutes, and seconds (e.g., 53°23′35″, duly qualified E else W for longitude, N else S for latitude. Where more precision is required than provided by this dissection, the figure for seconds can be expressed into fraction figures (e.g., 35.39″). Alternatively, for convenience in arithmetic operations, the angle can be expressed purely in degrees, to whatever decimal fractional place is appropriate. This usually written without the degree symbol last (e.g., 53.393°), but sometimes with the symbol immediately after the integer value (e.g. 53°.393).

In mathematics, angles are usually expressed in radians (symbol rad). The radian is defined as the angle subtended at the centre of a circle by a circumferential arc equal in length to the radius, hence (with tilde in suffix position indicating an incomplete number)

$$1 \text{ rad} = {}^{1}\!/_{2\pi} \text{ revolution} = {}^{360}\!/_{2\pi} \text{ degrees} = 57.295\,78_{\sim}{}^{\circ} = 57^{\circ}17'44.8_{\sim}{}''$$

$$360^{\circ} = 2\pi \text{ rad},\ 180^{\circ} = \pi \text{ rad},\ 90^{\circ} = {}^{\pi}\!/_{2} \text{ rad}.$$

The benefits of radians apply to much of the mathematics in cartography, for its facility in converting subtended angle to surface distance as well as in algebraic mathematical expressions. Despite being strange relative to maps, the radian is the preferred unit for most of this text, making facility with these conversions advantageous to the reader.

To minimize the need for repeated qualification, the convention in this book is that angles are represented algebraically by capital letters when the value is intended in degrees, in lower case letters when in radians. Greek letters (see Appendix A for full alphabet) are used generally for angles, with the particular examples

γ and Γ, ϕ and Φ, θ and Θ, ψ and Ψ, χ and X, λ and Λ, ξ and Ξ, ω and Ω,

the lower case in each pair implying radians, the upper case to the same angle expressed in degrees. Throughout all remaining text, it is assumed that giving the value of an angle in one form is readily interpretable in the other, e.g.

saying $\Lambda = -90$ implies $\lambda = -{}^{\pi}\!/_{2}$ and vice versa, $\Phi = 45$ implies $\phi = {}^{\pi}\!/_{4}$ and vice versa

The letter lambda (λ and Λ) is used routinely for longitudes, phi (ϕ and Φ) for latitudes.

TUTORIAL 4 VECTORS

In both the 2- and 3-dimensional co-ordinate systems, the line from O to P was addressed only indirectly, for its length or magnitude — a *scalar* or simple numerical quantity. Represented by ρ alone in the polar co-ordinates, this value is the same for all the points on any one circle centred on the origin, and likewise for all the points on any one spherical surface centred there. The line from O to P, however, can be seen to have a direction represented by the angular values in the polar co-ordinates, (i.e., θ in the 2-dimensional, λ and ϕ in the 3-dimensional). The line so regarded is termed a *vector*. To distinguish vectors from simple points they are in written in bold, upright, lower-case Roman letters. For example,

in 2-dimensional space $\mathbf{p} = (x, y)$,

in 3-dimensional space $\mathbf{p} = (X, Y, Z)$.

(While the concept of vector is extendable to ordered strings of any count of components, in this work only 2- and 3-dimensional spaces are of concern.}

The length of a vector, usually denoted by the modulus sidebars around its symbol, equals the square root of the sum of the squares of its co-ordinate elements. For example

in 2 dimensions: $|\mathbf{p}| = \sqrt{x^2 + y^2}$

in 3 dimensions: $|\mathbf{p}| = \sqrt{X^2 + Y^2 + Z^2}$

Vectors of length 1 are called *unit vectors.*

If $\mathbf{p}_a = (x_a, y_a)$ and $\mathbf{p}_b = (x_b, y_b)$ are two 2-dimensional vectors and δ is the (lesser) angle between them, it can be shown that

$$x_a x_b + y_a y_b = |\mathbf{p}_a||\mathbf{p}_b|\cos\delta.$$

When the two vectors are radii of a circle, this provides the length of the connecting arc via the angle between them.

If $\mathbf{p}_1 = (X_1, Y_1, Z_1)$ and $\mathbf{p}_2 = (X_2, Y_2, Z_2)$ are two 3-dimensional vectors, they define a plane. If δ is the (lesser) angle between $\mathbf{p}_1$ and $\mathbf{p}_2$ in that plane it can also be shown that

$$X_1X_2 + Y_1Y_2 + Z_1Z_2 = |\mathbf{p}_1||\mathbf{p}_2|\cos\delta.$$

If the two vectors are radii of a sphere, this provides the length of the connecting great-circle arc via the angle between them in their common plane.

The expression on the left of each of these two equations — a simple scalar — is called the *dot-product* or *scalar product* of the two vectors involved. One of two distinct forms of multiplication applicable to vectors, it is discussed further, along with the other (the *cross-product* or *vector product*), in Tutorial 21.

If s is any ordinary number, then the following arithmetic properties apply to vectors:

$$\mathbf{p}_1 + \mathbf{p}_2 = (X_1+X_2, Y_1+Y_2, Z_1+Z_2) = \mathbf{p}_2 + \mathbf{p}_1$$
$$s\,\mathbf{p}_1 = (sX_1, sY_1, sZ_1), \quad s\,\mathbf{p}_2 = (sX_2, sY_2, sZ_2).$$

If $\breve{\mathbf{i}}, \breve{\mathbf{j}}$ and $\breve{\mathbf{k}}$ are unit vectors lying respectively along the positive X, Y and Z half-axes, then

$$\mathbf{p}_1 = X_1\breve{\mathbf{i}} + Y_1\breve{\mathbf{j}} + Z_1\breve{\mathbf{k}}, \quad \mathbf{p}_2 = X_2\breve{\mathbf{i}} + Y_2\breve{\mathbf{j}} + Z_2\breve{\mathbf{k}}$$

Corresponding formulae to all the above apply to the 2-dimensional scene.

SURFACE DISTANCE

Great circles

It is relatively simple to show mathematically that the shortest surface path between two points on a sphere is along their common great circle. This holds not just for the east-west circumstance of the Equator and the north-south of the meridian pair (their *bimeridian*), but absolutely generally. There is, of course, a unique great circle through any two points on the surface that are not diametrically opposed — namely the intersection with the surface of the plane through those two points and the centre of the sphere[†]. Notably, the shortest path between two points with the same latitude is not along their common parallel, except for latitude 0° (the Equator) — all other parallels are lesser circles.

By having great circles appear on them as straight lines, some map projections are specifically advantageous in planning shortest travel routes. The only projection for which this is true everywhere, however, has gross distortion of scale. There are several projections for which the meridians are straight and to scale, and one projection in which all great circles through one point are straight and to scale. More generally, great-circle distance must be computed.

Great-circle arcs

Since circumferential length along any circle is simply the product of the subtended angle in radians and the radius of the circle, ascertaining the length of a great-circle arc is essentially a matter of determining that angle.

For a spherical Earth, the longitude and latitude, together with the radius, provide 3-dimensional polar co-ordinates for the point. Tutorial 3 provides the formulae for converting these to Cartesian form. Interpreting the Cartesian points as radial vectors of our sphere, the dot-product feature, covered in Tutorial 4, provides the cosine of the angle between these radial vectors — hence the angle itself. Finally, multiplication of that angle expressed in radians by the radius of the sphere gives the surface-arc distance between the two points. The method is illustrated overleaf for the distance between the main airports of Calgary and London — two points with essentially the same latitude (51°N) but over 113.5° (1.98 rad) apart in longitude, hence apart by more than 7650 km (or 4130 International Nautical Miles [INM] of 1852 m) along their common parallel. (A simpler formula for great-circle distance is shown late in Chapter 8.)

[†] For the ellipsoidal globe and the geoid itself the shortest path is not so simple. Termed the *geodesic*, it does not generally lie in a plane. However, the difference in length bewteen geodesic and great circle for a given pair of points on Earth is relatively minor.

TUTORIAL 5 COMPUTING SURFACE DISTANCE – 1

As described on the preceding page, the surface distance between any two points of known longitude and latitude can be calculated in four steps. The process is illustrated below by calculating the distance between Calgary International Airport (YYC) and London Heathrow Airport (LHR). The two airports have virtually the same latitude (51°N) but are sufficiently high in latitude that their common great circle arcs far to the north. (See, for instance, the map on page 83.)

Step 1. In three dimensions, convert polar co-ordinates to Cartesian co-ordinates for the two points, keeping the radius of Earth as an algebraic variable R.

YYC, point $P_1 = (X_1, Y_1, Z_1)$ at longitude Λ_1 of 114°01′W, latitude Φ_1 of 51°07′N

$$\Lambda_1 = -114.0167_{\sim} \qquad \Phi_1 = +51.1167_{\sim}$$
$$\sin\Lambda_1 = -0.913\,427_{\sim} \qquad \sin\Phi_1 = +0.778\,426_{\sim}$$
$$\cos\Lambda_1 = -0.407\,002_{\sim} \qquad \cos\Phi_1 = +0.627\,737_{\sim}$$

therefore,

$$X_1 = R\cos\Phi_1\cos\Lambda_1 = -0.255\,490_{\sim}\,R$$
$$Y_1 = R\cos\Phi_1\sin\Lambda_1 = -0.573\,392_{\sim}\,R$$
$$Z_1 = R\sin\Phi_1 = +0.778\,426_{\sim}\,R$$

LHR, point $P_2 = (X_2, Y_2, Z_2)$ at longitude Λ_2 of 0°27′W, latitude Φ_2 of 51°28′N

$$\Lambda_2 = -0.45 \qquad \Phi_2 = +51.4667_{\sim}$$
$$\sin\Lambda_2 = -0.007\,854_{\sim} \qquad \sin\Phi_2 = +0.782\,246_{\sim}$$
$$\cos\Lambda_2 = +0.999\,969_{\sim} \qquad \cos\Phi_2 = +0.622\,970_{\sim}$$

therefore,

$$X_2 = R\cos\Phi_2\cos\Lambda_2 = +0.622\,951_{\sim}\,R$$
$$Y_2 = R\cos\Phi_2\sin\Lambda_2 = -0.004\,893_{\sim}\,R$$
$$Z_2 = R\sin\Phi_2 = +0.782\,246_{\sim}\,R$$

Step 2. Take the dot-product of our two Cartesian-expressed vectors:

$$X_1X_2 = -0.159\,16_{\sim}\,R^2$$
$$Y_1Y_2 = +0.002\,81_{\sim}\,R^2$$
$$Z_1Z_2 = +0.608\,92_{\sim}\,R^2$$

$$X_1X_2 + Y_1Y_2 + Z_1Z_2 = +0.452\,57_{\sim}\,R^2$$

Step 3. Because both vectors are of length R, the angle between is

$$\arccos(+0.452\,57_{\sim}) = 63.091_{\sim}° = 1.1012_{\sim}\text{ rad}$$

this being the angle subtended at Earth's centre by the arc YYC – LHR

Step 4. Multiply the angle in radians by R (taken as 6371 km)

Distance between YYC and LHR = $1.1012_{\sim} \times 6371$ km = $7015._{\sim}$ km = $3788._{\sim}$ INM.

THE PLATE-CARRÉE PROJECTION

Concept

Every point on Earth other than the Poles (which have uniqueness in latitude but undefined longitudes) has a unique numerical identity in terms of longitude and latitude that can expressed as a number pair (Λ, Φ) or (λ, ϕ) where

$$-180 < \Lambda \leq +180 \text{ and } -90 < \Phi < +90,$$

Conversely, all such co-ordinates represent a unique spot on Earth.

By interpreting these numbers as 2-dimensional Cartesian co-ordinates (Tutorial 1) through multiplying each by some unit of linear measure s, we can directly obtain a map. The choice of multiplier dictates the span of the resulting map in each direction. If we choose $s = 2$ mm per degree as the multiplier, we obtain a whole-world map spanning 36 cm by 72 cm. As cartographic convention puts North generally at the top of a map, with longitude running left-to-right and latitude running bottom-to-top, we need to relate longitude to the x variable and latitude to the y variable to meet standard conventions for 2-dimensional plotting. We then obtain a map of points (x, y) using the simple equations

$$x = s\Lambda, \quad y = s\Phi$$

The Poles are represented by bounding lines at top and bottom. The result, using the reduced value of $s = 0.37$ mm per degree, is illustrated below.

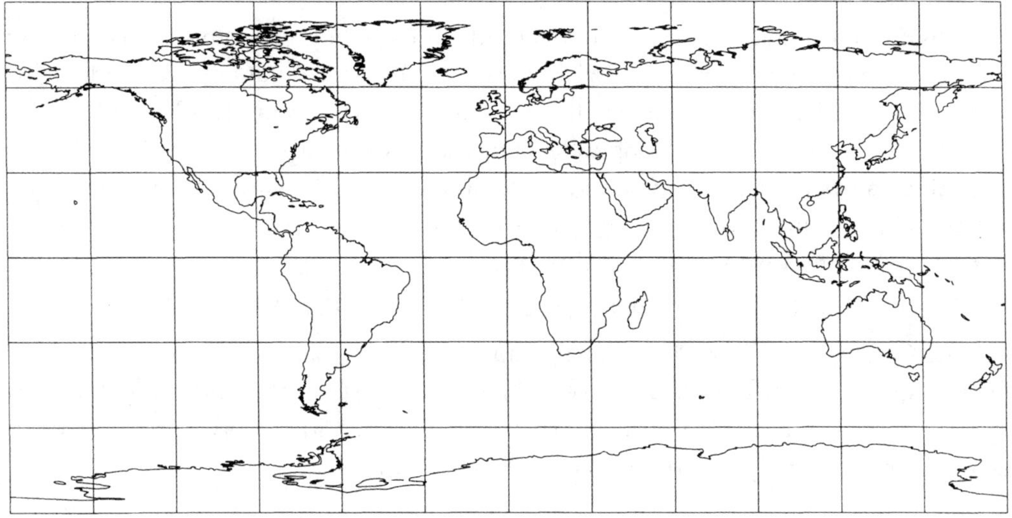

Exhibit 1–2: Plate-Carrée projection — whole World.
Centred on 0° (the Greenwich meridian).
Scale 1:300 000 000 along Equator.

Anticipating further developments, including the special problems of orienting a map centred on a Pole, we elaborate this mapping process by introducing *intermediate variables* labelled u and v to capture the valuations before expressing the result in familiar mathematical co-ordinates. We set

$$u = s\Lambda, \quad v = s\Phi$$

then convert u and v to x and y. To position North at the top, we convert them straightforwardly by equating x with u and y with v. This two-stage process allows rotation of the formulated map onto the map paper by any desired angle.

The expression of mapping variables x and y in terms of two intermediate variables that are functions of the longitude and latitude is the essence of map projecting. The intermediate variables may be Cartesian, as with u and v here, else 2-dimensional polar co-ordinates as used in the next chapter. In this projection, one intermediate variable is a function solely of longitude, the other a function solely of latitude. Many other notable map projections are similarly simple, though with more elaborate expressions connecting each geographical variable to its intermediate one. More generally, however, each intermediate variable is a function of both longitude and latitude.

This most basic of projections has been in use for about 2000 years[†], though it bears the relatively modern name **Plate-Carrée** projection (from the French descriptive *flat, square*); ancient use provided a map covering the Mediterranean area centuries before longitude was measured in the modern fashion. In Chapter 4 we will find that it is identically the Equidistant Cylindrical projection, by which name it will be generally identified in later chapters.

The formulae given above, using longitude in its standard Greenwich-based form, result in a map centred on the Greenwich meridian. To centre it at any longitude Λ_0 we need to re-express longitude relative to the desired central meridian. This basically means subtracting the Λ_0 value from all longitude values, after which we must subtract 360° from all resulting values over +180° and add 360° to all under −180° to bring values within the standard range. This gives us a relative longitude that we can distinguish with an underscore where necessary. Succinctly, we put

$$\underline{\Lambda} = (\Lambda - \Lambda_0) \pm 360 \qquad \text{such that} \quad -180 < \underline{\Lambda} \leq +180.$$

The Plate-Carrée centred on 150°E is illustrated opposite, again with $s = 0.37$ mm per degree to give the stated scale. Henceforth, we will assume that any differencing of longitudes for the purpose of producing such relative longitude involves this correction. This applies more usually to expressions in radians, putting

[†] By Marinus of Tyre [2000a]

$$\underline{\lambda} = (\lambda - \lambda_0) \pm 2n\pi \qquad \text{such that} \quad -\pi < \underline{\lambda} \le +\pi$$

However, in certain circumstances those bounds are adjusted to keep attached to more significant components on a world map associated lesser components lying beyond those bounds (e.g., to include in Eurasia the easternmost islands of Russia that reach to 175°W and to include in North America the Aleutian chain that reaches to 175°E).

Although most texts routinely use the expression in brackets to stress the differencing from the Greenwich value, reduction of clutter in the multitudinous formulae in this work favours the use of the single graphic λ. While the underlined form will be used at times, usually plain unadorned λ will be used to mean longitude in its chosen relative context rather than the Greenwich-based value. Where necessary the term *relative longitude* will be included nearby to make this clear.

Longitude and latitude are variables belonging to Earth, while x and y are variables belonging to the map. In future, to help with this dichotomy, variables of the map will be adorned with overscript dots (e.g., $\dot{x}$ and $\dot{y}$) in their equations and any valuations. When we later transform variables of one map into those of a derived map, we will use multiple dots as in $\ddot{x}$ and $\ddot{y}$ and even $\dddot{x}$ and $\dddot{y}$. However, such adornment will not be applied to those letters when used merely to refer to the axes in a general geometric sense. Besides $(\dot{x}, \dot{y})$ etc., to denote a point, curly brackets are used to denote the set of points forming the map (e.g., $\{\dot{x}, \dot{y}\}$). The overscript dots will be applied similarly to the intermediate variables of the mapping, with those already introduced to give $\dot{u}$ and $\dot{v}$.

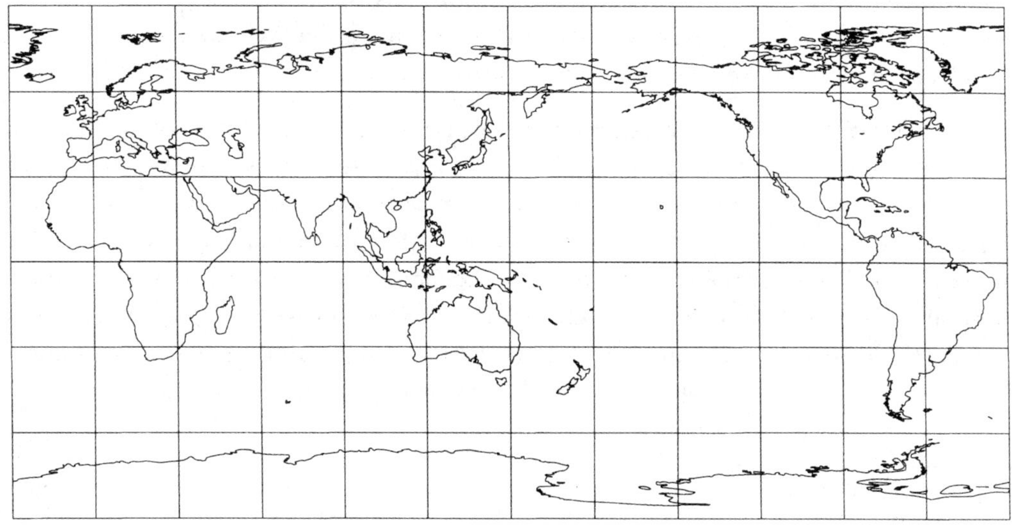

Exhibit 1–3: Plate-Carrée projection – whole World.
Centred on 150°E (meridian through Canberra).
Scale 1:300 000 000 along Equator.

SCALES

Background

Our Plate-Carrée maps cite a scale, specifically the ratio 1:300 000 000[†], meaning that distances on the map are $^{1}/_{300\,000\,000}$ those on Earth. Therefore 1 cm on the map equals 3000 km on Earth, which is an alternative means of expressing scale used on some maps. A narrow barred line is another means for showing scale.

However expressed, any scale applies precisely only to a limited extent, and often is grossly misleading – what appears as a clearly stated fact can be a major fiction. The scale cited above for the Plate-Carrée applies precisely along the Equator and along all meridians (drawn and not drawn). But it obviously does not hold true along any parallels other than the Equator. Because the length on the globe of the parallel for latitude ϕ is only $\cos\phi$ times that of the Equator, the map has 1 cm along the parallel representing only $3000\cos\phi$ km. At 60° this equals 1500 km. The scale progressively increases as unsigned latitude increases, to become infinite at each Pole.

The larger the second number is in the ratio expression, the smaller the fraction and the scale of the map are. Thus maps described as *small scale* typically cover larger areas. Generally, the greater the area covered, the greater the variation of scale is from the nominal. Cartographic authorities take great precautions to keep a tight rein on the variation of scale across their large-scale maps — a variation of 0.01% would be regarded as large. For individual maps covering large proportions of Earth, however, there is no way to keep anything like such rein. The ubiquitous Mercator map presents an even worse picture than the Plate-Carrée. Its escalation toward the Poles applies to meridians as well as parallels, putting the Polar regions out of reach.

Henceforth in this text, expression of scale is omitted from maps shown. This is not in rejection of scales, but because the maps shown — mostly whole-world — are essentially illustrations that can convey their messages adequately without reference to scale. The recurrent use of the familiar outline of world coastlines, plus selected meridians and parallels, illustrates not only variations in linear scale but also the consequent impact on areal scale and shapes. Collectively, these variations in length, area, and angle are referred to as *distortion*. The phenomenon is discussed repeatedly as new projections are

---

[†] Numbers are written in this book in international format, with the period as the decimal point but without commas. Both integer and fractional parts are punctuated by spaces into triples of digits, except that such spacing is omitted generally when the string is only four digits long.
Note, however, that numbers used in the UTM-UPS co-ordinate scheme of Chapter 18 are not punctuated at all, while those cited relative to U.S. patents and similar purposes are accorded the style used for the circumstance.

introduced, notably in Chapter 4 and Chapter 5. Then, after gaining use of differential geometry, it is covered more comprehensively in Chapter 13.

It is appropriate to have scale shown on the vast majority of maps, using any of the means mentioned above. The numeric expression of scale has clear advantages over the constructed bar, but it has disadvantages, too — particularly now with the ready availability of photocopiers with size-changing facility. Any copying with change of size makes the numeric statement invalid, but maintains equal validity for the bar. Anybody using such a technique, whether for serious report or student assignment, must remove the original numeric statements.

With its Poles represented by the full lines along the top and bottom of the encasing rectangle and correct scale along all meridians, the Plate-Carrée provides ready means of determining the distance of any point from either Pole. In Chapter 8, this is exploited to provide like facility for measuring distance from any given point.

TUTORIAL 6 THE TRANSCENDENTAL NUMBER π

The special number π (pi) is the proportionality factor of the circle — the ratio of its circumference to its diameter and of its area to one quarter the area of the square within which it is precisely inscribed. A precise valuation for π, albeit with very slow convergence, is given by the simple expression

$$\pi = 4\left(1 - \tfrac{1}{3} + \tfrac{1}{5} - \tfrac{1}{7} + \tfrac{1}{9} - \ldots\right)$$

Evaluation gives

$$\pi = 3.141\,592\,653\,590\,\sim$$

The number π is *irrational* (i.e., it cannot be expressed precisely as the quotient of two integers). Further, it is *transcendental*, which means it is not the solution of any polynomial equation of finite length that has only integer coefficients and integer powers of the variable.

Cartography is replete with this special number. It pervades most calculations involving circles and spheres, and, through the former giving rise to the angular measure radian with 2π to the revolution, it occurs in many trigonometrical expressions.

Any real number can, of course, be approximated as closely as desired with rational numbers. For π, the rational $^{22}/_{7}$ is the most familiar. Some of the historic examples, with discrepancy from true value, are:

3	= 3.0	−4.51~%
$^{256}/_{81}$	= 3.160 493 8~	+0.602~%
$^{25}/_{8}$	= 3.125	−0.528~%
$^{157}/_{50}$	= 3.14	−0.044 6~%
$^{22}/_{7}$	= 3.142 857 1~	+0.040 2~%
$^{355}/_{113}$	= 3.141 592 9~	+0.000 008 49~%

Another approximant is the irrational

$\sqrt{10}$	= 3.162 277 7~	+0.66~%

TUTORIAL 7 LOGARITHMS – NATURAL AND OTHER

For any positive numbers N and b, if $b^n = N$ then n is said to be the *logarithm* of N to the base b, usually written

$$n = \log_b N$$

Inversely, N is said to be the *antilogarithm* of n (briefly, antilog n).

The advantage of logarithms is the facilitation of multiplication and exponentiation, including the derivation of square and other roots — notably non-integral roots. The key advantage comes from multiplication of numbers equating with addition of indices — that is, with addition of their associated logarithms (i.e., if $M = b^m$ and $N = b^n$ then $M{\cdot}N = b^{m+n}$ and $\log_b (M{\cdot}N) = m+n$).

Until the emergence late in the 20th Century of the electronic calculator, logarithms were the essential means for any extensive calculation with many-digit numbers, including in cartography. Even with the powerful assistance of the computer, logarithms remain the means for elaborate root taking, using the fact:

$$\text{if } N = b^n \text{ and } r \text{ is positive, then } N^r = b^{n \cdot r}$$
$$\text{and } \sqrt[r]{N} = N^{1/r} = b^{n/r}$$

Because of our decimal number system, manual calculation is most convenient with 10 as the base. Such logarithms are usually written without explicit citation of the base, simply as

$$n = \log N$$

Of special convenience in mathematical developments beyond that for simple arithmetic are logarithms to the base

$$e = 1 + \frac{1}{1!} + \frac{1}{2!} + \frac{1}{3!} + \frac{1}{4!} + \frac{1}{5!} + \ldots$$

where ! represents the mathematical factorial expression (i.e., $1! = 1$, $2! = 2 \times 1$, $3! = 3 \times 2 \times 1$, etc). Evaluation gives e = 2.718‿; see further in Tutorial 16.

John Napier is credited with inventing the technique for converting multiplicative problems into additive ones, etc., by using logarithms. He achieved this by using a graphical technique with one axis the normal uniformly spaced arithmetic scale, and the other one with progressively shortening steps that we would now call logarithmic. His base was apparently very large, but mistakenly it has been firmly attached to logarithms to the base e, in the form of *napierian logarithms.* Preferably called *natural logarithms,* these can be distinguished by the inclusion of the symbol for the base number, but are now routinely accorded their own abbreviation $\ln N = \log_e N$

Logarithms to base 10 were first noted by Henry Briggs, who made the first pertinent tables, therefore this style is sometimes called *briggsian logarithms.* Because

$$\log (10^c{\cdot}N) = \log 10^c + \log N = c + \log N$$

all table-look-up needs with base 10 can be met easily by just the logarithms for numbers between 1 and 10, giving log values between 0 and 1.

The always positive fractional part is called the *mantissa.* Factor c is called the *characteristic* (written with a bar above when negative).

CARTOGRAPHIC GRIDS AND GRIDLINES

Graticules and grids

The Plate-Carrée has both parallels and meridians as straight parallel lines, the two sets being mutually orthogonal. This attribute, which holds also for the most familiar map (the Mercator), makes it a 'true' cylindrical projection. Many maps, however, have either parallels else meridians curved (although on many the curvature is not discernible to the human eye), and most projections have both curved. Whatever the shapes, the network of parallels and meridians is the *graticule.* Whichever parallels and meridians we choose to show on the map is arbitrary. The more that are shown, the greater the facility will be to relate between a longitude and latitude value pair and a point on the map — the fewer that are shown, the greater the clarity will be of the map itself. As discussed in the next chapter, however, a graticule was an essential component in most map preparation prior to the use of computers and its choice was influential over the precision of the map; it was appropriate, therefore, to state the graticule used in the preparation (which was usually that shown on the map).

The square graticule of the Plate-Carrée simultaneously provides a grid for positioning points. Conversion between longitude and latitude and co-ordinates $(\dot{x},\dot{y})$ is simple, but such is far from the case for most projections. There is an inherent desire, however, for a simple rectilinear scheme of identifying points on maps in both civilian and military usage. Since international adoption in 1983, this is now met by the Universal Transverse Mercator (UTM) (described in Chapter 18). Initiated by the U.S. military in the 1940s, this subsequently displaced the U.S. National State Plane Coordinate System of the 1930s, the British National grid of 1945, and various other national grids. In UTM (as in those earlier schemes) a point on the map is identified by an easting co-ordinate and a northing co-ordinate that together allow easy placement via a marked grid.

Except where they are mutually orthogonal straight lines (which is not the case with UTM), parallels and meridians are at variance with any rectangular grid, increasingly so away from the central zone of the map. Grids are usually set individually in restricted settings, reset over larger regions (with UTM they apply zonally about 300 km either side of a straight central meridian).

Grids that still survive by remaining as the underpinnings of land registration were used for the original surveying of much of North America. As an example, we peruse the most expansive — the Canadian Prairie scheme, which applies over a span of 2000 km from Winnipeg to the Rockies. It illustrates well the conflict between rectangularity and the longitude/latitude organization.

Its broad plan, similar to several in the U.S.A., was to divide the territory into 6-mile-by-6-mile square *townships*, each divided in turn into 36 1-mile squares called *sections*. Settlement was envisaged on the basis of one family per quarter-section (160 acres). An allowance 1 chain† wide was provided for north-south roads between every section and east-west between every two, making gross dimensions slightly more than the nominal.

The survey was seated on the border with the U.S.A. (49°N) and began at 97°30′W, with the numbering of the resulting perceived chequer board then running the unusual (but here natural) direction of north and west. On this board, each nominally 6-mile-by-6-mile township would be the intersection of a 6-mile-wide east-west band (called also a *township*) and a 6-mile-wide north-south band, called a *range*. As 6 miles equal about 5.4′ along every meridian, the township bands are effectively bands spanning that amount of latitude. However, while a span of 6 miles equals a like amount of longitude along the Equator, it exceeds 8′ of longitude at 49°N, is nearly 9′ by 54°, then almost 10′ at 58°N. On the curved surface, a north-south band cannot be both a uniform 6 miles wide and have both its sides run true north-south. The adopted compromise was to reset the longitudinal span repeatedly along the south-to-north traverse — specifically, every 24 miles, along the central parallel of four township bands. The land measurements along these *baselines* are correct, and are applied to the two bands either side by the drawing of true north-south lines from the baseline to effect the division into ranges. Clearly, along a parallel midway between baselines (e.g. that between ranges 2 and 3), the division lines coming from the south intersect the parallel a little east of the corresponding lines from the north. While it is a matter of planned discrepancy rather than error, such midway parallels are called *correction lines*. The discrepancy is less than 1% (less than a chain per mile), but it accumulates arithmetically, making the offset across the correction lines ever more as they reach westwards, and more so at higher latitudes.

To contain the offset within reasonable bounds, the scheme is reset every fourth degree of longitude, with a new meridian defined as the reference meridian, as follows:

2nd at 102°W, 3rd at 106°W, 4th at 110°W, 5th at 114°W, 6th at 118°W

Locations are then qualified as to the pertaining reference meridian. Thus, any position between 114°W and 110°W is stated as *west of the fourth*.

Clearly this scheme provides rectilinear co-ordinates down to the 6-mile-by-6-mile township, and that module can be further divided to carry the co-ordinate scheme down to whatever level of fineness is desired. Besides its use for land-title purposes, the scheme has been progressively applied to the labelling of roads and addresses over the wide geographical reach of the scheme.

---

† The quintessential Anglo-Saxon surveyor's unit; 1 chain = 4 rods = 22 yards = $\frac{1}{80}$ mile.

GEOGRAPHY VERSUS MATHEMATICS

Opposites meet

Cartographic science is heavily mathematical but built on a geographic foundation. Mathematics produces the projections that underlie the maps. So the two disciplines — mathematics and geography — work closely together. But, oddly, in their 2-dimensional notations they are substantially opposed.

Mathematics uses x, y to label its two variables and routinely (at least in modern Western practice} puts the line of increasing x pointing to the right with that for increasing y a right-angle away counter-clockwise, so upwards. Angles, which are measured from that rightwards-pointing line firstly towards the upwards line, are accordingly measured counter-clockwise.

The two key variables of geography are longitude (λ) and latitude (ϕ), which, in a typical map, also increase rightwards and upwards respectively. Geography, however, does not typically use them in that order but in the opposite order — namely as latitude then longitude (i.e., the paired variables are usually ϕ, λ). Except for the measurement of angles, including differentials of the variables, the difference is essentially pedantic. Ironically, this shows most in the very mathematics created through the need of cartography, namely *differential geometry*. Discussed in Chapter 12, this develops many formulae and tools embracing the representation of surfaces in general, and then applies them both to the globe and to maps.

To minimize difficulties for the reader, this work (as mentioned in its Preface) consistently uses the pairing longitude then latitude (i.e., λ, ϕ). The only adverse effect of this less-common practice is that many expressions have constituents swapped compared with the appearance in many other works.

One feature of the underlying conflict that gets early attention is azimuths — the standard bearing angle of cartography. They are measured from the axis representing increasing values of ϕ toward the axis representing increasing values of λ (i.e., clockwise from the meridian — the upward half-axis — while mathematics measures its angles counter-clockwise from the rightwards-pointing half-axis).

The one place where geographical practice is in angular tune with mathematics is at the North Pole, where longitude increases counter-clockwise. At the South Pole, the geographical angles of longitude become clockwise — i.e., the reverse.

TUTORIAL 8 ELLIPSES AND ELLIPSOIDS

Ellipse

Using the latter pair of axes and co-ordinates X, Y, Z of the 3-dimensional scene in anticipation of its use for the ellipsoid, the formula for an ellipse centred on the origin and passing through (±a, 0) and (0, ±b) is, for point $P = (Y, Z)$,

$$\frac{Y^2}{a^2} + \frac{Z^2}{b^2} = 1$$

Its area equals $\pi a b$. If $a = b$, the ellipse is a circle. If $a > b$, giving a true ellipse as illustrated below, then a is called the *semimajor axis*, b the *semiminor axis*.

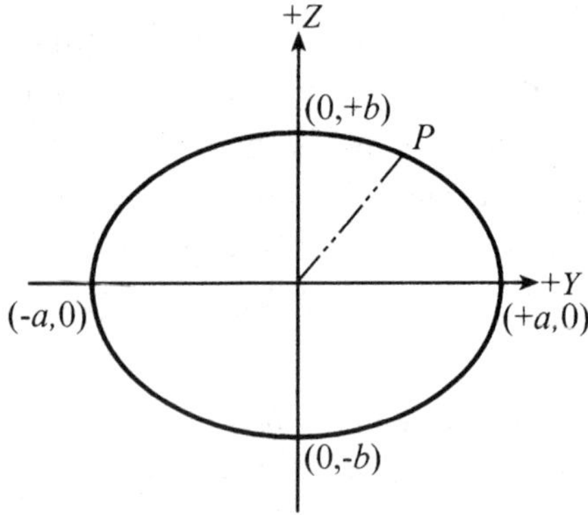

The distortion from circular of an ellipse is expressed in two ways, as

eccentricity $\varepsilon = \sqrt{\dfrac{a^2 - b^2}{a^2}}$

flattening $f = \dfrac{a - b}{a}$

Hence $\varepsilon^2 = f(2 - f) = 2f - f^2$

$$a^2\varepsilon^2 = a^2 - b^2 \quad \text{or} \quad b^2 = \left(1 - \varepsilon^2\right)a^2$$

allowing re-expression of the equation as

$$Y^2 + \frac{Z^2}{1 - \varepsilon^2} = a^2$$

Canonical form

Because for any angle θ

$$\frac{(a\sin\theta)^2}{a^2} + \frac{(b\cos\theta)^2}{b^2} = \sin^2\theta + \cos^2\theta = 1$$

the point $(a\sin\theta, b\cos\theta)$ must lie on the ellipse. If θ ranges over a revolution, it defines all points on the ellipse. This is called the *canonical form* of defining the ellipse.

The relationship of angle θ to the point is explained later in Tutorial 27. In the context of the globe the angle is known as the *parametric latitude*.

Ellipsoid

As a 3-dimensional object, the general ellipsoid has a third semiaxis in the third dimension. However, Earth is not so general. It is an *ellipsoid of revolution* — the axis of revolution is that of the minor semi-axis and the semiaxis of the third dimension is identical to the major semiaxis of the ellipse. Its equation is

$$\frac{X^2}{a^2} + \frac{Y^2}{a^2} + \frac{Z^2}{b^2} = \frac{X^2 + Y^2}{a^2} + \frac{Z^2}{b^2} = 1$$

or

$$X^2 + Y^2 + \frac{Z^2}{1 - \varepsilon^2} = a^2$$

Having three semiaxes equal produces a circular figure (i.e., the sphere).

The broader term *spheroid* is often used for the ellipsoid of revolution. Being regarded by this author as a looser term, it is abjured within this work.

CHAPTER 2

The Task in Hand: A globe as model and intermediary

Preamble

In developing the Plate-Carrée, a multiplier factor brought the numbers representing angles of longitude and latitude to linear measurements appropriately sized for the paper on which the map was to be plotted. In the sample maps shown in the preceding chapter, a scale of 1:300 000 000 is stated, indicating that distances on the map are only $\frac{1}{300\,000\,000}$ of the equivalent on Earth[†]. Virtually every map is a miniature — though not usually so gross a miniaturization — of the region it depicts. Such a scaling factor, therefore, is an essential of map production. But rather than repeatedly involve such a factor overtly, we shall create our own miniature model of Earth — i.e., a globe — and apply our mapping to this model, thus detaching discussion of scaling from the details of projecting. This will also allow us to affix a shape — we use a spherical globe for many chapters, then use an ellipsoidal globe before addressing the real shape of Earth. Globes also give us a privileged viewing perspective.

The trigonometry gets more complicated in this chapter, but nothing new is required in mathematics.

The maps in this chapter are again from just one projection; all are aerial views centred on the same spot.

[†] As discussed, any stated scale is true for only limited parts of any map.

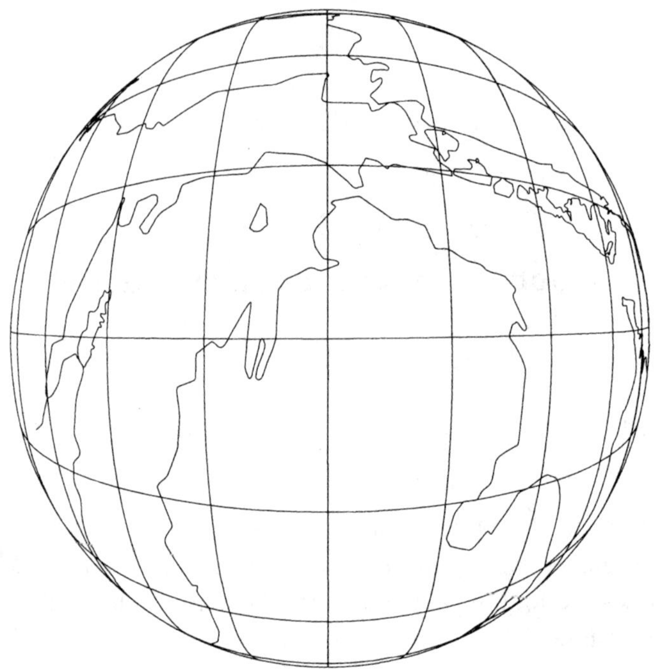

Exhibit 2–1: Earth seen from an airplane – the Aerial projection from an altitude of 12 km.
Focused at 85°W, 45°N, with 1° graticule.

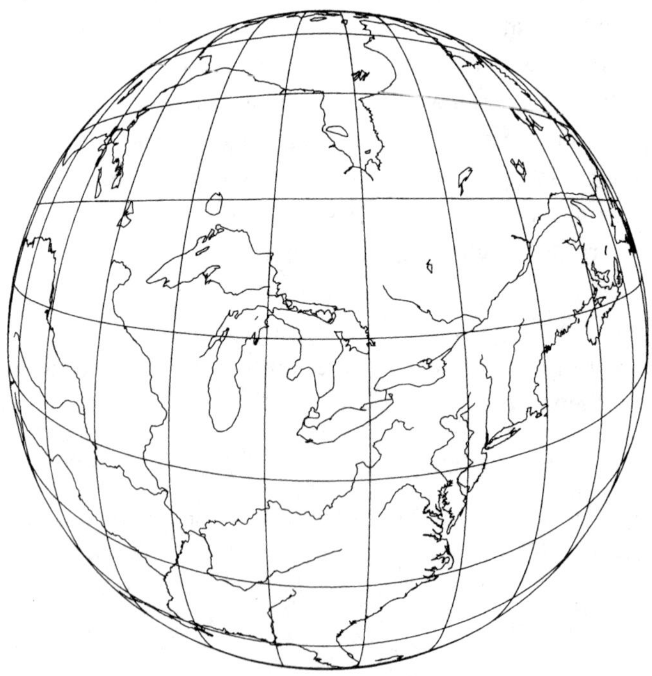

Exhibit 2–2: Earth seen from the Space Station – the Aerial projection from an altitude of 500 km.
Focused, as with Exhibit 2–1, at 85°W, 45°N, but with 5° graticule.

THE SCALE MODEL

Scale

The scaling of $^1/_{300\,000\,000}$ used for the Plate-Carrée maps in the preceding chapter (stated therein as "Scale 1:300 000 000") provided the multiplier that brought the massive radius of Earth down to a size befitting our map. As we have noted, any such scale stated on a map is only nominal, applying to some locations on the map but never precisely to all. In the case of the Plate-Carrée, the scale applies along all meridians but transversely only along the Equator. Every map, however, must have some assumed scale for its creation. More pedantically, because of the vagaries of representing an imprecise planet, every map must also be based on a shape together with some dimensions for Earth, most notably a representative for linear size. Were Earth spherical, the radius would be the most convenient — but, given the variation in the actual radius, a decision is required as to what radius value to use. For an ellipsoidal Earth the value of the eccentricity must augment a representative radius.

A representative radius term R_E naturally pervades most equations transforming geographical co-ordinates to mapping co-ordinates. A scaling factor must be incorporated in the process to achieve the objective of a map that can be plotted within the given confines of the paper. To keep equations as clear as possible, we devolve our mapping of Earth conceptually into two visual steps:

1, the creation of an appropriate 3-dimensional miniature of Earth — a *globe*.

2, a mapping of the globe without re-scaling.

The globe we make as our model of Earth is in the same linear proportion as we want for our map. Besides size, the shape of the globe is at our discretion. We have, and exploit, the option to have it truly spherical, perfectly ellipsoidal to any particular proportions, else any other shape. Indeed, we must decide on the shape before we decide on its exact size for any specified scaling. We can also use modified longitude and latitude values on our globe to accommodate other vagaries (e.g. geodesic versus geographic latitude already mentioned, and conformal latitude and others discussed in Chapter 17, where the technique of mapping ellipsoid onto sphere is developed).

Shape and size

As we have noted, Earth has a smaller diameter along the axis than across the Equatorial plane. The respective mean radii are, to the nearest metre, 6356.752 km and 6378.137 km in length. Between the Equator and the Poles there is a continuum of size between these two values. With an ellipsoidal globe, we can accommodate most of this variability — though only with much complication of the mathematics. With a spherical globe we must choose one specific radius to which to apply our scale factor. Because we are in three dimensions, and the Equatorial figure applies in two, it is appropriate to give that figure twice the weighting of the axial figure in any compromise computation of a single representative value. The primary mathematical choices are:

$$\text{the arithmetic mean} = \frac{6356.752_{\sim} + 2 \times 6378.137_{\sim}}{3}\ \text{km} = 6371.009_{\sim}\ \text{km}$$

$$\text{the geometric mean} = \sqrt[3]{6356.752_{\sim} \times 6378.137_{\sim}^{\,2}}\ \text{km} = 6371.007_{\sim}\ \text{km}$$

The latter is more appropriate but the two are trivially different for most mapping.

Where it is required that areas be consistently proportional between Earth and the map, the preferred choice as representative radius of Earth is the figure that would produce the *authalic sphere* — the sphere with surface area identical to that of Earth. Our geometric mean radius gives a sphere of volume equal to Earth's. The authalic radius, being derived from area, inherently involves a square root rather than the cube root. The surface area of even a representative ellipsoid, however, is not a simple computation. We develop in Chapter 17 spheres matching the ellipsoid in mathematical ways, including having equal surface area. Here it suffices to note that the authalic radius for Earth is very little different from the above figures, but that the maintenance of correct areas between ellipsoidal and spherical versions requires an adjustment to latitude of several arcminutes in middle latitudes between the two worlds (longitude remaining unaffected).†

Where it is required that surface shapes be consistent between Earth and map, the preferred choice as representative radius of Earth is the figure that would produce the *conformal sphere*. This too is developed in Chapter 17. Adopting a conformal sphere does change the representative radius significantly — indeed to one varying with chosen latitude of true scale (being about 6378 km in mid latitudes) — and requires adjustment to both longitude and latitude, the former varying with latitude, too.

In practice, all our developments of projections are algebraic rather than arithmetic, so the numeric values are not involved. For the illustrative maps, scaling is clearly a factor, but not one that matters. The message is in the shape of graticules and in the shape and areas of land masses defined by shorelines, hence there will be no statement of scale in any subsequent maps. The related equations will be developed using algebraic parameters. Until we turn from spherical world to ellipsoidal world in Chapter 17 the unadorned symbol R will denote the chosen radius of the spherical globe. For the rare calculation of real distances on a spherical Earth, the common rounded whole-kilometre figure of 6371 km suffices. Whatever single figure we use, we must remember that, while it is reasonably representative of mid latitudes, it is 14 km excessive when close to the Poles, and inadequate by 7 km when close to the Equator.

---

† The term *authalic* has been used to mean of equal-area at the detail level, but is not so used in this book. However, in the many projections that produce maps uniformly divided by meridians, being authalic in the overall sense implies that any lune defined by two meridians is of correct area. See Appendix C for further discussion of this term.

THE AERIAL PERSPECTIVE PROJECTION

A view from on-high

The progressive development and introduction of airplanes, rockets, satellites and space probes through the 20th Century allowed human beings to get far above Earth, and thereby see an extended part of its surface in a map-like visual image, even to see Earth as a sphere and its 2-dimensional image as a disc. A couple of centuries earlier, hot-air balloons allowed a much more limited aerial viewing. Before then, only steeply elevated escarpments and a few tall buildings allowed any looking down on the surface that man inhabited. Images from any distance have effectively been available, however, through the many centuries since globes were first created.

The picture in a viewer's eye, likewise on the film of a viewing camera, derives from the converging light rays from the area viewed. Clearly, the rays from the outer fringes are closer together for a given increment in distance from the central point — the point immediately underneath the viewer in gravitational terms, and equally the nearest point on the surface — but at the outer limits of the viewed area the rays are tangential to the spherical source and merge. The viewable area increases with distance from the observed globe, but always with progressive foreshortening away from the central point. It approximates a hemisphere as the distance from the globe increases, but can never become a hemisphere. This picture can be developed geometrically — the resulting map projection is called the **Aerial** or **Aerial Perspective** projection.

Let our viewing point be V and let F be the focal point on the globe's surface immediately below it, as illustrated in Exhibit 2–3 on page 31. If G is the centre of the globe, then the length of GF is R (the radius of the globe) and GFV is a straight line. The length of FV is the height of the viewing point above the surface, which we can express proportionally to the diameter of the globe as $h{\cdot}R$ (using the central dot to delineate two entities multiplied). As a 2-dimensional entity, our picture forms in a plane orthogonal to FV. While the picture forms close to point V, its internal pattern and proportions are exactly the same as were it the tangent plane through F; for convenience we use the latter as our mapping surface.

Let $P=(\lambda, \phi)$ be an arbitrary point on the globe. The corresponding point on the map, which we can label $\dot{P}$, is the point of interception of the line PV with our mapping plane. Let us label the angle at V between lines PV and FV as ξ and the angle at G between lines GP and GF as γ. Then, addressing triangle $\dot{P}VF$, we have

$$\tan\xi = \frac{\text{length of } \dot{P}F}{\text{length of } FV} = \frac{\dot{\rho}}{h R} \qquad \text{. . . . . . . (29a}$$

using $\dot{\rho}$ for the distance of our mapped point from the central point of the projection.

Now add a plane parallel to our mapping plane but passing through *P*, and let *F′* be its interception point with our central line *GF*. Clearly, since the length of both *GP* and *GF* equals *R*, triangle *PGF′* shows:

$$\text{length of } F'P = R\sin\gamma,$$
$$\text{length of } GF' = R\cos\gamma$$
$$\text{length of } F'F = R(1-\cos\gamma).$$

Addressing triangle *PVF′*, we then have

$$\tan\xi = \frac{\text{length of } F'P}{\text{length of } F'V} = \frac{\text{length of } F'P}{\text{length of } F'F + \text{length of } FV}$$
$$= \frac{R\sin\gamma}{R(1-\cos\gamma)+hR} = \frac{\sin\gamma}{h+1-\cos\gamma}$$

Connecting this with the expression in Equations 29a for the same tangent gives

$$\frac{\dot\rho}{hR} = \frac{\sin\gamma}{h+1-\cos\gamma}$$

Hence, noting that $h \neq 0$,

$$\dot\rho = \frac{hR\sin\gamma}{h+1-\cos\gamma} = \frac{R\sin\gamma}{1+h^{-1}(1-\cos\gamma)} \qquad \text{. . . . . (30a}$$

The geometric diagram, drawn to facilitate comprehension, has *h* as approximately 1 (i.e., *V* is about the same distance from *F* as is *G*). For aerial photography of Earth from an airplane, *h* is usually less than 0.001, the airplane being more likely below than above 6371 m altitude. For viewing from a satellite *h* ranges from about 0.1 to over 6 (for a geostationary satellite). For a view of Earth from a space probe *h* is unlimited.

Let $\dot\theta$ be the counter-clockwise rotational angle within the tangent plane of $\dot P$ about some chosen reference plane through *G* and *F* (hence also through *F′*). Then, seeing *F* as the origin for orthogonal axes with the *x* axis in that reference plane, interpreting $\dot\rho$ and $\dot\theta$ as polar co-ordinates provides Cartesian co-ordinates (Tutorial 1):

$$\dot x = \dot\rho\cos\dot\theta$$
$$\dot y = \dot\rho\sin\dot\theta \qquad \text{. . . . . (30b}$$

which provide a consistent mapping of point *P* on the globe into point $\dot P$ in the tangent plane. Substituting from Equation 30a gives:

$$\dot x = \frac{R\sin\gamma\cos\dot\theta}{1+h^{-1}(1-\cos\gamma)}$$
$$\dot y = \frac{R\sin\gamma\sin\dot\theta}{1+h^{-1}(1-\cos\gamma)} \qquad \text{. . . . . (30c}$$

If a Pole is the focal point of our projection (hence centre of this map), then γ is the polar distance (see page 9 for explanation) of *P* and all radial lines are (parts of) meridians. The angle $\dot\theta$ in the mapping plane, now identical to the rotational angle of *P* about the same chosen reference plane through F' and F, equals in magnitude the difference in longitude between the chosen meridian and that of point P. If F is the North Pole and we take the Greenwich Meridian as our radial line, then $\dot{\theta}$ is precisely the Greenwich longitude of P. However, we have two versions: the northern version with $\dot{\theta}$ identical to the longitudinal difference and a southern version in which the sign is reversed (longitude increasing clockwise at the South Pole, contrary to mathematical convention). If λ_{+x} is the longitude chosen to lie along the $+x$ half-axis and anchor our map, then

$$\dot{\theta} = \oplus(\lambda - \lambda_{+x}) \qquad \text{. . . . . (31a}$$

where the symbol $\oplus$ is to be positive for the northern version, negative for the southern version.[†]

Substituting from Equation 31a converts Equation 30b to:

$$\begin{aligned} \dot{x} &= \dot{\rho}\cos\oplus(\lambda - \lambda_{+x}) &&= +\,\dot{\rho}\cos(\lambda - \lambda_{+x}) \\ \dot{y} &= \dot{\rho}\sin\oplus(\lambda - \lambda_{+x}) &&= \oplus\dot{\rho}\sin(\lambda - \lambda_{+x}) \end{aligned} \qquad \text{. . . . . (31b}$$

If $\lambda_{+x} = {}^{\pi}\!/_{2}$ to put 90°E along the $+x$ half-axis (as is often used for Pole-centred maps) then:

$$\begin{aligned} \dot{x} &= +\,\dot{\rho}\cos(\lambda - \tfrac{\pi}{2}) &&= +\dot{\rho}\sin\lambda \\ \dot{y} &= \oplus\dot{\rho}\sin(\lambda - \tfrac{\pi}{2}) &&= -\oplus\dot{\rho}\cos\lambda \end{aligned} \qquad \text{. . . . . (31c}$$

Alternatively, if λ_{-y} is the longitude chosen to lie along the $-y$ half-axis,

$$\dot{\theta} = \oplus(\lambda - \lambda_{-y}) - \frac{\pi}{2}$$

and Equation 30b becomes:

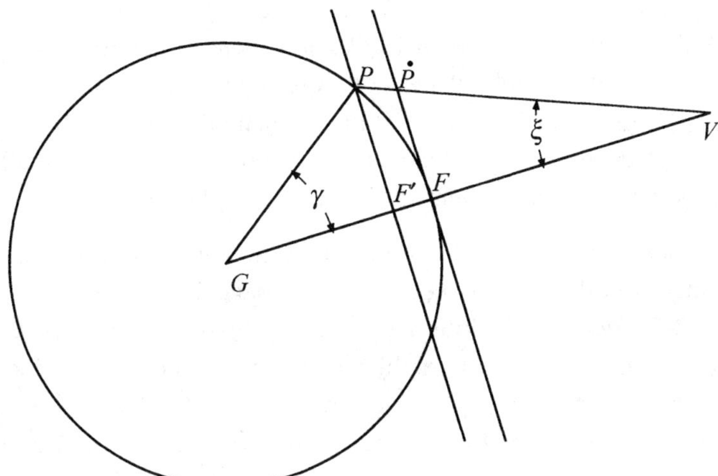

Exhibit 2–3: The Aerial projection – geometry.

[†] We adopt this symbol for such dichotomous use throughout this work. In this elementary example, it is a matter of which Pole is the centre of the map; more generally the sign is determined according to which hemisphere contains the central point, even without the map necessarily containing a Pole.

$$\begin{aligned}\dot{x} &= \dot{\rho}\cos\left(\oplus(\lambda-\lambda_{-y})-\tfrac{\pi}{2}\right) = +\dot{\rho}\sin\oplus\left(\lambda-\lambda_{-y}\right) = \oplus\dot{\rho}\sin\left(\lambda-\lambda_{-y}\right)\\ \dot{y} &= \dot{\rho}\sin\left(\oplus(\lambda-\lambda_{-y})-\tfrac{\pi}{2}\right) = -\dot{\rho}\cos\oplus\left(\lambda-\lambda_{-y}\right) = -\dot{\rho}\cos\left(\lambda-\lambda_{-y}\right)\end{aligned} \qquad \text{(32a}$$

Returning to Equation 30a and substituting $\gamma = (\pi/2 - |\phi|) = (\pi/2 - \oplus\phi)$ gives

$$\dot{\rho} = \frac{R\cos\oplus\phi}{1+h^{-1}(1-\sin\oplus\phi)} = \frac{R\cos\phi}{1+h^{-1}(1-\oplus\sin\phi)} \qquad \text{(32b}$$

Substituting from Equations 31a & 32b, Equation 30c gives the plotting equations for the Aerial or Aerial Perspective projection[†] at Simple aspect (i.e., centred on a Pole — see Chapter 3) as:

$$\begin{aligned}\dot{x} &= +\frac{R\cos\phi\cos(\lambda-\lambda_{+X})}{1+h^{-1}(1-\oplus\sin\phi)}\\ \dot{y} &= \oplus\frac{R\cos\phi\sin(\lambda-\lambda_{+X})}{1+h^{-1}(1-\oplus\sin\phi)}\end{aligned} \qquad \text{(32c}$$

For a general central point F the formulation is more complicated. Anticipating the general equations obtained in Chapter 8, where F would be the auxiliary pole $P_N = (\lambda_N, \phi_N)$ in our devised auxiliary scheme, the plotting equations for the Aerial projection at Oblique aspect (i.e., not centred on a true Pole) are:

$$\begin{aligned}\dot{x} &= \frac{+R\cos\phi\sin(\lambda-\lambda_N)}{1+h^{-1}(1-\cos\gamma)}\\ \dot{y} &= \frac{\mp R\left\{\cos\phi_N\sin\phi-\sin\phi_N\cos\phi\cos((\lambda-\lambda_N)\right\}}{1+h^{-1}(1-\cos\gamma)}\end{aligned} \qquad \text{(32d}$$

where the $\cos\gamma$ term is retained for ease of expression, its formulation being

$$\cos\gamma = \sin\phi_N\sin\phi + \cos\phi_N\cos\phi\cos(\lambda-\lambda_N)$$

These equations give us an initial glimpse at the mathematics of projective mapping using plane geometry and trigonometry. Many familiar projections, as normally used, have simpler plotting equations — but others are far more complicated.

There are four illustrations of the Aerial projection at Oblique aspect, all identically focused at a point midst the Great Lakes. Those on page 26 contrast the images from an airplane (Exhibit 2–1) and from the International Space Station (Exhibit 2–2). Exhibit 2–4 is the view from a height that encompasses the whole North American continent. Exhibit 2–5, on page 35, carries the viewing point out to the Moon. At the latter distance (350 000 km) the view is virtually identical with that from infinity, which is the

† The Aerial Perspective projection is a form of the General Perspective (Azimuthal) projection discussed in Chapter 3, with point V below the surface — i.e., distinctly not aerial. With aerial photography and similar Earth-observing imaging, the imaging plane is typically tilted relative to the line through the centre of Earth. That form of the projection is distinctively called the *Tilted Perspective* projection then the normal form emphatically called the *Vertical Perspective* projection. The Tilted projection and its challenges are covered in Chapter 19.

Orthographic projection — discussed in the Chapter 3, where it is illustrated centred on a Pole.†

Distortion

It is clear from the geometry depicted in Exhibit 2–3 that the scale decreases along radial lines emanating from the centre of the map. So, unlike with the Plate-Carrée, the scale that is true at the central point is never true along meridians. When viewing from a finite height, any circle centred on the focal point maps into a circle of lesser radius hence of diminished scale. This applies to all parallels. Viewing from infinity, however, produces a circle of identical radius — then, any circle centred on the focal point has correct scale on the map, including along parallels with a map focused on a Pole. Otherwise, the scale is not true for this projection except at the central point. A means of quantifying such linear distortion is shown in Chapter 4, after development of the full repertoire of literally projected projections and their algebraic kin.

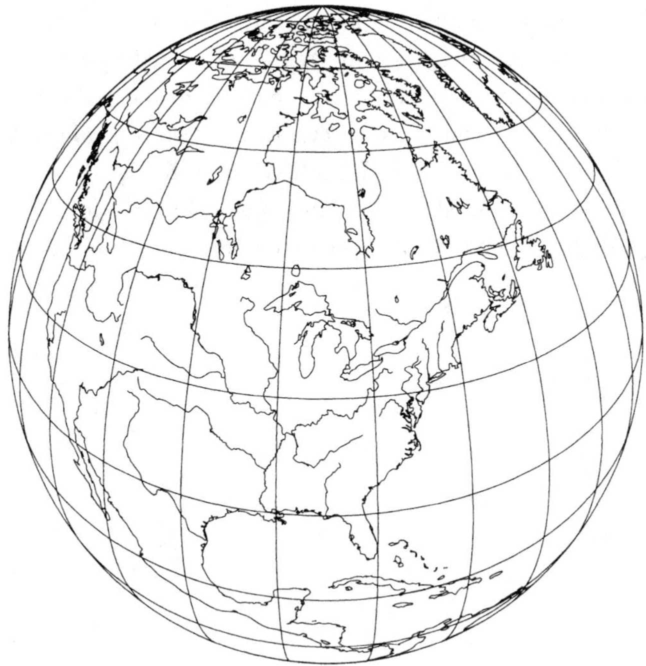

Exhibit 2–4: Earth seen from inner space – the Aerial projection from an altitude of 3000 km. Focused, as with Exhibit 2–1, at 85°W, 45°N, but with 10° graticule.

† The four maps in this chapter include major rivers to help provide realism. In later chapters some show major lakes but generally coastlines alone suffice for illustrating projections while minimizing clutter.

AZIMUTHS AND AZIMUTHALS

Linguistic parent and child

In the Aerial projection, the progressive foreshortening along lines radiating from the central point of our map (i.e., along meridians if that point is a Pole) results in the sight line becoming tangential to the circle; the foreshortening becomes infinite, the map reaches an absolute limit. Thus, this projection cannot go beyond a hemisphere for one map, and generally falls well short of that. While this foreshortening distorts the map significantly, indeed severely towards its natural edge, there is no disturbance of angles between directional lines emanating from the central point. If it is a Pole, all the points on any meridian lie on a corresponding line on the map. More generally, all the points lying on any directional line from the central point lie likewise on the map, and any two such lines on the map retain the same mutual angle as their original lines did on the globe. The term *azimuth* is used for the clockwise angle from North of any direction – (e.g. the direction East would be said to have an azimuth of 90°) — so the Aerial projection is described as *maintaining azimuthal angle* at its central point.

The Aerial projection originated as an eye view, but was developed using a plane to produce the intermediate variables, en route to the necessarily planar map. In Chapter 3 we introduce the cylinder and the cone as alternatives to the plane for the intermediate step. Such projections are duly qualified as cylindrical else conical (or conic). Projections that, like the Aerial, use a plane should be called distinctively *planar* projections. In the past they have widely been called *zenithal*, but now they are universally referred to as being *azimuthal*[†] projections because all inherently maintain azimuthal angle at the central point.

[†] The three classificatory words azimuthal, cylindrical and conical occur profusely through this book. When used as adjectives within a proper name they are capitalized — e.g., the Central Cylindrical projection. They are also used capitalized as nouns, to refer to defined projections collectively — e.g., the Cylindricals. In general use they are not capitalized.
The proper names of projections are often written without their common word — e.g., the Mercator rather than the Mercator projection, the Central Cylindrical rather than the Central Cylindrical projection.

GRATICULES

Lines before the map

With the Plate-Carrée, lines of both longitude and latitude appear as straight lines. Further, they are mutually orthogonal and form a grid that can be used as the basis for rectangular co-ordinates. With the Aerial projection, those lines are generally curved so there is no consequent rectangular grid. The inclusion in the presented map of such lines — collectively called the graticule — is still vital for the recognition and placement of places by longitude and latitude values. Equally important is that until the development of computer-based map production, the graticule signified the precision of the map. Placement of points was effected by individual calculation for points on the graticule but by interpolation generally elsewhere. Thus, the fineness of the graticule controlled the precision of the map, and a statement of the graticule used was a relevant footnote to the printed map.

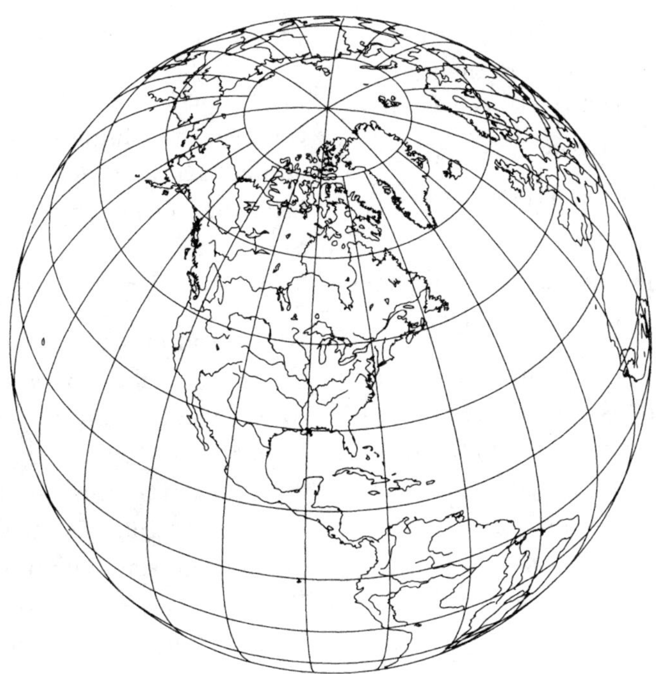

Exhibit 2–5: Earth seen from the Moon – the Aerial projection from an altitude of 350 000 km. Focused, as with Exhibit 2–1, at 85°W, 45°N, but with 15° graticule.

TUTORIAL 9 SINES, COSINES, ETC., – GENERALIZED

Tutorial 2 extended the definitions of sine and cosine for angle Θ beyond the 90° of Tutorial 1 and defined four related functions. With angle Θ anchored to the x axis and increasing as point $P = (x, y)$ gyrates counter-clockwise about the origin, those geometric definitions for sine and cosine were:

$$x = \rho\cos\Theta \quad \text{or} \quad \cos\Theta = \frac{x}{\rho}$$

$$y = \rho\sin\Theta \quad \text{or} \quad \sin\Theta = \frac{y}{\rho}$$

Because ρ is always positive, we can easily see that:

$$\sin(-\Theta) = -\sin\Theta,\quad \sin(90-\Theta) = +\sin(90+\Theta)$$

$$\cos(-\Theta) = +\cos\Theta,\quad \cos(90-\Theta) = -\cos(90+\Theta)$$

With tan, cot, csc and sec expressible in terms of sin and cos, we can see that:

$$\tan(-\Theta) = -\tan\Theta,\quad \tan(90-\Theta) = -\tan(90+\Theta)$$

$$\cot(-\Theta) = -\cot\Theta,\quad \cot(90-\Theta) = -\cot(90+\Theta)$$

$$\csc(-\Theta) = -\csc\Theta,\quad \csc(90-\Theta) = \csc(90+\Theta)$$

$$\sec(-\Theta) = +\sec\Theta,\quad \sec(90-\Theta) = -\sec(90+\Theta)$$

Clearly, adding one revolution to an angle leaves any function unchanged in value.

Alternatively to their geometric definitions, functions sin and cos can be defined algebraically as infinite series. These are more simply expressed with angles measured in radians, as:

$$\sin\theta = 0 + \frac{\theta^1}{1!} - \frac{\theta^3}{3!} + \frac{\theta^5}{5!} - \frac{\theta^7}{7!} \cdots$$

$$\cos\theta = 1 - \frac{\theta^2}{2!} + \frac{\theta^4}{4!} - \frac{\theta^6}{6!} + \frac{\theta^8}{8!} \cdots$$

where the exclamation mark indicates the *factorial function* — i.e., the product of all integers up to the stated value,

e.g. $3! = 1\times 2\times 3,\quad 4! = 1\times 2\times 3\times 4$

These series hold for any value of the angular variable. Although not obvious, the impacts of the right angle, the quadrant and the revolution are identical to those shown in the geometrical context. The following equalities still apply:

$$-\sin(-\theta) = \sin\theta = \cos\left(\tfrac{\pi}{2}-\theta\right) = +\cos\left(\theta-\tfrac{\pi}{2}\right)$$

$$\cos(-\theta) = \cos\theta = \sin\left(\tfrac{\pi}{2}-\theta\right) = -\sin\left(\theta-\tfrac{\pi}{2}\right)$$

$$-\tan(-\theta) = \tan\theta = \cot\left(\tfrac{\pi}{2}-\theta\right) = -\cot\left(\theta-\tfrac{\pi}{2}\right)$$

$$\sec(-\theta) = \sec\theta = \csc\left(\tfrac{\pi}{2}-\theta\right) = -\csc\left(\theta-\tfrac{\pi}{2}\right)$$

$$-\csc(-\theta) = \csc\theta = \sec\left(\tfrac{\pi}{2}-\theta\right) = +\sec\left(\theta-\tfrac{\pi}{2}\right)$$

The effect of adding any number of complete revolutions to the angle (i.e., adding $2n\pi$ rad to θ for any integer n) is to leave the value of any function unchanged, as in

$$\sin(2n\pi+\theta) = \sin\theta,\quad \cos(2n\pi+\theta) = \cos\theta,$$

The graph of $\sin\theta$ gives a widely known shape, called the *sinusoidal*. The graph for $\cos\theta$ is identical except for being offset in angular value by one right angle, making the y axis pass through the curve at its maximal height. For both, the values of the function stay within the bounds of ±1. The graphs for each of the other four functions head to infinity in every quadrant.

PART B

A Spherical World

Until the development of modern science and associated instruments showed in the 17th Century that it is otherwise, Earth was assumed to be a sphere. We have seen that the nominal surface we use — MSL — has Polar radii greater by over 20 km than its Equatorial radius. That is only by about 0.5 %, however, which is less than the accuracy of scale required for most maps. For maps of the world and of large regions, the accuracy of a spherical world is completely adequate. Further, even where the greater accuracy of the ellipsoid is ultimately required, the applicable cartographic principles can be learned much more easily on a spherical basis, the sphere being very much simpler than the ellipsoid mathematically. In this Part B, which forms three quarters of the book, we develop our science on the basis of a truly spherical world. In doing so we cover virtually all pertinent principles. Besides allowing us to study many projection ideas with less complication than would otherwise pertain, we find later that the spherical world can provide an intermediate stage in the development of projections using an ellipsoidal world.

TUTORIAL 10 TRIGONOMETRIC FORMULAE

The Pythagorean formula $\sin^2\theta + \cos^2\theta = 1$ given in Tutorial 2 can be developed by manipulation, including application to the right-angled halves and to the whole of an isosceles triangle, to give a great variety of relational formulae. The following is a selection of use in cartography for any angles α and β.

Function-product formulae

$$2\sin\alpha\sin\beta = \cos(\alpha-\beta) - \cos(\alpha+\beta)$$
$$2\sin\alpha\cos\beta = \sin(\alpha+\beta) + \sin(\alpha-\beta)$$
$$2\cos\alpha\sin\beta = \sin(\alpha+\beta) - \sin(\alpha-\beta)$$
$$2\cos\alpha\cos\beta = \cos(\alpha-\beta) + \cos(\alpha+\beta)$$

Function-sum and -difference formulae

$$\sin\alpha + \sin\beta = 2\sin\tfrac{1}{2}(\alpha+\beta)\cos\tfrac{1}{2}(\alpha-\beta)$$
$$\sin\alpha - \sin\beta = 2\cos\tfrac{1}{2}(\alpha+\beta)\sin\tfrac{1}{2}(\alpha-\beta)$$
$$\cos\alpha + \cos\beta = 2\cos\tfrac{1}{2}(\alpha+\beta)\cos\tfrac{1}{2}(\alpha-\beta)$$
$$\cos\alpha + \cos\beta = -2\sin\tfrac{1}{2}(\alpha+\beta)\sin\tfrac{1}{2}(\alpha-\beta)$$
$$\tan\alpha + \tan\beta = \frac{\sin(\alpha+\beta)}{\cos\alpha\cos\beta}$$
$$\tan\alpha - \tan\beta = \frac{\sin(\alpha-\beta)}{\cos\alpha\cos\beta}$$
$$\cot\alpha + \cot\beta = \frac{\sin(\alpha+\beta)}{\sin\alpha\sin\beta}$$
$$\cot\alpha - \cot\beta = \frac{\sin(\beta-\alpha)}{\sin\alpha\sin\beta}$$

Power formulae

$$\sin^2\alpha = \tfrac{1}{2}(1-\cos 2\alpha)$$
$$\cos^2\alpha = \tfrac{1}{2}(1+\cos 2\alpha)$$
$$\tan^2\alpha = \frac{1-\cos 2\alpha}{1+\cos 2\alpha} = \frac{1}{\cot^2\alpha}$$

Angle-sum and -difference formulae

$$\sin(\alpha+\beta) = \sin\alpha\cos\beta + \cos\alpha\sin\beta$$
$$\sin(\alpha-\beta) = \sin\alpha\cos\beta - \cos\alpha\sin\beta$$
$$\cos(\alpha+\beta) = \cos\alpha\cos\beta - \sin\alpha\sin\beta$$
$$\cos(\alpha-\beta) = \cos\alpha\cos\beta + \sin\alpha\sin\beta$$
$$\tan(\alpha+\beta) = \frac{\tan\alpha + \tan\beta}{1-\tan\alpha\tan\beta}$$
$$\tan(\alpha-\beta) = \frac{\tan\alpha - \tan\beta}{1+\tan\alpha\tan\beta}$$
$$\cot(\alpha+\beta) = \frac{\cot\beta\cot\alpha - 1}{\cot\beta + \cot\alpha}$$
$$\cot(\alpha-\beta) = \frac{\cot\beta\cot\alpha + 1}{\cot\beta - \cot\alpha}$$
$$\sin\left(\alpha+\tfrac{\pi}{2}\right) = +\cos\alpha$$
$$\sin\left(\alpha-\tfrac{\pi}{2}\right) = -\cos\alpha = -\sin\left(\tfrac{\pi}{2}-\alpha\right)$$
$$\cos\left(\alpha+\tfrac{\pi}{2}\right) = -\sin\alpha$$
$$\cos\left(\alpha-\tfrac{\pi}{2}\right) = +\sin\alpha = +\cos\left(\tfrac{\pi}{2}-\alpha\right)$$
$$\sin(\alpha+\beta)\sin(\alpha-\beta) = \sin^2\alpha - \sin^2\beta = \cos^2\beta - \cos^2\alpha$$
$$\cos(\alpha+\beta)\cos(\alpha-\beta) = \cos^2\alpha - \sin^2\beta = \cos^2\beta - \sin^2\alpha$$

Double-angle formulae

$$\sin 2\alpha = 2\sin\alpha\cos\alpha = \frac{2\tan\alpha}{1+\tan^2\alpha}$$
$$\cos 2\alpha = \cos^2\alpha - \sin^2\alpha = 2\cos^2\alpha - 1 = 1 - 2\sin^2\alpha = \frac{1-\tan^2\alpha}{1+\tan^2\alpha}$$
$$\tan 2\alpha = \frac{2\tan\alpha}{1-\tan^2\alpha}$$
$$\cot 2\alpha = \frac{\cot^2\alpha - 1}{2\cot\alpha}$$

CHAPTER 3

Shine a Light: Literal projections

Preamble

The word *projection* relates to the concept equivalent of shining a light from inside a transparent globe and catching the cast image on a piece of paper lying flat against else curled around the surface of the sphere. To obtain a map of familiar style, the paper would have to be translucent, for it would be the outside rather than inside surface that would have our map (unless we made our globe a rotational inversion of Earth). Using (in a darkroom) photographic film rather than the paper, simply place an electric bulb inside the globe, hold a flat piece of positive film against the globe at our point of primary interest, and expose it for the appropriate time. The developed film, viewed from the other side, provides a map that is precisely correct at the prime point, but increasingly distorted in scale as we move away from it. The projector light could be at various locations — within and beyond our globe — while the paper can be curled about the globe as a cylinder else as a cone. In this chapter, we consider such literal projecting, producing what are often called *perspective projections* (a term applied, as we have seen, also to the Aerial projection for which it might be more appropriate), successively for flat, cylindrical, and conical film.

Except for a few theorems and formulae of trigonometry, this chapter brings nothing new mathematically.

The need to observe the variation of spacing of parallels but not of meridians in the maps of this chapter results in unbalanced graticules, being usually 15° for parallels but 30° for meridians.

TUTORIAL 11 PLANE-TRIANGLE FORMULAE

The basic trigonometric functions were defined in Tutorial 2, each as a ratio of two sides of a right-angled triangle. There and in Tutorial 9, various relational formulae were developed applicable to angles in general, but only relating the various functions amongst themselves. There are also derivable formulae relating the functions and sides of a general triangle.

Let *ABG* be any such triangle, with sides of length *a*, *b*, *g* and opposite angles α, β, γ, as illustrated.

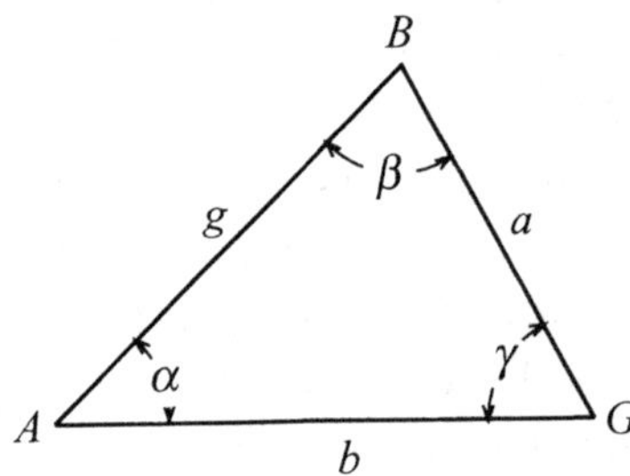

Further, let

$$s = \tfrac{1}{2}(a + b + g)$$

and let

$$r = \sqrt{\frac{(s-a)(s-b)(s-g)}{s}},$$

which can be demonstrated as equalling the radius of the triangle's inscribed circle. The area of the triangle then equals

$$rs = \sqrt{s(s-a)(s-b)(s-g)}$$
$$= \tfrac{1}{2}bg\sin\alpha = \tfrac{1}{2}ga\sin\beta = \tfrac{1}{2}ab\sin\gamma$$

and the following formulae apply. (See Tutorial 10 for other trigonometric formulae.)

Law of sines

$$\frac{a}{\sin\alpha} = \frac{b}{\sin\beta} = \frac{g}{\sin\gamma}$$

Law of cosines

$$a^2 = b^2 + g^2 - 2bg\cos\alpha$$
$$\text{or } \cos\alpha = \frac{b^2 + g^2 - a^2}{2bg}$$
$$b^2 = g^2 + a^2 - 2ga\cos\beta$$
$$\text{or } \cos\beta = \frac{g^2 + a^2 - b^2}{2ga}$$
$$g^2 = a^2 + b^2 - 2ab\cos\gamma$$
$$\text{or } \cos\gamma = \frac{a^2 + b^2 - g^2}{2ab}$$

Law of tangents

$$\frac{b-g}{b+g} = \frac{\tan\frac{1}{2}(\beta-\gamma)}{\tan\frac{1}{2}(\beta+\gamma)}$$
$$\frac{g-a}{g+a} = \frac{\tan\frac{1}{2}(\gamma-\alpha)}{\tan\frac{1}{2}(\gamma+\alpha)}$$
$$\frac{a-b}{a+b} = \frac{\tan\frac{1}{2}(\alpha-\beta)}{\tan\frac{1}{2}(\alpha+\beta)}$$

Half-angle formulae

$$\tan\tfrac{1}{2}\alpha = \frac{r}{s-a},\ \tan\tfrac{1}{2}\beta = \frac{r}{s-b},\ \tan\tfrac{1}{2}\gamma = \frac{r}{s-g}$$
$$\cot\tfrac{1}{2}\alpha = \frac{s-a}{r},\ \cot\tfrac{1}{2}\beta = \frac{s-b}{r},\ \cot\tfrac{1}{2}\gamma = \frac{s-g}{r}$$

and:

$$\sin\tfrac{1}{2}\alpha = \sqrt{\frac{(s-b)(s-g)}{bg}},\ \cos\tfrac{1}{2}\alpha = \sqrt{\frac{s(s-a)}{bg}}$$
$$\sin\tfrac{1}{2}\beta = \sqrt{\frac{(s-g)(s-a)}{ga}},\ \cos\tfrac{1}{2}\beta = \sqrt{\frac{s(s-b)}{ga}}$$
$$\sin\tfrac{1}{2}\gamma = \sqrt{\frac{(s-a)(s-b)}{ab}},\ \cos\tfrac{1}{2}\gamma = \sqrt{\frac{s(s-g)}{ab}}$$

DEVELOPABLE SURFACES

Flattenable surfaces

The cylinder and the cone were previously referred to as planes curled around our globe. For map projecting the actual process is the reverse — we project onto the cylinder else the cone then uncurl the curved surface to obtain a plane. In this sense, the two — cylinder and cone — are referred to as *developable surfaces*, or surfaces that, unlike other curved surfaces, can be physically manoeuvred to form a plane without any distortion between elements on the curved surface. The plane is also usually included under the rubric of developable surfaces, although it is already a plane. Projection directly onto a plane is termed *azimuthal*. Projection onto the cylinder and cone are called *cylindrical* and *conical* respectively. We use the term *mode* generically for the three.

In this chapter, we have the developable surface in bare contact with the surface of the globe, such that straight lines within the developable surface that contact the globe are tangential to it. In Chapter 5, we have them intersecting the globe — referred to as being *secantal*. Anticipating that, we refer to the developments in this chapter (and in the next) as being in *tangential* form. Thus, in this chapter the plane shares a single *contact point* with the globe and the others a *contact circle* — the cylinder with a great circle, the cone with a lesser circle. When developed, the contact circle becomes the *contact line*. Because the point is a degenerate form of circle and line, we can refer to all as having a contact circle and (when developed) a contact line.

The complexity of the geometrical development depends on the positioning of the developable surface relative to the axis (hence to the longitude/latitude grid) of the globe, which is referred to as *aspect*. For all three modes the simplest development occurs when the mapping surface — cylinder, cone else azimuthal plane — is symmetrical about the axis. This is called for each its *Simple aspect*. At any other position the mapping is said to be at *Oblique aspect*, though gyrated 90° from Simple the aspect is usually called *Transverse aspect* or *Equatorial aspect* (when Oblique may be used exclusive of such). The Simple aspect for an Azimuthal, being centred on a Pole, is also called *Polar aspect*; the Simple aspect for a Cylindrical, being in contact with the Equator, is also called *Equatorial aspect*. Through Chapter 7 we address only the Simple aspect for each projection. Elaboration to the oblique occurs with that for other projections in Chapter 8. At Simple aspect the azimuthal plane is in contact with a Pole (a 90° parallel), the cylinder in contact with the Equator (the 0° parallel) and the cone in contact with some other parallel.

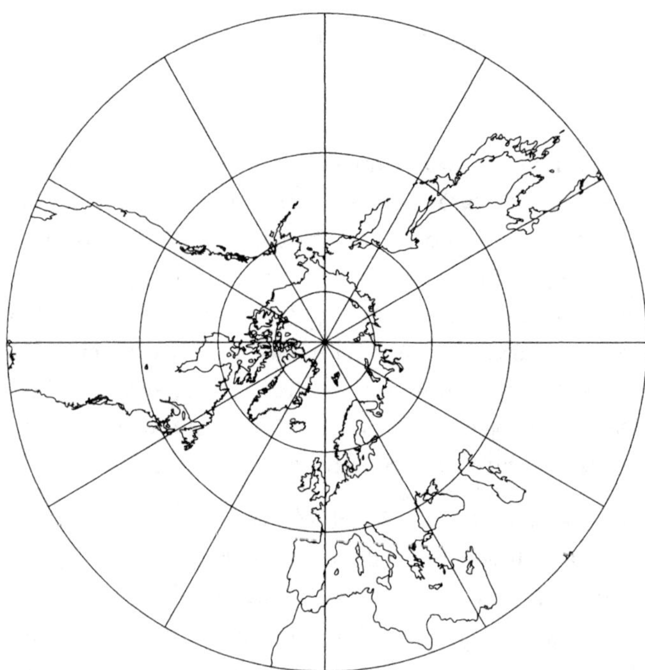

Exhibit 3–1: The Gnomonic or Central Azimuthal projection at Simple aspect.
North Pole to 30°N

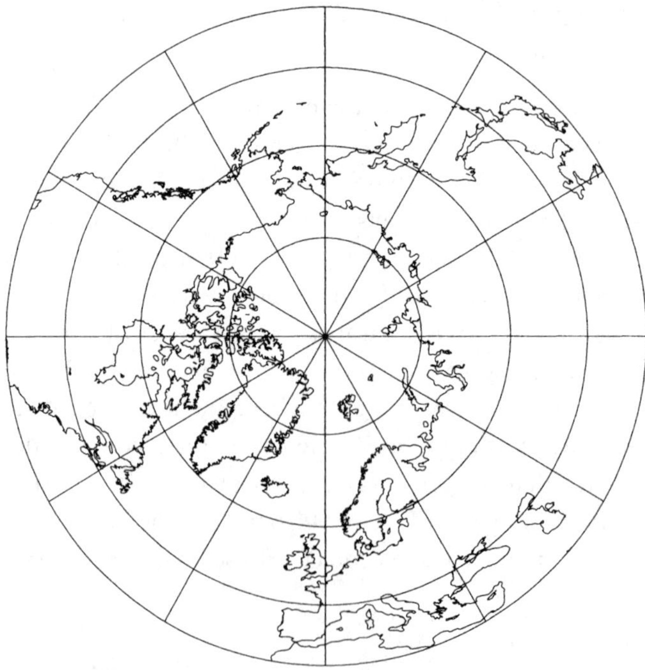

Exhibit 3–2: The Orthographic Azimuthal projection at Simple aspect.
North Pole to 30°N

PERSPECTIVE AZIMUTHAL PROJECTIONS

Basics

Azimuthal projections are based on projecting directly onto a flat plane, a plane contacting the globe at one point for the tangential form. Simple aspect occurs when the mapping surface and globe are co-axial — (i.e., the contact point, which becomes the focal point of our map, is a Pole) — hence also said to be at Polar aspect.

The Aerial projection of the preceding chapter is of this nature, though its projective lines meet at a point above the mapping plane. Here we deal with projective lines radiating from an illuminating projection point below the plane, either within the globe else beyond its confines. Specifically the projection point is on the perpendicular to the mapping plane at its point of contact with the globe. At Simple aspect, this means on Earth's axis — extended as necessary. The process is inherently similar to that for the Aerial. Given a projection point we obtain a formula for the distance $\dot{\rho}$ from the focal point to a general point and augment this with an unaffected angle to obtain polar co-ordinates for plotting. At Polar aspect, this measurement is along a meridian — independent of the choice of meridian, it provides a complete circumpolar solution. Obviously, such a projecting point lies in the plane of every great circle through the focal point. Hence every such great circle shows as a straight line on the map and, conversely, every straight line through the focal point represents (part of) a great circle.

The azimuthal map

As with the Aerial projection, the distance from the focal point to any specific point on the map depends, with Simple (or Polar) aspect, purely on the latitude of the specific point, while its exact position is set by a rotational angle $\dot{\theta}$ that depends purely on the longitude of the point (and some reference longitude).

The generic plotting equations for all azimuthal maps at Simple aspect are as shown with the Aerial projection as Equations 30b then, depending on which longitude is to be along a specific plotting half-axis, Equations 31b, 31c & 32a.

Every azimuthal map at Simple aspect has the meridians radiating out from the focal point, the parallels forming circles centred on the focal point. Any two meridians are at mutual angle equal to their difference in longitude, hence bimeridians appear as straight lines passing through the focal point. The spacing of the parallels varies with the projection, indeed characterizes the projection; the two maps opposite illustrate this with the upper one having radial divergence, the lower convergence. Azimuthal maps have an obvious merit when depicting circumpolar regions, but are widely used for other regions.

Projecting onto the azimuthal plane

The **Central Azimuthal** projection — usually called the **Gnomonic** projection[†] — has the projection point at the centre of the globe. It is illustrated below for one cross-sectional plane, with sample projection rays shown as dotted lines. Here and in succeeding geometric illustrations we label the centre of the globe (the *geocentre*) as G, the focal point as F, and the general point as $P = (\lambda, \phi)$. To distinguish the mapping plane forming the developable surface, it is shown in the style of a traditional cartographic *neatline*, with mapped points identified on its outer side, adorned with their dot overscript. That neatline carries the mappings of points on the bimeridian of the cross-sectional plane. The dashed line (altered to have interspersed dots where coincident with a projection ray) is Earth's axis and the double-dotted line represents the Equator. Focal point F is thus a Pole (though readily seen as the North Pole, our development allows it to be the South Pole). We denote by γ the angle subtended at G by the chord PF, which angle is identically the polar distance of point P (i.e., of latitude $|\phi|$).

As the length of $G\dot{F}$ (identically that of GF) is R, we get

$$\dot{\rho} = \text{length } \dot{F}\dot{P} = R\tan\gamma \qquad \text{........ (44a}$$

Inherently positive angle γ is the complement of the unsigned latitude ϕ, hence

$$\dot{\rho} = R\tan\gamma = R\tan\left(\tfrac{\pi}{2} - |\phi|\right) \; = R\cot|\phi| \qquad \text{....... (44b}$$

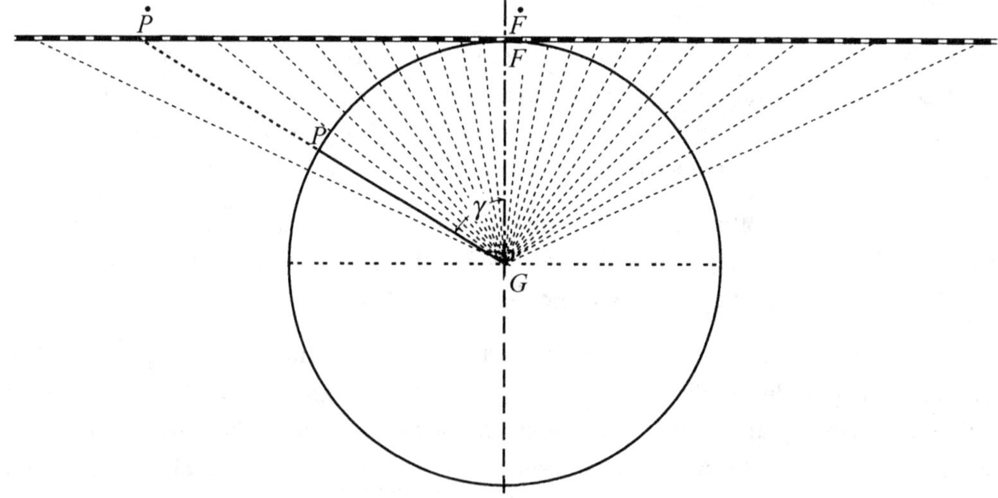

Exhibit 3–3: The geometry of central projecting.

[†] From *gnomon*, a name for the central shadow-casting rod of a sundial, < Latin < Greek: γνωμον *inspector*.

which is infinite for $\phi=0$, precluding that value (when the projecting rays would be parallel with the developable surface). With latitudes south of the Equator being negative, this becomes

$$\dot{\rho} = R\cot|\phi| = R\cot(\oplus\phi) \quad = \oplus R\cot\phi \quad \text{for } \phi \neq 0 \qquad \text{(45a}$$

Substitution in any of Equations 31b, 31c & 32a gives plotting equations for the tangential Central Azimuthal or Gnomonic projection at Simple aspect. Choosing the longitude to lie along the $-y$ half-axis, therefore using Equations 32a, gives

$$\begin{aligned} \text{for } \phi \neq 0: \quad \dot{x} &= \oplus\dot{\rho}\sin(\lambda - \lambda_{-y}) &&= +R\cot\phi\sin(\lambda - \lambda_{-y}) \\ \dot{y} &= -\dot{\rho}\cos(\lambda - \lambda_{-y}) &&= -\oplus R\cot\phi\cos(\lambda - \lambda_{-y}) \end{aligned} \qquad \text{(45b}$$

It is obvious from the Exhibit 3–3 that the increment in $\dot{\rho}$ for a given decrement in latitude escalates rapidly as γ increases (i.e., as $|\phi|$ decreases) tending to infinity as $|\phi|$ approaches zero. That escalation is illustrated in the table immediately below, applied to a *unit globe* (i.e., with $R=1$). Anticipating other projecting points and other modes, the angle is equated with a more general angle, and the complement of this angle relative to the right angle is also shown. The growing escalation limits the utility of the Gnomonic to distinctly less than one hemisphere. Because the projection point is the centre of the globe and hence in all globe-bisecting planes, however, it has the distinct and unique utility of all sections of great circles projecting into straight lines and of all straight lines on the map being projections of arcs of great circles. Otherwise, the projection finds little use.

	90°	80°	70°	60°	50°	40°	30°	20°	10°	0°
γ	0°	10°	20°	30°	40°	50°	60°	70°	80°	90°
$\dot{\rho}$	0.0000	0.1763	0.3640	0.5774	0.8391	1.1918	1.7321	2.7475	5.6712	∞
difference	0.1763	0.1876	0.2134	0.2617	0.3527	0.5403	1.0154	2.9238	∞	

Central projecting: spacing of parallels for unit globe.

The **Stereographic Azimuthal** projection has projection from the diametrically opposite point[†] of the globe relative to the focal point — i.e., the other Pole for Polar aspect. The relevant geometry is illustrated overleaf in Exhibit 3–4, with V denoting the projection point and ξ the angle subtended at that point by the chord PF. An ancient theorem of geometry [1956a] proves that the angle subtended at the centre by any chord of a circle is twice the angle subtended at any point on the circumference by that chord. Applying it to chord FP and point V on the circumference shows that

$$\dot{\rho} = \text{length } F\dot{P} = 2R\tan\xi \quad = 2R\tan\frac{\gamma}{2} \qquad \text{(45c}$$

[†] The word for the diametrically opposite point is its *antipodes* else *antipodean point*; some islands in the Pacific ocean that are opposite a spot in the Cherbourg peninsula of northern France are called the Antipodes islands. More generally, 'the Antipodes' implies just the opposite side of the world.

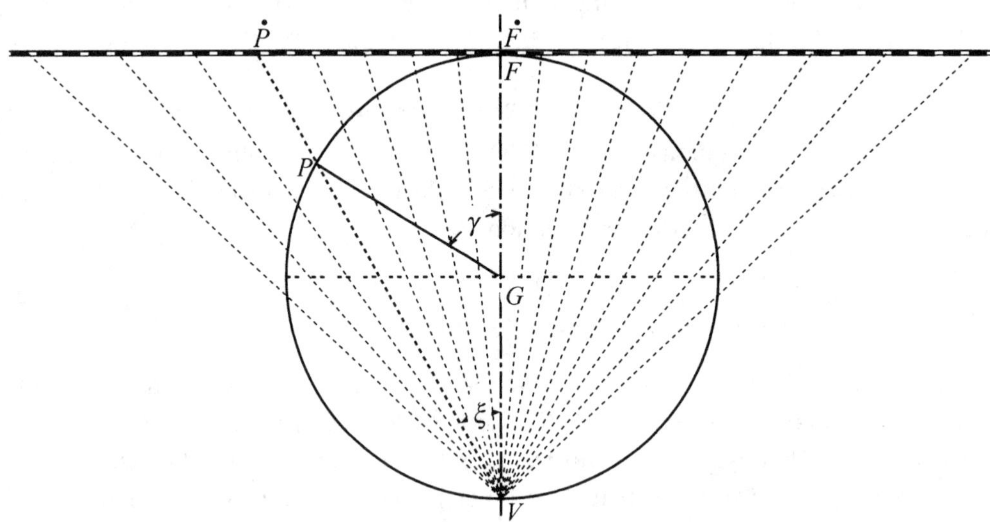

Exhibit 3–4: The geometry of stereographic projecting.

Relating γ to latitude as previously done (which applies to any mapping in azimuthal mode at Simple aspect), this becomes

$$\dot{\rho} = 2R\tan\frac{1}{2}\gamma = 2R\tan\frac{1}{2}\left(\frac{\pi}{2} - |\phi|\right) = 2R\tan\frac{1}{2}\left(\frac{\pi}{2} - \oplus\phi\right) = 2R\tan\left(\frac{\pi}{4} - \frac{\oplus\phi}{2}\right) \quad \ldots \text{(46a}$$

which clearly extends correctly from one Pole across the Equator into the other hemisphere. Substitution in any of Equations 31b, 31c & 32a gives plotting equations for the tangential Stereographic Azimuthal or simply **Stereographic** projection at Simple aspect. Choosing the longitude to lie along the $-y$ half-axis, using Equations 32a gives:

$$\begin{aligned} \dot{x} &= \oplus\dot{\rho}\sin(\lambda - \lambda_{-y}) &&= \oplus 2R\tan\left(\frac{\pi}{4} - \frac{\oplus\phi}{2}\right)\sin(\lambda - \lambda_{-y}) \\ \dot{y} &= -\dot{\rho}\cos(\lambda - \lambda_{-y}) &&= -2R\tan\left(\frac{\pi}{4} - \frac{\oplus\phi}{2}\right)\cos(\lambda - \lambda_{-y}) \end{aligned} \quad \ldots\ldots\ldots \text{(46b}$$

Although the Stereographic can map more than one hemisphere, it also ultimately tends to infinity, now as the other Pole is approached. On the way, the mapping of higher latitudes in that other hemisphere becomes grossly distorted. Besides often being used for maps of Polar regions, however, the projection gains much wider use (for smaller regions) because it has the distinct advantage among azimuthal maps of being conformal — i.e., it maintains shape in a local sense — a matter discussed in the Chapter 4.

Values of $\dot{\rho}$ for intervals of 10° and the difference between adjacent values are shown for the unit globe immediately next. In the primary hemisphere, the spacing of parallels escalates, but much more modestly than for the Central.

	90°	80°	70°	60°	50°	40°	30°	20°	10°	0°
γ	0°	10°	20°	30°	40°	50°	60°	70°	80°	90°
$\dot{\rho}$	0.0000	0.1750	0.3527	0.5359	0.7279	0.9326	1.1547	1.4004	1.6782	2.0
difference		0.0875	0.0888	0.0916	0.0960	0.1023	0.1110	0.1229	0.1389	0.1609

Stereographic projecting: spacing of parallels for unit globe.

Thirdly, we consider having the projector at the technically impossible point of infinity, making the rays parallel and all impacting our mapping surface at right angles, hence the name **Orthographic Azimuthal** projection. From the geometry shown below, we see that

$$\dot{\rho} = \text{length } \dot{F}\dot{P} = R \sin\gamma \qquad \text{. . . . . (47a}$$

Again, relating γ to latitude in the standard azimuthal manner, this becomes

$$\dot{\rho} = R \sin\gamma = R \cos|\phi| \quad = R \cos \oplus\phi \quad = R \cos \phi \qquad \text{. . . . . (47b}$$

Substitution in any of Equations 31b, 31c & 32a gives plotting equations for the tangential Orthographic Azimuthal or simply **Orthographic** projection at Simple aspect. Choosing 90°E to lie along the $+x$ half-axis, therefore using Equations 31c, gives:

$$\begin{aligned} \dot{x} &= \ \ + \dot{\rho} \sin\lambda &&= \ \ + R\cos\phi \sin\lambda \\ \dot{y} &= -\oplus \dot{\rho} \cos\lambda &&= -\oplus R\cos\phi \cos\lambda \end{aligned} \qquad \text{. . . . . (47c}$$

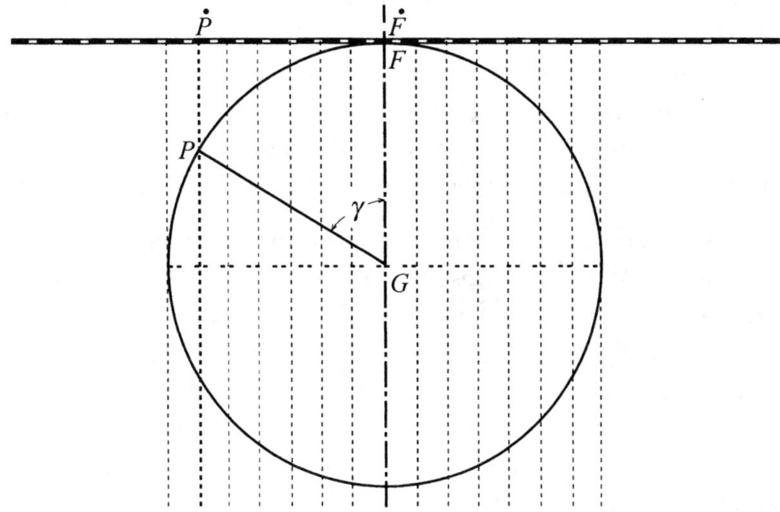

Exhibit 3–5: The geometry of orthographic projecting.

The geometry visibly shows that the orthographic form brings a progressive narrowing of the spacing of parallels, tending to zero as the Equator is approached. Technically, it projects the entire globe onto the mapping surface, but it overlays the second hemisphere on the first, so it cannot be used beyond the Equator. Were one to literally project to form the map, one would have to have just one hemisphere rather than a whole globe. Again, values of $\dot{\rho}$ for intervals of 10° and the difference between adjacent values are shown for the unit globe immediately below. Unlike for the Central and Stereographic projections, the spacing shrinks as we move away from the contact line.

	90°		80°		70°		60°		50°		40°		30°		20°		10°		0°
γ	0°		10°		20°		30°		40°		50°		60°		70°		80°		90°
$\dot{\rho}$	0.0000		0.1736		0.3420		0.5000		0.6428		0.7660		0.8660		0.9397		0.9848		1.0
difference		0.1736		0.1684		0.1580		0.1428		0.1233		0.1000		0.0737		0.0451		0.0152	

Orthographic projecting: spacing of parallels for unit globe.

Orthographic projecting makes all circles on the globe's surface appear on the map as ellipses, including as true circles those in a plane parallel to the mapping surface and as straight lines all great circles passing through the focal point. When focused on a Pole, the Orthographic projection not only maps parallels into true circles, but those circles are equal in size to the parallels. Hence, we can say that the scale of the map holds true along all parallels. (This illustrates well the difference that can exist on any map between scales in different directions at a given point, for the above table shows that uniformity of scale does not apply radially along the meridians relative to the focal point.)

The Gnomonic and Orthographic projections are illustrated on page 42. The latter, in Exhibit 3–2, shows as a complete hemisphere with increasing crowding towards the edge — a circle of identical radius to that of the globe. The Gnomonic map in Exhibit 3–1 has the spacing of parallels escalating away from the Pole, to be infinite in size were it carried to the Equator; it is shown down to latitude 45°, where it reaches the same diameter as the Orthographic and to which latitude it provides a reasonable map.

Each of the described projections — the Gnomonic, the Stereographic and the Orthographic — is merely a specific case of a general arrangement. The projection point could be anywhere along the axis. We now look at the **General Perspective Azimuthal** projection, which allows the projection point to be any distance hR from point F (i.e., from the contact point of globe and mapping plane, measured positively into the globe — e.g. $h = +2$ for the Stereographic). The relevant geometry, again with V denoting the projection point and ξ the angle subtended at that point by the chord PF, is illustrated opposite in Exhibit 3–6, with h somewhat less than 2.

Bisecting triangle GFP into mirror-image right-angled triangles shows that:

$$\text{length of } FP = 2R\sin\frac{\gamma}{2}$$

$$\text{angle } VFP = \text{angle } GFP = \frac{1}{2}(\pi - \gamma) = \frac{\pi}{2} - \frac{\gamma}{2}$$

$$\text{angle } FPV = \pi - (\text{angle } VFP + \text{angle } PVF) = \pi - \left\{\left(\frac{\pi}{2} - \frac{\gamma}{2}\right) + \xi\right\} = \frac{\pi}{2} + \frac{\gamma}{2} - \xi$$

By the Law of Sines (see Tutorial 11, page 40) applied to triangle *VFP*,

$$\frac{\text{length of } FV}{\sin(\text{angle } FPV)} = \frac{\text{length of } FP}{\sin(\text{angle } FVP)} \quad \text{or} \quad \frac{hR}{\sin\left(\frac{\pi}{2} + \frac{\gamma}{2} - \xi\right)} = \frac{2R\sin\frac{\gamma}{2}}{\sin\xi}$$

Hence, using an angle-difference formula from Tutorial 10 (page 38),

$$hR\sin\xi = 2R\sin\frac{\gamma}{2}\sin\left(\frac{\pi}{2} + \frac{\gamma}{2} - \xi\right) \quad = 2R\sin\frac{\gamma}{2}\cos\left(\frac{\gamma}{2} - \xi\right)$$

$$\sin\xi = 2h^{-1}\sin\frac{\gamma}{2}\cos\left(\frac{\gamma}{2} - \xi\right) \quad = 2h^{-1}\left(\sin\frac{\gamma}{2}\right)\left(\cos\frac{\gamma}{2}\cos\xi + \sin\frac{\gamma}{2}\sin\xi\right)$$

$$= h^{-1}\left(2\sin\frac{\gamma}{2}\cos\frac{\gamma}{2}\cos\xi + 2\sin\frac{\gamma}{2}\sin\frac{\gamma}{2}\sin\xi\right) \quad = h^{-1}\sin\gamma\cos\xi + 2h^{-1}\sin^2\frac{\gamma}{2}\sin\xi$$

So
$$\left\{1 - 2h^{-1}\sin^2\frac{\gamma}{2}\right\}\sin\xi = h^{-1}\sin\gamma\cos\xi$$

$$\tan\xi = \frac{h^{-1}\sin\gamma}{1 - 2h^{-1}\sin^2\frac{\gamma}{2}} = \frac{\sin\gamma}{h - 2\sin^2\frac{\gamma}{2}} = \frac{\sin\gamma}{h - (1 - \cos\gamma)} = \frac{\sin\gamma}{h - 1 + \cos\gamma} \quad \ldots\ldots \text{(49a}$$

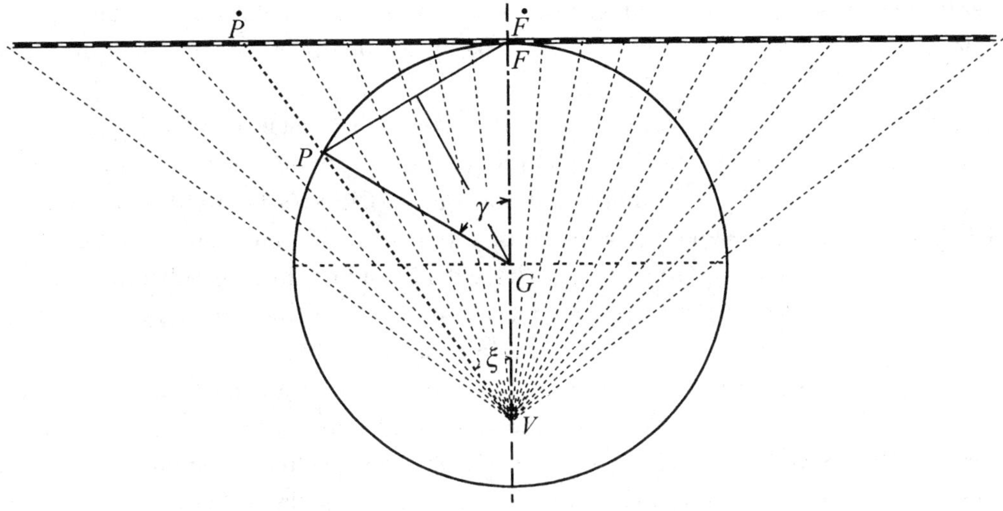

Exhibit 3–6: The geometry of general perspective projecting.

Hence,

$$\dot{\rho} = \text{length } \dot{F}\dot{P} = hR\tan\xi \quad = \frac{hR\sin\gamma}{h-1+\cos\gamma} \quad = \frac{R\sin\gamma}{1-h^{-1}(1-\cos\gamma)} \qquad \text{. . . . . (50a}$$

Converting from γ to ϕ as previously, this becomes

$$\dot{\rho} = \frac{hR\cos\phi}{h-1+\oplus\sin\phi} \quad = \frac{R\cos\phi}{1-h^{-1}(1-\oplus\sin\phi)} \qquad \text{. . . . . . . (50b}$$

Substitution in any of Equations 31b, 31c & 32a gives plotting equations for the tangential General Perspective Azimuthal projection at Simple aspect (see footnote on page 32 regarding nomenclature). Choosing the longitude to lie along the $-y$ half-axis, therefore using Equations 32a, gives:

$$\begin{aligned} \dot{x} &= \oplus\dot{\rho}\sin(\lambda-\lambda_{-y}) &&= \oplus\frac{R\cos\phi}{1-h^{-1}(1-\oplus\sin\phi)}\sin(\lambda-\lambda_{-y}) \\ \dot{y} &= -\dot{\rho}\cos(\lambda-\lambda_{-y}) &&= -\frac{R\cos\phi}{1-h^{-1}(1-\oplus\sin\phi)}\cos(\lambda-\lambda_{-y}) \end{aligned} \qquad \text{. . . . . . . (50c}$$

Except for the opposite sign of h, the final form of Equation 50a is identically that of Equation 30a and Equations 50b identically Equations 32b. Equation 50c and Equation 32c differ only by choice of fixed longitude. All those earlier equations were obtained for the Aerial projection. Not surprisingly, we see that other projection as a further generalization of this new projection, encompassed by the new equations if we invert the signage of h in the other, making $h<0$ for aerial points.

Equations 50b and 50c cover the preceding perspective projections with $h=1$ for the Gnomonic, $h=2$ for the Stereographic, and (despite the infinite not being accommodated in the development) with $h=\infty$ for the Orthographic. Various people have advocated other finite values in pursuit of reduced distortion in some particular way, including:

- P. de la Hire — $h = 2+\sqrt{2}/2 = 2.707_{\sim}$ to make the Equator twice the length of the 45° parallel
- P. A. Parent — $h = 2.594_{\sim}$ for minimally distorted radial distance to the Equator
- A. R. Clarke — $h = 2.35$ to 2.65 including 2.4 to provide the 99° radial span embraced by twilight — **Clarke's Twilight** projection
- H. F. Gretschel — $h = 1+\frac{1}{2}(1+\sqrt{5}) = 2.6180_{\sim}$ to give true meridional scale at the Equator
- P. Fischer — $h = (\pi/2-1)^{-1} = 2.752_{\sim}$ to have the Equator at its true radial distance
- J. Lowry — $h = 2.6858_{\sim}$ the geometric mean for having true radial distance to each of the 18 parallels 0°, 5°, . . . 85°
- A. L. Nowicki — $h = 2.537\,48_{\sim}$ to provide the slightly greater-than-a-hemisphere surface of the Moon visible from Earth

Further choices have been advocated for the General Perspective when enhanced by the secantal approach introduced in Chapter 5. Distortion is the subject of Chapter 13. Optimizing distortion is the subject of Chapter 14.

PERSPECTIVE CYLINDRICAL PROJECTIONS

Basics

Cylindrical projections are based on the globe fitting within a cylinder, with the contact line being any great circle and the projection point in the plane of the contact line. For the central cylindrical projection the projection point is fixed at the geocentre. Otherwise the projection point circulates about the geocentre, mapping from one position all points on the further side of the globe in the plane through the projection point orthogonal to the contact plane. (As a literal projection, this means an infinite number of infinitesimally narrow collimated exposures.) We define the intermediate parameters $\dot{u}$ and $\dot{v}$ to be respectively the distance around the cylinder from some reference line orthogonal to the contact line and the orthogonal distance from the contact line to the point (as we did for the Plate-Carrée in Chapter 1). The value of the first parameter clearly depends on the choice of reference line. These parameters form a system of rectilinear co-ordinates.

Simple aspect for a cylindrical projection occurs when the contact line is the Equator. Then, the projection point is in the Equatorial Plane and, for a given point, in the meridional plane that contains the said point. We look at the geometry for each projection similarly to the method for azimuthal projecting — working in a 2-dimensional setting to obtain the formula for the distance $\dot{v}$ of any point on a specific meridian from its contact point on the Equator. With projecting in the plane of the meridian, all the points on any one meridian fall on a ray of the cylinder (i.e., a straight line along the surface), so parameter $\dot{v}$ is a function of latitude only (dependent on the particular projection). Parameter $\dot{u}$, which expresses the distance of a point around the curve of the cylinder, then depends purely and proportionally on the longitude of the global point relative to some reference meridian. Because the formula for $\dot{v}$ holds regardless of actual meridian, it provides a complete around-the-world solution.

The cylindrical map

Developing the cylinder into a map requires cutting the mapping surface lengthwise (not necessarily on a straight line), then flattening. Obviously, one cuts it approximately (if not precisely) on the opposite side to a chosen central meridian of longitude Λ_0 or λ_0. We again, as explained in Chapter 1, interpret lambda to mean longitude relative to the chosen central meridian. Because the length of the Equator on the map is identically the $2\pi R$ that it is on the globe, and traversing along it is proportional to longitudinal difference, we get

$$\dot{u} = R\lambda \qquad \text{. . . . . (51a}$$

Both $\dot{u}$ and $\dot{v}$ have positive and negative values, each ranging symmetrically in absolute value on either side of zero — the one about the central meridian, the other about the Equator. We choose these two lines as our axes — the central meridian as the y axis, the Equator as the x axis — then equate $\dot{x}$ with $\dot{u}$ and $\dot{y}$ with $\dot{v}$. (We do otherwise for the transverse case, and generally incorporate intermediate parameters when using differential geometry in later chapters. Hence, we deploy intermediate parameters here, as we did in Chapter 1 for the Plate-Carrée, although we could have gone directly to the final map co-ordinates.) The longitude/latitude grid on any cylindrical map at Simple aspect is thus rectangular (as shown below), with uniformly spaced straight verticals for

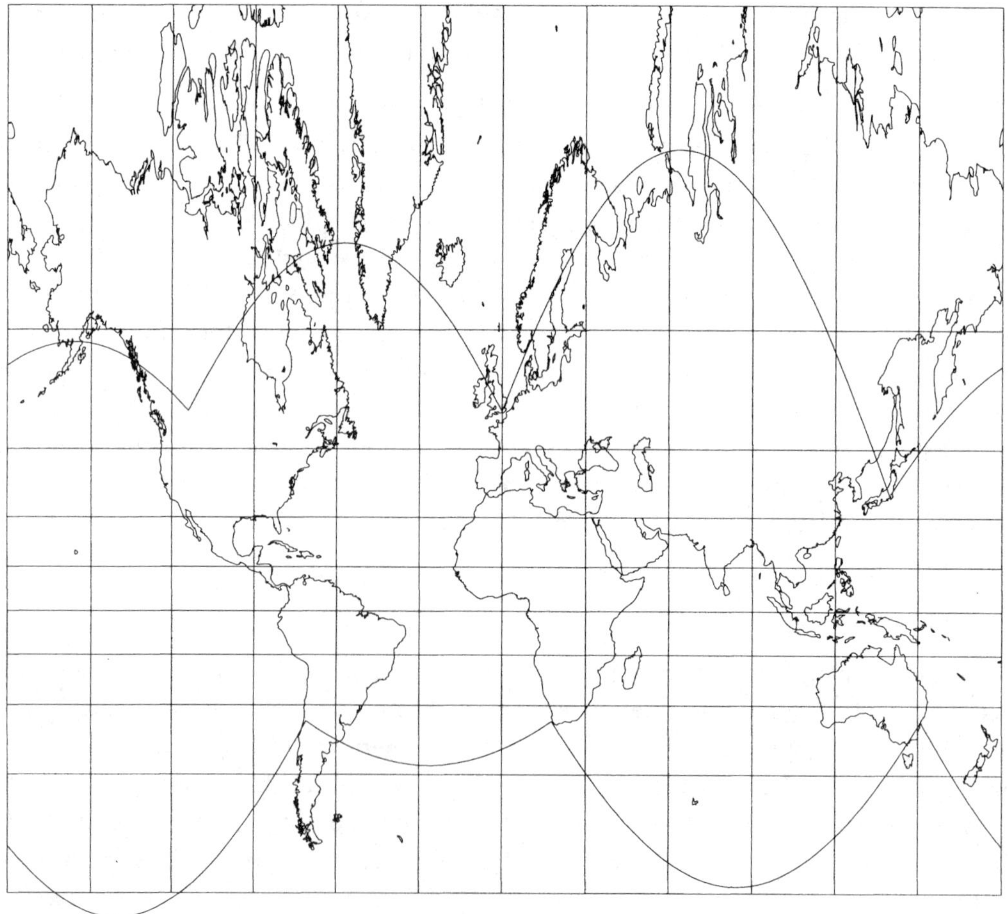

Exhibit 3–7: Central Cylindrical projection at Simple aspect.
Cut off at 75°N and 67°30'S.
With great-circle routes as detailed for Exhibit 4–3 on page 82.

the meridians, while parallels are straight horizontal lines spaced according to the projection. As the length of parallels on the globe are proportional to the cosine of the latitude, the uniformity of spacing meridians forces growing expansion of scale along the parallels of the map. The following table shows, for steps of 10°, the cosine then its reciprocal for each angle, the latter representing the resulting expansion of scale — becoming infinite for each Pole, where a proper line, however short, represents a point.

latitude:	0°	10°	20°	30°	40°	50°	60°	70°	80°	90°
cos	1.0000	0.9848	0.9397	0.8660	0.7660	0.6428	0.5000	0.3420	0.1736	0.0000
$\cos^{1}$	1.0000	1.0154	1.0642	1.1547	1.3054	1.5557	2.0000	2.9238	5.7588	∞

Cylindrical projections: scaling of meridians.
(Tangential form at Simple aspect: see text above for explanation.)

We now address the parameter $\dot{v}$ that defines the spacing of the parallels. The 2-dimensional geometry of these cylindrical projections is essentially the same as for the Azimuthals, except for the contact of mapping surface and globe being at the Equator rather than a Pole and with the projection point being always in the Equatorial plane rather than always on the axis of the globe. The result of that difference is just to make the angle γ of the earlier illustrations identically the latitude rather than the polar distance.

Assuming that the central meridian lies along the y axis with North at the top of the map, the generic plotting equations for the cylindrical mode at Simple aspect are:

$$\dot{x} = \dot{u} = R\lambda$$
$$\dot{y} = \dot{v} \qquad \text{. . . . . (53a}$$

with λ being the relative longitude.

Projecting onto the cylinder

The diagrams and equations developed for projecting onto the azimuthal plane apply similarly to the cylindrical situation. The cylinder simultaneously makes contact with opposite sides of the globe — specifically at the Equator for Simple aspect — but we consider projection onto one side only, as illustrated overleaf in Exhibit 3–8. More particularly, as with the azimuthal, we base our development on projection onto one meridional line — here a single meridian rather than the abutting ends of two. All of the four geometric diagrams developed for azimuthal projecting can be used for cylindrical projecting if the double-dotted lines are interpreted as Earth's axis, the dashed lines as the Equator, with distance $\dot{\rho}$ now variable $\dot{v}$. Angle γ then becomes the latitude.

To have the Earth's axis in the familiar vertical position we need only rotate a diagram clockwise 90°. Exhibit 3–11 is just the earlier Exhibit 3–6 for general perspective projecting rotated by a right angle. As before the projecting point is some way within the globe, but could be beyond the further side of the globe so beyond the further side of the cylinder (hence the omission of that component). We turn, however, first to the specific choices of central, stereographic, and orthographic projecting. In adapting the equations developed for azimuthal projecting, we must substitute parameter symbol $\dot{v}$ for $\dot{\rho}$.

For the **Central Cylindrical** projection, Equation 44a translates to

$$\dot{v} = R \tan\phi \qquad \text{(54a)}$$

which self-corrects to negative value for negative latitude. The plotting equations for the tangential Central Cylindrical projection at Simple aspect are thus:

$$\begin{aligned} \dot{x} &= \dot{u} = R\lambda \\ \dot{y} &= \dot{v} = R\tan\phi \end{aligned} \qquad \text{(54b)}$$

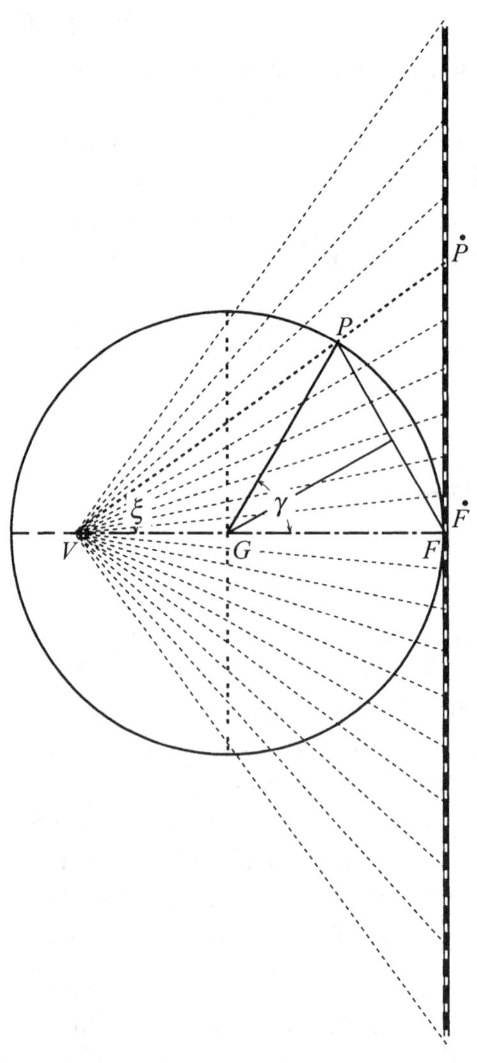

Exhibit 3–8: The geometry of general perspective projecting – cylindrical version.

As with the Central Azimuthal, the spacing of parallels for the Central Cylindrical projection escalates rapidly as we go away from our contact line — here the Equator — identically per degree in stepping away from the Equator as in stepping away from the Pole in the Azimuthal (shown in the table on page 45). Neither Pole can appear on the map and regions close to the Pole become vastly distorted, but the projection can provide a useful map up to about 45° from the Equator. It is illustrated in Exhibit 3–7 on page 52.

For the **Stereographic Cylindrical** projection, projected from the antipodal point, Equation 45c translates to

$$\dot{v} = 2R \tan\frac{\phi}{2} \qquad \text{. . . . . (55a}$$

which again self-corrects for hemispheric signs. The plotting equations for the tangential Stereographic Cylindrical projection at Simple aspect, which is very rarely seen, are thus:

$$\begin{aligned} \dot{x} &= \dot{u} = R\lambda \\ \dot{y} &= \dot{v} = 2R \tan\frac{\phi}{2} \end{aligned} \qquad \text{. . . . . (55b}$$

Again, the spacing for each step away from the Equator is identical with that for each equal angular step away from the Pole in the corresponding Azimuthal projection, shown in the table on page 47. Given that there is no call to make the angle between projection ray and the axial line of the projecting greater than 45° for a hemisphere, the Stereographic Cylindrical projection has a relatively subdued distortion even when used for a whole hemisphere. Its sparse use reflects its lack of advantages compared with rivals that present their own advantages.

For the **Orthographic Cylindrical** projection, with its parallel rays, Equation 47a translates to

$$\dot{v} = R \sin\phi \qquad \text{. . . . . (55c}$$

which again self-corrects for hemispheric signs. The plotting equations for the tangential Orthographic Cylindrical projection at Simple aspect (used to a limited extent despite severe distortion in the form of latitudinal compaction towards the Poles) are thus:

$$\begin{aligned} \dot{x} &= \dot{u} = R\lambda \\ \dot{y} &= \dot{v} = R \sin\phi \end{aligned} \qquad \text{. . . . . (55d}$$

These produce a map of height $2R$. With the length of the Equator $2\pi R$ this gives the map an area of $4\pi R^2$, which is identical with that of our globe. Chapter 4 shows that this equality of area between map and globe does not apply just overall but also applies to any defined section (e.g., the area of the section delimited by any two meridians and any two parallels). Thus, the projection is an equal-area projection, serving a purpose discussed generally in Chapter 4. Though the compaction of parallels in higher latitudes is extreme as illustrated in Exhibit 4–3 on page 82 (to be opposite a contrasting map) for many purposes (e.g., world forests, population densities) the highest latitudes are of little consequence, so the extreme distortion can often be ignored.

As with the Azimuthals, each of these three projections in the cylindrical mode is a specific case of the **General Perspective Cylindrical** projection. Equation 50a translates to

$$\dot{v} = \frac{hR\sin\phi}{h-1+\cos\phi} = \frac{R\sin\phi}{1-h^{-1}(1-\cos\phi)} \qquad \text{(56a}$$

where h again is such that hR is the distance of the projection point from the contact point of the globe and the mapping surface, measured positively into the globe. The plotting equations for the tangential General Perspective Cylindrical projection at Simple aspect are thus:

$$\dot{x} = \dot{u} = R\lambda$$
$$\dot{y} = \dot{v} = \frac{R\sin\phi}{1-h^{-1}(1-\cos\phi)} \qquad \text{(56b}$$

In the infinite range of choices for projection depth, only the single choice of infinity — the Orthographic Cylindrical projection — finds much use, and even that is both modest and usually in the secantal form described in Chapter 5 rather than the tangential. The most used projection is a cylindrical projection but not a literal projection; it, the Mercator, is developed in Chapter 4.

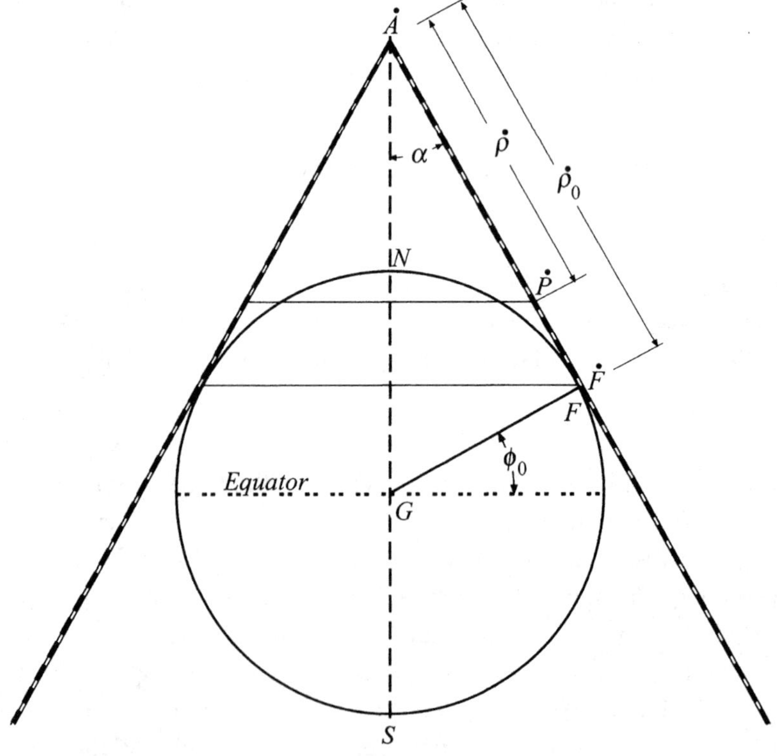

Exhibit 3–9: The cross-section of cone and globe.

PERSPECTIVE CONICAL PROJECTIONS

Basics

Conical projections are based on fitting the globe within a hollow circular cone, the contact line being a circle fixed in size by the taper of the cone. The narrower the cone is the greater the contact circle will be. At the extreme of narrowness the cone becomes a cylinder and the contact circle a great circle with the conical projection merging into the cylindrical. The more open the cone is the smaller the contact circle; at the extreme the cone becomes a plane and the contact circle a point with the conical projection merging into the azimuthal. Exhibit 3–9 provides a cross-section of a cone sitting on the globe, expressed in the style adopted earlier in this chapter. The developable surface now appears symmetrically on each side of the globe. To maintain the established notation (and anticipate the development of the cone into a map), we adorn points and related measurements on the cone with the dot overscript.

The taper of a cone is characterized by the ratio of its cross-sectional radius to the length of slope from its apex. This ratio, obviously identical for any distance from the apex of a given cone, is called the *constant of the cone* and usually denoted by c. If α is the half-angle within the cone at the apex (i.e., the angle between the side and the axis of symmetry of the cone), then

$$c = \sin\alpha \qquad \text{. . . . . (57a}$$

Hence, at distance $\dot\rho$ along the slope from the apex, the cross-section has

$$\text{radial length} = c\,\dot\rho, \quad \text{circumference} = 2\pi c\,\dot\rho \qquad \text{. . . . . . (57b}$$

The diagram puts the cone co-axial with the globe — i.e., at Simple aspect. The contact line is thus a parallel — denoted as latitude ϕ_0 with $\dot\rho_0$ its distance down slope from the apex. As with the Azimuthals, we have two positions available — a northern version as illustrated opposite and a southern version with the cone inverted. Conical projections are rarely other than in this northern version at Simple aspect. From triangle *GAF*, we can see that angle *FGA* forms the geometric complement of both angle α and angle ϕ_0. Hence those two angles are identical in magnitude. Allowing that southern latitudes are negative but c must be positive, Equation 57a revises for contact latitude ϕ_0 to

$$c = \sin|\phi_0| = \oplus\sin\phi_0 \qquad \text{. . . . . (57c}$$

using our conditional sign symbol. Triangle *GAF* then shows

$$\dot\rho_0 = R\cot|\phi_0| = R\cot\oplus\phi_0 = \oplus R\cot\phi_0 = R\frac{\cos\phi_0}{\oplus\sin\phi_0} = R\frac{\cos\phi_0}{c} \qquad \text{. . . . . . . (57d}$$

which is positive in both versions. As all projecting will be relative to *F*, but the intermediate parameters based on point $\dot A$, this particular length and its formula become pervasive in the plotting formulae for conical projections.

As with the cylindrical mode, we project all the points on any one meridian from a projection point situated within the plane holding that meridian, at a specific distance along the line through the geocentre from the focal point (but see further discussion below). As with Cylindricals, the projection point circulates about the axis of the cone except for the central projection. Clearly, the mapped points from any one meridian are on a ray of the cone (i.e., a straight line through the apex).

The conical map

To develop a cone into a map requires, as with a cylinder, cutting it lengthwise so as to be able to flatten the surface. Again, we assume a central longitude and cut down a ray on the opposite side. The resulting flat map has the shape of a sector of a disc with centre $\dot{A}$ — the apex of the erstwhile cone — now forming the apical point of the sector. Unless $c < \frac{1}{2}$, the sector is not less than half the disc, so the crucial point is not visually an apex. The ray carrying the mapped points of any one meridian becomes a radial line of the sector and the angle at the apical point between any two rays is fraction c of the angle at

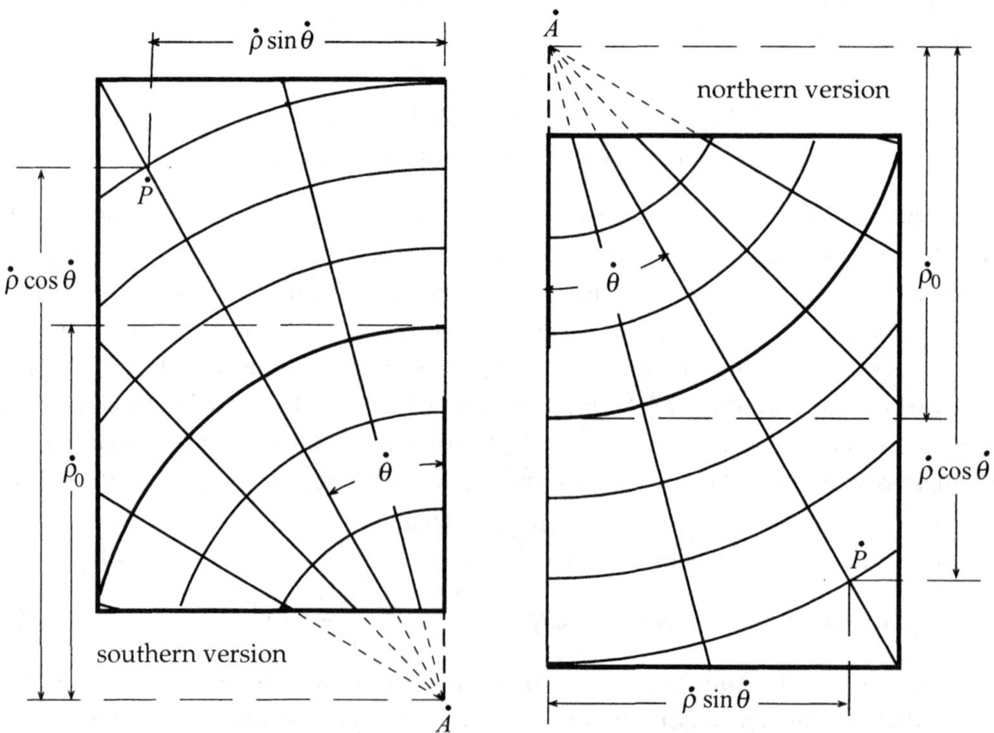

Exhibit 3–10: The Conical map – generic graticule plus dimensioning

the Pole between their respective meridians. That is, having λ mean longitude relative to the central meridian and denoting by $\dot{\theta}$ the counter-clockwise apical angle on the map from the central meridian to that for longitude λ,

$$\dot{\theta} = \oplus c\lambda \qquad \text{(59a)}$$

allowing for the clockwise pattern of longitudes at the South Pole.

The points on the cone a given distance from its apex form a circle on the cone, with circumferential length $2\pi c$ times that distance. On the flattened surface, they form an incomplete circular arc centred on the apical point with radius equal to the given distance and of length equal to that circumferential length. Thus the arc is proportion c of the full circumference of a circle of that radial length. The set of points on the contact line, which must be a parallel, is referred to as the *standard parallel* on both globe (being a full circle) and map (being a circular arc of radius $\dot{\rho}_0$ and length $2\pi c\,\dot{\rho}_0$, centred on $\dot{A}$).

If we posit a projection that maps global point $P = (\lambda, \phi)$ onto point $\dot{P}$ distance $\dot{\rho}$ down the cone, then $\dot{P}$ and all points of identical latitude must lie on that circular arc of radius $\dot{\rho}$ centred on $\dot{A}$. Hence, all parallels within the map are concentric circular arcs, their spacing depending on the particular projection. The pattern is illustrated in Exhibit 3–10 opposite. These are limited to mid latitudes; however, a conical map can readily include a Pole, and readily span even from there to the other hemisphere — as is visible in the diagram overleaf. It should be noted, however, that unless the projection point is on the axis of the globe (which implies being the geocentre given our presumptions), any conical mapping of a Pole is onto a proper arc, not a point.

Putting the central meridian on the y axis, the two variables $\dot{\rho}$ and $\dot{\theta}$ thus provide polar co-ordinates relative to $\dot{A}$ and the central meridian for the mapping. The two diagrams in Exhibit 3–10 show that angle $\dot{\theta}$ measures counter-clockwise from the $-y$ axial direction for a cone contacting the northern hemisphere and likewise but from the $+y$ axial direction for a cone contacting the southern hemisphere.

The apical point about which the angular parameter is measured (and from which the distance too) is rarely chosen to be the origin for plotting purposes. More usually that origin is some distance away, at a latitude of focal interest — typically that of the standard parallel. Setting $\dot{A} = (0, \dot{y}_a)$, the general plotting equations become, using Equation 59a:

$$\dot{x} = \oplus\dot{\rho}\sin\theta \quad = \oplus\dot{\rho}\sin(\oplus c\lambda) \quad = \dot{\rho}\sin c\lambda$$
$$\dot{y} = \dot{y}_a - \oplus\dot{\rho}\cos\dot{\theta} \quad = \dot{y}_a - \oplus\dot{\rho}\cos(\oplus c\lambda) \quad = \dot{y}_a - \oplus\dot{\rho}\cos c\lambda$$

If we do put the origin at the intersection with the y axis of the standard parallel, then $\dot{y}_a = \oplus\dot{\rho}_0$ and (using Equation 57d) the plotting equations become:

$$\dot{x} = \dot{\rho}\sin c\lambda$$
$$\dot{y} = \oplus\dot{\rho}_0 - \oplus\dot{\rho}\cos c\lambda \quad = R\cot\phi_0 - \oplus\dot{\rho}\cos c\lambda \qquad \text{(59b)}$$

Projecting onto the cone

While the whole range of perspective distances can again be considered, only the Central projection has received any significant recognition. Ambiguously called the **Perspective Conic** projection, it is more often used in the secantal form discussed in Chapter 5 rather than tangentially as illustrated and considered here.

Except for the central case, conical perspective projecting presents choices of alignment not occurring with azimuthal and cylindrical modes, but the routine choice is to set the viewpoint on the perspective line from the contact point through the geocentre. As with Cylindricals, we find that the algebra developed for Azimuthals translates easily to the new situation by the substitution of angles. Exhibit 3–11 below illustrates the new

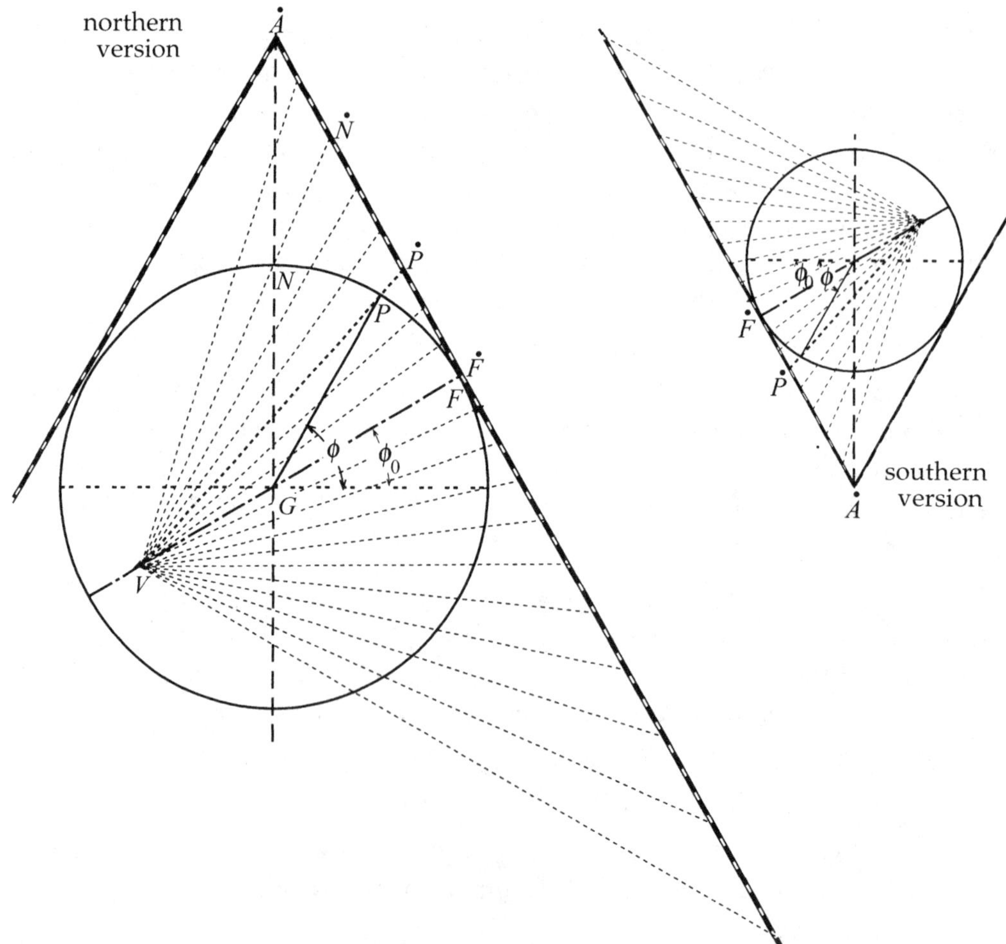

Exhibit 3–11: The geometry of general perspective projecting – conical version.

geometry, with the southern version in miniature alongside the northern version. The diagram for the latter is essentially identical to those in Exhibits 3–6 and 3–8, except for the focal projection line being at angle ϕ_0 to the Equator. For the northern version, γ of the earlier developments becomes $(\phi - \phi_0)$. Using Equation 50a, with hR again the distance of the projection point from the contact point, measured positively into the globe, we get

$$\text{offset of } \dot{P} \text{ from } \dot{F} = \frac{hR\sin(\phi-\phi_0)}{h-1+\cos(\phi-\phi_0)} = \frac{R\sin(\phi-\phi_0)}{1-h^{-1}\{1-\cos(\phi-\phi_0)\}}$$

which is negative for points further from $\dot{A}$ than $\dot{F}$, positive otherwise. The same holds for the southern version if the angle difference has its sign inverted. Incorporating our conditional sign symbol we then get

$$\dot{\rho} = \text{length of } \dot{A}\dot{P} = \rho_0 - \frac{R\sin\oplus(\phi-\phi_0)}{1-h^{-1}\{1-\cos\oplus(\phi-\phi_0)\}}$$

$$= \oplus R\cot\phi_0 - \frac{\oplus R\sin(\phi-\phi_0)}{1-h^{-1}\{1-\cos(\phi-\phi_0)\}} \qquad \text{. . . . . (61a}$$

$$= \oplus R\left(\cot\phi_0 - \frac{\sin(\phi-\phi_0)}{1-h^{-1}\{1-\cos(\phi-\phi_0)\}}\right)$$

Taking particular values 1, 2, and ∞ for the flexible intermediate parameter h we get:

for the **Central Conical** — projected from the centre of the globe

$$\dot{\rho} = \dot{\rho}_0 - \oplus R\tan(\phi-\phi_0) \qquad = \oplus R\{\cot\phi_0 - \tan(\phi-\phi_0)\} \qquad \text{. . . . . (61b}$$

for the **Stereographic Conical** — projected from the antipodal point on the focal line

$$\dot{\rho} = \dot{\rho}_0 - \oplus 2R\tan\frac{(\phi-\phi_0)}{2} \qquad = \oplus R\left\{\cot\phi_0 - 2\tan\frac{(\phi-\phi_0)}{2}\right\} \qquad \text{. . . . . (61c}$$

for the **Orthographic Conical** — with all projection lines parallel to focal line VF

$$\dot{\rho} = \dot{\rho}_0 - \oplus R\sin(\phi-\phi_0) \qquad = \oplus R\{\cot\phi_0 - \sin(\phi-\phi_0)\} \qquad \text{. . . . . (61d}$$

Substituting any of these values for $\dot{\rho}$ in Equations 59b gives the relevant plotting equations. The general plotting equations for the tangential General Perspective Conic projection at Simple aspect are:

$$\dot{x} = \oplus R\left(\cot\phi_0 - \frac{\sin(\phi-\phi_0)}{1-h^{-1}\{1-\cos(\phi-\phi_0)\}}\right)\sin c\lambda$$

$$\dot{y} = R\cot\phi_0 - R\left(\cot\phi_0 - \frac{\sin(\phi-\phi_0)}{1-h^{-1}\{1-\cos(\phi-\phi_0)\}}\right)\cos c\lambda \qquad \text{. . . . . (61e}$$

The Central Conical — the only literal conical projection that maps the Pole into a point — is shown overleaf; it illustrates the fact that a Conical map can span from one Pole to the other hemisphere. With its standard parallel set at 40°N the illustration has $c = 0.6428_{\sim}$ making the angle at the apex for the illustrated complete revolution around the globe approximately 231°. The illustrative generic Conical maps shown on page 58 and the Equidistant projection (developed in Chapter 4) on page 110, are, having the Pole as an arc, more representative of Conical projections than the illustration below (though, by having their parallels evenly spaced, they are not widely representative).

Conical projections have the same pattern of spacing of parallels and geometric limits as those of the other two modes, but are measured relative to the standard parallel ϕ_0 — hence in terms of $\gamma=\oplus(\phi-\phi_0)$ rather than the polar distance else latitude. Every Conical map can cover the whole of the hemisphere containing its standard parallel, and it is quite common for them to straddle the Equator. The maximum coverage is:

$$\text{for } \Phi_0 > 0: \quad +90° \text{ to } (\Phi_0 - 90)°$$
$$\text{for } \Phi_0 < 0: \quad (90 + \Phi_0)° \text{ to } -90°$$

As noted, the cone stands between the cylinder at one extreme and the azimuthal plane at the other and offers a continuum of choice in that range. Conical maps typically offer advantages over the two alternative modes only if they are well away from those extremes. In practice, that means at Simple aspect, when the standard parallel is in the mid latitudes, and then usually exclusive of both Poles.

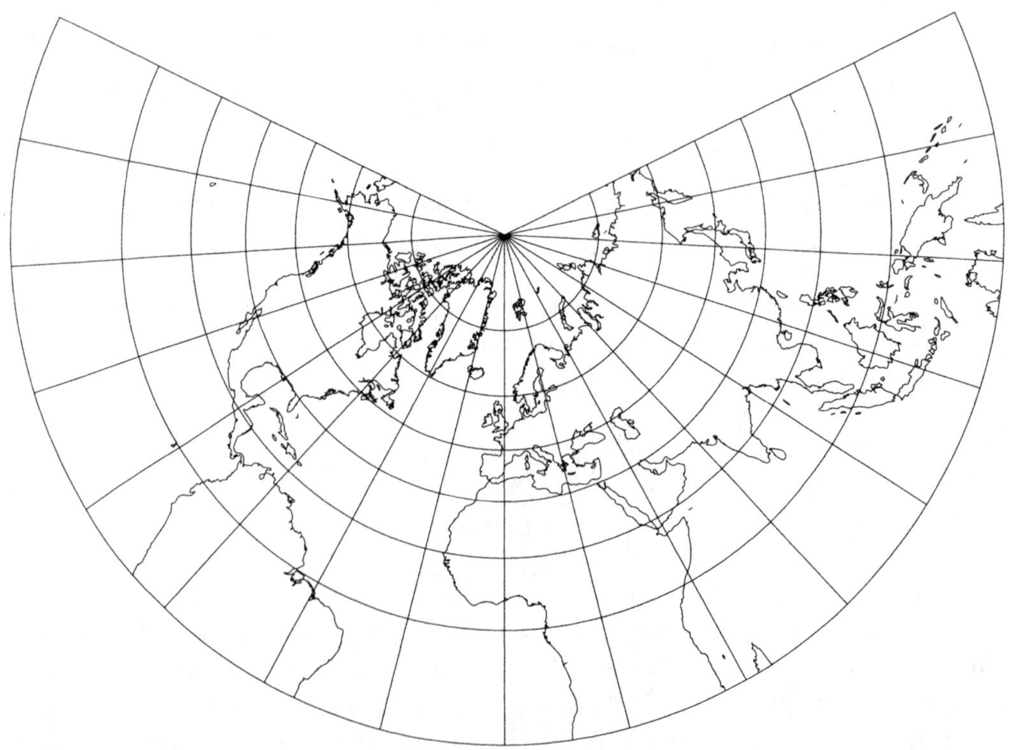

Exhibit 3–12: Central Conical projection.

COMPOUND PROJECTING

Basics

As mentioned in Chapter 2, before the advent of machine calculators (notably the electronic computer), graticules provided the basic network for drawing maps. Points in general were then placed by interpolation. For literal projections, the accuracy of the map thus depended on how many geometric lines one chose to use. For the many other projections covered here in forthcoming chapters, the equivalent task was the more laborious one of calculating the transformation. To reduce this burden, various attempts were made at creating maps via repeated projecting. The typical detailed cartography book illustrated such methods until the proliferation of computers in the 1970s (see, for example, 1962d and 1973a). While not fitting our shine-a-light literal rubric, such methods fall fully within the *perspective* description.

In his extensive survey of Russian cartographic developments, D. H. Maling [1960a] describes the multiple perspective method of M. D. Solov'ev. In effect, Solov'ev projects first onto a double globe, then projects from there to the azimuthal plane. Taking the diagram for stereographic projecting (Exhibit 3–4), we draw a circle of radius $2R$ centred on V, hence also passing through F. The point of intersection of the projective ray through P with this circle becomes the source point for another perspective transformation in the context of that greater circle. If the second transformation is stereographic (viewed from the further edge of the larger circle), then the angle offset of the ray through P is half what it is at V, while the diameter is doubled to $4R$. Hence, from Equation 45c we get

$$\dot{\rho} = 4R\tan\frac{\gamma}{4} \quad = 4R\tan\left(\frac{\pi}{8} - \frac{\oplus\phi}{4}\right)$$

If the second transformation is orthographic, then the angle offset of the radius to P from the pertinent centre V is half what it is at G, while the radius is doubled to $2R$. Hence, from Equation 47a we get

$$\dot{\rho} = 2R\sin\frac{\gamma}{2} \quad = 2R\sin\left(\frac{\pi}{4} - \frac{\oplus\phi}{2}\right) \qquad \text{. . . . . (63a}$$

This simply derived formula proves to maintain equality of areas between the original globe and the azimuthal map, giving the Equal-area Azimuthal projection of Chapter 4.

These two formulae represent particular cases of the general formulae

$$\dot{\rho} = nR\tan\frac{\gamma}{n} \quad \text{and} \quad \dot{\rho} = nR\sin\frac{\gamma}{n}$$

which provide for $n = 1$ the Gnomonic and Orthographic projections respectively and for $n = 2$ the Stereographic and Equal-area Azimuthal projections respectively.

TUTORIAL 12 MENSURATION FORMULAE

Circle

For a circle or circular disc of radius r:

$$\text{circumference} = 2\pi r = 6.2832_{\sim} r$$

$$\text{enclosed area} = \pi r^2 = 3.1416_{\sim} r^2$$

For a sector of such a circle subtending angle $\delta < \pi$ at the centre (as illustrated below):

$$\text{arc length: } s = r\delta$$

$$\text{chord length: } c = 2r\sin\frac{\delta}{2}$$

$$\text{chord to centre: } d = r\cos\frac{\delta}{2}$$

$$\text{chord to arc: } h = r - d = r\left(1 - \cos\frac{\delta}{2}\right)$$

Sphere

For a sphere of radius r:

$$\text{surface area} = 4\pi r^2 = 12.5664_{\sim} r^2$$

$$\text{volume} = \tfrac{4}{3}\pi r^3 = 4.1888_{\sim} r^3$$

For a segment of such a sphere defined by parallel planes distance h apart (as illustrated below),

$$\text{curved surface area} = 2\pi rh = 6.2832_{\sim} rh$$

which holds for any such planes, including the tangential.

The surface area of a wedge of any sphere having central angle α equals the proportion $\alpha/2\pi$ of the whole. The same proportionality to the area of the whole segment applies to the intersection of wedge and segment.

Cone

For a right circular cone of base radius r and slant height s with α the half-angle at the apex (as illustrated below):

$$\text{apical half-angle } \alpha = \arcsin\frac{r}{s}$$

$$\text{circumference of base} = 2\pi r$$

$$\text{area of curved surface} = \pi r s$$

$$\text{characterizing constant } c = \sin\alpha$$

the last equalling the slope of the side relative to the axis.

If the cone is cut straight down from the apex and unrolled as is done to develop it as a flat map surface, the result is a sector of a circular disc of radius s having a sectorial perimeter length of $2\pi r$. Hence,

$$\text{central angle} = \frac{2\pi r}{2\pi s}\text{rad} = \frac{r}{s}\text{rad}$$

The area of the curved surface of the cone

$$= \text{area of segment} = \frac{r}{s}\pi s^2 = \pi r s$$

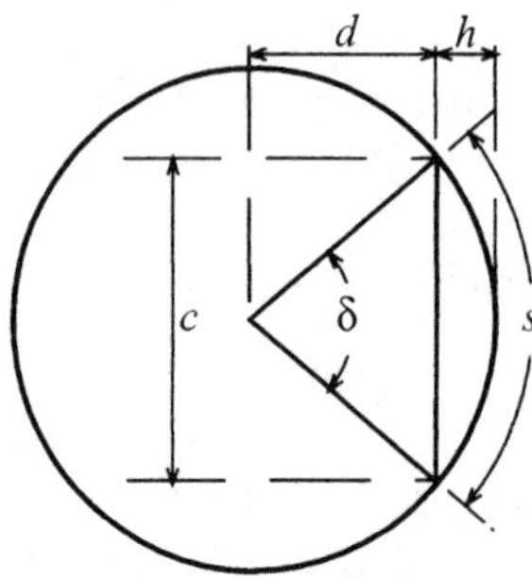

Circle (of radius r)

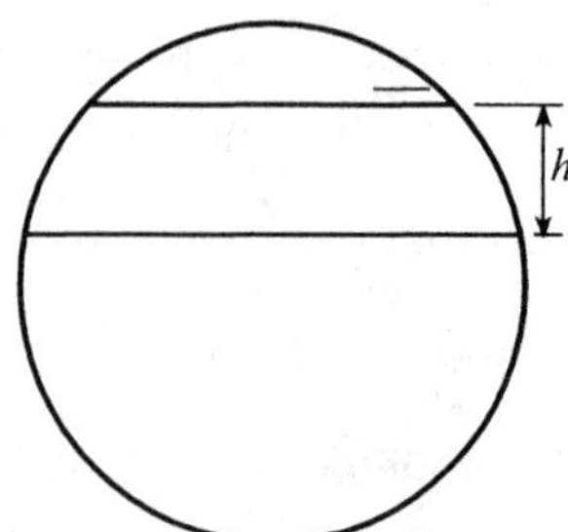

Sphere (of radius r)

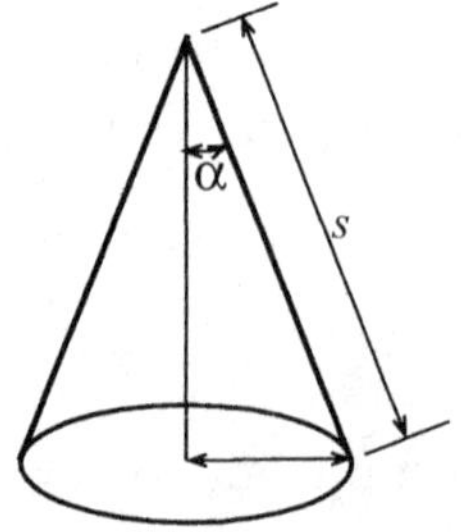

Cone (of base radius r)

CHAPTER 4

'True' but Synthetic: Equidistant and other algebraic variants

Preamble

Chapter 3 introduced three modes of developable surface and established the style and formulae for the resulting maps. At Simple aspect, all had parallels as concentric circular arcs and meridians as straight lines radiating from that centre (allowing infinity for radial length) — because one intermediate parameter was a multiple of relative longitude, the other a function of just latitude. For this characteristic, we accord the parenthesized adjective *'true'*. Only a few of those projections give maps that find significant use. In this chapter, we address utility then establish formulae. Although no longer literal projections, that noun is used throughout cartography for any formula producing maps. In this chapter we stay within the realm of 'true' projections, synthesizing projections algebraically in each mode that are:

equidistant (have uniformity of scale along meridians and the Equator),
equal-area (have area on the map of any object equal its area on the globe),
conformal (preserves angles and local shapes from the globe to the map),
The three attributes are mutually exclusive.

This chapter introduces integration and differentiation (i.e., calculus).

For the reason stated in Chapter 3 as well as for consistency, the maps in this chapter are presented with the same graticule as in that preceding chapter — i.e., with a 15° spacing of parallels and 30° spacing of meridians.

TUTORIAL 13 DIFFERENTIATION

Differentiation is a process of mathematical analysis that, for a given function, derives a second function expressing the rate of change of the value of the original one. Let $y = f(x)$ be the original function (e.g. a function that shows for time point x the cumulative distance travelled by a car). If the rate of change of y (i.e., the speed in our example) is steady between time points x and $x + \Delta x$, then the change Δy in y equals $\{f(x + \Delta x) - f(x)\}$, so

$$\text{rate of change} = \frac{\Delta y}{\Delta x} = \frac{f(x + \Delta x) - f(x)}{\Delta x}$$

If the rate of change is not steady, this formula can still give a fair estimate of the rate of change at time point x if the time interval Δx is small, depending on how far from steady the rate is at that value of x.

The (first) *derivative* of $y = f(x)$ with respect to x is defined as the limit of this quotient, for the general point x, as Δx tends to zero — if such limit exists. That is, for example,

$$\text{rate of change at } x = \lim_{\Delta x \to 0} \frac{f(x + \Delta x) - f(x)}{\Delta x}$$

Such derivative can be subjected to the same treatment giving, if its limit exists, a second derivative (the acceleration in our example), and so on. The successive derivatives of $y = f(x)$ are denoted in this work by the styles:

$$\frac{dy}{dx} \text{ and } \frac{df}{dx} \text{ and } f^{(1)}$$

$$\frac{d^2y}{dx^2} \text{ and } \frac{d^2f}{dx^2} \text{ and } f^{(2)}$$

Perhaps surprisingly, many mathematical functions have those limits, i.e., many are *differentiable*, and have derivatives also expressible as differentiable functions. Most notable is the function e^x, where e is the special value 2.718~; its derivative is identical with itself as, therefore, are all its further derivatives (see Tutorial 16). Clearly the derivative of a constant (e.g. $y = f(x) = 6$) is zero, since it is unchanging. The derivative of the simple power x^n for any n equals nx^{n-1}. Tutorial 15 lists the first derivatives for numerous functions pertinent to cartography.

The derivative of the sum of two differentiable functions equals the sum of their derivatives. Let $f(x)$ and $g(x)$ be two such functions. Then,

$$\frac{d(f+g)}{dx} = \frac{df}{dx} + \frac{dg}{dx}$$
$$= f^{(1)} + g^{(1)}$$

For the derivative of the product of two differentiable functions we get

$$\frac{d(f \cdot g)}{dx} = f\frac{dg}{dx} + g\frac{df}{dx}$$
$$= f \cdot g^{(1)} + f^{(1)} \cdot g$$

Since the derivative of any constant k is zero, it follows that

$$\frac{d(k \cdot g)}{dx} = k\frac{dg}{dx} + 0 = k \cdot g^{(1)}$$

If x is itself a function of another variable z (e.g., $x = g(z)$) then

$$\frac{df}{dz} = \frac{df}{dx}\frac{dx}{dz} = \frac{df}{dx}\frac{dg}{dz}$$

ATTRIBUTES

Equidistant, equal-area, conformal

To this point we have concentrated on the basic requirement of flattening the sphere. Usefulness, however, favours a map meeting some criteria concerning scale variation, consistency of areas, consistency of local shapes, and so on — the applicable criteria depending on the intended purpose of the map. In this chapter we will look at the three prime requirements, for maps that are equidistant, for maps that are equal-area, and for maps that are conformal. We will find that each attribute is mutually exclusive of the others — a given map has at most one of these attributes (and many much-used maps have none). Uniformity of scale along meridians and the Equator makes *equidistant* maps of special use for planning transportation and other communications. Equality of areas between globe and map (*equal-area*) is of value for showing comparative areal extents of things (e.g. the amount of forest in different countries, in which circumstance consistency of shape and linear scale can be compromised). Consistency of angles and local shapes from globe to map renders *conformality* the best attribute when trying to picture regions of Earth; it is the most valued of all attributes for regional and more local mapping in general. We also look at *loxodromes* — lines on a map having consistent azimuthal bearing. Having straight loxodromes was of vital importance in medieval times, because of facilitating ocean navigation, and remains of value; it is a subset of conformality.

Target and strategy

Our target is to produce maps in each mode that are unchanged in general appearance from those of the preceding chapter but which have revised spacings of the parallels to provide the wanted attributes. For appearance, we must keep one intermediate parameter a multiple of relative longitude and the other a function of just latitude (which we call 'true' projection}. The latter parameter, which controls the spacing of parallels, is differently labelled as a Cartesian co-ordinate in the Cylindricals versus a polar co-ordinate in the other two modes. For convenience we will refer to it as the *flexible parameter*. Our strategy is then to establish the separate algebraic formulae this parameter must meet in each mode to provide each attribute. Note that despite discarding projective geometry, every map-producing transformation in cartography is a projection.

Projections that do not have any of these (or other) specific attributes are often referred to in other works as *conventional*. Those already described are mostly so classified. Three of them, however, are not such — the Plate-Carrée, the Stereographic (Azimuthal), and the Orthographic Cylindrical are equidistant, conformal, and equal-area respectively. This terminological dichotomy is not used in this book.

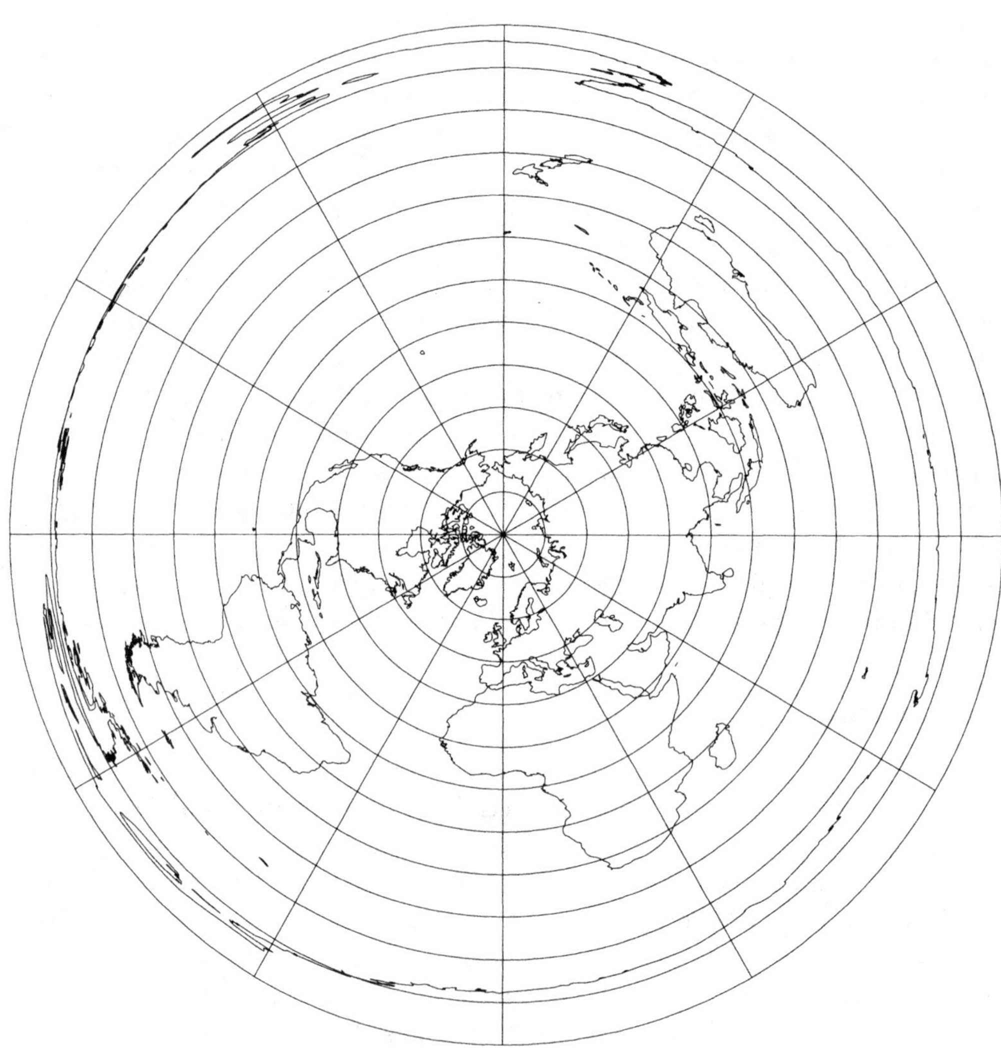

Exhibit 4–1: The Equidistant Azimuthal projection at Simple aspect – whole world.
On such a map, the straight line from the centre to any other point is of correct azimuth at the centre, but also of length proportional to the distance between those two points.
Hence, the concentric rings comprise points equidistant from the centre.
With the 10° graticule, each step outwards equals 1111 km.

EQUIDISTANT PROJECTIONS

Concept

Whether azimuthal, cylindrical else conical, all the projections developed to this point have correct scale (i.e., 1:1 relative to the globe) at their contact point or line — a specific parallel at Simple aspect. For Azimuthals, it is a point, so scale has only spot relevance. For Cylindricals the scale applies along the whole length of the Equator for Cylindricals, for Conicals along the whole standard parallel. Our tables of spacing of parallels on pages 45, 47, and 48, however, show that the spacing varies markedly, implying the scale along the meridians varies likewise. We have remarked earlier that scale cannot be held uniform across a whole map. The best that can be achieved is to hold scale uniform everywhere across the map for one intermediate parameter plus along two lines orthogonal to that at most. We now explore having scale correct along all meridians, while maintaining its correctness along the contact line/point, and develop equidistant projections for all three of our mapping modes. They have meridional lengths on the map between any two latitudes equal to the corresponding length on the globe. In Chapter 5 we expand our development to have correct scale along two parallels rather than one.

Having true scale along all meridians is synonymous with having all parallels correctly spaced. We thus create equidistant projections for each of our three modes by re-spacing as necessary.

Result and examples

Because the meridional length on the globe between any two latitudes is simply R times the latitudinal difference (in radians), we get:

for the **Equidistant Azimuthal** (also **Postel**) projection, differencing from the Pole,

$$\dot{\rho} = R\gamma \quad = R\left(\frac{\pi}{2} - \oplus\phi\right) \qquad \text{(69a}$$

for the **Equidistant Cylindrical** (identically the Plate-Carrée of Chapter 1) projection,

$$\dot{v} = R\,\phi \qquad \text{(69b}$$

for the **Equidistant Conical** projection (compare with Equation 61a),

$$\dot{\rho} = \dot{\rho}_0 - \oplus R\,(\phi - \phi_0) = \oplus R\{\cot\phi_0 - (\phi - \phi_0)\} = \oplus R\{\phi_0 + (\cot\phi_0) - \phi\} \qquad \text{(69c}$$

For each, the other parameter remains a simple multiple of longitude. Plotting equations are obtained by substituting in any of Equations 31b, 31c & 32a, in Equations 53a and in Equations 59b respectively. The Azimuthal is illustrated on the facing page and on page 188 (at oblique aspect). The Cylindrical is illustrated on pages 15 and 17. The Conical is illustrated on pages 58 and 110 (in secantal form, as described in Chapter 5).

Equidistant maps have obvious advantages for perceiving distance along the meridians (at Simple aspect) while suffering distortion transversely, rendering them useful for restricted regions. It is an equidistant azimuthal projection that is set into the floor of the Paris Observatory, effected in the 1670s by Jean Dominique Cassini. At Simple aspect, the Equidistant Azimuthal projection is good for general use when restricted to Polar regions. The Equidistant Conical maps are good for regional maps of modest coverage — e.g., smaller countries and individual states of the U.S.A. The Equidistant Cylindrical (identically the Plate-Carrée, our first projection) is of comparable utility when applied close to the Equator; the simplest of all projections, it was often used in past centuries. When turned from Simple aspect to some oblique position (Chapter 8), both azimuthal and cylindrical modes carry the facility to wider scenes.

The transverse distortion increases progressively as we go away from the contact line, without limit. For the Equidistant Cylindrical, the distortion with latitude is the same as for the literal projections because the meridians are parallel. For the Equidistant Azimuthal and Equidistant Conical, the converging/diverging of the meridians makes the distortion dependent on the spacing of the latitudes, hence specific to the projection. For the Azimuthal, for example, the length of the Equator on the equidistant map at Simple aspect equals

$$2\pi\left(R\left(\tfrac{\pi}{2}-0\right)\right)=\pi^2 R=9.8696_{\sim}R$$

against a distance on the globe of $2\pi R=6.2832_{\sim}R$. This represents an increase in length on the map of over 50%. For latitudes in the further hemisphere, the map continues to increase the length of parallels while the globe shrinks, so scale increases in a rapidly escalating manner to become infinite at the further Pole. That point is represented on our map by a circle twice as long as its Equator.

Notwithstanding such gross distortion of scale, the Equidistant Azimuthal is of considerable value even for a whole-globe map when turned from Simple aspect to some oblique position (developed in Chapter 8) focused on a place of interest. Being an Azimuthal, the shortest route to any other place is a straight line of correct azimuthal bearing at the focus. Because it is equidistant, the line is uniformly proportional to distance for all such routes (see page 188). The projection has particular value with air travel, giving a clear picture of the shortest route from its centre to anywhere else, and of the distance involved — a facility enhanced by the ability to embrace the whole world. The map is thus ideal for any travel office, preferably with concentric distance rings for increments of distance added.

EQUAL-AREA PROJECTIONS

Concept

The concept of equal-area is independent of shape — we merely ask that any object on the globe becomes an object of identical area on the map. A symmetric selection on the globe can become quite skewed on the map and a selection with straight sides can become a circle. With careful design it can retain much of its shape from globe to map, but it cannot fully maintain its shape. It is impossible to achieve equal-area without some distortion of shape. This is not to say that on a given map the existence of one implies that the other is grossly discrepant, but, for small-scale maps of large regions, such implication is usual. Because the spacing of parallels in 'true' projections in azimuthal, cylindrical, and conical modes can be varied arbitrarily, and the spacing of the straight meridians is proportional to longitude difference as it is on the globe, equal-area for all three modes can be achieved easily without violating the 'true' status. Working between a pair of meridians, our method is to find formulae for the spacing of the parallels so that the area between any two is identical with that on the globe. We can obviously deal with the area between an arbitrary parallel and a fixed one. If we get equal area between any arbitrary one and a common fixed one, then, by addition else subtraction of the values for two different arbitrary parallels, we get equal area between the two. Given that area is essentially the product of measurements along orthogonal lines, once equal-area is achieved, it can usually be maintained by reciprocal changes of scale along orthogonal directions. Much later, within the context of differential geometry, we apply such a rule. Here we look to direct comparison of areas (accommodating as we do the surface curvature across any region on the globe versus the flatness of the same region on the map).

The term *authalic,* already introduced relative to the whole Earth and globe, is often regarded as synonymous with equal-area. Seen by this author as not etymologically appropriate, it is restricted in this work to regional applicability. The terms *equivalent, equi-areal* and *orthambedic* may be deemed more appropriate, but are not used in this work either.

Result and examples

Equality of area between globe and map for any entity — e.g. an island, a continent, a lake, a country, another politically defined region — will occur if it holds true between any arbitrary pairs of parallels and of meridians. On all our maps so far, the meridians have been spaced uniformly relative to their spacing on the globe, so that we can take any one pair of meridians, including the two versions of any single one (i.e., the whole encirclement of the globe).

The surface area of a sphere between two parallel planes (see Tutorial 12, page 64) is the product of the spacing of the planes and the circumference of the complete sphere. For our globe, the last item equals $2\pi R$ so the surface area between the planes of any two latitudes ϕ and ϕ_i is

$$2\pi R \left| R\sin\phi - R\sin\phi_i \right| = 2\pi R^2 \left| \sin\phi - \sin\phi_i \right|$$

If ϕ_i is the Equator, this becomes

$$2\pi R^2 \left| \sin\phi \right| \qquad \text{. . . . . (72a}$$

If ϕ_i is a Pole, it becomes

$$2\pi R^2 \left(1 - \left| \sin\phi \right| \right)$$

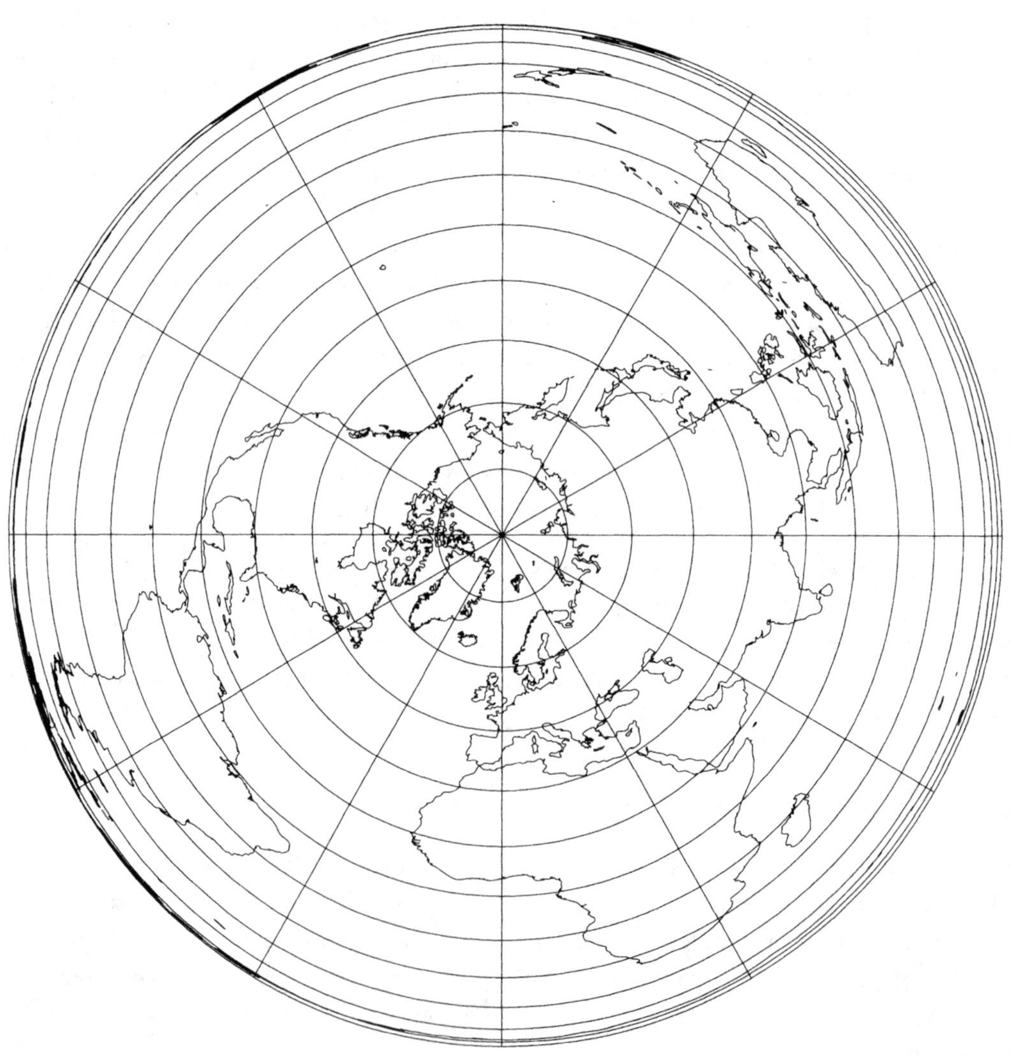

Exhibit 4–2: The (Lambert) Equal-area Azimuthal projection at Simple aspect – whole world.

The surface area of a sphere between latitudes ϕ and ϕ_i and longitudes λ_1 and $\lambda > \lambda_1$ is

$$\frac{\lambda - \lambda_1}{2\pi} 2\pi R^2 \left|\sin\phi - \sin\phi_i\right| \quad = R^2(\lambda - \lambda_1)\left|\sin\phi - \sin\phi_i\right| \qquad \text{........ (73a}$$

Developing the formulae for equal-area projecting in any mode requires establishing the area therein for a latitudinal span from ϕ to ϕ_i and longitudinal span λ to $\lambda > \lambda_1$, then equating it to the area on the globe shown in this equation. A map that meets these conditions could be shrunken in one direction and expanded orthogonally thereto without violating them, provided the product of the two multipliers equals 1. In this chapter, however, we insist on true scale for the contact latitude — Pole, Equator, else other standard parallel.

We again address the azimuthal, cylindrical, and conical modes in that order, via the intermediate parameters. In all three modes, the longitude-based parameter is a linear multiple of longitude and the area contained between two meridians is proportional to their difference in longitude-based parameter relative to that for the (complete) Equator.

For any **Azimuthal** at Simple aspect, the area between latitude ϕ and the Pole is that of the circular disk of radius $\dot\rho$ i.e., $\pi\dot\rho^2$ and the span of $\dot\theta$ for the Equator is 2π. Hence the area between the parallels for latitudes ϕ and ϕ_i and the meridians for longitudes λ_1 and $\lambda > \lambda_1$ equals (noting Equation 31a)

$$\frac{\dot\theta - \dot\theta_1}{2\pi}\left|\pi\dot\rho^2 - \pi\dot\rho_i^{\,2}\right| \quad = \frac{\dot\theta - \dot\theta_1}{2\pi}\pi\left|\dot\rho^2 - \dot\rho_i^{\,2}\right| \quad = \oplus\frac{\lambda - \lambda_1}{2}\left|\dot\rho^2 - \dot\rho_i^{\,2}\right|$$

Equating this with the corresponding area on the globe (Equation 73a), we get

$$\oplus\frac{(\lambda - \lambda_1)}{2}\left|\dot\rho^2 - \dot\rho_i^{\,2}\right| = \oplus(\lambda - \lambda_1)R^2\left|\sin\phi - \sin\phi_i\right|$$

$$\left|\dot\rho^2 - \dot\rho_i^{\,2}\right| = 2R^2\left|\sin\phi - \sin\phi_i\right|$$

Because the sign for the expression within the modulus parentheses on the left is always the opposite of that on the right, we can write this more simply as

$$\dot\rho^2 - \dot\rho_i^{\,2} = 2R^2(\sin\phi_i - \sin\phi)$$

This is the general formula for equal-area of an azimuthal map. Taking the North Pole as our fixed latitude and, using general formulae from Tutorials 9 and 10, we get

$$\dot\rho^2 = 2R^2(1 - \sin\phi) \quad = 2R^2\left\{1 - \cos\left(\frac{\pi}{2} - \phi\right)\right\} \quad = 2R^2\left\{2\sin^2\frac{1}{2}\left(\frac{\pi}{2} - \phi\right)\right\} \qquad \text{...... (73b}$$

Hence, inverting signs for the southern cone, we get

$$\dot\rho = R\sqrt{2(1 - \sin\oplus\phi)} \quad = 2R\sin\left(\frac{\pi}{4} - \frac{\oplus\phi}{2}\right) \qquad \text{........ (73c}$$

This formula for $\dot\rho$ is not identical with (or equivalent to) any of those already obtained in the previous chapter for literal projections (but is identical to the unnamed projection of Equation 63a that was created by compound projecting). Substitution in any of Equations 31b, 31c & 32a provides plotting equations for the **Equal-area Azimuthal** projection. The projection was first shown by J. H. Lambert of Alsace with six others in his grand opus [1772a] and is commonly prefixed with his name; it is illustrated facing the preceding page in Exhibit 4–2. Taken from Simple to Equatorial aspect, it forms the basis of the Hammer, Nordic and Briesemeister projections of Chapter 10.

For any **Cylindrical** at Simple aspect, the area between latitude ϕ and the Equator is that of the rectangle $\dot{v}$ high by the length of the Equator (which is $2\pi R$). Hence, the area between the parallels for latitudes ϕ and ϕ_i and the meridians for longitudes λ_1 and $\lambda > \lambda_1$ equals (noting Equation 51a)

$$\frac{\dot{u}-\dot{u}_1}{2\pi R}\,2\pi R\left|\dot{v}-\dot{v}_i\right| = \left(\dot{u}-\dot{u}_1\right)\left|\dot{v}-\dot{v}_i\right| = R\left(\lambda-\lambda_1\right)\left|\dot{v}-\dot{v}_i\right|$$

Equating this with the corresponding area on the globe (Equation 73a), we get

$$R\left(\lambda-\lambda_1\right)\left|\dot{v}-\dot{v}_i\right| = R^2\left(\lambda-\lambda_1\right)\left|\sin\phi-\sin\phi_i\right|$$

$$\left|\dot{v}-\dot{v}_i\right| = R\left|\sin\phi-\sin\phi_i\right|$$

As the sign for the expression within the modulus parentheses on the left is always the same as that on the right, we can write this more simply as

$$\dot{v}-\dot{v}_i = R\left(\sin\phi-\sin\phi_i\right)$$

Taking the Equator as our fixed latitude ϕ_i, we get

$$\dot{v} = R\sin\phi \qquad \text{. . . . . (74a}$$

which, for having equal area, must hold for all ϕ. If we look back to Equation 55d, we find that this is precisely the formula of the Orthographic Cylindrical, which we saw had an overall area equal to that of the whole globe, but which we can now see is comprehensively an equal-area projection. Shown as his cylindrical equal-area map by Lambert in his grand opus [1772a] so again commonly prefixed with his name, the **(Lambert) Equal-area Cylindrical** projection is illustrated in Exhibit 4–3 on page 82, in visual contrast with the Mercator projection shown on the facing page.

For any **Conical** at Simple aspect, the area between latitude ϕ and the apex of the cone is that of a sector with central angle $c2\pi$ of the circular disk of radius $\dot{\rho}$, i.e., $c\pi\dot{\rho}^2$; the span of $\dot{\theta}$ for the Equator is $c2\pi$. Hence, the area between the parallels for latitudes ϕ and ϕ_i and the meridians for longitudes λ_1 and $\lambda > \lambda_1$ equals (noting Equation 59a)

$$\frac{\dot{\theta}-\dot{\theta}_1}{c2\pi}\left|c\pi\dot{\rho}^2 - c\pi\dot{\rho}_i^{\,2}\right| = \frac{\oplus c\lambda - \oplus c\lambda_1}{c\,2\pi}\,c\pi\left|\dot{\rho}^2-\dot{\rho}_i^{\,2}\right| = \oplus c\,\frac{\lambda-\lambda_1}{2}\left|\dot{\rho}^2-\dot{\rho}_i^{\,2}\right|$$

Equating this with the corresponding area on the globe (Equation 73a), we get

$$\oplus c\,\frac{\left(\lambda-\lambda_1\right)}{2}\left|\dot{\rho}^2-\dot{\rho}_i^{\,2}\right| = \left(\lambda-\lambda_1\right)R^2\left|\sin\phi-\sin\phi_i\right|$$

$$c\left|\dot{\rho}^2-\dot{\rho}_i^{\,2}\right| = \oplus 2R^2\left|\sin\phi-\sin\phi_i\right|$$

Because the sign for the expression within the modulus parentheses on the left is always the opposite of that on the right, we can write this more simply as

$$\dot{\rho}^2 - \dot{\rho}_i^{\,2} = \oplus\frac{2R^2}{c}(\sin\phi_i - \sin\phi)$$

or . . . . (75a

$$\dot{\rho}^2 = \dot{\rho}_i^{\,2} + \oplus\frac{2R^2}{c}(\sin\phi_i - \sin\phi)$$

In either form, this is the general formula for equal-area on a conical map, applying to any two latitudes. We take the latitude of the standard parallel as our fixed latitude — i.e., $\phi_i = \phi_0$, hence $\sin\phi_i = \sin\phi_0$ and $\dot{\rho}_i = \dot{\rho}_0$. Using Equations 57c & 57d we then get

$$\begin{aligned}\dot{\rho}^2 &= (\oplus R\cot\phi_0)^2 + \oplus\frac{2R^2}{\oplus\sin\phi_0}(\sin\phi_0 - \sin\phi)\\ &= \frac{R^2}{\sin^2\phi_0}\left\{\cos^2\phi_0 + 2\sin\phi_0(\sin\phi_0 - \sin\phi)\right\}\\ &= \frac{R^2}{\sin^2\phi_0}(\cos^2\phi_0 + 2\sin^2\phi_0 - 2\sin\phi_0\sin\phi)\end{aligned}$$

$$\dot{\rho} = \frac{\oplus R}{\sin\phi_0}\sqrt{(1 + \sin^2\phi_0 - 2\sin\phi_0\sin\phi)} \qquad \text{. . . . . (75b}$$

which is necessarily positive.† This formula for $\dot{\rho}$ is not identical with (or identically equivalent to) any of those obtained earlier for Conicals (Equations 61a, 61b, 61c, 61d & 69c) so must be added to our collection. Substitution in Equations 59b provides plotting equations for the tangential Equal-area Azimuthal projection. The projection is usually named after its developer, H. C. Albers [1805a]. Because the name is also applied to a form using two standard parallels, this version is named more precisely the **Albers with one standard parallel** or the **1-standard-parallel Albers** projection; the map is illustrated only in 2-standard form, on page 116. Lambert included a restricted version of this projection in his grand opus [1772a]; it was a form with two standard parallels, restricted to having a Pole as one standard.

This development of equal-area projections has been within the context of the 'true' forms of azimuthal, cylindrical, and conical projecting, with the developable surface applied tangentially. Chapter 5 varies this by applying the developable surfaces as secants rather than tangents (e.g., getting the Albers with two standard parallels). More significantly in terms of visible results, we develop further equal-area projections of a cylindrical and conical nature in Chapter 6 in which, while retaining the feature of parallels being arcs of concentric circles, we no longer have meridians as straight lines — we're no longer 'true'.

† Check: as ϕ_0 tends to the right angle, hence $\sin\phi_0$ to $\oplus 1$, this approaches the formula for the Equal-area Azimuthal in Equations 73c.

TUTORIAL 14 INTEGRATION

Integration is a process of mathematical analysis that, for a given function, derives a second function expressing the cumulative effect of the original one over an interval of its underlying variable.

Letting $y = g(x)$ be the original function, we have a continuity of values over some interval a to $b>a$ (e.g. a function that shows the current speed of a car for time point x). The cumulative effect of continuing speed is to increase the distance travelled by the car. It can be shown that, for any function, the cumulation equals the area under the graph† of the function between those values.

Ascertaining that area by ordinary geometric means is difficult for a general function. Clearly, it can be approximated by sample values taken over small sub-intervals, with the accuracy of the result depending on the vagaries of the function and the size of the sub-intervals. If the interval a to b is divided into n sub-intervals of size Δx, (i.e., $b=a+n\Delta x$), demarcated by the successive points

$$a = x_0, x_1, x_2, .. x_{n-1}, x_n = b$$

then, if g_i for $i = 1, 2, . n$ is a value in the interval x_{i-1} *to* x_i, the area can be approximated by

$$(g_1 + g_2 + g_3 ... + g_n)\Delta x = \sum_{i=1}^{n} g_i \Delta x$$

More particularly, we define

$$\overline{S}_n = \sum_{i=1}^{n} \overline{g}_i \Delta x \text{ and } \underline{S}_n = \sum_{i=1}^{n} \underline{g}_i \Delta x$$

where upper and lower bars for the g_i terms indicate the maximum and the minimum values of g in the interval x_{i-1} to x_i, respectively. Then, clearly

$$\overline{S}_n \geq \text{area} \geq \underline{S}_n$$

If these two sums converge to a common limit S as n tends to infinity, for example,

$$\lim_{n\to\infty} \overline{S}_n = \lim_{n\to\infty} \underline{S}_n = S$$

then we have a value for the area. The result is denoted using a stylized S as

$$\int_a^b g \, dx \text{ or } \int_a^b g(x) \, dx$$

for the area under the graph of $g(x)$ in the interval a to b of variable x. It is called the *integral* of the function over that interval.

The value for this integral over a specific interval of any function is purely a number. If the function $g(x)$ is identically the first derivative $f^{(1)}(x)$ of some function $f(x)$ (see Tutorial 13), however, then it can be shown that, for any a and $b>a$,

$$\int_a^b g \, dx = f(b) - f(a) = [f(x)]_a^b$$

the last expression being the conventional way to indicate the preceding difference expression. More generally and regardless of interval, if $g(x) = f^{(1)}(x)$, we write

$$\int g \, dx = f(x) + k$$

where k is a constant. Since the derivative of any constant is zero, the function $g(x)$ is simultaneously the first derivative of any composite function $f(x) + k$.

† The area computed is actually that between the graph and the x axis, with negative value for area below that axis.

CONFORMAL PROJECTIONS

Concept

Conformality of map with globe relates to the consistency of the azimuthal angle between the two, and manifests itself to the observer primarily as a maintaining of local shape from globe to map — hence, its alternate label *orthomorphic*. We address first the angle between meridian and parallel — it is everywhere on the globe a right angle, and must be likewise on the map. This is so for all 'true' azimuthal, cylindrical and conical projections, with their straight meridians intersecting parallels that are either orthogonal straight lines else circular arcs centred on the intersection point of the meridional lines. It can also be so for projections with curved meridians, as long as that additional curvature is consistently orthogonal to the parallels. At this stage, however, we address only the 'true' style.

Having meridians everywhere orthogonal to the parallels is a minimal requirement for — but by no means the definition of — conformality. To consider the general angle, let us assume that that minimal requirement is met, then take a general vector not coincident with either a meridian else a parallel, and resolve it into the equivalent two vectors along those distinctive directions. They define a rectangle, the diagonal of which is the general vector. On the map, the two constituent vectors will remain orthogonal, but their length will depend on the scale applying in each direction. If the scale along the meridian differs from that along the parallel, their mutual rectangle will differ from that of the globe. Hence the diagonal, which represents the mapped version of the general vector, will be at a different angle to the meridian and parallel than it was on the globe. Thus, we can see that a further requirement for conformality is that the linear scale at every point must be identical along meridian and parallel. Conversely, if we meet this condition, then angles are maintained everywhere, making it a sufficient condition for conformality. (The escalation of scale compromises the retention of shape, however, keeping that aspect of conformality a relatively local phenomenon.)

Result and examples

For all three developable surfaces, the scale along the meridians depends on the particular projection formula. In azimuthal and conical modes, the scale along parallels is also dependent on the particular projection formula. In cylindrical mode, because each parallel is the same length as the Equator, the scale along the parallels is a fixed function of latitude, independent of the particular projection. Again, we establish the algebraic formulae that a projection must meet to provide the required attribute — now conformality — then derive the formula that must apply to the flexible intermediate parameter. The equations that we must construct then rearrange, however, involve more than simple rearrangement; they involve mathematical integration (see Tutorial 14).

To compare scales, we consider the lengths of small elements along meridian and parallel at point $P = (\lambda, \phi)$. If $\Delta\lambda$ and $\Delta\phi$ represent small intervals of longitude and latitude at P, then the corresponding lengths along parallel and meridian on the globe at P are $(R\cos\phi)\,\Delta\lambda$ and $R\,\Delta\phi$ — the cosine term reflecting the diminished radius of the lesser circle forming the parallel relative to that of a great circle. Identical scale along parallel and meridian at any point implies

$$\frac{\text{length along meridian on map}}{\text{length along meridian on globe}} = \frac{\text{length along parallel on map}}{\text{length along parallel on globe}}$$

or

$$\frac{\text{length along meridian on map}}{\text{length along parallel on map}} = \frac{\text{length along meridian on globe}}{\text{length along parallel on globe}} \qquad \ldots\ldots (78a)$$

$$= \frac{R\,\Delta\phi}{(R\cos\phi)\Delta\lambda} = \frac{\Delta\phi}{\cos\phi\,\Delta\lambda}$$

TUTORIAL 15 INTEGRALS AND DERIVATIVES

The following lines show a miscellany of functions for which the integrals and derivatives are analytically expressible (the constant of integration being omitted). The terms within the integral expressions are the derivatives of the functions to the right. Reciprocally, the functions to the right are the results of the integrations on the left.

(a) $\int k\,\mathrm{d}\alpha \qquad = k\alpha$

(b) $\int \alpha^{-1}\,\mathrm{d}\alpha \qquad = \ln\alpha$

(c) $\int \alpha^{k}\,\mathrm{d}\alpha$ for $k \neq -1 \qquad = (k+1)^{-1}\alpha^{k+1}$

(d) $\int \mathrm{e}^{k\alpha}\,\mathrm{d}\alpha \qquad = k^{-1}\,\mathrm{e}^{k\alpha}$

(e) $\int \sin k\alpha\,\mathrm{d}\alpha \qquad = -k^{-1}\cos k\alpha$

(f) $\int \cos k\alpha\,\mathrm{d}\alpha \qquad = k^{-1}\sin k\alpha$

(g) $\int \left(\sin^{-1} k\alpha\right)\mathrm{d}\alpha \qquad = k^{-1}\ln\left(\sin^{-1} k\alpha - \cot k\alpha\right) \qquad = k^{-1}\ln\tan\left(\frac{k\alpha}{2}\right)$

(h) $\int \left(\cos^{-1} k\alpha\right)\mathrm{d}\alpha \qquad = k^{-1}\ln\left(\cos^{-1} k\alpha + \tan k\alpha\right) \qquad = k^{-1}\ln\tan\left(\frac{\pi}{4} + \frac{k\alpha}{2}\right)$

(i) $\int \left(\sin^{-2} k\alpha\right)\mathrm{d}\alpha \qquad = -k^{-1}\cot k\alpha$

(j) $\int \left(\cos^{-2} k\alpha\right)\mathrm{d}\alpha \qquad = k^{-1}\tan k\alpha$

This provides an equation that expresses $\Delta\dot{\rho}$ hence $\dot{\rho}$ for azimuthal and conical modes, and $\Delta\dot{v}$ hence $\dot{v}$ for the cylindrical mode, in terms of $\Delta\phi$ hence ϕ. It should be noted that this condition says nothing about the actual scale. Indeed any map that meets these conditions can be shrunken or expanded without violating them, as long as the change is applied equally to both rectilinear axes. In this chapter, however, we retain true scale for the contact latitude — Pole, Equator else other standard parallel — again leaving such changes for later.

We begin with the cylindrical mode, which, with its rectangularity, is the simplest.

For any **Cylindrical** at Simple aspect, the length on the map along a parallel for any interval $\Delta\lambda$ of longitude is independent of latitude. It is everywhere identical to the corresponding length $R\Delta\lambda$ along the Equator. Applying Equation 78a gives

$$\frac{\Delta\dot{v}}{\Delta\dot{u}} = \frac{\Delta\dot{v}}{R\Delta\lambda} = \frac{\Delta\phi}{\cos\phi\,\Delta\lambda}$$

$$\Delta\dot{v} = \frac{R\,\Delta\phi}{\cos\phi} \quad = R\,\Delta\phi\cos^{-1}\phi$$

Clearly, the increment in $\dot{v}$ for a given increment in latitude progressively increases as we move away from the Equator, from 1 to infinity at the Pole (a table of actual values is given on page 83).

Our formula for $\dot{v}$ must be such as provides for this escalation, becoming itself infinite at either Pole. It can be obtained by addition, over the range of latitude concerned, of all the little increments applicable at each latitude. The number of these is infinite, so impractical by simple addition. The problem can be resolved easily by the technique of integration, which belongs to the realm of calculus. Before employing this technique (which, along with calculus in general, was invented only in the later 1600s by Gottfried Leibniz and Isaac Newton), we will explore using simple addition, as Gerardus Mercator must have done in the 1560s when he established this projection.

We could estimate $\dot{v}$ for any latitude by applying the reciprocal cosine for half that latitude uniformly, as a multiplying factor to the span of latitudes. If we take latitude 60°, for example, we could use the factor for 30°. If we use the latitude value in degrees as a suffix to label the length to that latitude, but ensure that the increment in latitude is expressed in radians, then we would get as our estimate

$$\dot{v}_{60} = R\,\frac{60\pi}{180}\cos^{-1}\left(\frac{60^\circ}{2}\right) = R\,\frac{60\pi}{180} \times 1.1547_{\sim} \quad = 1.2092_{\sim}R$$

Since the expansion factor escalates with latitude, applying the value for the half-latitude clearly underestimates the value for $\dot{v}$. This is most obvious for the Pole, for which $\dot{v}$ is infinity but the factor of the half-latitude (45°) has the very modest value $1.4142_{\sim}$, giving a derived estimate $\dot{v}_{90} = 2.2214_{\sim}R$.

The rate of escalation of the factor also escalates as latitude increases (as does the rate of increase in that rate, and so on). For lower latitudes, this simple estimate is reasonable, but it quickly becomes unreasonable, and increasingly so, as we depart further from the Equator. Better estimates can be obtained by fragmenting the overall interval into sub-intervals, applying within each the factor for its central latitude, then adding the values for each fragment. To get a good estimate would require fragmenting into single degrees, or so. As a simple illustration, however, we will just fragment the 60° problem into three bands of 20° each. We then get

$$\dot{v}_{60} = R\frac{60°\pi}{180°}\frac{1}{3}\left\{\cos^{-1}10° + \cos^{-1}30° + \cos^{-1}50°\right\} \quad = 1.3006_{\sim}R$$

This equation can be expressed more neatly using the distinctive mathematical symbol Σ for repeated summation. Generalizing to n segments and for any latitude (expressed conveniently in degree form as Φ) gives the general estimating formula

$$\dot{v} = R\frac{\Phi\pi}{180°}\frac{1}{n}\sum_{i=1}^{n}\cos^{-1}\left\{\left(i - \tfrac{1}{2}\right)\frac{\Phi}{n}\right\}$$

The larger the value n, the better is the estimate. For 60°, 6 bands of 10° gives the estimate $1.3126_{\sim}R$, 15 bands of 4° gives $1.3163_{\sim}R$, and so on, with the change from one to the next getting progressively smaller. What we want is the value when n is so large that increases in it make no discernible difference at the level of precision sought. Such a process of repeated approximation with convergence to a finite value can be pursued to obtain the value of our variable for each latitude. It is laborious, at least without modern calculators, but at least one computation serves all points at a given latitude (including those of like latitudinal magnitude in the other hemisphere). Later, with other projections, we will be left to resort to repeated approximation to solve our problem. Here, however, we can apply mathematical *integration*. Integration represents the limit of this summation as the underlying incremental steps $\Delta\dot{v}$ and $\Delta\phi$ become infinitesimal and tend to zero, to become the differential elements $d\dot{v}$ and $d\phi$. Substituting these in our last equation (after expressing it in radian rather than degree terms) then applying the integral (represented by a very elongated and stylized S for summation) we get

$$\int d\dot{v} = R\int \cos^{-1}\phi \, d\phi$$

Effecting the integration gives

$$\dot{v} = R\ln\left\{\tan\left(\frac{\pi}{4} + \frac{\phi}{2}\right)\right\} + k_0$$

where k_0 is an arbitrary constant. Knowing $\dot{v}$ to be zero for zero latitude, we get

$$0 = \dot{v}_0 = R\ln\left\{\tan\left(\tfrac{\pi}{4} + \tfrac{0}{2}\right)\right\} + k_0 \quad = R\ln\left\{\tan\tfrac{\pi}{4}\right\} + k_0$$
$$= R\ln 1 + k_0 = R\cdot 0 + k_0 = k_0$$

which shows that, for our situation, the k_0 must be zero. Hence, our formula must be

$$\dot{v} = R \ln\left\{\tan\left(\frac{\pi}{4} + \frac{\phi}{2}\right)\right\} \quad \text{. . . . . (81a}$$

The value at 60° is 1.3170~ R, which is close to that obtained above with segments of 4°.

For any southern latitude, the value must obviously be the same in magnitude as for the corresponding northern latitude, but of opposite sign. For negative ϕ we have

$$-\ln\left\{\tan\left(\frac{\pi}{4} + \frac{|\phi|}{2}\right)\right\} = -\ln\left\{\tan\left(\frac{\pi}{4} - \frac{\phi}{2}\right)\right\} = \ln\left\{\tan^{-1}\left(\frac{\pi}{4} - \frac{\phi}{2}\right)\right\} = \ln\left\{\tan\left(\frac{\pi}{4} + \frac{\phi}{2}\right)\right\}$$

Equation 81a thus holds identically for Southern latitudes.

Using a power formula from Tutorial 10 this is expressible without fractional angles as,

$$\dot{v} = R \ln\left\{\tan\left(\frac{\pi}{4} + \frac{\phi}{2}\right)\right\} = \frac{R}{2}\ln\left\{\tan^2\left(\frac{\pi}{4} + \frac{\phi}{2}\right)\right\} = \frac{R}{2}\ln\frac{1-\cos\left(\frac{\pi}{2}+\phi\right)}{1+\cos\left(\frac{\pi}{2}+\phi\right)} = \frac{R}{2}\ln\frac{1+\sin\phi}{1-\sin\phi}$$

$$= \frac{R}{2}\ln\frac{(1+\sin\phi)^2}{1-\sin^2\phi} = \frac{R}{2}\ln\left\{\frac{1+\sin\phi}{\cos\phi}\right\}^2 = R\ln\frac{1+\sin\phi}{\cos\phi} = R\ln\{\sec\phi + \tan\phi\} \quad \text{. . (81b}$$

However expressed, this condition is clearly different from any previous one, so must be added to our repertoire. Substituted in Equations 53a, it provides the plotting equations for the tangential **Conformal Cylindrical** or **Mercator** projection at Simple aspect — the most famous of all projections. Introduced by Mercator[†] in 1569[‡], well before calculus became a known technique, it is shown as Exhibit 4–4 on page 83. Later in this chapter we discuss a particular attribute of it that made it of extreme value from its earliest times, when European exploration around the world was at its peak. This value extends partially into current times, though its commonness today is an unfortunate carryover of past glory.

The growing spacing between parallels away from the Equator of the Mercator projection, necessary to maintain conformality, is shown numerically as well as graphically on page 83, the table showing, for the unit globe and for each of ten selected latitudes, the distance of map parallels from the Equator and the increment for each from its preceding value. It shows the spacing from 70° to 80° as four times that from 0° to 10°. Comparison with the tabulated figures on page 48 for the orthographic spacing shows a stark contrast with what we have just found is the Equal-area Cylindrical (shown overleaf facing the Mercator), the maintenance of equal-area requiring the reciprocal effect of shrinking spacing as one approaches a Pole. The areal exaggeration of Mercator is well demonstrated by noting the actual areas of Greenland, Australia, South America and

[†] The Latinized pen name of Gerhard Kramer of Flanders, who also introduced the name *atlas* for a collection of maps, many of which he produced in subsequent years [1949a]

[‡] Though recent research indicates such a projection was developed by Etzlaub in 1511 [1976a].

Africa, which are 2.2, 7.7, 17.9 and 30.3 million square kilometres respectively. The Equal-area Cylindrical map gives an image consistent with those real numbers, though Greenland is somewhat blurry. It depicts clearly the hugeness of Africa, however, not only relative to Greenland but compared with North America, for instance.

Various efforts have been made to reduce the accelerated expansion of the Mercator map while retaining its general appearance, which many people seem to see as correct — i.e., as closely representing Earth. The Miller Cylindrical projection, described in Chapter 10, is the most common. Exhibit 10–8 on page 246 shows that it fits the whole world within the dimensions of the world to ±80° in the Mercator. Though retaining the general appearance of the Mercator, the Miller loses both the conformality of its predecessor and the special characteristic discussed a little later under the Loxodromes rubric.

The vertical exaggeration in the Mercator map shows up with the plots of great circle routes between points in higher latitudes. The maps below in Exhibit 4–3 and opposite in Exhibit 4–4 show, as curved arcs, the great circle interconnections of Tokyo, Calgary, and London and of Sydney, Santiago, and Cape Town. As can be seen, these loop into latitudes well beyond those of the cities involved. Despite the loops being so long on the Mercator map, they are the shortest real routes. In Exhibit 4–3, those of the northern hemisphere are virtually lost in the compaction of higher latitudes

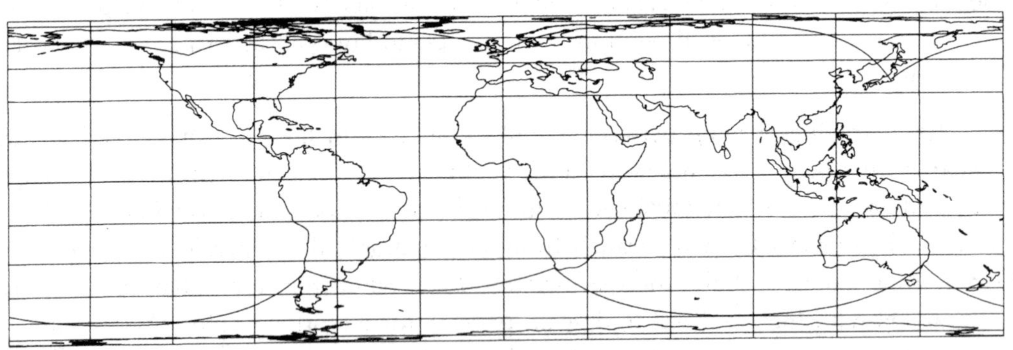

Exhibit 4–3: Orthographic Cylindrical projection – an equal-area projection.
(Lambert Cylindrical Equal-area projection.)
With great-circle routes Calgary – London – Tokyo – Calgary
[114°W, 51°N; 0°, 51°N; 140°E, 36°N; 114°W, 51°N]
and Santiago – Cape Town – Sydney – Santiago
[70°W, 34°S; 18°E, 34°S; 151°E, 34°S; 70°W, 34°S].

characteristic of that projection, but those in southern ones, which are at lower latitudes, remain clear. In Exhibit 3–7 (the Central Cylindrical projection, page 52) the exaggeration is greater than in Mercator. These great-circle routes are shown in all further cylindrical projections.

lat	0°		10°		20°		30°		40°		50°		60°		70°		80°		90°
distce	0.0000		0.1754		0.3564		0.5493		0.7629		1.0107		1.3170		1.7354		2.4362		∞
diffce		0.1754		0.1810		0.1929		0.2136		0.2478		0.3063		0.4184		0.7008		∞	

Mercator projection: spacing of parallels for unit globe
(Tangential form at Simple aspect.)

Both the azimuthal and conical modes of projection can be addressed for conformality in like manner to the cylindrical, except for being elaborated to accommodate the dependence of their latitudinal scales on the flexible parameter.

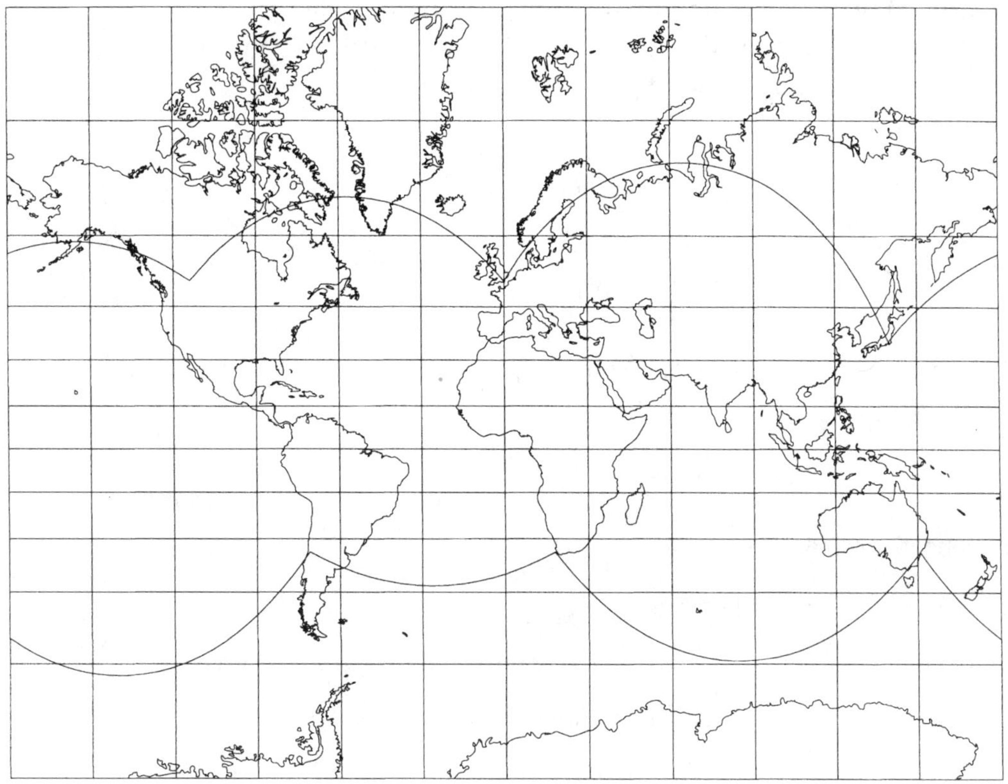

Exhibit 4–4: The Mercator — the Conformal Cylindrical projection — at Simple aspect.
Cut off at 75°N and 67°30′S.
With great-circle routes as detailed for Exhibit 4–3 opposite.

For any **Conical** at Simple aspect, the length of parallel for longitudinal difference $\Delta\lambda$ involves the constant of the cone, and is $c\,\dot{\rho}\Delta\lambda$. The incremental change in length along the meridian on the map for change $\Delta\phi$ in latitude is $-\oplus\Delta\dot{\rho}$, so applying Equation 78a gives

$$\frac{-\oplus\Delta\dot{\rho}}{c\,\dot{\rho}\Delta\lambda} = \frac{\Delta\phi}{\cos\phi\,\Delta\lambda} \qquad \text{or} \qquad \frac{-\oplus\Delta\dot{\rho}}{\dot{\rho}} = \frac{c\,\Delta\phi}{\cos\phi}$$

Taking this to the infinitesimal gives

$$-\oplus\dot{\rho}^{-1}\mathrm{d}\dot{\rho} = c\cos^{-1}\phi\;\mathrm{d}\phi \qquad \text{. . . . . (84a}$$

Integrating gives

$$-\oplus\int\dot{\rho}^{-1}\mathrm{d}\dot{\rho} = c\int\cos^{-1}\phi\;\mathrm{d}\phi$$

The right-hand side of this equation is familiar to us, integrating to a logarithm. The left-hand side also integrates to a logarithm, giving overall

$$-\oplus\ln\dot{\rho} = c\ln\left\{\tan\left(\frac{\pi}{4}+\frac{\phi}{2}\right)\right\} - \oplus\ln k$$

for some non-zero constant *k*, where, anticipating our next steps, we have chosen to express the constant of integration also in logarithm form and particularly signed. Having a logarithm limits *k* to being strictly positive, but its logarithm can be any finite number, so adopting the logarithm does not constrain the basic value. Noting that

$$-\oplus\ln\left\{\tan\left(\frac{\pi}{4}+\frac{\phi}{2}\right)\right\} = \ln\left\{\tan^{-\oplus 1}\left(\frac{\pi}{4}+\frac{\phi}{2}\right)\right\}$$

$$= \ln\left\{\tan\left(\frac{\pi}{4}-\frac{\oplus\phi}{2}\right)\right\}$$

and repositioning the signs, we get

$$\ln\dot{\rho} = c\,\ln\left\{\tan\left(\frac{\pi}{4}-\frac{\oplus\phi}{2}\right)\right\} + \ln k$$

$$= \ln\left\{\tan\left(\frac{\pi}{4}-\frac{\oplus\phi}{2}\right)\right\}^{c} + \ln k$$

$$= \ln\left(k\left\{\tan\left(\frac{\pi}{4}-\frac{\oplus\phi}{2}\right)\right\}^{c}\right)$$

Exponentiating converts this to

$$\dot{\rho} = k\left\{\tan\left(\frac{\pi}{4}-\frac{\oplus\phi}{2}\right)\right\}^{c} \qquad \text{. . . . . (84b}$$

which provides a general statement for conformality on a 'true' conical map.

Having true scale along the standard parallel ϕ_0 demands (see Equation 57d)

$$\dot{\rho}_0 = \oplus R\cot\phi_0$$

Inserting these values in our equation gives

$$\oplus R\cot\phi_0 = \dot{\rho}_0 = k\left\{\tan\left(\frac{\pi}{4}-\frac{\oplus\phi_0}{2}\right)\right\}^{c}$$

$$k = \frac{\oplus R \cot \phi_0}{\left\{\tan\left(\frac{\pi}{4} - \frac{\oplus \phi_0}{2}\right)\right\}^c} \quad \text{........ (85a}$$

Inserting the indicated value in Equation 84b and substituting the related value for c gives

$$\dot{\rho} = \frac{\oplus R \cot \phi_0}{\left\{\tan\left(\frac{\pi}{4} - \frac{\oplus \phi_0}{2}\right)\right\}^c} \left\{\tan\left(\frac{\pi}{4} - \frac{\oplus \phi}{2}\right)\right\}^c$$

$$= \frac{\oplus R \cot \phi_0}{\left\{\tan\left(\frac{\pi}{4} - \frac{\oplus \phi_0}{2}\right)\right\}^{\oplus \sin \phi_0}} \left\{\tan\left(\frac{\pi}{4} - \frac{\oplus \phi}{2}\right)\right\}^{\oplus \sin \phi_0}$$

This is often written, more neatly mathematically, as

$$\dot{\rho} = \oplus R \cot \phi_0 \left\{\frac{\tan\left(\frac{\pi}{4} - \frac{\oplus \phi}{2}\right)}{\tan\left(\frac{\pi}{4} - \frac{\oplus \phi_0}{2}\right)}\right\}^{\oplus \sin \phi_0} \quad \text{........ (85b}$$

For repeated computation, the constant k should obviously be evaluated once for any given standard parallel using Equation 85a, then applied repetitively to the initial form of the formula in Equation 84b.

We thus have, by substitution in Equation 59b, plotting formulae for the tangential **Conformal Conical** projection at Simple aspect or the **Lambert Conic** projection with one standard parallel at Simple aspect, the latter name (like Albers for the Equal-area Conical) being also applied to the form with two standard parallels. Yet another projection in Lambert's grand opus [1772a], it is illustrated in its 2-standard form in Exhibit 5–10 on page 120.

The **Azimuthals** represent limiting cases of the conical, with $c = 1$. Equation 84b then becomes the simpler

$$\dot{\rho} = k\left\{\tan\left(\frac{\pi}{4} - \frac{\oplus \phi}{2}\right)\right\} \quad \text{........ (85c}$$

which provides a general expression for conformality on any 'true' azimuthal map. If $k = 2R$ the equation becomes

$$\dot{\rho} = 2R \tan\left(\frac{\pi}{4} - \frac{\oplus \phi}{2}\right) \quad \text{........ (85d}$$

which is identically Equation 46a for the flexible parameter of the (tangential) Stereographic Azimuthal projection. Hence the Stereographic projection, often called the **Lambert Conformal Azimuthal** projection and illustrated in Exhibit 4–5 overleaf, provides the tangential conformal azimuthal projection at Simple aspect.

ROTATION

Changing what is at the top (of the map)

North is generally at the top of a map, but having a Pole within a map generally precludes that, while various special maps want it otherwise. The simple transformation:

$$\begin{aligned} \ddot{x} &= \dot{x}\cos\chi - \dot{y}\sin\chi \\ \ddot{y} &= \dot{x}\sin\chi + \dot{y}\cos\chi \end{aligned} \qquad \text{(86a}$$

provides a counter-clockwise rotation equal to angle χ without change of origin.

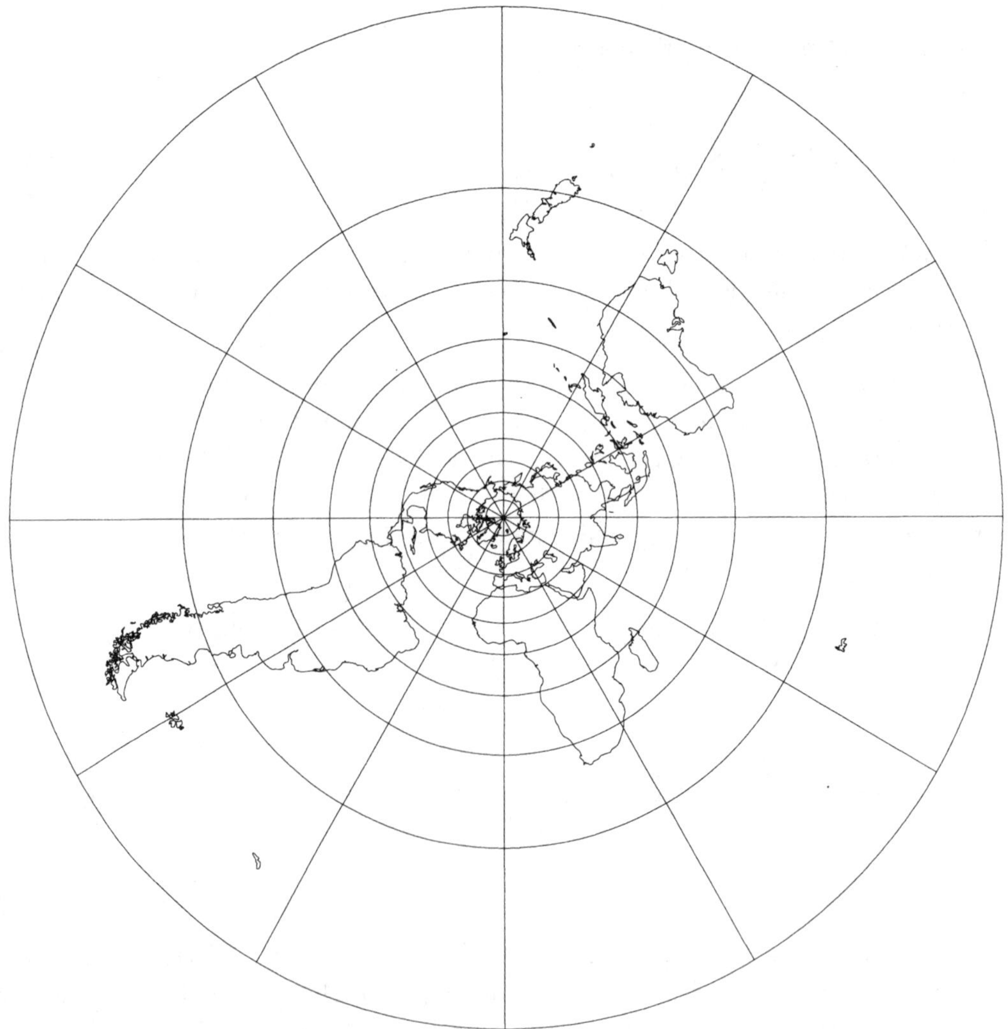

Exhibit 4–5: The Stereographic projection at Simple aspect – North Pole to 60°S.

LOXODROMES

A matter for navigation

Loxodromes (*rhumb lines* to the navigator) are lines of constant bearing. That is, they represent the successive locations were one to travel/navigate maintaining one specific compass bearing†. Every meridian is a loxodrome; persistent travel in a north else south direction from any point will take the traveller to a succession of positions all on the meridian of the beginning point, until a Pole is reached (at which point there is no further scope to travel in the chosen direction). The Equator is a loxodrome too; travel due east else due west from any point on the Equator will produce a succession of positions all on the Equator. The same is true of every parallel, but of no other circles or closed loops on the globe's surface. If any direction other than one of the four cardinal points of the compass is taken and maintained precisely, the result on Earth — and likewise in a simulation sense on the globe — is a line spiralling towards one of the Poles, passing through all longitudes without limit but never quite getting to the Pole in finite distance.

Except along bimeridians and the Equator, no loxodrome corresponds to the shortest route between two points. Assuming possession of data on the magnetic variation along the route, however, plus sufficiently close approximation of position while travelling to know the applicable variation, loxodromes provide the simplest route (ignoring intervening natural hazards). The discrepancy between loxodromic distance and great-circle distance (the shortest) increases with latitude and travel distance, but even in mid latitudes the discrepancy is insufficient to outweigh the convenience over significant distances. To use constant bearing for a trip, it suffices to know the bearing of the specific destination at the departure point. This could be provided by tables but is more conveniently ascertained from a map. While loxodromes of any shape could be drawn on the map connecting preset points, such map-reading is preferably achieved by making all loxodromes straight lines; then, any one that is wanted can be drawn and its bearing angle measured. Except for the four cardinal points of the compass, this is quite challenging on almost any map projection. At modest latitudes, and for shorter distances, the Plate-Carrée provides a reasonable approximation. It was widely used for navigational maps for the Mediterranean sea in medieval times (usually inscribed with travel lines between ports and called *Portolan charts*). But one somewhat older projection — the Mercator — uniquely meets the need. On it, every loxodrome is a straight line, every straight line is a loxodrome, and the compass bearing of that line is identically the azimuth angle relative to any meridian. Nothing could be simpler.

† I.e., constant bearing relative to the geographic pole, nowadays observable via a gyrocompass. One could have projections based on a grid that had the magnetic poles as a basis for their graticules, in which case bearing would be constant for a magnetic compass. In practice, the past use was with a magnetic compass corrected for the progressively changing magnetic variation — a data feature usual on navigational maps.

Clearly, having this distinctive attribute requires meridians and parallels to be straight lines, and the map overall to be conformal — hence the inherent uniqueness of the Mercator map. Developed in the 1560s in Holland as the various European nations were expanding their travel and trade then their colonization and empires across the Atlantic, Indian, and other oceans, this projection was of enormous political significance. Most of that travel was in modest latitudes, with only a trivial amount much beyond 55° of latitude. Hence, the escalating distortion of scale was relatively contained. Except in the tropics, the extended distance along the loxodrome is excessive for long-distance travel; the length of the loxodrome spanning many degrees of longitude in less modest latitudes can be a significant extension of the shortest route. Once applicable only to transocean sailing, that extends now to transcontinental flying. This difference shows conspicuously with travel between Europe and the Pacific Rim, for which the shortest route may well be literally over the North Pole (a point impossible to portray on Mercator). Travel between most westerly cities of North America and northern Europe, however, is invariably over central or northern Greenland — far away from their loxodromes. The situation may be illustrated by travel between Calgary (Alberta) and London, both cities at about 51°N but separated by 114° of longitude. Hence, the loxodrome is along the parallel with a length of 8000 km. The great-circle route, taken in essence by direct flights, crosses the 70° parallel and is barely 7000 km (see Tutorial 5 on page 14 for calculation).

Thus, the convenience of constant bearing can have a significant price. The answer is to segment the great-circle route of longer journeys, using a Gnomonic or appropriate equidistant Azimuthal map. The nodal points between segments are then transcribed to a Mercator map using longitude and latitude, and the bearing measured for each segment.

Simple consideration of a Mercator map (e.g., Exhibit 4–4) shows that for two points $P_1 = (\lambda_1, \phi_1)$ and $P_2 = (\lambda_2, \phi_2)$ not on the same parallel, if β is the bearing at P_1 of the loxodrome to P_2 and s is the length of the loxodrome track between P_1 and P_2, then

$$\tan\beta = \frac{\lambda_2 - \lambda_1}{\ln\tan\left(\frac{\pi}{4} - \frac{\phi_2}{2}\right) - \ln\tan\left(\frac{\pi}{4} - \frac{\phi_1}{2}\right)} \qquad \text{(88a}$$

$$s = R\frac{\phi_2 - \phi_1}{\cos\beta} \qquad \text{(88b}$$

Other projections have been developed to address specifically the loxodrome scene, most notably the Loximuthal (see page 154). Samuel Herrick [1944c] argued that any conformal map can serve navigational purposes adequately if equipped with a grid and corrective data equivalent to that provide widely for magnetic variation.

Loxodromes have in past times also been called *Mercator tracks*.

DISTORTION OF SCALE

Quantifying the variation

Variation of scale along meridians is illustrated by many preceding map exhibits in this chapter and others in Chapter 3. Some variations are shown quantitatively in tabular samples of the spacing of parallels. In the extreme, the scale becomes infinitely inflated, resulting in the Polar regions being beyond the limits of any practical Mercator map and the Equator likewise beyond the Gnomonic at Simple aspect. We now derive formulae that establish the variation of scale along both meridians and parallels applying to our projections. Before doing so, we should note that, while the scale along meridians generally varies, the scale is uniform along any parallel in any of our projections because the parallel is a circular arc and the positioning of longitude along it is a linear multiple of the longitudinal angle. We denote:

${}_{\mathrm{m}}\dot{m}$ = relative local scale along meridian

${}_{\mathrm{t}}\dot{m}$ = relative local scale transversely to meridian (effectively along parallel)

The Cylindricals provide the simplest beginning. In every one, the length of all parallels is the same. The Equator is of true length, so the others are necessarily relatively exaggerated. We know that the true length of each parallel is proportional to the cosine of its latitude. For a globe of radius R that means $2\pi R \cos\phi$. Hence, since the length on the map is consistently the $2\pi R$ of the Equator, we have

$${}_{\mathrm{t}}\dot{m} = \frac{2\pi R}{2\pi R\cos\phi} = \frac{1}{\cos\phi} \qquad \text{. . . . . (89a}$$

We used this very factor to establish the Mercator projection where, to achieve conformality, we made the scaling along the meridians identical at any point to this exaggerated scaling along the intersecting parallel. We thus know the scale variation along the meridians for this particular projection, and we might note that we obtained the spacing of any parallel from the Equator by integrating this trigonometric function — i.e., because integration and differentiation are mutually inverse processes, the local variation of scale is the derivative of the function $\dot{v}$ representing accumulated distance. We will find that differentiation of this function provides the means for establishing scale variation along the meridians generally.

For any increment $\Delta\phi$ in latitude along a meridian on the globe of radius R, the distance along the meridian is increased by $R\,\Delta\phi$. Along a derived map with formula $\dot{v} = \dot{v}(\phi)$, the equivalent increase equals $\dot{v}(\phi + \Delta\phi) - \dot{v}(\phi)$. Hence, the relative local scale at latitude ϕ is

$$\lim_{\Delta\phi \to 0} \frac{\dot{v}(\phi + \Delta\phi) - \dot{v}(\phi)}{R\Delta\phi} = \frac{1}{R} \lim_{\Delta\phi \to 0} \frac{\dot{v}(\phi + \Delta\phi) - \dot{v}(\phi)}{\Delta\phi}$$

The latter lim term is identically the definition of the derivative of $\dot{v}$ relative to ϕ, hence

$$_{\mathrm{m}}\dot{m} = \frac{1}{R}\frac{\mathrm{d}\dot{v}}{\mathrm{d}\phi} \qquad \text{(90a}$$

We can then apply this general result to all the cylindrical projections developed to this point to see the local scale along the meridians at any latitude. To do this, we need to differentiate each formula for $\dot{v}$, using established formulae for the differentiation of various routine mathematical functions shown in Tutorial 15 (page 78). For the five distinct cylindrical projections using Equations 89a & 90a, we get:

Central Cylindrical (Equation 54a)

$$_{\mathrm{m}}\dot{m} = \frac{1}{R}\frac{\mathrm{d}}{\mathrm{d}\phi}(R\tan\phi) \qquad = \cos^{-2}\phi$$

Stereographic Cylindrical (Equation 55a)

$$_{\mathrm{m}}\dot{m} = \frac{1}{R}\frac{\mathrm{d}}{\mathrm{d}\phi}\left(2R\tan\frac{\phi}{2}\right) \qquad = \cos^{-2}\frac{\phi}{2}$$

Orthographic Cylindrical (Equation 55c) — an equal-area projection

$$_{\mathrm{m}}\dot{m} = \frac{1}{R}\frac{\mathrm{d}}{\mathrm{d}\phi}(R\sin\phi) \qquad = \cos\phi$$

Equidistant Cylindrical (Equation 69b)

$$_{\mathrm{m}}\dot{m} = \frac{1}{R}\frac{\mathrm{d}}{\mathrm{d}\phi}(R\phi) \qquad = 1$$

Conformal Cylindrical (Mercator) (Equation 81a)

$$_{\mathrm{m}}\dot{m} = \frac{1}{R}\frac{\mathrm{d}}{\mathrm{d}\phi}\left(R\ln\left\{\tan\left(\frac{\pi}{4}+\frac{\phi}{2}\right)\right\}\right) \qquad = \cos^{-1}\phi$$

The last expression is identical to that for the transverse distortion shown on page 91, as needed for conformality. The expression for the Orthographic is the reciprocal of that, as required for an equal-area projection.

These formulae for the cylindrical projections are evaluated for steps of 10° on page 97.

For Azimuthals and Conicals the equivalent function $\dot{\rho}$ defines the length of the meridians. It differs only by increasing as $\oplus\phi$ decreases. Thus, we get

$$_{\mathrm{m}}\dot{m} = \frac{1}{R}\lim_{\Delta\phi\to 0}\frac{\dot{\rho}(\phi)-\dot{\rho}(\phi+\Delta\phi)}{\Delta\phi} \quad = \frac{-\oplus 1}{R}\frac{\mathrm{d}\dot{\rho}}{\mathrm{d}\phi} \qquad \text{(90b}$$

Unlike the Cylindricals, however, both Azimuthals and Conicals have parallels of varying length, depending on this function $\dot{\rho}$. Specifically, the length of the parallel for latitude ϕ is $2\pi c\dot{\rho}$, where c = 1 for any Azimuthal and $c = \sin\oplus\phi_0$ for a cone with standard parallel ϕ_0. Comparing this with the true length of the parallel of $2\pi R\cos\phi$ shows

$$_{t}m = \frac{2\pi c\dot{\rho}}{2\pi R\cos\phi} = \frac{c}{R\cos\phi}\dot{\rho} \qquad \text{. . . . . (91a}$$

Applying Equations 90b & 90a with c= 1 to the five distinct Azimuthals, we get Central Azimuthal (Equation 45a):

$$_{m}\dot{m} = \frac{-\oplus 1}{R}\frac{\mathrm{d}}{\mathrm{d}\phi}(\oplus R\cot\phi) = -\left\{\frac{-1}{\sin^2\phi}\right\} = \frac{1}{\sin^2\phi}$$

$$_{t}\dot{m} = \frac{1}{R\cos\phi}(\oplus R\cot\phi) = \frac{\oplus 1}{\sin\phi}$$

Stereographic Azimuthal (Equation 46a) — a conformal projection (see Equation 84a):

$$_{m}m = \frac{-\oplus 1}{R}\frac{\mathrm{d}\dot{\rho}}{\mathrm{d}\phi} = \frac{1}{R}\frac{\dot{\rho}}{\cos\phi} = \frac{1}{R\cos\phi}2R\tan\left(\frac{\pi}{4} - \frac{\oplus\phi}{2}\right) = \frac{2R}{\cos\phi}\tan\left(\frac{\pi}{4} - \frac{\oplus\phi}{2}\right)$$

$$_{t}m = \frac{1}{R}\frac{\dot{\rho}}{\cos\phi} = \frac{1}{R\cos\phi}2R\tan\left(\frac{\pi}{4} - \frac{\oplus\phi}{2}\right) = \frac{2R}{\cos\phi}\tan\left(\frac{\pi}{4} - \frac{\oplus\phi}{2}\right)$$

Orthographic Azimuthal (Equation 47b):

$$_{m}\dot{m} = \frac{-\oplus 1}{R}\frac{\mathrm{d}}{\mathrm{d}\phi}(R\cos\phi) = -\oplus\{-\sin\phi\} = \oplus\sin\phi$$

$$_{t}\dot{m} = \frac{1}{R\cos\phi}(R\cos\phi) = 1$$

Equidistant Azimuthal (Equation 69a):

$$_{m}\dot{m} = \frac{-\oplus 1}{R}\frac{\mathrm{d}}{\mathrm{d}\phi}\left(R\left(\frac{\pi}{2} - \oplus\phi\right)\right) = -\oplus\{-\oplus 1\} = 1$$

$$_{t}\dot{m} = \frac{1}{R\cos\phi}\left(R\left(\frac{\pi}{2} - \oplus\phi\right)\right) = \left(\frac{\pi}{2} - \oplus\phi\right)\frac{1}{\cos\phi}$$

Equal-area Azimuthal (Equation 73c):

$$_{m}\dot{m} = \frac{-\oplus 1}{R}\frac{\mathrm{d}}{\mathrm{d}\phi}\left(R\sqrt{2(1-\sin\oplus\phi)}\right) = -\oplus\left\{\frac{1}{2\sqrt{2(1-\sin\oplus\phi)}}2(-\oplus\cos\oplus\phi)\right\} = \frac{\cos\phi}{\sqrt{2(1-\sin\oplus\phi)}}$$

$$_{t}\dot{m} = \frac{1}{R\cos\phi}\left(R\sqrt{2(1-\sin\oplus\phi)}\right) = \frac{\sqrt{2(1-\sin\oplus\phi)}}{\cos\phi}$$

where the identity of value for the two terms within the conformal projection and their reciprocity within the equal-area projection can again be observed — as we will find again with the Conicals. These formulae for Azimuthals are evaluated for steps of 10° on page 95.

Applying Equations 90b & 91a with $c = \sin\phi_0$ to the six distinct Conicals i.e., with

$$_{t}m = \frac{c}{R\cos\phi}\dot{\rho} = \frac{\oplus\sin\phi}{R\cos\phi}\dot{\rho}$$

we get

Central Conical (Equation 61b):

$$ {}_{\mathrm{m}}\dot{m} = \frac{-\oplus 1}{R}\frac{\mathrm{d}}{\mathrm{d}\phi}\left(\oplus R\left\{\cot\phi_0 - \tan(\phi-\phi_0)\right\}\right) \qquad = \cos^{-2}(\phi-\phi_0) $$

$$ {}_{\mathrm{t}}\dot{m} = \frac{\oplus \sin\phi_0}{R\cos\phi}\left(\oplus R\left\{\cot\phi_0 - \tan(\phi-\phi_0)\right\}\right) \qquad = \sin\phi_0\left\{\cot\phi_0 - \tan(\phi-\phi_0)\right\}\cos^{-1}\phi $$

Stereographic Conical (Equation 61c):

$$ {}_{\mathrm{m}}\dot{m} = \frac{-\oplus 1}{R}\frac{\mathrm{d}}{\mathrm{d}\phi}\left(\oplus R\left\{\cot\phi_0 - 2\tan\frac{\phi-\phi_0}{2}\right\}\right) \qquad = \cos^{-2}\frac{(\phi-\phi_0)}{2} $$

$$ {}_{\mathrm{t}}\dot{m} = \frac{\oplus \sin\phi_0}{R\cos\phi}\left(\oplus R\left\{\cot\phi_0 - 2\tan\frac{\phi-\phi_0}{2}\right\}\right) \qquad = \sin\phi_0\left\{\cot\phi_0 - \tan\frac{\phi-\phi_0}{2}\right\}\cos^{-1}\phi $$

Orthographic Conical (Equation 61d):

$$ {}_{\mathrm{m}}\dot{m} = \frac{-\oplus 1}{R}\frac{\mathrm{d}}{\mathrm{d}\phi}\left(\oplus R\left\{\cot\phi_0 - \sin(\phi-\phi_0)\right\}\right) \qquad = \cos(\phi-\phi_0) $$

$$ {}_{\mathrm{t}}\dot{m} = \frac{\oplus \sin\phi_0}{R\cos\phi}\left(\oplus R\left\{\cot\phi_0 - \sin(\phi-\phi_0)\right\}\right) \qquad = \sin\phi_0\left\{\cot\phi_0 - \sin(\phi-\phi_0)\right\}\cos^{-1}\phi $$

Equidistant Conical (Equation 69c):

$$ {}_{\mathrm{m}}\dot{m} = \frac{-\oplus 1}{R}\frac{\mathrm{d}}{\mathrm{d}\phi}\left(\oplus R\left\{\phi_0 + (\cot\phi_0) - \phi\right\}\right) = -\{-1\} \qquad = 1 $$

$$ {}_{\mathrm{t}}\dot{m} = \frac{\oplus \sin\phi_0}{R\cos\phi}\left(\oplus R\left\{\phi_0 + \cot\phi_0 - \phi\right\}\right) \qquad = \sin\phi_0\left\{\phi_0 + \cot\phi_0 - \phi\right\}\cos^{-1}\phi $$

Equal-area Conical (Equation 75b):

$$ {}_{\mathrm{m}}\dot{m} = \frac{-\oplus 1}{R}\frac{\mathrm{d}}{\mathrm{d}\phi}\left(\frac{\oplus R}{\sin\phi_0}\sqrt{(1+\sin^2\phi_0 - 2\sin\phi_0\sin\phi)}\right) \qquad = \frac{\cos\phi}{\sqrt{(1+\sin^2\phi_0 - 2\sin\phi_0\sin\phi)}} $$

$$ {}_{\mathrm{t}}\dot{m} = \frac{\oplus \sin\phi_0}{R\cos\phi}\left(\frac{\oplus R}{\sin\phi_0}\sqrt{(1+\sin^2\phi_0 - 2\sin\phi_0\sin\phi)}\right) \qquad = \frac{\sqrt{(1+\sin^2\phi_0 - 2\sin\phi_0\sin\phi)}}{\cos\phi} $$

Conformal Conical (Equation 85b, noting it is from integration of Equation 84a):

$$ {}_{\mathrm{m}}\dot{m} = \frac{-\oplus 1}{R}\left(-c\dot{\rho}\cos^{-1}\phi\right) = \frac{\cos\phi_0\,\dot{\rho}\cos^{-1}\phi}{R} \qquad = k\cos\phi_0\left\{\tan\left(\frac{\pi}{4} - \frac{\oplus\phi}{2}\right)\right\}^{\oplus\sin\phi_0}\cos^{-1}\phi $$

$$ {}_{\mathrm{t}}\dot{m} = \frac{\oplus \sin\phi_0}{R\cos\phi}\left(\oplus Rk\cot\phi_0\tan^{\oplus\sin\phi_0}\left(\frac{\pi}{4} - \frac{\oplus\phi}{2}\right)\right) \qquad = k\cos\phi_0\left\{\tan\left(\frac{\pi}{4} - \frac{\oplus\phi}{2}\right)\right\}^{\oplus\sin\phi_0}\cos^{-1}\phi $$

$$ \text{where } k^{-1} = \left\{\tan\left(\frac{\pi}{4} - \frac{\oplus\phi_0}{2}\right)\right\}^{\oplus\sin\phi_0} $$

Following the interleaved Tutorial 16, this chapter concludes, as Endnote A, with synopses of the projection formulae plus sample evaluations of numerical results, on facing pages for each of the three modes. These formulae for Conicals are evaluated on page 99.

TUTORIAL 16 THE TRANSCENDENTAL NUMBER e

As discussed in Tutorial 7, the special number e in mathematics that forms the base of the natural logarithms is defined as

$$e = 1 + \frac{1}{1!} + \frac{1}{2!} + \frac{1}{3!} + \frac{1}{4!} + \frac{1}{5!} + \ldots$$

where the symbol ! represents the mathematical factorial expression (i.e., 2! equals the product 2 x 1, 3! = 3 x 2 x 1, etc).

Like the number π, e is irrational, i.e., it cannot be expressed precisely as the quotient of two integers and, further, it is transcendental — i.e., it is not the solution of any polynomial equation of finite length that has only integer coefficients and integer powers of the variable.

Evaluating this series gives

$$e = 2.718\,281\,828\,459 \ldots$$

Unlike π, e is not routinely approximated to with rational numbers.

As a base for logarithms, e is awkward compared with 10 for desktop use. It and so-called natural logarithms (Tutorial 7) occur naturally in differentiation and integration.

The arithmetic series given above is obtained by putting $x = 1$ in the *exponential series*

$$e^x = 1 + \frac{x^1}{1!} + \frac{x^2}{2!} + \frac{x^3}{3!} + \frac{x^4}{4!} + \frac{x^5}{5!} + \ldots$$

Simple differentiation (with respect to x) of this series produces the same series; hence the series is its own derivative and its own integral.

This series has coefficients and powers in common with those for sin and cos in Tutorial 9, but is devoid of the alternating signs that occur in those series. If we postulate an entity i such that $i^2 = -1$, however, then put $x = ib$, we get

$$e^{ib} = 1 + \frac{(ib)^1}{1!} + \frac{(ib)^2}{2!} + \frac{(ib)^3}{3!} + \frac{(ib)^4}{4!} + \frac{(ib)^5}{5!} + ..$$

$$= 1 + \frac{ib^1}{1!} - \frac{b^2}{2!} - \frac{ib^3}{3!} + \frac{b^4}{4!} + \frac{ib^5}{5!} + ..$$

$$= \left\{1 - \frac{b^2}{2!} + \frac{b^4}{4!} - ..\right\} + i\left\{\frac{b^1}{1!} - \frac{b^3}{3!} + \frac{b^5}{5!} - ..\right\}$$

$$= \cos b + i \sin b$$

$$e^{-ib} = \left\{1 - \frac{b^2}{2!} + \frac{b^4}{4!} - ..\right\} - i\left\{\frac{b^1}{1!} - \frac{b^3}{3!} + \frac{b^5}{5!} - ..\right\}$$

$$= \cos b - i \sin b$$

Hence,

$$\cos b = \frac{1}{2}\left\{e^{ib} + e^{-ib}\right\}, \quad \sin b = \frac{1}{2}\left\{e^{ib} - e^{-ib}\right\}$$

Since

$$e^b = \left\{1 + \frac{b^2}{2!} + \frac{b^4}{4!} + ..\right\} + \left\{\frac{b^1}{1!} + \frac{b^3}{3!} + \frac{b^5}{5!} + ..\right\}$$

$$e^{-b} = 1 - \frac{b^1}{1!} + \frac{b^2}{2!} - \frac{b^3}{3!} + \frac{b^4}{4!} - \frac{b^5}{5!} + \ldots$$

$$= \left\{1 + \frac{b^2}{2!} + \frac{b^4}{4!} + ..\right\} - \left\{\frac{b^1}{1!} + \frac{b^3}{3!} + \frac{b^5}{5!} + ..\right\}$$

we can (discarding the entity i) define the similar *hyperbolic functions*. These are:

$$\cosh b = 1 + \frac{b^2}{2!} + \frac{b^4}{4!} + .. = \frac{1}{2}\left\{e^b + e^{-b}\right\}$$

$$\sinh b = \frac{b^1}{1!} + \frac{b^3}{3!} + \frac{b^5}{5!} + .. = \frac{1}{2}\left\{e^b - e^{-b}\right\}$$

along with corresponding functions tanh, etc. The distinct mutual relationships of these functions are given in Tutorial 26 (on page 289).

Although there is no real number that is negative when squared, the imagined entity i is a potent intermediate tool for solving many real mathematical problems via complex numbers — the subject of Tutorial 23 (on page 220).

ENDNOTE A: SUMMARY OF FORMULAE AND SCALE FACTORS

Azimuthal: developable surface is a plane.

Simple aspect: surface is orthogonal to (hence, symmetric about) globe's axis.
Tangential: surface touches but does not intersect globe's surface; contact is a point.
At Simple aspect contact point is one of the geographic Poles.
Hence, on map:
meridians are straight lines through Pole, intersecting at true longitudinal angles,
parallels are concentric circles, centred on Pole, spaced according to projection,
meridians and parallels meet orthogonally, everywhere.
Convention: Pole is at origin point for plotting of map, i.e., at $(0, 0)$,
chosen meridian lies along the x axis.
Intermediate parameters (with $\oplus$ negative when focused on South Pole, else positive):

$$\dot{\theta} = \oplus\lambda = \frac{\oplus\pi\Lambda}{180}$$

$\dot{\rho} = \dot{\rho}(\phi)$, as defined for projection.

Plotting basis: $x = \rho\cos\theta,\ y = \rho\sin\theta$

<u>Central Azimuthal</u> (<u>Gnomonic</u>) — projected from the centre of the globe
From geometry,

$$\dot{\rho} = \oplus R\cot\phi \qquad = \oplus R\cot\Phi$$

<u>Stereographic</u> Azimuthal — projected from the opposite point on the globe's surface
From geometry,

$$\dot{\rho} = 2R\tan\left(\frac{\pi}{4} - \frac{\oplus\phi}{2}\right) \qquad = 2R\tan\left(45 - \frac{\oplus\Phi}{2}\right)$$

a conformal projection

<u>Orthographic</u> Azimuthal — projected from infinity — projection lines orthogonal to plane
From geometry,

$$\dot{\rho} = R\cos\phi \qquad = R\cos\Phi$$

<u>Equidistant Azimuthal</u> — i.e., distance along meridians same as on globe

$$\rho = R\left(\frac{\pi}{2} - \oplus\phi\right) \qquad = \frac{R\pi(90 - \oplus\Phi)}{180}$$

(Lambert) <u>Equal-area Azimuthal</u>
From geometry of areas,

$$\dot{\rho} = R\sqrt{2(1 - \oplus\sin\phi)} \qquad = R\sqrt{2(1 - \oplus\sin\Phi)}$$

(Lambert) <u>Conformal Azimuthal</u> — see Stereographic above

Scaling factors at selected latitudes for various tangential Azimuthals at Simple aspect so true scale at Pole only.

	Central	Stereographic (conformal)	Orthographic	Equidistant	Equal-area	Conformal (see Stereo...)
90°						
along meridian	1	1	1	1	1	
along parallel	1	1	1	1	1	
80°						
along meridian	1.0310	1.0077	0.9848	1	0.9962	
along parallel	1.0154	1.0077	1	1.0051	1.0038	
70°						
along meridian	1.1325	1.0311	0.9397	1	0.9848	
along parallel	1.0642	1.0311	1	1.0206	1.0154	
60°						
along meridian	1.3333	1.0718	0.8660	1	0.9659	
along parallel	1.1547	1.0718	1	1.0472	1.0353	
50°						
along meridian	1.7040	1.1325	0.7660	1	0.9397	
along parallel	1.3054	1.1325	1	1.0861	1.0642	
40°						
along meridian	2.4203	1.2174	0.6428	1	0.9063	
along parallel	1.5557	1.2174	1	1.1392	1.1034	
30°						
along meridian	4	1.3333	0.5	1	0.8660	
along parallel	2	1.3333	1	1.2092	1.1547	
20°						
along meridian	8.5486	1.4903	0.3420	1	0.8192	
along parallel	2.9238	1.4903	1	1.3001	1.2208	
10°						
along meridian	33.1634	1.7041	0.1736	1	0.7660	
along parallel	5.7588	1.7041	1	1.4178	1.3054	
0°						
along meridian	∞	2	0	1	0.7071	
along parallel	∞	2	1	1.5708	1.4142	

Cylindrical: developable surface is a cylinder.

Simple aspect: surface is symmetric about globe's axis, i.e., is co-axial with globe.
Tangential: surface touches, does not intersect globe's surface; contact is a great circle.
At Simple aspect contact line is the Equator.
Hence, on map: Equator is a straight line,
meridians are parallel straight lines, orthogonal to Equator, spaced uniformly,
parallels are parallel straight lines, spaced by projection symmetrically about Equator,
meridians and parallels meet orthogonally, everywhere.
Convention: Equator is along x axis, with east in positive direction,
chosen central meridian along the y axis, with north in positive direction.
Intermediate parameters:

$$\dot{u} = R\lambda = \frac{\pi\Lambda}{180},$$

$\dot{v} = \dot{v}(\phi)$ = as defined for projection

Plotting basis $\dot{x} = u$, $\dot{y} = v$

Central Cylindrical — projected from the centre of the globe
From geometry,

$$\dot{v} = R\tan\phi \qquad = R\tan\Phi$$

Stereographic Cylindrical — projected from the opposite point on the globe's surface
From geometry,

$$\dot{v} = 2R\tan\frac{\phi}{2} \qquad = 2R\tan\frac{\Phi}{2}$$

Orthographic Cylindrical — projected from infinity — projection lines orthogonal to plane
From geometry,

$$\dot{v} = R\sin\phi \qquad = R\sin\Phi$$

an equal-area projection

Equidistant Cylindrical — i.e., distance along meridians same as on globe

$$\dot{v} = R\,\phi \qquad = \frac{R\pi\Phi}{180}$$

(Lambert) Equal-area Cylindrical see Orthographic above.

Conformal Cylindrical (Mercator)
By equiscaling

$$\dot{v} = R\,\ln\left\{\tan\left(\frac{\pi}{4} + \frac{\phi}{2}\right)\right\} \qquad = R\,\ln\left\{\tan\left(45 + \frac{\Phi}{2}\right)\right\}$$

Scaling factors at selected latitudes for various tangential Cylindricals at Simple aspect so true scale at Equator only.

	Central	Stereographic	Orthographic (equal-area)	Equidistant	Equal-area (see Ortho...)	Conformal (Mercator)
90°						
along meridian	∞	2	0	1		∞
along parallel	∞	∞	∞	∞		∞
80°						
along meridian	33.1634	1.7041	0.1736	1		5.7588
along parallel	5.7588	5.7588	5.7588	5.7588		5.7588
70°						
along meridian	8.5486	1.4903	0.3420	1		2.9238
along parallel	2.9238	2.9238	2.9238	2.9238		2.9238
60°						
along meridian	4	1.3333	0.5	1		2
along parallel	2	2	2	2		2
50°						
along meridian	2.4203	1.2174	0.6428	1		1.5557
along parallel	1.5557	1.5557	1.5557	1.5557		1.5557
40°						
along meridian	1.7040	1.1325	0.7660	1		1.3054
along parallel	1.3054	1.3054	1.3054	1.3054		1.1034
30°						
along meridian	1.3333	1.0718	0.8660	1		1.1547
along parallel	1.1547	1.1547	1.1547	1.1547		1.1547
20°						
along meridian	1.1325	1.0311	0.9397	1		1.0642
along parallel	1.0642	1.0642	1.0642	1.0642		1.0642
10°						
along meridian	1.0310	1.0077	0.9848	1		1.0154
along parallel	1.0154	1.0154	1.0154	1.0154		1.0154
0°						
along meridian	**1**	**1**	**1**	**1**		**1**
along parallel	**1**	**1**	**1**	**1**		**1**

Conical: developable surface is a circular cone, apex *A*, taper characterized by constant *c*.

Simple aspect: surface is symmetric about globe's axis, i.e., cone is co-axial with globe.
Tangential: surface touches, does not intersect globe's surface; contact is a lesser circle.
At Simple aspect contact line is the standard parallel, denoted $\phi_0 = \arcsin c$.
Hence, on map: standard parallel is a circular arc, centred on the Pole at *A*,
meridians are straight lines through *A*, intersecting there at *c* times longitudinal angles,
parallels are circular arcs, all centred on *A*, spaced according to projection,
meridians and parallels meet orthogonally, everywhere.
Convention: chosen central meridian along the *y* axis,
chosen latitude ϕ_a passes through origin, usually $\phi_a = \phi_0$.
Intermediate parameters (with ⊕ negative for cone on South Pole, else positive):

$$\dot{\theta} = \oplus c\,\lambda \quad = \oplus c\frac{\pi\Lambda}{180}$$

$\dot{\rho} = \dot{\rho}(\phi)$, as defined for projection, mostly relative to $\rho_0 = \rho(\phi_0) \quad = \oplus R \cot\phi_0$

Plotting basis: $x = \oplus\rho\sin\theta,\ \ y = \rho_a - \oplus\rho\cos\theta$

Central Conical — projected from the centre of the globe
From geometry,

$$\dot{\rho} = \oplus R\{\cot\phi_0 - \tan(\phi - \phi_0)\} \qquad = \oplus R\{\cot\Phi_0 - \tan(\Phi - \Phi_0)\}$$

Stereographic Conical — projected from the opposite point on the globe's surface
From geometry,

$$\dot{\rho} = \oplus R\left\{\cot\phi_0 - 2\tan\frac{(\phi - \phi_0)}{2}\right\} \qquad = \oplus R\left\{\cot\Phi_0 - 2\tan\frac{(\Phi - \Phi_0)}{2}\right\}$$

Orthographic Conical — projected from infinity — projection lines orthogonal to plane
From geometry,

$$\dot{\rho} = \oplus R\{\cot\phi_0 - \sin(\phi - \phi_0)\} \qquad = \oplus R\,\{\cot\Phi_0 - \sin(\Phi - \Phi_0)\}$$

Equidistant Conical — i.e., distance along meridians same as on globe

$$\dot{\rho} = \oplus R\,\{(\cot\phi_0) + \phi_0 - \phi\} \qquad = \oplus R\,\{(\cot\Phi_0) + \Phi_0 - \Phi\}$$

Equal-area Conical (Albers) with 1 standard parallel
From geometry of areas,

$$\dot{\rho} = \frac{\oplus R}{\sin\phi_0}\sqrt{(1 + \sin^2\phi_0 - 2\sin\phi_0\sin\phi)} \qquad = \frac{\oplus R}{\sin\Phi_0}\sqrt{(1 + \sin^2\Phi_0 - 2\sin\Phi_0\sin\Phi)}$$

(Lambert) Conformal Conic with 1 standard parallel
By reciprocity of scaling factors,

$$\dot{\rho} = \frac{\oplus R\cot\phi_0}{\left\{\tan\left(\frac{\pi}{4} - \frac{\oplus\phi_0}{2}\right)\right\}^{\sin\phi_0}}\left\{\tan\left(\frac{\pi}{4} - \frac{\oplus\phi}{2}\right)\right\}^{\oplus\sin\phi_0} \qquad = \frac{\oplus R\cot\Phi_0}{\left\{\tan\left(45 - \frac{\oplus\Phi_0}{2}\right)\right\}^{\sin\phi_0}}\left\{\tan\left(45 - \frac{\oplus\Phi}{2}\right)\right\}^{\oplus\sin\Phi_0}$$

Scaling factors at selected latitudes for various tangential Conicals at Simple aspect so true scale at standard parallel only — here 40°.

	Central	Stereographic	Orthographic	Equidistant	Equal-area (Albers)	Conformal (Lambert)
90°						
along meridian	2.4203	1.2174	0.6428	1	—	—
along parallel	1.5557	∞	∞	∞	∞	∞
80°						
along meridian	1.7041	1.1325	0.7660	1	0.4527	1.5047
along parallel	1.3054	3.0642	2.0321	1.8272	2.2089	1.5047
70°						
along meridian	1.3333	1.0718	0.8660	1	0.7552	1.1987
along parallel	1.1547	1.7362	1.3001	1.2557	1.3242	1.1987
60°						
along meridian	1.1325	1.0311	0.9397	1	0.9131	1.0730
along parallel	1.0642	1.3054	1.0924	1.0833	1.0951	1.0730
50°						
along meridian	1.0311	1.0077	0.9848	1	0.9821	1.0163
along parallel	1.0154	1.1043	1.0181	1.0172	1.0182	1.0163
40°						
along meridian	**1**	**1**	**1**	**1**	**1**	**1**
along parallel	**1**	**1**	**1**	**1**	**1**	**1**
30°						
along meridian	1.0311	1.0077	0.9848	1	0.9867	1.0147
along parallel	1.0154	0.9495	1.0134	1.0141	1.0135	1.0147
20°						
along meridian	1.1325	1.0311	0.9397	1	0.9524	1.0587
along parallel	1.0642	0.9358	1.0492	1.0540	1.0500	1.0587
10°						
along meridian	1.3333	1.0718	0.8660	1	0.9028	1.1348
along parallel	1.1547	0.9528	1.1042	1.1196	1.1077	1.1348
0°						
along meridian	1.7041	1.1325	0.7660	1	0.8412	1.2509
along parallel	1.3054	1	1.1792	1.2148	1.1898	1.2509
–10°						
along meridian	2.4203	1.2174	0.6428	1	0.7698	1.4218
along parallel	1.5557	1.0822	1.2779	1.3475	1.2990	1.4218
–20°						
along meridian	4	1.3333	0.5	1	0.6903	1.6739
along parallel	2	1.2101	1.4076	1.5315	1.4486	1.6739
–30°						
along meridian	8.5486	1.4903	0.3420	1	0.6040	2.0561
along parallel	2.9238	1.4043	1.5820	1.7914	1.6557	2.0561
–40°						
along meridian	33.1634	1.7041	0.1736	1	0.5119	2.6665
along parallel	5.7588	1.7041	1.8264	2.1716	1.9535	2.6665

TUTORIAL 17 CIRCULAR-CURVE FITTING

Two points define a unique straight line. Similarly, any three points not on a straight line define a unique circle. In cartography there is a frequent need to establish formulae for circular arcs symmetric about one axis and not crossing the other.

We first address an arc for parallels, therefore symmetric about the y axis. Given points $A_1 = (-a, y_a)$, $B = (0, y_b)$, and $A_2 = (+a, y_a)$, we seek the radius r of the arc passing through them and the location of its centre (necessarily on the y axis). We assume $a > 0$ and that y_a and y_b have the same sign. The diagram also has $y_a > y_b$.

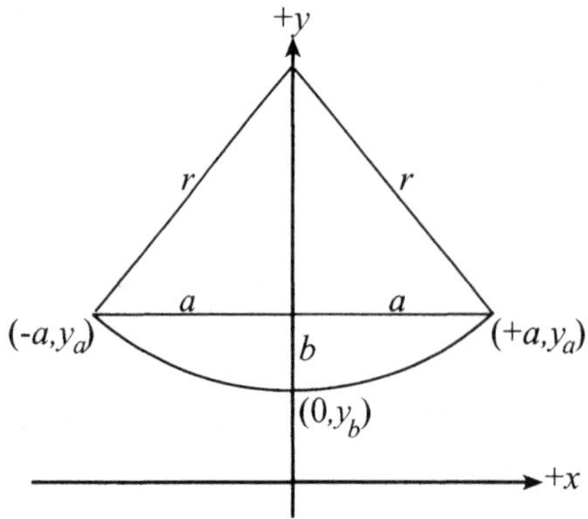

By Pythagoras, we have

$$r^2 = a^2 + (r-b)^2 = a^2 + r^2 - 2rb + b^2$$

$$0 = a^2 - 2rb + b^2$$

$$r = \frac{a^2+b^2}{2b} = \frac{1}{2}\left(\frac{a^2}{b} + b\right) = \frac{a}{2}\left(\frac{a}{b} + \frac{b}{a}\right)$$

Circular arcs for parallels

Accommodating the inverted signs for an inverted image, we have

$$b = |y_a - y_b|$$

$$r = \frac{a}{2}\left(\frac{a}{|y_a - y_b|} + \frac{|y_a - y_b|}{a}\right)$$

The centre of the circle providing the arc is at $(0, c)$, where $c = y_b + r$ for arcs concave upwards and $c = y_b - r$ for arcs concave downwards. The formula for the circle is

$$x^2 + \{y - c\}^2 = r^2$$

or

$$y = c \pm \sqrt{r^2 - x^2}$$

where for a fractional arc as illustrated the $\pm$ takes the sign of $(y_b - y_a)$.

Circular arcs for meridians

For meridians, we interchange co-ordinates — i.e., given points $A_1 = (x_a, -a)$, $B = (x_b, 0)$, and $A_2 = (x_a, +a)$ we get:

$$b = |x_a - x_b|$$

$$r = \frac{a}{2}\left(\frac{a}{|x_a - x_b|} + \frac{|x_a - x_b|}{a}\right)$$

The centre of the circle providing the arc is at $(c, 0)$, where $c = x_b + r$ for arcs concave toward the right, $c = x_b - r$ for arcs concave toward the left. The formula for the circle is

$$x = c \pm \sqrt{r^2 - y^2}$$

where, for a fractional arc, the $\pm$ takes the sign of $(x_b - x_a)$.

Circular arcs for point-Polar meridians

For point-polar meridians, the Poles are on the y axis, so likewise are the end points of the arc. Hence, $x_a = 0$ and:

$$b = |x_b|$$

$$r = \frac{a}{2}\left(\frac{a}{b} + \frac{b}{a}\right)$$

CHAPTER 5

Slicing into the Globe: Secantal projections

Preamble

For all projections covered in preceding chapters, the scale for the map is defined from the single contact parallel of each. That scale holds along all parallels in the Orthographic Azimuthal, but along no other parallel in the other projections. In the equidistant projections, the same scale holds along all meridians, but it holds along no meridian in any of the others. Limited applicability of the scale is unavoidable for any projection. We can improve that facet a little, however, and significantly reduce the rapidity of change from nominal by adopting a secantal approach — slicing our developable surface into our globe, geometrically. When carried out using a literal secant, the results are of little practical value. As a base for synthesis in the manner of Chapter 4, however, the method — now only notionally secantal — provides projections of vital importance.

No new mathematics is needed for this chapter.

The maps continue the unbalanced graticule of the preceding two chapters — i.e., with a 15° spacing of parallels and 30° spacing of meridians.

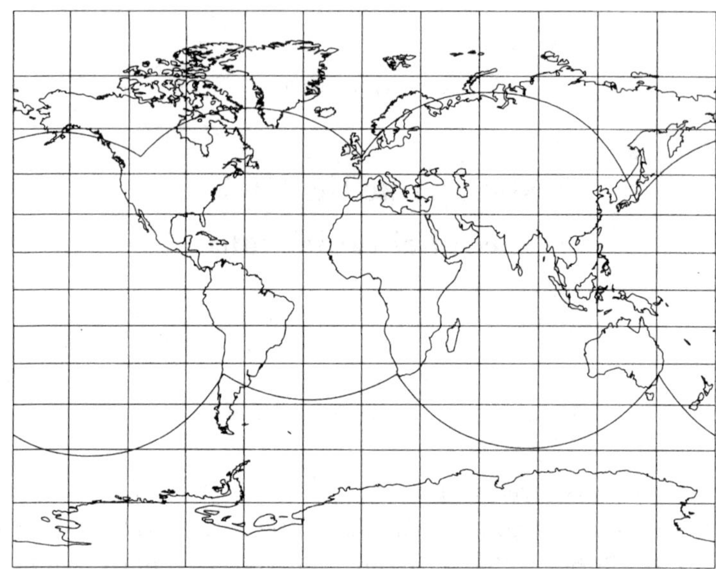

Exhibit 5–1: The Gall Stereographic projection —
the Stereographic Cylindrical projection with standard parallels set at ±45° latitude.
With great-circle routes as detailed for Exhibit 4–3 on page 82.

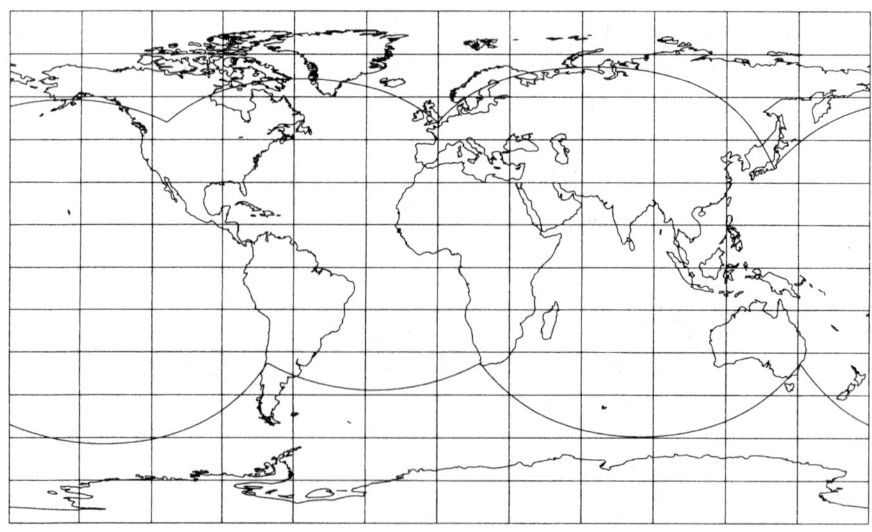

Exhibit 5–2: The Equirectangular projection —
the Equidistant Cylindrical projection with two standard parallels.
Standard parallels set at ±30° latitude.
With great-circle routes as detailed for Exhibit 4–3 on page 82.

GEOMETRY

Basics

The word *secant* relates to a straight line intersecting a circle or other curve, with the section between the intersections being a *chord*. The adjective *secantal* is used here to mean a mapping plane that intersects the globe, literally in a geometric sense else notionally within our algebra. All of our three developable surfaces — azimuthal, cylindrical, and conical — can be placed slicing through the globe rather than external to it. Assuming Simple aspect, the result for the azimuthal mode is a single lengthy non-trivial contact parallel, for the cylindrical mode two mirror-image latitudes straddling the Equator, for the conical mode two distinct contact parallels (one of which could be a Pole, one of which could be the Equator). The contact parallels, being intersections between mapping plane and globe, have true scale on the map. They are called *standard parallels.*

Exhibit 5–3 below illustrates the geometry and derives the trigonometry for the secant. For visual convenience, it has a large secantal angle, of size 2ψ. The diagram shows:

- the orthogonal distance from centre of the globe to the secant is $R\cos\psi$
- the continuation of that line to the circumference line has length $R(1-\cos\psi)$
- the length of the chord is $2R\sin\psi$

The secantal technique can be applied to all forms of literal projecting but, because of the consequent shrinkage along meridians to chordal lengths within/between the lines

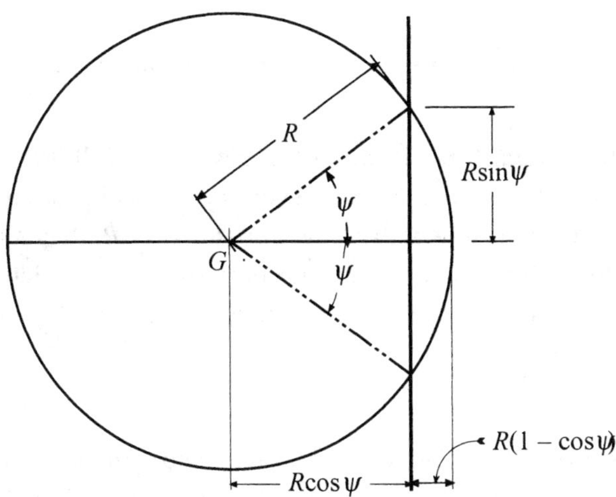

Exhibit 5–3: The secant: geometry and trigonometry.

of true scale, there is scope neither for being equal-area nor for being equidistant. Nor again is there in fact scope for conformality for a literally secantal azimuthal projection, although that is less obvious. Thus, projected secantally the Stereographic Azimuthal loses its conformality and the Orthographic Cylindrical loses its equal-area status. Resorting to algebra for re-spacing of parallels (in the manner of Chapter 4) allows synthesis of notionally secantal versions of both, retaining their respective attributes. As is shown below, however, there cannot be a secantal equidistant azimuthal projection.

The pattern of circular arcs with a common centre for parallels and straight lines converging to that centre for meridians is retained for all the secantal projections. Hence they all qualify as 'true' projections. An essential feature of being 'true' is that each of the two intermediate variables is a function of just one of the basic variables longitude and latitude, and that the variable using longitude is a linear multiple of (relative) longitude — features assumed in several algebraic steps in our developments below.

Except with orthogonal projecting, the effect of the secantal approach is to reduce the scale along parallels and hence the width of the map, and not only to moderate the rate of change (from true) of the scale along the meridians but to invert that effect between the lines of true scale. For the azimuthal mode, the reduction of width applies equally to all directions. We denote by σ the resulting relative scale at the focal point(s).

For an **azimuthal** projection at Simple aspect with standard parallel ϕ_1, we have the secantal semi-angle

$$\psi = \frac{\pi}{2} - \oplus\phi_1$$

True scale occurs along the circle of radius $\dot{\rho}_1$ and circumference $2\pi\dot{\rho}_1$ that represents the parallel of length $2\pi R\cos\phi_1$ on the globe. Hence,

$$\dot{\rho}_1 = R\cos\phi_1 \qquad \text{. . . . . (104a}$$

For a **cylindrical** projection at Simple aspect with standard parallels $\pm\phi_1$, we have

$$\psi = \phi_1$$

assuming $\phi_1 > 0$. The standard parallels must, if covering a full 360° of longitude, be of length $2\pi R\cos\phi_1$ to be of true scale. Because every cylindrical projection has all parallels of the same length, and the true length of the Equator is $2\pi R$, they suffer with standard parallels $\pm\phi_1$ a multiplicative lateral shrinkage of $\cos\phi_1$, giving relative scale along the Equator of $\sigma = \cos\phi_1$. Hence, the generic intermediate parameters [Equations 53a] revise to:

$$\begin{aligned} \dot{u} &= \sigma R\lambda = \cos\phi_1 R\lambda \\ \dot{v} &= \dot{v}(\phi) \end{aligned} \qquad \text{. . . . . (104b}$$

For a **conical** projection at Simple aspect we denote the two standard parallels ϕ_1 and ϕ_2 with ϕ_1 the one closer to the Equator. The secantal semi-angle is then half their difference, i.e.,

$$\psi = \oplus\frac{\phi_1 - \phi_2}{2} \qquad \text{. . . . . (105a}$$

As previously shown, the angle between any two meridians on the conical map is some constant $c < 1$ times the corresponding angle on the globe. This, the constant of the notional cone, is reduced by going secantal, producing a narrowing or lateral shrinkage of the map.

For a span of $\Delta\lambda$ in longitude (measured in radians, as routinely implied by the notation), the length along parallel ϕ_i is $(R\cos\phi_i)\Delta\lambda$ on the globe and $c\,\dot{\rho}_i\,\Delta\lambda$ on the map. To have true scale along standard parallel ϕ_1 we must, therefore, have

$$c\,\dot{\rho}_1\,\Delta\lambda = (R\cos\phi_1)\Delta\lambda$$

Dividing through by $\Delta\lambda$, and repeating the process for the other standard parallel ϕ_2, gives

$$c\,\dot{\rho}_1 = R\cos\phi_1 \quad \text{and} \quad c\,\dot{\rho}_2 = R\cos\phi_2$$

Hence,

$$\frac{R\cos\phi_1}{\dot{\rho}_1} = c = \frac{R\cos\phi_2}{\dot{\rho}_2} \qquad \text{. . . . . (105b}$$

So

$$\frac{\dot{\rho}_1}{\dot{\rho}_2} = \frac{\cos\phi_1}{\cos\phi_2}$$

Differencing the two map variables then gives

$$\dot{\rho}_1 - \dot{\rho}_2 = \dot{\rho}_1\left\{1 - \frac{\dot{\rho}_2}{\dot{\rho}_1}\right\} = \dot{\rho}_1\left\{1 - \frac{\cos\phi_2}{\cos\phi_1}\right\} = \dot{\rho}_1\left\{\frac{\cos\phi_1 - \cos\phi_2}{\cos\phi_1}\right\} \qquad \text{. . . . (105c}$$

and

$$\dot{\rho}_2 - \dot{\rho}_1 = \dot{\rho}_2\left\{1 - \frac{\dot{\rho}_1}{\dot{\rho}_2}\right\} = \dot{\rho}_2\left\{1 - \frac{\cos\phi_1}{\cos\phi_2}\right\} = \dot{\rho}_2\left\{\frac{\cos\phi_2 - \cos\phi_1}{\cos\phi_2}\right\}$$

These equations hold for all secantal 'true' conical projections.

For each of the three modes the generic plotting basis (i.e., expressed in terms of the intermediate parameters) for Simple aspect remains unchanged (see Equations 32a, 53a & 59b). We again denote by λ_{+y} the longitude chosen to lie along the $+y$ half-axis for azimuthal maps and by λ_0 the central longitude assumed to lie along the y axis for cylindrical and conical maps. We then set the origin co-ordinates for Azimuthals corresponding to the Pole, for Cylindricals to $(\lambda_0, 0)$, and for Conicals to (λ_0, ϕ_0), where ϕ_0 is now arbitrarily chosen, usually as the arithmetic mean of the two standard latitudes.

We now explore the effect of using a secantal situation, literally, for the literal projections of Chapter 3. Then, using the synthetic algebraic manner of Chapter 4, we develop equidistant, equal-area and conformal projections, calling them secantal although they are only notionally so.

SECANTAL LITERAL PROJECTIONS

Orthographic projections

Because the projective rays of any orthographic projecting are parallel lines, moving the mapping plane along the central ray has no impact on the map, hence there is no change to scale along any meridian. For Azimuthals, with the scale true along all parallels, there is no impact at all. For Cylindricals and Conicals there is a migration of true scale from the single contact parallel of the tangential to the two contact parallels. With the scale on these parallels enlarged for both modes in the tangential map, the result is shrinkage of the map transverse to the central longitude. For example, were the Orthographic Cylindrical to be recast secantally with contact parallels at ±20° (where the tangential map had a scale $\cos^{-1} 20°$ times nominal) then the scale along all parallels would be reduced by the inverse of this, i.e., multiplied by $\cos 20° = 0.9397_{\sim}$, with corresponding reduction in the width of the map. Such directional shrinkage obviously removes any equal-area or conformal status that applied to the tangential (though it could feasibly introduce it where previously not existing). However, in the one mode that is used secantally — the cylindrical — an inverse expansion of the map along the meridians can obviously counter the shrinkage to maintain equal-area; this is discussed below.

Non-orthographic perspective projections

For perspective projection from a finite point, changing from tangential to secantal affects the scaling along both meridians and parallels. Assuming the radiative point is beyond the chord, the scale along meridians is reduced between the old contact parallel and the new, but enlarged beyond the latter (see Exhibit 5–4).

With point V distance hR below the surface point F we get

$$\text{length of } \dot{F}V = \text{length of } FV = R\big(h - (1 - \cos\psi)\big)$$
$$= R(h - 1 + \cos\psi)$$

Despite moving the mapping plane, the earlier derived Equation 49a still holds, i.e.,

$$\tan\xi = \frac{\sin\gamma}{(h - 1 + \cos\gamma)}$$

Therefore, by simple trigonometry,

$$\text{length of } \dot{F}\dot{P} = R(h - 1 + \cos\psi)\tan\xi = R\frac{(h - 1 + \cos\psi)\sin\gamma}{(h - 1 + \cos\gamma)} \qquad \ldots\ (106a$$

For any secantal **Azimuthal,** γ is the polar distance of P and ψ is also the polar distance of the chosen standard latitude ϕ_1. Using our conditional sign symbol relative to the Pole that is the centre of our map this becomes

$$\dot{\rho} = \text{length of } \dot{F}\dot{P} = R\frac{(h-1+\oplus\sin\phi_1)\cos\phi}{(h-1+\oplus\sin\phi)}$$

Substitution in any of Equations 31b, 31c & 32a gives plotting equations for the secantal General Perspective Azimuthal projection at Simple aspect. Choosing the longitude to lie along the $+x$ half-axis, therefore using Equations 31b, gives:

$$\dot{x} = +R\frac{(h-1+\sin\phi_1)\cos\phi}{(h-1+\sin\phi)}\cos(\lambda-\lambda_{+x})$$
$$\dot{y} = \oplus R\frac{(h-1+\sin\phi_1)\cos\phi}{(h-1+\sin\phi)}\sin(\lambda-\lambda_{+x}) \quad \text{. . . . . (107a}$$

With $h=2$ this becomes the secantal Stereographic (Azimuthal) projection. This erstwhile conformal projection, however, loses that vital attribute when moving from tangential to secantal when done literally, so loses most of its value. A notionally secantal form, developed later, takes its place and a significant place in mapping.

For any secantal **Cylindrical**, Equation 106a holds but with γ as the latitude of point P and ψ the latitude ϕ_1 of the northern standard parallel. With the negative values on the other side of the central ray automatically covered by the sine and cosine terms, we get

$$\dot{v} = R\frac{(h-1+\cos\phi_1)\sin\phi}{(h-1+\cos\phi)}$$

Using Equations 104b, the plotting equations for the secantal General Perspective Cylindrical projection at Simple aspect become:

$$\dot{x} = \dot{u} = \sigma R\lambda \quad = \cos\phi_1 R\lambda$$
$$\dot{y} = \dot{v} \quad = R\frac{(h-1+\cos\phi_1)\sin\phi}{(h-1+\cos\phi)} \quad \text{. . . . . (107b}$$

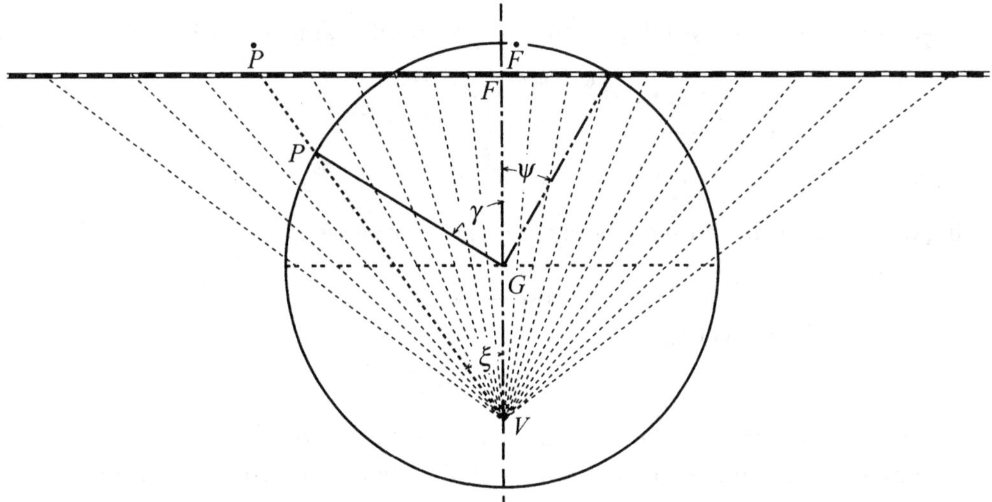

Exhibit 5–4: The geometry of general perspective projecting – secantal version.

Despite lacking any special attributes, the secantal Stereographic Cylindrical projection, with $h=2$, has found use. Its plotting equations are:

$$\dot{x} = \dot{u} = \sigma R\lambda = \cos\phi_1 R\lambda$$

$$\dot{y} = \dot{v} = R\frac{(1+\cos\phi_1)\sin\phi}{(1+\cos\phi)} \qquad \text{. . . . . (108a}$$

Various choices of ϕ_1 have been adopted with the projection, for example:

Gall Stereographic of 1855	$\phi_1 = 45°$	$\sigma = \cos\phi_1 = 0.7071_{\sim}$
an unnamed Russian projection of 1929	$\phi_1 = 55°$	$\sigma = \cos\phi_1 = 0.5736_{\sim}$
B.S.A.M. of 1937	$\phi_1 = 35°$	$\sigma = \cos\phi_1 = 0.8660_{\sim}$

The latter two were used in atlases published by the U.S.S.R. The Gall Stereographic projection, illustrated on page 102 in Exhibit 5–1, has found various usages.

Henry James and A. R. Clarke chose a more distant viewpoint than the stereographic, to produce maps that comfortably embraced more than a hemisphere with modest overall distortion. Seeking to reach at Simple aspect as far as the further Tropic, James [1860a] adopted $h=2.5$ (reckoned by Clarke [1862b] to be better at $h=2.368$); at Oblique aspect (Chapter 8) centred at 15°E on the Tropic of Cancer, it shows all the major land masses. **Clarke's Twilight** projection [ibid] was designed to reach only 18° beyond the hemisphere, equal to the amount recognised as in twilight by astronomers. He had $h=2.4$ and a secantal half-angle of 21.$_{\sim}$° (i.e., latitude 69.$_{\sim}$° for Simple aspect).

For any secantal **Conical**, with standard parallels ϕ_1 and ϕ_2 and assuming that the radiant point is on the line that bisects the angle between those two,

$$\psi = \oplus\frac{\phi_2 - \phi_1}{2}$$

$$\gamma = \oplus\phi - \oplus\frac{\phi_2 + \phi_1}{2} = \oplus(\phi - \phi_0) \quad \text{where} \quad \phi_0 = \frac{\phi_2 + \phi_1}{2} \qquad \text{. . . . . (108b}$$

So Equation 106a together with Equation 57d gives for the familiar northern version

$$\dot{\rho} = \dot{\rho}_0 - \frac{\left\{h - 1 + \cos\left(\frac{\phi_2 - \phi_1}{2}\right)\right\}\sin\left(\frac{\phi - \phi_0}{2}\right)}{\left\{h - 1 + \cos\left(\frac{\phi - \phi_0}{2}\right)\right\}}$$

Adding our conditional sign symbol for the inverted cone we get

$$\dot{\rho} = \oplus R\left\{\cot\phi_0 - \frac{\left\{h - 1 + \cos\left(\frac{\phi_2 - \phi_1}{2}\right)\right\}\sin\left(\frac{\phi - \phi_0}{2}\right)}{\left\{h - 1 + \cos\left(\frac{\phi - \phi_0}{2}\right)\right\}}\right\} \qquad \text{. . . . . (108c}$$

The plotting equations for the secantal General Perspective Conical projection are obtained by substitution from Equation 108c in Equations 59b. Any Pole included in the resulting map is a point if it is a standard parallel — otherwise it is a circular arc.

NOTIONALLY SECANTAL EQUIDISTANT PROJECTIONS

Basis

Equidistant tangential projections are achieved by the simple expedient of making every meridian of true scale. In the secantal context, being equidistant implies true scale along the standard parallel(s) and along all meridians. For the azimuthal mode, we find this is infeasible for any proper secant. For the other two modes, it is feasible and a simple adaptation of the tangential case. For both, the distance on the map between the two standard parallels cannot be that of the chord. It is, necessarily, identically the length of the circumferential curve between the ends of the chord.

The Equidistant Azimuthal projection

Equidistance for the azimuthal mode at Simple aspect requires (see Equation 69a)

$$\dot{\rho} = R\left(\tfrac{\pi}{2} - \oplus\phi\right)$$

for any latitude ϕ. Equating, for the specific latitude ϕ_1, with Equation 104a gives

$$R\left(\tfrac{\pi}{2} - \oplus\phi_1\right) = R\cos\phi_1$$

$$\tfrac{\pi}{2} - \oplus\phi_1 - \cos\phi_1 = 0$$

which limits the choices for ϕ_1. Indeed the equation is true solely for the value $\pi/2$ for $\oplus\phi_1$ i.e., with contact parallel a Pole, which is the tangential case. In other words, as already asserted, there is no feasible equidistant azimuthal projection that is properly secantal — i.e., has true scale along a non-Polar parallel.

The Equidistant Cylindrical projection

Equidistance for the cylindrical mode requires (see Equation 69b)

$$\dot{v} = R\phi$$

for any latitude ϕ. Applying the lateral shrinkage factor $\sigma = \cos\phi_1$ gives, for the secantal Equidistant Cylindrical projection at Simple aspect with standard parallels $\pm\phi_1$, the plotting equations

$$\begin{aligned} \dot{x} = \dot{u} = \sigma R\lambda &= R(\cos\phi_1)\lambda \\ \dot{y} = \dot{v} &= R\phi \end{aligned} \qquad \text{. . . . . (109a}$$

Called also the **Rectangular, Equirectangular, and La Carte Parallelogrammatique** projection, it is illustrated with $\phi_1 = 30°$ in Exhibit 5–2 (page 102). The projection is among the very oldest, having been used by Marinus of Tyre two millennia ago [see 2000a]; his standard parallel through the island of Rhodes was about 36°N, giving a lateral shrinkage of about 4:5. The version with $\phi_1 = 45°$ introduced by James Gall [1855a] has the distinctive name **Gall Isographic** projection, with lateral-shrinkage factor $0.7071_{\sim} = \cos 45°$.

The Equidistant Conical projection

The situation for equidistance for the conical mode is similar to that for the cylindrical, but complicated by the curvature of the parallels. Putting ϕ_1 for ϕ_0 in Equation 69c, which specifies the condition for equidistant scaling along the meridians in conical mode, gives

$$\dot{\rho} - \dot{\rho}_1 = \oplus R\,(\phi_1 - \phi) \quad \text{or} \quad \dot{\rho} = \dot{\rho}_1 + \oplus R\,(\phi_1 - \phi) \qquad \text{. . . . . (110a}$$

Substituting ϕ_2 for the general ϕ, and equating with Equation 105c gives

$$\dot{\rho}_1 \left\{ \frac{\cos\phi_1 - \cos\phi_2}{\cos\phi_1} \right\} = \dot{\rho}_1 - \dot{\rho}_2 = \oplus R\,(\phi_2 - \phi_1)$$

Hence,

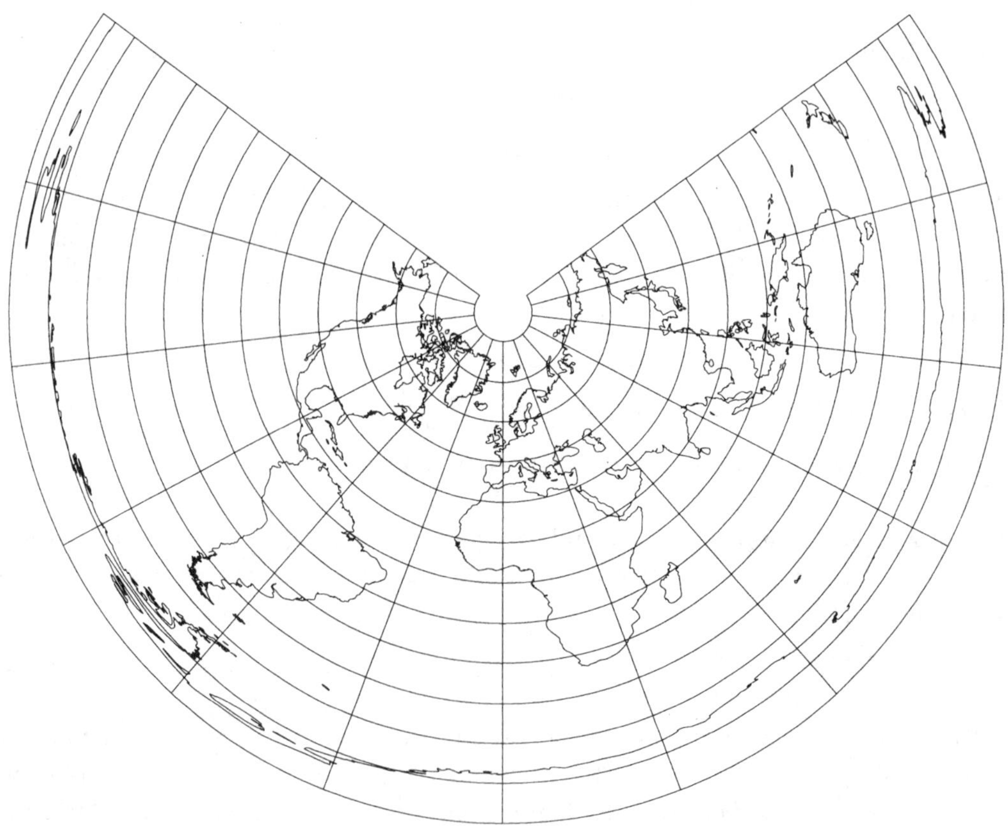

Exhibit 5–5: The Equidistant Conical projection with two standard parallels – whole world.
Standard parallels set at 30°N and 60°N.

$$\dot{\rho}_1 = \oplus \frac{R(\phi_2 - \phi_1)\cos\phi_1}{\cos\phi_1 - \cos\phi_2} \qquad \text{(111a}$$

Substituting for ρ_1 in Equation 110a gives

$$\begin{aligned}\dot{\rho} &= \oplus \frac{R(\phi_2 - \phi_1)\cos\phi_1}{\cos\phi_1 - \cos\phi_2} + \oplus R(\phi_1 - \phi) \\ &= \oplus R\left\{\frac{(\phi_2 - \phi_1)\cos\phi_1}{\cos\phi_1 - \cos\phi_2} + (\phi_1 - \phi)\right\} \qquad \text{(111b} \\ &= \oplus R\left\{\frac{(\phi_2 - \phi_1)\cos\phi_1 + \phi_1(\cos\phi_1 - \cos\phi_2)}{\cos\phi_1 - \cos\phi_2} - \phi\right\}\end{aligned}$$

That is

$$\dot{\rho} = \oplus R\left\{\frac{\phi_2\cos\phi_1 - \phi_1\cos\phi_2}{\cos\phi_1 - \cos\phi_2} - \phi\right\} \qquad \text{(111c}$$

Substituting from Equation 111a in Equation 105b, we get

$$c = \frac{R\cos\phi_1}{\dot{\rho}_1} = \oplus \frac{R\cos\phi_1(\cos\phi_1 - \cos\phi_2)}{R(\phi_2 - \phi_1)\cos\phi_1} = \oplus \frac{\cos\phi_1 - \cos\phi_2}{(\phi_2 - \phi_1)} \qquad \text{(111d}$$

Substituting this in the first line of Equation 111b gives

$$\dot{\rho} = R\left\{c\cos\phi_1 + \oplus(\phi_1 - \phi)\right\} \qquad \text{(111e}$$

which simplifies our general formula, since c is constant across our map.

The plotting equations for the secantal Equidistant Conical projection at Simple aspect with standard parallels ϕ_1 and ϕ_2 are obtained by substituting in Equations 59b for c as defined by Equation 111d and for $\dot{\rho}$ as defined in either of Equation 111c & 111e, then for $\dot{\rho}_0$ by substitution for variable ϕ in the obtained formula for $\dot{\rho}$ the latitude of the point on the central meridian chosen to be at the origin.

The projection is illustrated in Exhibit 5–5 opposite with standard parallels at 30°N and 60°N.

Called also the **Simple Conic** projection and simply just the **Conic** projection, this too is about 2000 years old, created rudimentarily by Cladius Ptolemy. Medieval mapmakers rendered various improvements and many cartographers have endeavoured to develop criteria for optimal choice of standard parallels — a matter of obvious concern in all these secantal projections. J. N. de l'Isle, to whom Snyder [1979b] attaches credit for the modern form with a map of Russia in 1745, chose (at Simple aspect) ¼ of the latitudinal span in from top and bottom; C. H. Deetz and O. S. Adams [1921f] chose ⅙ and A. R. Hinks [1912b] chose ⅐. Others have developed mathematical formulae beyond such simple rules to determine the optimal positioning, as discussed in Chapter 14.

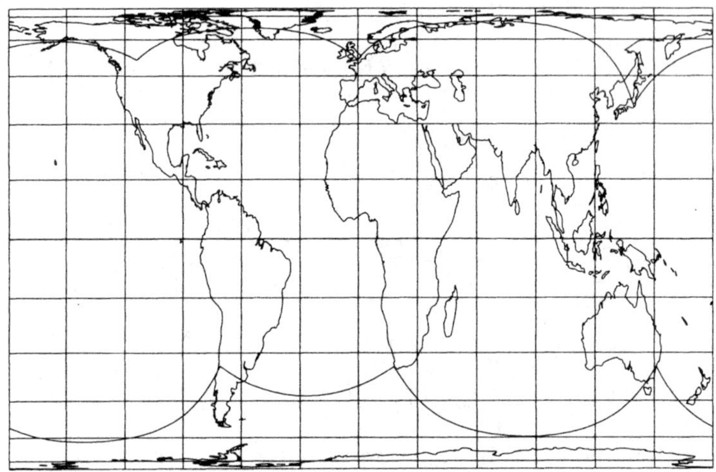

Exhibit 5–6: The Gall Orthographic (or Peters) projection –
the secantal Equal-area Cylindrical with standard parallels at ±45°.
Shows selected great-circle routes as detailed for Exhibit 4–3 on page 82.

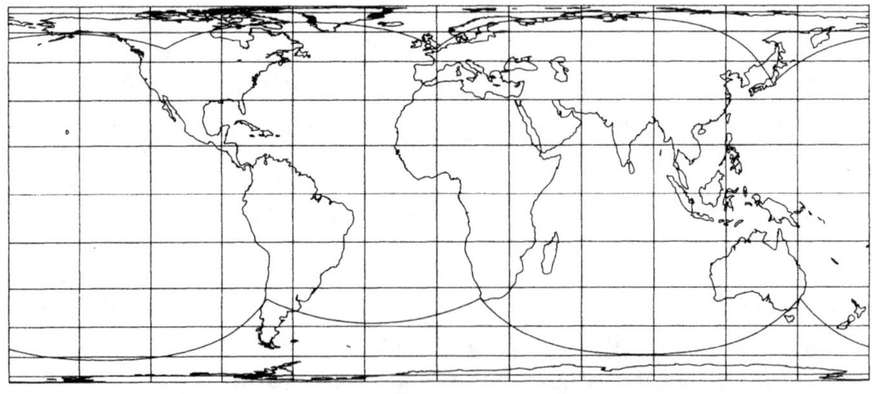

Exhibit 5–7: The Behrmann projection –
the secantal Equal-area Cylindrical with standard parallels at ±30°.
Shows selected great-circle routes as detailed for Exhibit 4–3 on page 82.

NOTIONALLY SECANTAL EQUAL-AREA PROJECTIONS

Basis

We use formulae of the preceding chapter to develop equal-area equations when the true scale is not at the tangential contact. For the azimuthal mode, we find this is infeasible. For the other two modes it is feasible and a simple adaptation of the tangential case. For both, with the narrowing of the space between meridians at the more central latitudes, the distance on an equal-area map between the two standard parallels cannot be that of the chord. Both require expansion along meridians to offset the lateral contraction.

The Equal-area Azimuthal

Equal-area for the azimuthal mode requires (see Equation 73c)

$$\dot{\rho}^2 = 2R^2(1 - \oplus\sin\phi)$$

for any latitude ϕ. Equating for the specific latitude ϕ_1 with Equation 104a gives

$$2R^2(1 - \oplus\sin\phi_1) = \left(R\cos\phi_1\right)^2 \quad = R^2\cos^2\phi_1$$

so

$$2(1 - \oplus\sin\phi_1) = \cos^2\phi_1 \qquad = 1 - \sin^2\phi_1$$

I.e.

$$\sin^2\phi_1 - \oplus 2\sin\phi_1 + 1 = 0$$

This equation has the single solution $\sin\phi_1 = 1$. Within our context this implies $\oplus\phi_1 = {}^{\pi}\!/_{2}$. Hence, the contact parallel is a Pole and the mapping plane is tangential. In other words, as already asserted, there is no feasible equal-area azimuthal projection that is properly secantal.

The Equal-area Cylindrical

Equal-area for the cylindrical mode in the tangential form occurred with the Orthographic projection. As we have noted above, however, equal-area status was lost with the secantal form of that literal projection because of uni-directional shrinkage. Equal-area status can easily be restored, by inverse expansion of the rectangular map along the meridians. Indeed, given orthogonal axes and non-zero σ, the mapping:

$$\begin{aligned} \ddot{x} &= \sigma\dot{x} \\ \ddot{y} &= \sigma^{-1}\dot{y} \end{aligned} \qquad \text{. . . . . (113a}$$

of any equal-area map $\{\dot{x},\dot{y}\}$ produces another equal-area map $\{\ddot{x},\ddot{y}\}$. This transformation, applying reciprocal stretching and shrinking orthogonally to any map, maintains any equal-area attribute possessed by that map; its essence is to relocate the lines of true scale. It is a modification of general applicability.

Secantal Cylindricals have a lateral shrinkage by multiplicative factor $\cos\phi_1$. Applied to the Orthographic Cylindrical projection with $\sigma = \cos\phi_1$, Equation 113a re-establishes the equal-area attribute. From Equations 104b and 55c the plotting equations for the secantal Equal-area Cylindrical projection at Simple aspect with standard parallels $\pm\phi_1$ are:

$$\dot{x} = \dot{u} = \cos\phi_1 \, R\,\lambda$$
$$\dot{y} = \dot{v} = \cos^{-1}\phi_1 \, R \sin\phi \qquad \text{. . . . . (114a}$$

Several particular values of ϕ_1 have found advocates, each version of the projection typically accorded the advocate's name. Example projections include:

Gall Orthographic [1855a]	$\phi_1 = 45°$	$\sigma = \cos\phi_1 = 0.7071_\sim$
Behrmann Cylindrical Equal-area [1910a]	$\phi_1 = 30°$	$\sigma = \cos\phi_1 = 0.8660_\sim$
Edwards [1953a]	$\phi_1 = 37.4°$	$\sigma = \cos\phi_1 = 0.7944_\sim$

The **Peters** projection, announced in 1967 and widely promoted for world maps, particularly relative to the less-developed world, appears to be identically the Gall Orthographic, which is illustrated in Exhibit 5–6 on page 112 [for discussion of dispute see 1980a, 1980b]. As an equal-area map, this projection has been heavily promoted for political reasons as a rectangular map that avoids the exaggeration of the more-developed countries relative to the mostly Equatorial less-developed countries that occurs with Mercator. It has the irony, however, of being close to conformal for the developed countries while producing a somewhat unbecoming droopiness for the Equatorial regions.

After (purely cartographic) analysis seeking the least overall angular distortion within the confines of the equal-area cylindrical style, Walter Behrmann promoted the use of ±30° as standard parallels. It, too, is illustrated on page 112 (Exhibit 5–7). The Behrmann has the same scale as the Gall Orthographic along the parallels of true scale, but having them at lower latitudes results in it having less lateral shrinkage and less vertical expansion than the Gall Orthographic map. Although visibly distorting the highest latitudes much more than does the Gall, the Behrmann would seem a better political choice, since it maintains shapes much better in the lowest latitudes. (Some other standard parallel between the two could be even better, e.g. $\arccos 0.8 = 36°52._\sim'$.)

Trystan Edwards advocated keeping the ratio of linear distortion along the meridian to that along the parallel at any point within the range 4:3 to 3:4 for a maximal distance from the Equator. That gives standard parallels at ±37°24′.

Both Chapter 6 and Chapter 10 show alternative equal-area maps used on a whole-world basis.

The Equal-area Conical

Equal-area for a conical projection with constant c occurs if and only if Equation 75a holds true for any two latitudes. Applying it to our two standard parallels by substituting ϕ_1 for ϕ and ϕ_2 for ϕ_i gives

$$\dot{\rho}_1^{\,2} - \dot{\rho}_2^{\,2} = \oplus\frac{2R^2}{c}(\sin\phi_2 - \sin\phi_1)$$

Equation 57d, which holds for any latitude of true scale on any Conical map, similarly gives

$$\dot{\rho}_1^{\,2} - \dot{\rho}_2^{\,2} = \frac{R^2}{c^2}\left(\cos^2\phi_1 - \cos^2\phi_2\right)$$

for any latitude ϕ_1 that is a standard parallel. Equating these two equations gives

$$\oplus c^2\, 2R^2\left(\sin\phi_2 - \sin\phi_1\right) = c\, R^2\left(\cos^2\phi_1 - \cos^2\phi_2\right)$$

Hence,

$$\begin{aligned} c &= \oplus\frac{(\cos^2\phi_2 - \cos^2\phi_1)}{2(\sin\phi_1 - \sin\phi_2)} = \oplus\frac{(\sin^2\phi_1 - \sin^2\phi_2)}{2(\sin\phi_1 - \sin\phi_2)} \\ &= \oplus\frac{(\sin\phi_1 + \sin\phi_2)}{2} \end{aligned} \qquad \text{(115a}$$

producing a derived value for the constant c applicable to equal-area with standard parallels ϕ_1 and ϕ_2.

A cone cutting the globe at two separate parallels cannot meet this condition. It is progressively approximated as the two parallels become closer, to be met as they merge — when the constant of the cone becomes the sine of the single standard parallel.

Equal area in a conical projection with one standard parallel requires meeting Equation 75a for any latitude relative to the standard, and the same must hold here relative to either standard parallel. (With the equal-area condition already met between the two, meeting relative to either means meeting relative to the other.) We still have

$$\dot{\theta} = \oplus c\left(\lambda - \lambda_i\right)$$

but there is no actual cone.

Substituting ϕ_1 for ϕ_i in our equal-area formula in Equation 75a, then the above value for constant c, we get

$$\begin{aligned} (c\dot{\rho})^2 &= (c\dot{\rho}_1)^2 + \oplus c 2R^2(\sin\phi_1 - \sin\phi) \\ &= (R\cos\phi_1)^2 + (\sin\phi_1 + \sin\phi_2)R^2(\sin\phi_1 - \sin\phi) \end{aligned} \qquad \text{(115b}$$

Substituting our derived value for c on the right-hand side and simplifying gives

$$\begin{aligned} (c\,\dot{\rho})^2 &= R^2\{\cos^2\phi_1 + (\sin\phi_1 + \sin\phi_2)(\sin\phi_1 - \sin\phi)\} \\ &= R^2\{\cos^2\phi_1 + \sin^2\phi_1 + \sin\phi_2\sin\phi_1 - (\sin\phi_1 + \sin\phi_2)\sin\phi\} \end{aligned}$$

Taking the (positive) square root overall, we get as a general formula

$$\dot{\rho} = \frac{R}{c}\sqrt{1 + \sin\phi_1 \sin\phi_2 - (\sin\phi_1 + \sin\phi_2)\sin\phi}$$
$$= \frac{\oplus 2R}{\sin\phi_1 + \sin\phi_2}\sqrt{1 + \sin\phi_1 \sin\phi_2 - (\sin\phi_1 + \sin\phi_2)\sin\phi} \quad \text{. . . . . . (116a}$$

Plotting equations for the secantal Equal-area Conical projection at Simple aspect are obtained by substitution in Equations 59b for c as defined by Equation 115a, for $\dot{\rho}$ as defined by Equation 116a then for $\dot{\rho}_0$ by substitution for variable ϕ in the obtained formula for $\dot{\rho}$ the latitude of the point on the central meridian chosen to be at the origin. It is usually named after the promoter [1805a] as the **Albers projection with two standard parallels** or **2-standard-parallel Albers** projection. Snyder, however, points out [1978d] that it was well presented 60 years previously by de l'Isle. Illustrated immediately below, this projection has found widespread use in atlases, especially for regional thematic maps.

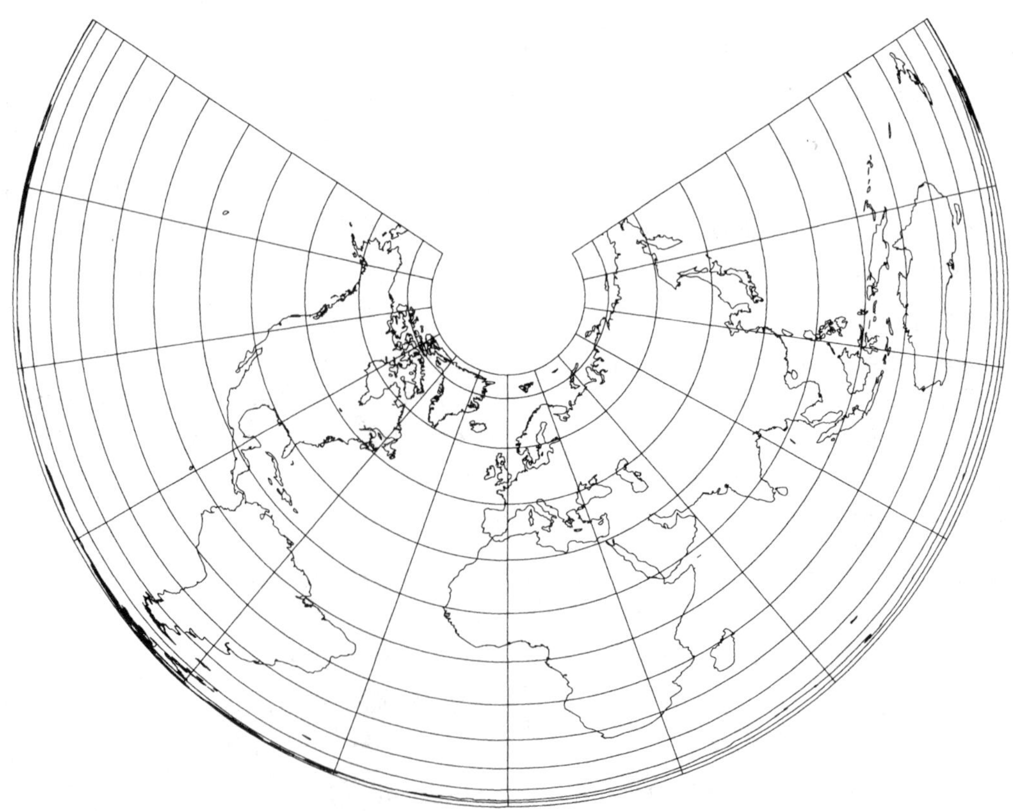

Exhibit 5–8: The Albers projection with two standard parallels — the secantal Equal-area Conical projection.
Standard parallels set at 30°N & 60°N.

For the two standard parallels, Equation 116a simplifies to

$$\dot{\rho}_i = \frac{R\cos\phi_i}{c} = \oplus\frac{2R\cos\phi_i}{(\sin\phi_1 + \sin\phi_2)} \qquad \text{for } i = 1,2$$

A special version is formed when one of the standard parallels is a Pole (making that geographical point a spot on the map). With $\oplus\Phi_2 = 90°$, hence $\sin\phi_2 = \oplus 1$, we get

$$c = \frac{(1 + \oplus\sin\phi_1)}{2}$$

Addressing the term within the square root in Equation 116a, noting that the square of our conditional sign symbols equals +1, we get

$$\begin{aligned} 1 + \sin\phi_1\sin\phi_2 - (\sin\phi_1 + \sin\phi_2)\sin\phi &= 1 + \oplus\sin\phi_1 - (\sin\phi_1 + \oplus 1)\sin\phi \\ &= 1 + \oplus\sin\phi_1 - \sin\phi_1\sin\phi - \oplus\sin\phi \\ &= 1 + (\oplus\sin\phi_1) - (\oplus\sin\phi_1)(\oplus\sin\phi) - (\oplus\sin\phi) \\ &= (1 + \oplus\sin\phi_1)(1 - \oplus\sin\phi) \end{aligned}$$

Hence for this special form Equation 116a reduces

$$\dot{\rho} = \frac{2R}{1 + \oplus\sin\phi_1}\sqrt{(1 + \oplus\sin\phi_1)(1 - \oplus\sin\phi)} = 2R\sqrt{\frac{(1 - \oplus\sin\phi)}{(1 + \oplus\sin\phi_1)}} \quad \text{. . . . . . . (117a}$$

This particular form, presented (in northern version only) by Lambert [1772a] some 30 years prior to introduction of the general form, is usually called the **Lambert Equal-area Conic** projection. It is illustrated immediately below.

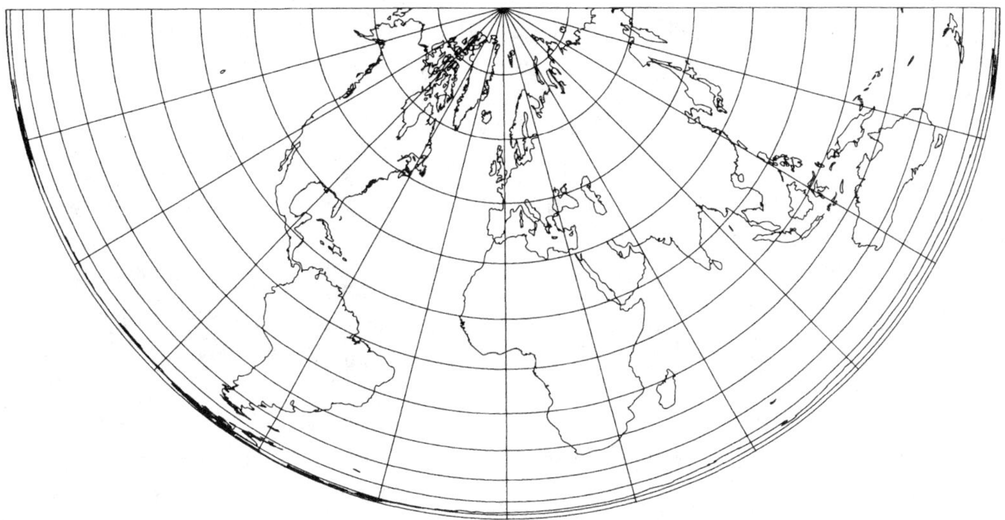

Exhibit 5–9: The (Lambert) Equal-area Conic projection,
a version of Albers projection with one of its two standard parallels at a Pole.
Standard parallels set at 0° and 90°N.

NOTIONALLY SECANTAL CONFORMAL PROJECTIONS

Basis

We again use formulae of the preceding chapter to develop pertinent equations, now for conformality. Since conformality merely requires, at each point on the map, that the meridian and parallel are mutually orthogonal and that the scale in every direction is the same, the lack of any absolute numeric demand avoids compromising the azimuthal mode — so all three modes are algebraically feasible. Clearly the resulting maps cannot change their shapes or internal proportions from those obtained in the tangential form. Adopting a notionally secantal approach merely changes the parallel(s) having true scale — the standard parallels. Effectively, the plotting equations for each mode become multiplied in both axes by the shrinkage factor.

The Conformal Azimuthal

The conformal tangential Azimuthal is the Stereographic projection, the key formula for which is Equation 46a. That proved to be a particular form of a general formula for conformality of an Azimuthal map given as Equation 85c, with unrestricted constant k. The value *2R* for this constant gave the tangential Stereographic, with its true scale at the Pole. If we write $k = \sigma 2R$ in Equation 85c it becomes

$$\dot{\rho} = \sigma 2R \tan\left\{\frac{\pi}{4} - \frac{\oplus\phi}{2}\right\} \qquad \text{(118a}$$

with σ the relative scale at the focal point. Since any circumference is multiple 2π of its radius, and the radius of parallel ϕ_1 on the globe is $R\cos\phi_1$, the relative scale along that parallel in the tangential projection is

$$\frac{1}{R\cos\phi_1} 2R\tan\left\{\frac{\pi}{4} - \frac{\oplus\phi_1}{2}\right\} = \cos^{-1}\phi_1\, 2\tan\left\{\frac{\pi}{4} - \frac{\oplus\phi_1}{2}\right\}$$

Equating this with σ^{-1} and applying to Equation 118a gives true scale at latitude ϕ_1 and relative scale σ at the focal point. Then

$$\dot{\rho} = \frac{\cos\phi_1}{2\tan\left\{\frac{\pi}{4} - \frac{\oplus\phi_1}{2}\right\}} 2R\tan\left\{\frac{\pi}{4} - \frac{\oplus\phi}{2}\right\} = R\cos\phi_1 \frac{\tan\left\{\frac{\pi}{4} - \frac{\oplus\phi}{2}\right\}}{\tan\left\{\frac{\pi}{4} - \frac{\oplus\phi_1}{2}\right\}} \qquad \text{(118b}$$

Substitution in any of Equations 31b, 31c & 32a gives plotting equations for the re-scaled Conformal Azimuthal projection, known in one specific re-scaling as the **Universal Polar Stereographic** projection (discussed further in Chapter 18). The appearance is unchanged from that of the tangential version on page 86.

The Conformal Cylindrical

To obtain a notionally secantal conformal cylindrical projection we again change the size of the map to relocate true scale. In this case, there are two linear intermediate variables to be multiplied, necessarily by the identical value, namely the factor $\sigma = \cos\phi_1$. From Equation 81a and Equations 104b the plotting equations become:

$$\dot{x} = \dot{u} = \sigma R \lambda = R\cos\phi_1 \lambda$$

$$\dot{y} = \dot{v} = \sigma R \ln\left\{\tan\left(\frac{\pi}{4} + \frac{\phi}{2}\right)\right\} = R\cos\phi_1 \ln\left\{\tan\left(\frac{\pi}{4} + \frac{\phi}{2}\right)\right\} = \frac{R\cos\phi_1}{2}\ln\left\{\frac{1+\sin\phi}{1-\sin\phi}\right\} \quad . . . \text{(119a}$$

The map appearance is unchanged from that of the tangential version on page 83 (the Mercator); it is merely re-scaled.

Equations 114a are equivalent to transforming the tangential formula of Equation 74a by:

$$\ddot{x} = \sigma\dot{x}$$

$$\ddot{y} = \sigma\dot{y}$$

This transformation, applying identical stretching orthogonally to any map, maintains any conformality possessed by that map; its essence is to relocate the lines of true scale. It is thus a modification of general applicability, which can be accommodated directly by including the scaling factor σ directly into plotting equations as done above.

The Conformal Conical

Conformality for a conical projection with constant c occurs if and only if Equation 84b holds true for every latitude. Applying it successively to our two standard latitudes ϕ_1 and ϕ_2 then taking the quotient gives

$$\frac{\rho_1}{\rho_2} = \frac{\left\{\tan\left(\frac{\pi}{4} - \frac{\oplus\phi_1}{2}\right)\right\}^c}{\left\{\tan\left(\frac{\pi}{4} - \frac{\oplus\phi_2}{2}\right)\right\}^c} = \left\{\frac{\tan\left(\frac{\pi}{4} - \frac{\oplus\phi_1}{2}\right)}{\tan\left(\frac{\pi}{4} - \frac{\oplus\phi_2}{2}\right)}\right\}^c$$

Equation 57b, which holds for any latitude of contact of cone and globe — hence for any latitude of true scale on any conical map — similarly gives

$$\frac{\rho_1}{\rho_2} = \frac{\cos\phi_1}{\cos\phi_2}$$

Hence, conformality with standard parallels ϕ_1 and ϕ_2 implies

$$\frac{\cos\phi_1}{\cos\phi_2} = \left\{\frac{\tan\left(\frac{\pi}{4} - \frac{\oplus\phi_1}{2}\right)}{\tan\left(\frac{\pi}{4} - \frac{\oplus\phi_2}{2}\right)}\right\}^c$$

Taking logarithms to bring the constant c down from exponent status, gives

$$\ln\left(\frac{\cos\phi_1}{\cos\phi_2}\right) = c\ \ln\left\{\frac{\tan\left(\frac{\pi}{4} - \frac{\oplus\phi_1}{2}\right)}{\tan\left(\frac{\pi}{4} - \frac{\oplus\phi_2}{2}\right)}\right\} \quad = c\ \ln\left\{\tan\left(\frac{\pi}{4} - \frac{\oplus\phi_1}{2}\right)\tan\left(\frac{\pi}{4} + \frac{\oplus\phi_2}{2}\right)\right\}$$

Hence,

$$c = \frac{\ln\left(\cos\phi_1 \sec\phi_2\right)}{\ln\left\{\tan\left(\frac{\pi}{4} - \frac{\oplus\phi_1}{2}\right)\tan\left(\frac{\pi}{4} + \frac{\oplus\phi_2}{2}\right)\right\}} \qquad \text{. . . . . (120a}$$

and also, identically,

$$c = \frac{\ln\left(\cos\phi_2 \sec\phi_1\right)}{\ln\left\{\tan\left(\frac{\pi}{4} - \frac{\oplus\phi_2}{2}\right)\tan\left(\frac{\pi}{4} + \frac{\oplus\phi_1}{2}\right)\right\}} \qquad \text{. . . . . (120b}$$

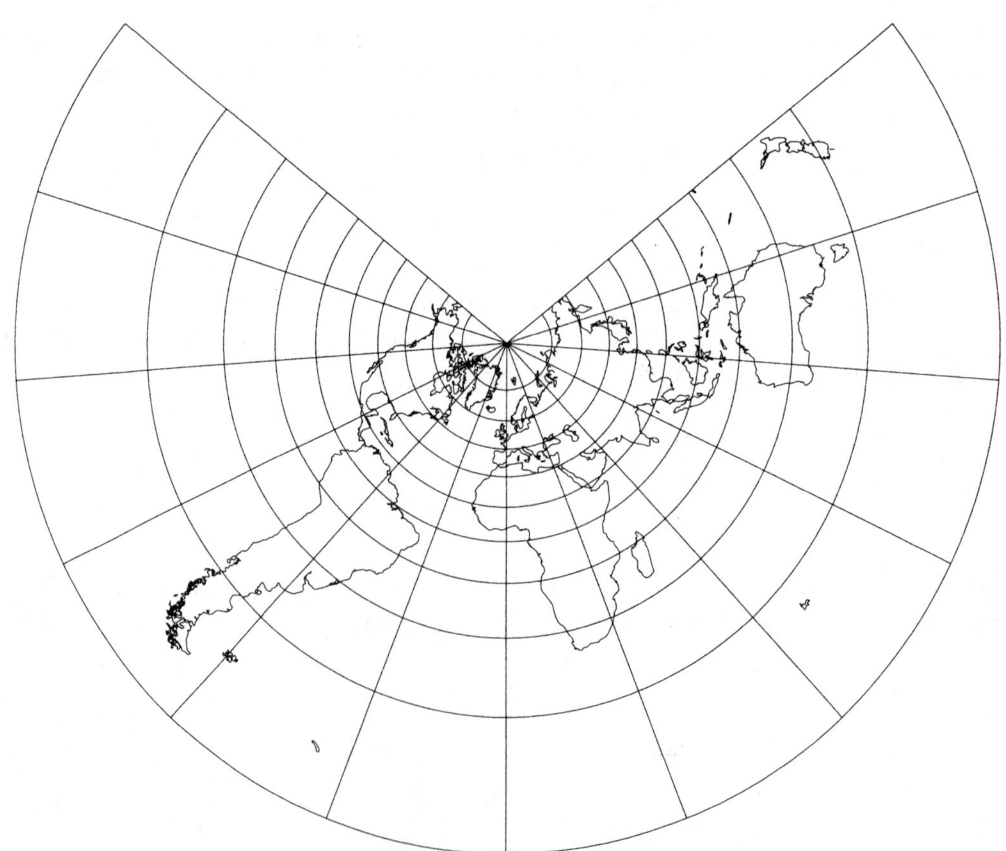

Exhibit 5–10: The (Lambert) Conformal Conic projection at Simple aspect – North Pole to 60°S.
Standard parallels set at 30°N and 60°N.
Cut off at 65°S

producing a derived value for the constant c applicable to conformality with standard parallels ϕ_1 and ϕ_2.

Being conformal with those standard parallels further implies the scale on the map be true along the meridians at the points of intersection with those two parallels, i.e., instantaneously at latitudes ϕ_1 and ϕ_2. Hence, (see Equation 84a)

$$\frac{\mathrm{d}\dot{\rho}}{\mathrm{d}\phi} = \frac{-\oplus c\dot{\rho}}{R\cos\phi} = \oplus 1$$

separately for both $\phi = \phi_1$ and $\phi = \phi_2$, so

$$c\dot{\rho}_1 = R\cos\phi_1 \quad \text{and} \quad c\dot{\rho}_2 = R\cos\phi_2 \qquad \text{. . . . . (121a}$$

Equation 84b then gives

$$\dot{\rho} = k\left\{\tan\left(\frac{\pi}{4} - \frac{\oplus\phi}{2}\right)\right\}^c \qquad \text{. . . . . (121b}$$

Evaluating for the two standard parallels then substituting from Equations 121a gives

$$k = \frac{R\cos\phi_1}{c\left\{\tan\left(\frac{\pi}{4} - \frac{\oplus\phi}{2}\right)\right\}^c} \quad \text{and} \quad k = \frac{R\cos\phi_2}{c\left\{\tan\left(\frac{\pi}{4} - \frac{\oplus\phi}{2}\right)\right\}^c} \qquad \text{. . . . . . (121c}$$

Substituting these twin formulae in Equation 121b gives

$$\dot{\rho} = \frac{R\cos\phi_1}{c}\left\{\frac{\tan\left(\frac{\pi}{4} - \frac{\oplus\phi}{2}\right)}{\tan\left(\frac{\pi}{4} - \frac{\oplus\phi_1}{2}\right)}\right\}^c \quad \text{and} \quad \dot{\rho} = \frac{R\cos\phi_2}{c}\left\{\frac{\tan\left(\frac{\pi}{4} - \frac{\oplus\phi}{2}\right)}{\tan\left(\frac{\pi}{4} - \frac{\oplus\phi_2}{2}\right)}\right\}^c \qquad \text{. . . . . (121d}$$

which establishes alternative identically valued formulae for $\dot{\rho}$ in the Conformal Conical projection with the two standard parallels ϕ_1 and $\phi_2 > \phi_1$.

Because the general variable ϕ occurs in only one place in this complicated formula, it is common practice to keep all the other terms consolidated into one distinct factor, i.e., keep the factor k (which needs to be evaluated, from either of the formulae in Equations 121c, only once for a given map), using the value of c defined by Equation 120a else 120b as a precursor then applying it to Equation 121b.

The plotting equations for the notionally secantal conformal conical projection at Simple aspect, usually known as the **Lambert Conformal Conic projection with two standard parallels,** are obtained by substitution in Equations 59b for c as defined in either of Equations 120a & 120b and for $\dot{\rho}$ as defined in Equation 121d, then for $\dot{\rho}_0$ by substitution for variable ϕ in the obtained formula for $\dot{\rho}$ the latitude of the point on the central meridian chosen to be at the origin. The first projection in Lambert's grand opus [1772a], it is illustrated opposite. **Ney's** projection was a variant akin to that for Lambert's Equal-area Conic, with one of its standard parallels set barely short of the Pole at 89°58′58″, the other at 71° else 74°.

THE IMPACT ON SCALE

Basic impact

For both cylindrical and conical modes, a secantal approach brings shrinkage of scale along lines parallel to the new standard lines for lines between them, and an increase of scale along such lines outside them. The change along those lines — parallels for Simple aspect — depends purely on angle spanned by the secant and of that of the general line from the nearest standard parallel. Transversely, the equal-area map has compensating enlargement. The conformal clearly must have identical shrinkage relative to the tangential case else the conformality is lost (effectively the place of true scale alone is changed but not the picture).

The crucial value of the secantal projections is to allow a range of scale variation both up and down, and thus opportunity to have maps of greater coverage for a given percentage variation from stated scale.

Balancing the impact

In its tangential form, the conformal Cylindrical is the Mercator projection, presented in the preceding chapter, where the escalation of scale for the projection at each multiple of 10° in latitude is tabulated on page 83. At the first step, it is merely about 1.5% but by 60° it is double, and on the way to being infinite at a Pole. Indeed it is almost meaningless to express a scale for such a map (or, for that matter, a linear scale for the equal-area Cylindrical).

If we re-express the scale as applying at some non-zero latitude, then the escalation of scale is moderated outside the two standard latitudes at the price of a de-escalation between them. The factor involved is the cosine of the chosen latitude. If 20° is the standard, then, since $\cos 20° = (1.0642_{\sim})^{-1}$, the de-escalation becomes a maximum of nearly 6.5% at the Equator, while a like reduction occurs in all the escalation figures.

The table opposite gives, for each of a selection of standard parallels, the revised numerical values at each multiple of 10° in latitude plus some smaller ones, with the Mercator values again in the right-most column for the choice of zero as standard latitude (i.e., the tangential case). Values within the standard parallel are shown in inverse form to allow ready appreciation of the percentage effect. With 20° as standard, the escalation still exceeds 8.5% by 30°, but a map limited to about ±25° latitude stays within the accuracy of ±6.5% applicable between the standard parallels.

This is a very limited band compared with the typical appearance of a Mercator map. As we shall find in Chapter 17, however, this famous projection is now very widely used for bands far narrower than the above, to provide the maps accurate to within less than one part in a thousand that are required for surveying and related purposes. If we want accuracy better than 1 part in a thousand, then the latitude to be chosen as standard must have its tangential scale factor less than 1.001. With

$$(1.001)^{-1} = \cos 2.5626_{\sim}° \quad = \sqrt{(1.002_{\sim})^{-1}} \quad = \sqrt{\cos 3.6216_{\sim}°}$$

we have ±2°34.~′ as the maximum for the standard parallels and ±3°37.~′ as the maximal span for retention of the prescribed accuracy. This provides a straddle of the Equator of barely more than a 7°.

Alternatively, if we begin with a specified straddle, we can establish the figure for the optimal standard parallels together with the proportional variation of scale by reversing the sequence. For example, for a straddle reaching to ±30° we get

$$\sqrt{\cos 30°} = \sqrt{(1.1547_{\sim})^{-1}} \quad = (1.0075_{\sim})^{-1} \quad = \cos 21.4707_{\sim}°$$

showing ±21°28.~′ as the appropriate choice for the standard parallels, with ±7.5% as the consequent accuracy.

	---------- **Standard parallels**								----------
lat.	60°	50°	40°	30°	20°	10°	5°	2°	1°	0°
90°	∞	∞	∞	∞	∞	∞	∞	∞	∞	∞
80°	2.8794	3.7017	4.4115	4.9872	5.4115	5.6713	5.7369	5.7553	5.7579	5.7588
70°	1.4619	1.8794	2.2398	2.5320	2.7475	2.8794	2.9127	2.9220	2.9234	2.9238
60°	1	1.2856	1.5321	1.7321	1.8794	1.9696	1.9924	1.9988	1.9997	2
50°	$(1.2856)^{-1}$	1	1.1918	1.3473	1.4619	1.5321	1.5498	1.5548	1.5555	1.5557
40°	$(1.5321)^{-1}$	$(1.1918)^{-1}$	1	1.1305	1.2267	1.2856	1.3004	1.3046	1.3052	1.1034
30°	$(1.7321)^{-1}$	$(1.3473)^{-1}$	$(1.1305)^{-1}$	1	1.0851	1.1372	1.1503	1.1540	1.1545	1.1547
20°	$(1.8794)^{-1}$	$(1.4619)^{-1}$	$(1.2267)^{-1}$	$(1.0851)^{-1}$	1	1.0480	1.0601	1.0635	1.0640	1.0642
10°	$(1.9696)^{-1}$	$(1.5321)^{-1}$	$(1.2856)^{-1}$	$(1.1372)^{-1}$	$(1.0480)^{-1}$	1	1.0116	1.0148	1.0153	1.0154
5°	$(1.9924)^{-1}$	$(1.5498)^{-1}$	$(1.3004)^{-1}$	$(1.1503)^{-1}$	$(1.0601)^{-1}$	$(1.0116)^{-1}$	1	1.0032	1.0037	1.0038
2°	$(1.9988)^{-1}$	$(1.5548)^{-1}$	$(1.3046)^{-1}$	$(1.1540)^{-1}$	$(1.0635)^{-1}$	$(1.0148)^{-1}$	$(1.0032)^{-1}$	1	1.0005	1.0006
1°	$(1.9997)^{-1}$	$(1.5555)^{-1}$	$(1.3054)^{-1}$	$(1.1547)^{-1}$	$(1.0640)^{-1}$	$(1.0153)^{-1}$	$(1.0014)^{-1}$	$(1.0005)^{-1}$	1	1.0002
0°	$(2)^{-1}$	$(1.5557)^{-1}$	$(1.3054)^{-1}$	$(1.1547)^{-1}$	$(1.0642)^{-1}$	$(1.0154))^{-1}$	$(1.0038)^{-1}$	$(1.0006)^{-1}$	$(1.0002)^{-1}$	1

A summary of the scaling factors at selected latitudes for various standard parallels for the secantal Conformal Cylindrical (Mercator) projection at Simple aspect.

TUTORIAL 18 FUNCTIONS AND THEIR ROOTS

Any expression of the form

$a_n x^n + a_{n-1} x^{n-1} + \ldots + a_0$

with non-zero a_n is called an *n*th-order polynomial in x. Elements $a_n, a_{n-1}, \ldots a_0$ are called the *coefficients*. These can be numbers of any form, including complex numbers (the subject of Tutorial 23).

Polynomials of smaller orders have distinctive names, notably:

order 1 *linear*,
order 2 *quadratic*,
order 3 *tertiary* or *ternary*
order 4 *quaternary* or *quartic*

though all can be referred to by the number. Expressed as

$y = a_n x^n + a_{n-1} x^{n-1} + \ldots + a_0$

every polynomial defines a curve in 2-dimensional space, including a straight line for those of order 1.

Values of x that render the polynomial equal to zero are called its *roots*. Some may involve the imaginary entity i and the complex numbers mentioned in Tutorial 16 and discussed in later chapters. Every polynomial has, nominally, exactly as many roots as its order — though some can be identical. For polynomials of order less than 5 algebraic manipulation provides formulae for the roots. Polynomials of order 2 are much used in this work. The roots of

$$a_2 x^2 + a_1 x + a_0$$

are

$$x = \frac{-a_1 \pm \sqrt{a_1^2 - 4a_2 a_0}}{2a_2}$$

For polynomials of order greater than 4 approximation techniques must be used to ascertain the roots. Such techniques apply to functions of more general nature. In this work, those involving trigonometric functions of the variable are of particular concern. Typically these functions arise from a requirement to make one such general function equal another, but if the two are differenced as function f this corresponds to finding the roots of f. For example, given ϕ, find ψ such that

$$\pi \sin\phi = 2\psi + \sin 2\psi$$

i.e.

$$f = \pi \sin\phi - 2\psi - \sin 2\psi = 0$$

ψ now being the variable in place of x.

For value ψ_0 of ψ, the corresponding point on the curve of f is distance $f(\psi_0)$ from the x axis and, assuming the function differentiable there, the slope of the tangent at the point on the curve is $f^{(1)}(\psi_0)$. Hence, the point

$$\psi_1 = \psi_0 - \frac{f(\psi_0)}{f^{(1)}(\psi_0)}$$

is on the ψ axis and such that generally

$$|f(\psi_1)| < |f(\psi_0)|$$

putting ψ_1 closer to the root than ψ_0.

Given a suitable starting value ψ_1, repeated iteration in the form

$$\psi_{n+1} = \psi_n - \frac{f(\psi_n)}{f^{(1)}(\psi_n)} \quad \text{for } n = 0, 1, 2, \ldots$$

can reveal a root. Named the *Newton method* from its original application to polynomials, it is more commonly applied as the *Newton-Raphson method*.

CHAPTER 6

Turning from 'True': Pseudo... projections

Preamble

Whether projected geometrically else synthesized, all projections of the preceding chapters have produced 'true' maps within their respective modes. That is, at Simple aspect, they have straight meridians meeting at one common apical point (at the extreme, infinity) and parallels that are arcs of circles centred at that point — hence, meridians and parallels are everywhere mutually orthogonal. This appearance of meridians and parallels clearly applies unnecessary constraints to our mapping. We want our map to be as true as possible, in the general sense of that word, but flattening the spherical surface demands, as we have noted, some departure from the complete truth. In this chapter, we take a first step beyond such constraints. We allow meridians to be not orthogonal to the parallels and, more crucially although not necessarily, to be curved.

Although it delves more deeply into algebra, this chapter brings nothing new in mathematics.

The characterizing differences between the many projections in this chapter are largely the curvatures of the meridians. So the finer latitudinal graticule used in previous chapters is discarded in favour of a uniform one of 30°. Most of the maps are of equal-area projections, so the scaling is generally consistent with that attribute.

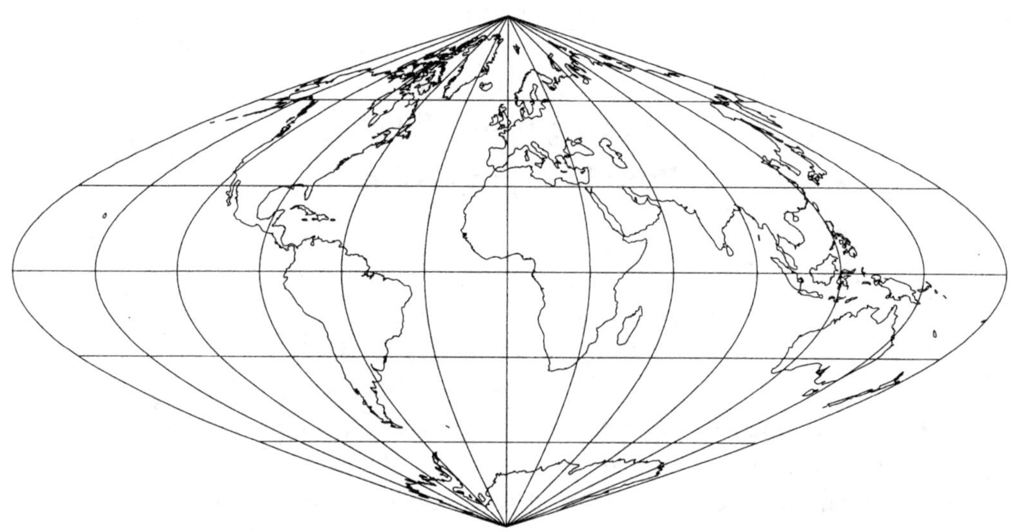

Exhibit 6–1: The Sinusoidal projection at Simple aspect.

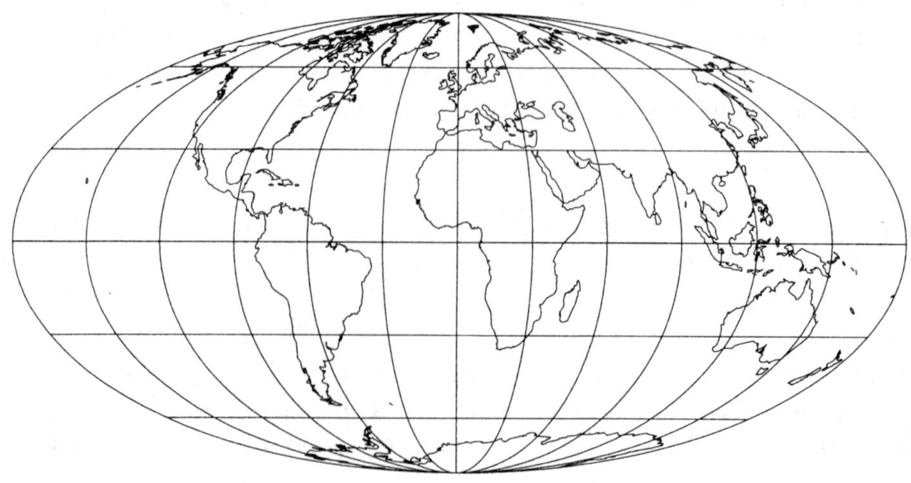

Exhibit 6–2: The Mollweide projection at Simple aspect.

THE APPROACH

Allowing curved meridians

The characteristics of maps from 'true' projections corresponded to having each of the two intermediate parameters a function of just one basic parameter — one of latitude alone and one of longitude alone — and having the latter a fixed linear multiple of longitudinal difference within any one map. We now make the multiplier of longitude vary with latitude while retaining the uniform spacing of meridians along each parallel. That is, we retain the linear involvement of longitude but introduce an additional multiplier term that is a function of latitude. Thus, writing λ for relative longitude:

$$\dot{\rho} = \dot{\rho}(\phi), \qquad \dot{\theta} = \theta^*(\phi)\,\lambda \quad \text{for Azimuthals} \qquad \text{(127a}$$

$$\dot{u} = u^*(\phi)\,R\,\lambda, \quad \dot{v} = \dot{v}(\phi) \qquad \text{for Cylindricals} \qquad \text{(127b}$$

$$\dot{\rho} = \dot{\rho}(\phi), \qquad \dot{\theta} = \theta^*(\phi)\,c\,\lambda \quad \text{for Conicals} \qquad \text{(127c}$$

where the asterisked terms are functions of ϕ alone. There is, therefore, no inflexible parameter across all projections within a mode.

The linear λ term causes all Cylindricals and Conicals to have uniform scale along all parallels. With the azimuthal mode, uniform spacing along parallels that remain full circles implies that any curvature of meridians in that mode must be radially uniform. Such shaping is left to Chapter 14, which shows the Wiechel projection as an example.

Cylindrical and conical maps, built around and symmetric about a straight central meridian, readily accommodate the approach. It is particularly effective for the cylindrical mode, allowing a reduction of the lengths of the parallels as latitude increases — even, optionally, making the poles† points on the map. This usually involves bending the meridians, although a tilting of straight meridians is allowable. The appearance changes drastically from the now-familiar array of rectangles, but continuing existence of all latitudinal parallels as parallel straight lines keeps some cylindrical styling for them.

Aided by the fact that the concentric circular arcs for the cylindrical mode are straight lines, there is unlimited choice for the spacing of the parallels and for the shape of the meridians. Many projections of this nature are used. Because the concentric circular arcs of the conical map are truly curved, however, there is much less opportunity to apply this technique to that mode (and only more subtle changes in appearance, relative to the 'true' Conical, when it is); there is just one such projection commonly used.

† In anticipation of having projections not at Simple aspect, hence with the poles of the projection not the geographic Poles and the equator of the projection not the true Equator, we now desist capitalizing those two words unless explicitly meaning the true geographic entities.

The prefix *pseudo* (meaning *false* in Latin) is added to the base modal term for those projections that maintain the quintessential characteristics of each class — parallels appearing as straight lines in the cylindrical and as concentric circular arcs in the conical, each with a straight central meridian. In this chapter, we explore this first step away from 'true', with many *pseudocylindrical* projections then one *pseudoconical*[†] projection.

Because the spacing of parallels and the scale along each parallel offer unlimited choices (even though the latter must be uniform), so does the shape of the meridians — hence much of the conceptual appearance of the map. Endless choice has produced a very great variety of pseudocylindrical projections in use in atlases and other forms, along with the one pseudoconical projection. The choice is often conditioned by mere appearance, but objective criteria — particularly being equal-area — are also addressed. (Having lost orthogonality of meridians and parallels, conformality is no longer feasible. Having concentric arcs but curved meridians precludes being equidistant.)

Since the longitude-dependent parameter for a given latitude is proportional to relative longitude, it follows that the ratio of the area between one meridian and the central meridian to that between any other meridian and the central meridian is just the ratio of their respective relative longitudes. Thus, even with curved meridians, the area of the band between two parallels is uniformly divided between the meridians, just as on the globe and with all the 'true' projections. Therefore, in equating areas to derive equal-area for pseudocylindrical and pseudoconical projections, we can compare areas between whichever two longitudes are convenient.

We can also see that the length along a parallel from the central meridian, ($R\cos\phi$ if true scale) must on the map change with latitude inversely with the rate of change of the new factor. That is, with λ as relative longitude,

$$\dot{x} = \frac{R\lambda\cos\phi}{\dfrac{\mathrm{d}\dot{v}}{\mathrm{d}\phi}}$$

I.e., relative to Equation 127b,

$$\dot{u}^* = \frac{\cos\phi}{\dfrac{\mathrm{d}\dot{v}}{\mathrm{d}\phi}}$$

As usual, we assume the Simple aspect for all our developments, and use the standard plotting features for each mode. The many pseudocylindrical projections described individually are grouped firstly as those having the poles as points (*point-polar projections*) and then as those having lines for the poles (*flat-polar projections*).

[†] As with the words azimuthal, cylindrical and conical explained in the footnote on page 34, the words pseudocylindrical and pseudoconical are capitalized if used as adjectives within a proper name, and also as nouns to refer to defined projections collectively — e.g., the Pseudocylindricals. The same holds for the word pseudoazimuthal.

SYMMETRIC POINT-POLAR PSEUDOCYLINDRICAL PROJECTIONS

Opting for true scale along central meridian and all parallels

For our first pseudocylindrical projection we choose as a basis the 'true' Equidistant Cylindrical (so the equator is of correct scale and of length $2\pi R$, and the parallel straight lines representing the latitudinal parallels are spaced equidistantly along the central meridian, which must therefore also be of correct scale but of length πR). In addition to the detailed choices implicit with that basis and being point-polar, we further opt to have the spacing of the meridians correct along every parallel — i.e., the spacing from the fixed central meridian. Such a combination of choices clearly defines the exact shape of all meridians, and hence the precise form of our graticule. Having the spacing of parallels set by the underlying basis of the Equidistant Cylindrical, we already have the formula (Equation 69b) $\dot{v} = R\,\phi$. Development focuses on ascertaining the latitudinal factor in the formula for $\dot{u}$.

For correct scale, the spacing along any parallel from the central meridian to any other meridian must be R times the cosine of the latitude times the relative longitude (and precisely R times the relative longitude along the equator). So, the needed formula is

$$\dot{u} = (R\cos\phi)\,\lambda \qquad \text{. . . . . (129a}$$

Interestingly, this projection, which has correct scale along all parallels and the central meridian (although not along other meridians, as they now follow a curved path between the equispaced parallels), is also an equal-area projection. We prove this by finding the area between parallels ϕ_1 and ϕ_2, with the latter the larger, from the central meridian to any other. This is obtained by integration of the function in our formula, using formula (f) of Tutorial 15 (page 78), to give

$$\text{area} = \int_{\phi_1}^{\phi_2} R\cos\phi\; R\,\lambda\,\mathrm{d}\phi \quad = R^2\lambda\int_{\phi_1}^{\phi_2}\cos\phi\,\mathrm{d}\phi$$

$$= R^2\lambda\left[\sin\phi\right]_{\phi_1}^{\phi_2} \quad = R^2\lambda\,(\sin\phi_2 - \sin\phi_1)$$

Hence the area between any two longitudes λ_1 and λ_2, assuming $\lambda_1 < \lambda_2$, is

$$R^2(\lambda_2 - \lambda_1)(\sin\phi_2 - \sin\phi_1)$$

which has previously (Equation 73a) been shown to be the area of the corresponding panel on the globe. Thus, choosing to have correct scale along the parallels, along with our other choices, results in having an equal-area projection.

We could, alternatively, have chosen equal-area (represented by this expression) as our requirement, rather than the correct spacing along the parallels. Then, through differentiation of it, we would have obtained the formula of Equation 129a. With the value of that formula for a given longitude proportional to the cosine of latitude, the shape of each meridian is that of the cosine curve (relative to the vertical axis rather than the more familiar horizontal axis), ranging from a south pole at $-\pi/2$ to a north pole at $+\pi/2$. This is identically the familiar sine wave appearing between 0 and $+\pi$, prompting the common name **Sinusoidal** projection. First used by Jean Cossin of Dieppe in 1570 for a world map then by Jodocus Hondius for Africa and South America in a Mercator atlas early in the 17th Century, the latter led to it being called the **Mercator Equal-area** projection. Later that century promotion by Nicolas Sanson of France and then by John Flamsteed, the initial British Astronomer Royal, for star maps, led to the more enduring name of **Sanson-Flamsteed** projection. The plotting equations for the projection are:

$$\dot{x} = (R\cos\phi)\,\lambda \qquad \dot{y} = R\,\phi \qquad \text{. . . . . (130a}$$

Representative unit plotting values (i.e., for the unit globe) for the Sinusoidal, with longitudes relative to the chosen central meridian, are summarized next.

	$\dot{x}$							$\dot{y}$
	0°	±30°	±60°	±90°	±120°	±150°	±180°	all
±90°	0.000	0.000	0.000	0.000	0.000	0.000	0.000	1.571
±60°	0.000	0.262	0.524	0.785	1.047	1.309	1.571	1.047
±30°	0.000	0.453	0.907	1.360	1.814	2.267	2.721	0.524
0°	0.000	0.524	1.047	1.571	2.094	2.618,	3.142	0.000

each co-odinate to have the sign respectively of the relative longitude and of the latitude.

Unit plotting co-ordinates for the Sinusoidal projection
to 3 decimal places of approximation.

Clearly the arch of the curved meridians becomes progressively greater as they become increasingly distant from the central meridian, although they always cross the equator orthogonally. Hence, away from the equator and the central meridian, the panels of the longitude/latitude graticule (and every object mapped) become increasingly distorted angularly. Thus, for a panel in these outer zones, what is essentially rectangular (but convex) on the globe becomes something of a parallelogram on the flat map — a parallelogram with curved sides that precisely offset the loss of area deriving from the loss of surface convexity. As can be seen with the Mollweide in Exhibit 6–2 on page 126, any mapped entity that runs north-to-south can be turned almost halfway to being parallel to the equator if in high latitudes, and turned in opposite directions for opposite sides of our map. Thus, while able to depict reasonably the whole globe, the Sinusoidal has its depiction of a given place in higher latitudes greatly influenced by the choice of central meridian. By making the meridians meet at the poles, the Sinusoidal is much more compatible with our perceptions of Earth (and indeed with reality) than the 'true' Cylindricals, with their gross exaggeration of width near the poles. It also has the debatable value of pointed polar meetings of meridians — a common mental image of the Poles, although a geographical absurdity for meridians 180° apart. The sharpness of these pointed junctions, and the angular distortion it represents, present problems to the map's use; Chapter 7 discusses one way to alleviate this, while the projection developed next in this chapter removes the pointedness while maintaining equal-area, albeit at the price of even greater distortion.

Within the context of point-polar Pseudocylindricals, with correct scale along both the equator and the central meridian (implying equal spacing of parallels), the Sinusoidal is the result whether we choose correct scale along all parallels else equal-area else sinusoidal curves as the shape of the meridians. Any one of these implies the other two. We could still ignore all three and choose some other attribute while retaining the true scales, but, with conformality precluded, candidate attributes are scarce. So, we will now discard the true scales (specifically the equal spacing of parallels) then make choices. Generally, we can make two choices before constraining the map. For the initial development of the Sinusoidal it was equal spacing of parallels and true scale along them. Now we make neither of those; instead, we choose the shape of the meridians along with having equal-area. Were we to choose the sinusoidal shape we would be back to the Sinusoidal projection. Instead, we choose other shapes that offer visual advantages over the sine curves. Again, the choice is unlimited

Opting for elliptical meridians

Ellipses offer shapes that are inherently rounder than sinusoidal curves, hence potentially more tuned to an image of Earth. Prompted by the work of G. G. Schmidt [1803a], C. B. Mollweide [1805b] used ellipses in creating the second-oldest Pseudocylindrical. He set the central meridian at half the length of the equator and made the outer pair of meridians form an ellipse. He then made the overall size and set the spacing of parallels so as to provide equal-area. Since meridians are required to be equally spaced along all parallels in any Pseudocylindrical, and multiplying one co-ordinate of an ellipse produces another ellipse with common transverse axis, all the meridians become semi-ellipses through the common polar points. Those for ±90° relative longitude, having a mutual spacing equal to half the equator, must form a circle. Each meridian inside the circle has the poles as the end points of its major axis and each meridian outside has the poles as end points of its minor axis. Development will show that the two basic options define the spacing of the parallels, and, therefore, inherently define their lengths.

Let the length of the central meridian be $2r$, making the length of the equator $4r$. These two figures then define the dimensions of the bounding ellipse, giving its area as $2\pi r^2$ which for equal-area must equal the globe's total area of $4\pi R^2$. Equating these gives

$$r^2 = 2R^2, \text{ so } \quad r = \sqrt{2}\, R \approx 1.4142_{\sim} R \qquad \text{. . . . . (131a}$$

The length of the equator, at four times this measurement, is thus $4\sqrt{2}\,R \approx 5.6569_{\sim} R$ (compared with $2\pi R \approx 6.2832_{\sim} R$ for the Sinusoidal). The ±90° meridians form a circle of radius r and area πr^2, equalling half that of the globe.

To ascertain the thickness of the strip of that circle adjacent to the equator that is required to equal the area of the hemisphere from the equator to parallel ϕ we address the spacing of the parallels in the limited context of this central circle. The full area of the latter on the globe has already been established (Equation 72a) as $2\pi R^2 |\sin\phi|$. The area in the circle can be seen in the diagram below to be twice that of a triangle plus a sector. While it is the thickness of this slice that we need to find — a linear distance — we characterize the slice by the angle subtended at the centre of the circle by the arc segment of the slice. We denote this by ψ. (The value for ψ depends solely and directly on the latitude we are addressing, but ψ shouldn't be confused in any way with latitude, despite the similarity of this diagram to many others where the like angle is latitude. Those others are, of course, cross-sections of the globe; this is a circular disc on the map.)

Using this angle, we can readily see via the diagram that:

$$\text{area of sector} = \frac{\psi}{2\pi}\pi r^2 = \frac{1}{2}\psi r^2$$

$$\text{area of triangle} = \frac{1}{2}(r\sin\psi)(r\cos\psi) = \frac{r^2}{2}\sin\psi\cos\psi = \frac{r^2\sin 2\psi}{4}$$

by the first double-angle formula of Tutorial 10 (page 38). Adding twice these two areas and equating with the corresponding area within one hemisphere, we get

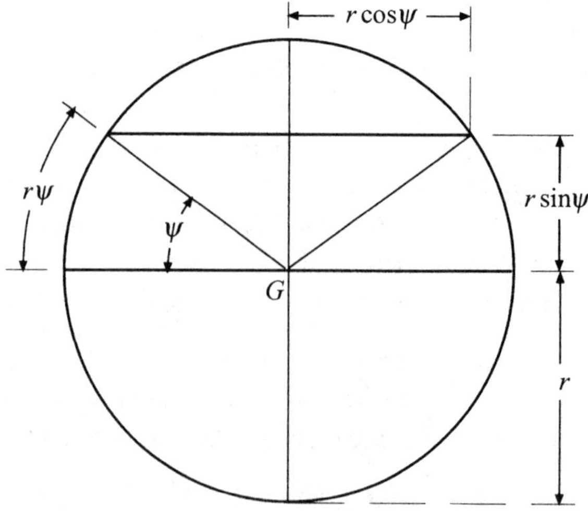

Exhibit 6–3: Geometry for the equal-area computation for the Mollweide projection.

$$\pi R^2 \sin\phi = \psi r^2 + \frac{1}{2} r^2 \sin 2\psi$$

Putting $\psi = \pi/2$ for a polar value gives $r^2 = 2R^2$, as we have already shown above. Substituting from Equation 131a for r^2 in this general equation gives us

$$\pi R^2 \sin\phi = 2R^2 \psi + R^2 \sin 2\psi$$

$$\pi \sin\phi = 2\psi + \sin 2\psi$$

We need to find the value of ψ in terms of ϕ. There is no analytic formula derivable from this to express ψ routinely for any ϕ; it can only be done by iterative approximation for each specific value of ϕ. To do this, we re-express the last equation as a function of variable ψ equated to zero and find its zeros for a given value of ϕ. The function becomes

$$f(\psi) = \pi \sin\phi - 2\psi - \sin 2\psi = 0 \qquad \text{. . . . . (133a}$$

which, mathematically, is transcendental but differentiable. The Newton-Raphson iterative approximation methodology (see Tutorial 18, page 124) can be used to find the zeros. The methodology takes any approximate solution ψ_n then finds a better approximation ψ_{n+1} using the proven iterative approximating formula for zeros of such differentiable functions

$$\psi_{n+1} = \psi_n - \frac{f(\psi_n)}{f^{(1)}(\psi_n)} \quad \text{for } n = 0, 1, 2, ..$$

where the denominator function is the first derivative of our special function. In our current case this is

$$f^{(1)}(\psi) = \frac{\mathrm{d}f}{\mathrm{d}\psi} = -2 - 2\cos 2\psi \quad = -2(1 + \cos 2\psi)$$

so

$$\psi_{n+1} = \psi_n + \frac{\pi \sin\psi_0 - 2\psi_n - \sin 2\psi_n}{2(1 + \cos 2\psi_n)}$$

The value of ϕ is a reasonable opening estimate ψ_0 for ψ, with convergence from there being rapid everywhere. Because the result for any latitude is universally true for any Mollweide map at Simple aspect, a single table of values can be constructed, requiring only multiplication by the scale factor to give the plotting co-ordinates. The plotting equations for the **Mollweide** (or **Babinet** or **Elliptical** or **Homalographic** or **Homolographic**) projection are:

$$\begin{aligned} \dot{x} &= \frac{2}{\pi}\sqrt{2}\,(\cos\psi)\, R\lambda &&= 0.900\ 32_{\sim} R(\cos\psi)\,\lambda \\ \dot{y} &= \sqrt{2}\, R \sin\psi &&= 1.414\ 21_{\sim} R \sin\psi \end{aligned} \qquad \text{. . . . . (133b}$$

The scale at the equator, as already seen, is diminished compared with the true scale (which applies with the Sinusoidal); near the poles it is exaggerated. This relative change is progressive from the equator to either pole. Clearly at some latitude, matched north and south of the equator, the scale along the parallel is true. Since the true total length of parallel ϕ is $2\pi R \cos\phi$, the latitudes of true scale are those that fit the equation

$$2\sqrt{2}\, R\cos\psi = \pi R\cos\phi$$

or

$$8\cos^2\psi = \pi^2\cos^2\phi = \pi^2\left(1-\sin^2\phi\right) \quad = \pi^2 - \pi^2\sin^2\phi$$
$$= \pi^2 - (2\psi + \sin 2\psi)^2$$

i.e., fit the equation

$$8\cos^2\psi + (2\psi + \sin 2\psi)^2 = \pi^2 = 9.8696_{\sim}$$

Solving for ψ then substituting in the earlier function gives $\Phi \approx \pm 40°44._{\sim}'$ for the parallels of correct scale. At all lower latitudes the scales along the parallels are less than the correct value; in all higher latitudes they are greater than the correct value. Since it is an equal-area projection, the opposite statement must apply to the spacing of parallels.

Representative plotting values for Mollweide for the unit globe, with longitudes relative to the chosen central meridian, are summarized just below. The result is illustrated in the lower map on page 126. The elliptical outline contrasts clearly with the pointed outline of the Sinusoidal above it. At the poles, all meridians in Mollweide end running parallel to the equator, meaning the angular distortion is maximal. Again we will look into alleviation of this distortion in the Chapter 7.

	$\dot{x}$							$\dot{y}$
	0°	±30°	±60°	±90°	±120°	±150°	±180°	all
±90°	0.000	0.000	0.000	0.000	0.000	0.000	0.000	1.326
±60°	0.000	0.305	0.610	0.915	1.223	1.525	1.830	1.097
±30°	0.000	0.431	0.862	1.294	1.726	2.156	2.587	0.603
0°	0.000	0.471	0.943	1.414	1.886	2.357	2.828	0.000

each co-ordinate to have the sign respectively of the relative longitude and of the latitude.

Unit plotting co-ordinates for the Mollweide projection
to 3 decimal places of approximation.

Multiplying one co-ordinate and dividing the other by one number would retain semi-ellipses for meridians and the equal-area attribute while changing the length of the parallels. Hence, it is feasible to have true scale at any chosen latitude. Lawrence Bromley [1965a] proposed using $\pi\sqrt{2}/4$ as the factor, to give the equator true scale. As discussed in Chapter 7, other values allow joining part of a Mollweide map with partial maps from various other equal-area projections to provide a composite map that is equal-area.

Several early cartographers used equally spaced elliptical meridians along with straight parallels but, forfeiting equal-area, spaced the parallels evenly. The second projections of Peter Apian [1551a, see also 1935a] and of Georges Fournier [1643a] are examples; the former has parallels equally spaced along the central meridian, the latter along the 90° meridians that formed the outer boundary of his hemispheric map. In 1975 Lawrence Fahey [see 1993a] proposed using the spacing of a 'true' secantal Cylindrical at approximately ±35°. Max Eckert [1906a] used elliptical meridians in two flat-polar projections. Later in this chapter we address elliptical meridians more generally.

Hybrid formulae of the Sinusoidal and Mollweide projections are discussed further below; composite maps involving them are described in the Chapter 7.

The Latvian cartographer R. V. Putnins used partial semi-ellipses for meridians in the first pair of three pairs of point-polar projections that he introduced together [1934a]. Putnins' general methodology for these six (labelled P_1 through P_6) and for six matched flat-polar versions (distinguished by the addition of a prime marker) is discussed in detail l in Endnote A to this chapter. All 12 projections make the equator twice the length of the central meridian, routinely expressed as $4\pi s$ and $2\pi s$, respectively, and all are authalic pole-to-pole between the meridians. Putnins designed a single meridional graticule for each pair, allowing spacing of parallels for equal-area; this is done in the second of each pair, the first having equally spaced parallels. The precise curves used for meridians in each pair are defined by the circumstances of the flat-polar versions (discussed on page 147 onwards). Endnote A provides a complete development for the semi-ellipse versions, through to the constants that show in the plotting equations. For the other projections the constant are explained but not developed.

Endnote A provides a full developemnt of the formulae for the partial semi-ellipses used for meridians in projections **Putnins P_1** and **Putnins P_2**. Equations 164a & 164b give the general equation and that for the outside meridians (i.e., $\lambda = \pm\pi$) as, respectively,

$$\frac{\dot{x}^2}{(2\lambda s)^2} + \frac{3\dot{y}^2}{(2\pi s)^2} = 1 \qquad \text{(135a}$$

$$\dot{x}^2 + 3\dot{y}^2 = (2\pi s)^2 \qquad \text{(135b}$$

Equally spaced parallels imply that $\dot{y} = 2s\phi$. Substituting the specific factor s_1 applicable to this pair of projections, then substituting this into the equation for the ellipses gives the plotting equations for the P_1 projection as:

$$\begin{aligned} \dot{x} &= 4s_1 R\left\{\sqrt{1-3\left(\tfrac{\phi}{\pi}\right)^2} - \tfrac{1}{2}\right\}\lambda &&= 1.89490\_R\left\{\sqrt{1-3\left(\tfrac{\phi}{\pi}\right)^2} - \tfrac{1}{2}\right\}\lambda \\ \dot{y} &= 2s_1 R\phi &&= 0.94745\_R\phi \end{aligned} \qquad \text{(135c}$$

using the formula for s_1 given in Equation 166b. The P_1 has true scale at $\Phi \approx \pm 29°46\_'$.

Using that same factor, the plotting equations for the equal-area P_2 projection are:

$$\begin{aligned} \dot{x} &= 4s_1 R\left(\cos\psi - \tfrac{1}{2}\right)\lambda &&= 1.89490\_R\left(\cos\psi - \tfrac{1}{2}\right)\lambda \\ \dot{y} &= 2\frac{\pi}{\sqrt{3}} s_1 R\sin\psi &&= 1.71848\_R\sin\psi \end{aligned} \qquad \text{(135d}$$

with ψ the solution of $2\psi + \sin 2\psi - 2\sin\psi = A_1 \sin\phi$

and factor A_1 again defined later, in Equation 166a. The P_2 has true scale at $\Phi \approx \pm 37°28\_'$.

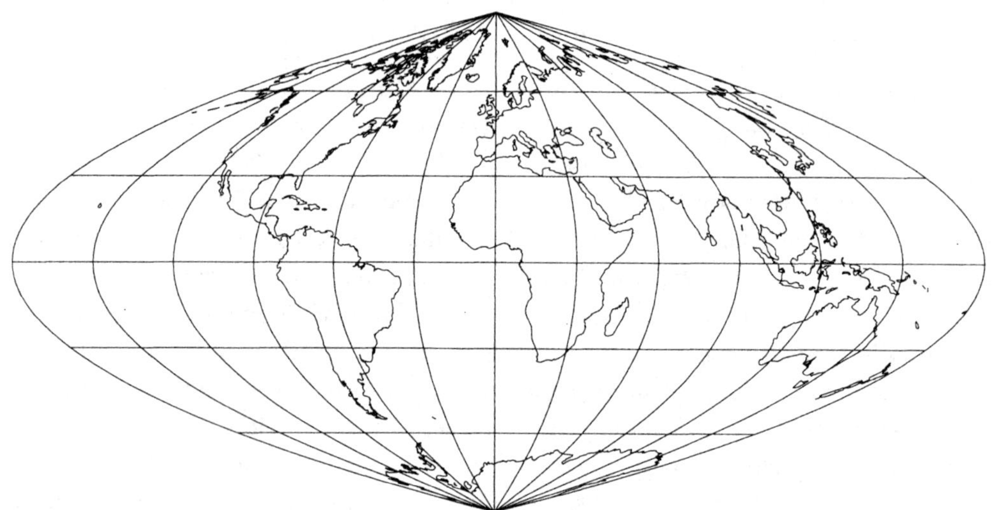

Exhibit 6–4: The Craster Parabolic or Putnins P_4 projection at Simple aspect.

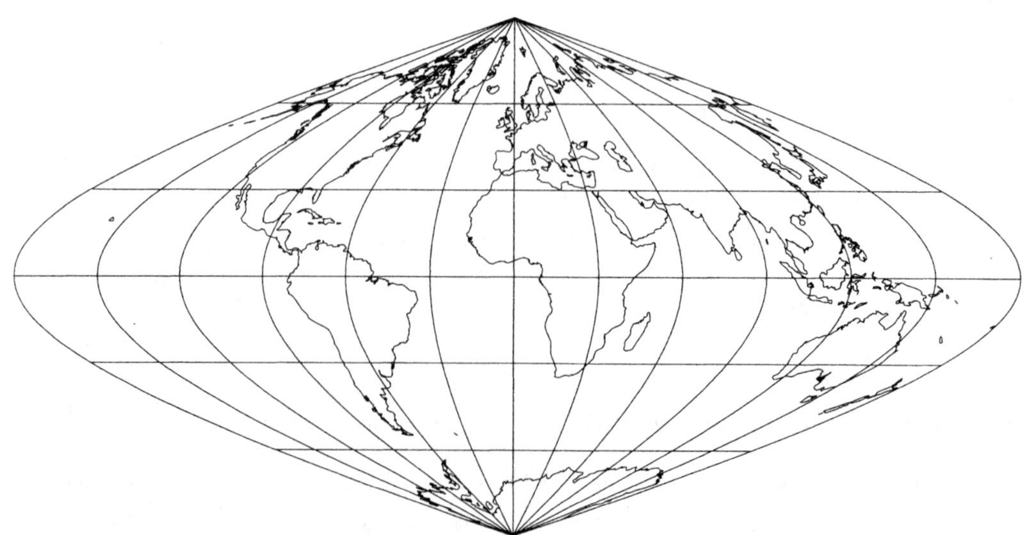

Exhibit 6–5: The Putnins P5 hyperbolic projection at Simple aspect.

Opting for parabolic meridians

The strategy of the Mollweide can be applied using any cartographically suitable shape for the meridians. Choosing the shape of a parabola means that the formula for each meridian derives from the general parabolic equation

$$y^2 = ax + b \qquad \text{. . . . . (137a}$$

for some a and b. As with the Mollweide, it opts for the central meridian to be half the length of the equator. Let these lengths again be $2r$ and $4r$, respectively (though no longer referring to any radius as such). For the meridian of relative longitude λ, passing through both poles and being proportionally spaced along the equator, we must have:

$$\dot{x} = 0 \quad \text{for } \dot{y} = \pm r$$

$$\dot{y} = 0 \quad \text{for } \dot{x} = \frac{\pm 2r}{\pi}\lambda$$

making Equation 137a into

$$\dot{y}^2 = -\frac{\pi r \dot{x}}{2\lambda} + r^2 \quad \text{or} \quad \dot{x} = \frac{2}{\pi r}\left(r^2 - \dot{y}^2\right)\lambda \;= \frac{2r}{\pi}\left\{1 - \left(\frac{\dot{y}}{r}\right)^2\right\}\lambda \qquad \text{. . . . . . (137b}$$

If we now address the left-most meridian (i.e., –180° relative longitude, for which $\lambda = -\pi$), the area between it and the central meridian, from the equator to any value y, is

$$\int_0^{\dot{y}} -\dot{x}\,\mathrm{d}\dot{y} = \int_0^{\dot{y}} \left\{\frac{2\lambda}{\pi r}\left(\dot{y}^2 - r^2\right)\right\}\mathrm{d}\dot{y} = \int_0^{\dot{y}} \left\{\frac{-2\pi}{\pi r}\left(\dot{y}^2 - r^2\right)\right\}\mathrm{d}\dot{y} = \int_0^{\dot{y}} \left\{\frac{2}{r}\left(r^2 - \dot{y}^2\right)\right\}\mathrm{d}\dot{y}$$

$$= \left[\frac{2}{r}\left(r^2\dot{y} - \frac{\dot{y}^3}{3}\right)\right]_0^y$$

$$= \frac{2}{r}\left(r^2\dot{y} - \frac{\dot{y}^3}{3}\right) \qquad = 2r^2\left(\frac{\dot{y}}{r} - \frac{\dot{y}^3}{3r^3}\right)$$

Taking $\dot{y} = +r$, this must equal half the area of a hemisphere of our globe, so

$$2r^2\left(\frac{r}{r} - \frac{r^3}{3r^3}\right) = \pi R^2$$

$$\frac{4r^2}{3} = \pi R^2 \quad \text{or} \quad r^2 = \frac{3\pi R^2}{4}$$

$$r = \sqrt{\frac{3\pi}{4}}\,R = \frac{\sqrt{3\pi}}{2}R \quad = 1.534\,99_{\smile}R$$

so the length of the equator, at four times this, is $6.139\,96_{\smile}R$, which is closer to the $6.283\,19_{\smile}R$ for the Sinusoidal than to the $5.656\,85_{\smile}R$ for Mollweide (reflecting the fact that the parabola, over this range, is closer to the sine wave than to an ellipse).

Substituting for r in the preceding general equation and equating that with half the area between the equator and latitude ϕ (Equation 73a) gives the corresponding value of $\dot{y}$ as the solution of

$$2r^2\left(\frac{\dot{y}}{r}-\frac{\dot{y}^3}{3r^3}\right)=\pi R^2\sin\phi \qquad =\frac{4}{3}r^2\sin\phi$$

i.e., of

$$3\left(\frac{\dot{y}}{r}-\frac{\dot{y}^3}{3r^3}\right)=3\left(\frac{\dot{y}}{r}\right)-\left(\frac{\dot{y}}{r}\right)^3 \qquad =2\sin\phi$$

This can be shown to be

$$\dot{y}=2r\sin\frac{\phi}{3}=\sqrt{3\pi}\;R\sin\frac{\phi}{3} \qquad =3.069\,98_{\sim}R\sin\frac{\phi}{3} \qquad \text{. . . . . (138a}$$

Substituting in the second of Equations 137b completes the plotting equations for the (**Craster**) **Parabolic** projection as:

$$\dot{x}=\frac{2}{\pi}\frac{\sqrt{3\pi}}{2}R\left(1-4\sin^2\frac{\phi}{3}\right)\lambda \qquad =\frac{\sqrt{3\pi}}{\pi}R\left(2\left(1-2\sin^2\frac{\phi}{3}\right)-1\right)\lambda$$

$$=\sqrt{\frac{3}{\pi}}\;R\left\{2\cos\frac{2\phi}{3}-1\right\}\lambda \qquad =0.977\,21_{\sim}R\left(2\cos\frac{2\phi}{3}-1\right)\lambda \qquad \text{. . . . . (138b}$$

Equating with the true length of a parallel and solving gives $\Phi\approx\pm36°46._{\sim}'$ as the latitudes of true scale. The Parabolic, which was introduced by J. E. E. Craster [1929b] then independently [1934a] as the **Putnins** $\mathbf{P_4}$, is illustrated (above a hyperbolic) on page 136.

Representative plotting values for the Parabolic for the unit globe, with longitudes relative to the chosen central meridian, are summarized in the following table.

	$\dot{x}$							$\dot{y}$
	0°	±30°	±60°	±90°	±120°	±150°	±180°	all
±90°	0.000	0.000	0.000	0.000	0.000	0.000	0.000	1.535
±60°	0.000	0.272	0.545	0.817	1.089	1.361	1.634	1.050
±30°	0.000	0.450	0.900	1.350	1.800	2.250	2.700	0.533
0°	0.000	0.512	1.023	1.535	2.047	2.558	3.070	0.000

each x and y to have the sign respectively of the relative longitude and of the latitude.

Unit plotting co-ordinates for the Parabolic projection
to 3 decimal places of approximation.

The associated **Putnins** $\mathbf{P_3}$ projection has the same layout for meridians as the P_4 but has the parallels spaced equally rather than for equal-area. Its plotting equations are:

$$\dot{x}=\sqrt{\frac{3}{\pi}}\left\{1-\left(\frac{2\phi}{\pi}\right)^2\right\}R\lambda \qquad =0.977\,205_{\sim}R\left\{1-\left(\frac{2\phi}{\pi}\right)^2\right\}\lambda$$

$$\dot{y}=\sqrt{\frac{3}{\pi}}\;R\,\phi \qquad =0.977205_{\sim}R\,\phi \qquad \text{. . . . . (138c}$$

The P_3 has true scale at $\Phi\approx\pm28°14._{\sim}'$.

Opting for hyperbolic meridians

A hyperbola has two separate curves, the pair forming mirror images about some straight *conjugate* axis, and each symmetric itself about a straight *transverse axis* orthogonal to the conjugate axis (each visually somewhat similar to the parabola).

For our purposes, we use part of one curve as one meridian, with the x axis as the conjugate axis. The formula for each meridian derives from the hyperbolic equation $x^2 - 3y^2 = a^2$ for some a, the hyperbola having to pass through the poles on the map for point-polar projections.

The **Putnins $\mathbf{P_5}$** and **Putnins $\mathbf{P_6}$** projections [1934a] are examples of this rarely used shape among point-polar projections. The former is illustrated in Exhibit 6–5 on page 136, where comparison with the P_4 above it shows the hyperbolic meridians to be very similar in appearance to parabolic meridians — just slightly more sharply curved near the equator. With A_5 and s_5 as defined within Endnote A in Equations 166d, the plotting equations for the P_5 projection with equispaced parallels are:

$$\dot{x} = 4s_5R\left\{2 - \sqrt{1 + 3\left(\frac{2\phi}{\pi}\right)^2}\right\}\lambda \quad = 1.013\,46_{\sim}R\left\{2 - \sqrt{1 + 3\left(\frac{2\phi}{\pi}\right)^2}\right\}\lambda \qquad \ldots\ldots \text{(139a}$$

$$\dot{y} = 4s_5R\,\phi \quad = 1.013\,46_{\sim}R\,\phi$$

The P_5 has true scale at $\Phi \approx \pm 21°14._{\sim}'$.

The plotting equations for the equal-area P_6 projection are:

$$\dot{x} = 4s_5R\left(2 - \sqrt{1 + \psi^2}\right)R\,\lambda \quad = 1.013\,46_{\sim}R\left\{2 - \sqrt{1 + \psi^2}\right\}\lambda \qquad \ldots\ldots \text{(139b}$$

$$\dot{y} = \frac{2\pi}{\sqrt{3}}s_5R\psi \quad = 0.919\,10_{\sim}R\,\psi$$

with $\psi = \sqrt{3}$ for $\phi = \pi/2$ otherwise the solution of

$$\left(4 - \sqrt{1 + \psi^2}\right)\psi - \ln\left(\psi + \sqrt{1 + \psi^2}\right) = A_5 \sin\phi$$

The P_6 has true scale at $\Phi \approx \pm 25°42._{\sim}'$.

Opting for higher-order functions shaping meridians

The **Adams Quartic** or **Quartic Authalic**[†] projection [1945a], originated by Karl Siemon [1937a], has meridians following a fourth-order (*quartic*) curve. Besides being point-polar, it opts for equal-area and true scale along the equator. The plotting equations are:

[†] The term *Authalic* in this name relates to being equal-area in a comprehensive detailed sense. As noted in Chapter 4, that term is used in this work (and generally today) only in a looser/broader context, in particular where areas are correct between meridians but not necessarily between parallels.

$$\dot{x} = R\,\frac{\cos\phi}{\cos\frac{\phi}{2}}\,\lambda \qquad \dot{y} = 2R\sin\frac{\phi}{2} \qquad \text{(140a)}$$

making the parallels get gradually closer together as latitude increases toward the poles. As can be seen from the latter equation, the full length of the central meridian between the poles is $4R\sin 45° = 2\sqrt{2}\,R \approx 2.828\,43_{\sim}R$ against $\pi R \approx 3.141\,59_{\sim}R$ were it of true scale. Therefore, its scale is $0.900\,32_{\sim}$ of true and its length equal to $0.450\,16_{\sim}$ the length of the equator. It is rarely used directly and is not illustrated here, but its flat-polar derivative — the McBryde-Thomas Flat-polar Quartic projection (Exhibit 6–14 on page 150) — is used more. (The Eckert-Greifendorff, illustrated on page 242, has quartic meridians coming to points for the poles, but curved, albeit almost imperceptibly curved parallels.)

H.-C.-A. de P. Foucault [1862a] spaced the parallels at the same proportions as with the Stereographic Cylindrical projection, then set their lengths to provide equal-area. The plotting equations for the **Foucault Stereographic Equivalent** projection are:

$$\dot{x} = 2\frac{1}{\sqrt{\pi}}R\cos\phi\cos^2\frac{\phi}{2}\lambda = 1.128\,38_{\sim}R\cos\phi\cos^2\frac{\phi}{2}\lambda \qquad \dot{y} = \sqrt{\pi}\,R\tan\frac{\phi}{2} = 1.772\,45_{\sim}R\tan\frac{\phi}{2} \qquad \text{(140b)}$$

The resulting sharply pointed map is shown on page 155.

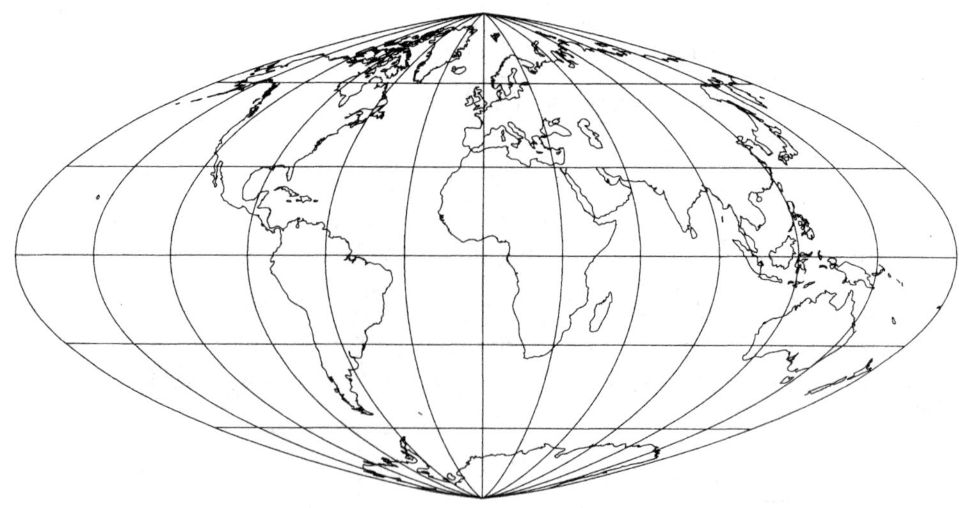

Exhibit 6–6: The Boggs Eumorphic projection at Simple aspect.

Opting for straight meridians

The simplest curve is the straight line. The 'true' Cylindricals had straight meridians, the Pseudocylindricals have license to bend those straight lines. However, the meridians of the 'true' Cylindricals were mutually parallel; our license includes the alternative of straight lines that are not parallel. Within the context of point-polar Pseudocylindricals, this means a diamond-shaped map, each half-meridian a straight line. However, the constriction of straight meridians converging to a point renders such projection, whatever its other attributes, virtually without utility. It occurs, for one hemisphere, in the Collignon, which is discussed and illustrated with other novelty maps in Chapter 14. Straight meridians produce a map of more potential utility in flat-polar style (see overleaf), but even then they find only marginal use.

Seeking good shape

The **Boggs Eumorphic** projection (from Greek for *good* and *shape*) [1929c] is a hybrid of two projections, the Sinusoidal and Mollweide, adapted to maintain equal-area and the 2:1 ratio of equator to central meridian. Its parallels are spaced at the arithmetic mean of the values for the two sources, then the lengths along the parallels are computed (hence shapes constructed) to maintain equal-area. (This is virtually the same as averaging the co-ordinates.) Finally, reciprocal multipliers are applied to make the distances from the origin to the intersection for ±90° longitude and latitude identical. The resulting map has, like the Sinusoidal (but not the Mollweide), a slope discontinuity at each pole, but it is only half as abrupt as in the former. The plotting equations, with Boggs's multipliers, are:

$$\dot{x} = 1.001\,379(2R)\frac{2\sqrt{2}\,\pi(\cos\phi\cos\psi)\,\lambda}{\left(2\sqrt{2}\,\cos\psi + \pi\cos\phi\right)} = \frac{2.002\,758\,R\lambda}{\sec\phi + 1.110\,72_{\sim}\sec\psi} \quad \text{. . . . (141a}$$

$$\dot{y} = 0.998\,622\,9\left(\frac{R}{2}\right)\left(\phi + \sqrt{2}\,\sin\psi\right) = 0.499\,311\,45\,R\left(\phi + \sqrt{2}\,\sin\psi\right)$$

with ψ the solution of Equation 133a.

The good appearance of the Eumorphic is visible from the map opposite. It may be seen as superior to both its progenitors but, like Mollweide, it has lost the true scale along the parallels and the ease of computation of the Sinusoidal. The scale is true at $\Phi \approx \pm 40°15._{\sim}'$.

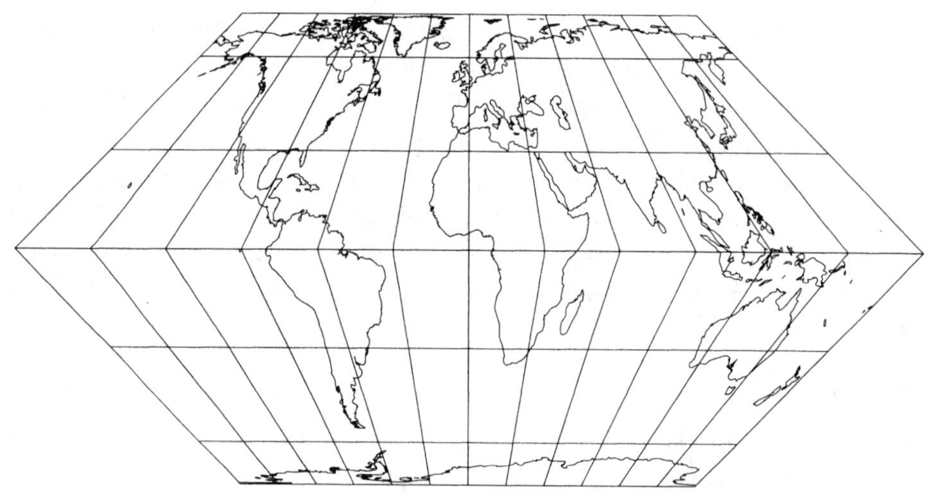

Exhibit 6–7: The Eckert II projection at Simple aspect.

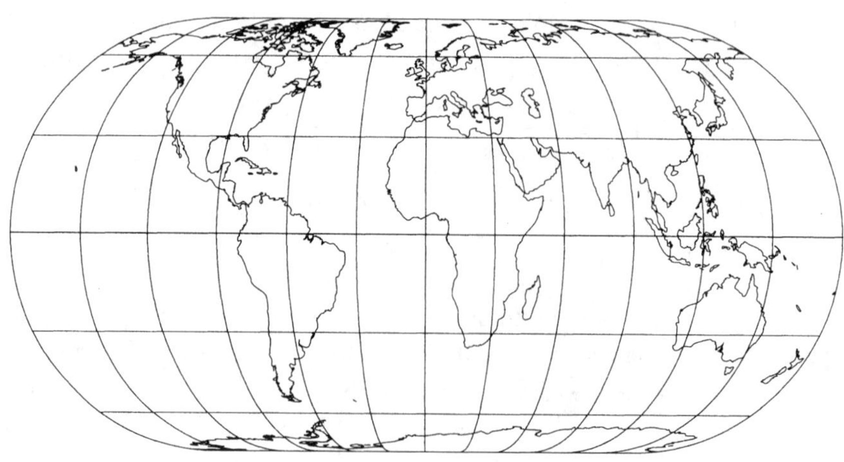

Exhibit 6–8: The Eckert IV projection at Simple aspect.

SYMMETRIC FLAT-POLAR PSEUDOCYLINDRICAL PROJECTIONS

Poles not as points

While getting away from equator-length polar parallels was part of the motivation for creating the Pseudocylindricals, mapping each pole into a point is not a necessity. We saw with the Sinusoidal that a point for each pole can produce perceptual problems. Further, the extremely narrow angle between outer meridians and parallels near the poles produces both gross distortion and some illegibility in a World map. In Mollweide, the meridians become ever closer to being parallel to the parallels.

One means of reducing this angular distortion is to break the World into adjacent fragments. This technique, producing interrupted versions, is discussed in Chapter 7. Another common means is to compromise by having the poles represented by lines of some markedly lesser length than the equator.

The **Trapezoidal** (or **Donis**) projection, which essentially dates back 2000 years, uses straight meridians intersecting one chosen parallel at true scale near each of the upper and lower limits of a regional map. The poles do not appear on the map, but the meridians as drawn would meet at some point well beyond the polar latitude. Ortelius [1570a; illustrated in 1973b; see also 1928c], following others, adopted curved meridians akin to the later Eckert IV (see Exhibit 6–8) but with inner ones meeting at a point and parallels equispaced along the central meridian. More typically, an explicit size is adopted for the length of the polar parallels, as well as of the central meridian — relative to the length of the equator. With this additional option, the variety of feasible projections is greatly expanded. The longitude-proportional spacing of meridians along parallels is applied to the polar parallels, too (although the scale there is infinite).

It is now convenient to look at alternatives grouped by creators rather than by shape of meridians.

The Eckert and Winkel projections

The use of lines for poles with Pseudocylindricals appears to have been pioneered by Max Eckert of Germany, who introduced six such projections [1906a] — a pair with straight meridians, a pair with semi-ellipses for meridians, a pair with sinusoidal meridians. In each pair, the first has equally spaced parallels, the other has equal-area (those two attributes being mutually exclusive except in the Sinusoidal). For all, the central meridian is kept to its proper proportion of half the length of the equator, and the lines for poles are made equal to the central meridian.

The **Eckert I** and **Eckert II** projections have straight meridians. The two provide very similar maps, the latter (shown as Exhibit 6–7 on page 142) having parallels that get noticeably closer together only as a pole is approached. While the graticule pattern is good within a hemisphere, the abrupt change of line across the equator tends to relegate the maps to mere novelty status.

The plotting equations for the Eckert I, with equally spaced parallels, are:

$$\dot{x} = 2\sqrt{\frac{2}{3\pi}}R\left(1-\frac{|\phi|}{\pi}\right)\lambda \qquad = 0.921\,318_{\sim}R\left(1-\frac{|\phi|}{3\pi}\right)\lambda$$
$$\dot{y} = 2\sqrt{\frac{2}{3\pi}}\,R\,\phi \qquad = 0.921\,318_{\sim}R\,\phi \qquad \text{. . . . . (144a}$$

which gives true scale at $\Phi \approx \pm 47°10._{\sim}'$.

The plotting equations for the equal-area Eckert II are:

$$\dot{x} = +2R\frac{\sqrt{4-3\sin|\phi|}}{\sqrt{6\pi}}\lambda \qquad = +\,0.460\,66_{\sim}R\sqrt{4-3\sin|\phi|}\,\lambda$$
$$\dot{y} = \frac{\phi}{|\phi|}\sqrt{\frac{2\pi}{3}}\,R\left(2-\sqrt{4-3\sin|\phi|}\right) \qquad = \frac{\phi}{|\phi|}1.447\,20_{\sim}R\left(2-\sqrt{4-3\sin|\phi|}\right) \qquad \text{. . . . (144b}$$

which gives true scale at $\Phi \approx \pm 55°10._{\sim}'$.

The **Eckert III** and **Eckert IV** projections use semi-ellipses for meridians, so avoid abrupt changes of line across the equator. The semi-ellipses range from the straight central meridian only to a semicircle at ±180° relative longitude. This choice of shaping results in the maps having a curved outline that some see as attractive in appearance, offsetting the adversity of extended poles. Using semi-ellipses makes the map similar to a Mollweide map. Indeed the pattern of meridians is exactly that of the inner hemisphere of that preceding projection, except for being moved outward by a multiple of relative longitude.

The plotting equations for the Eckert III, with equally spaced parallels, are:

$$\dot{x} = \frac{2R}{\sqrt{4\pi+\pi^2}}\left\{1+\sqrt{\left(1-\left[\frac{2\phi}{\pi}\right]^2\right)}\right\}\lambda \qquad = 0.422\,24_{\sim}R\left\{1+\sqrt{\left(1-\left[\frac{2\phi}{\pi}\right]^2\right)}\right\}\lambda$$
$$\dot{y} = \frac{4}{\sqrt{4\pi+\pi^2}}R\,\phi \qquad = 0.844\,76_{\sim}R\,\phi \qquad \text{. . . (144c}$$

which gives true scale at $\Phi \approx \pm 35°58._{\sim}'$.

Development for the spacing of parallels in the equal-area Eckert IV follows the same path as for the Mollweide. Allowing for the intrusion of a rectangle of width $2r$ between the semicircular parts of the Mollweide diagram on page 132, equating with the full area between the equator and latitude ϕ on the globe gives

$$2\pi R^2 \sin\phi = 2\left\{\frac{1}{2}\psi r^2 + \frac{1}{2}(r\sin\psi)(r\cos\psi)\right\} + 2r^2\sin\psi$$
$$= r^2\{\psi + \sin\psi\cos\psi + 2\sin\psi\}$$

The polar value $\phi = {}^{\pi}\!/_2 = \psi$ gives

$$r = \sqrt{\frac{2\pi}{2+\frac{\pi}{2}}}\,R \quad = 1.326\,50\_R$$

while rationalization of the overall general equation gives

$$2\psi + \sin 2\psi + 4\sin\psi = (4+\pi)\sin\phi \qquad \text{. . . . . (145a}$$

This can be solved for ψ in terms of a given value of ϕ by the Newton-Raphson method.

The plotting equations for the equal-area Eckert IV, with ψ defined by Equation 145a, are:

$$\dot{x} = \frac{2R}{\sqrt{4\pi+\pi^2}}(1+\cos\psi)\lambda \quad = 0.422\,24\_R\,(1+\cos\psi)\lambda$$
$$\dot{y} = \frac{2\pi}{\sqrt{4\pi+\pi^2}}R\sin\psi \quad = 1.326\,50\_R\sin\psi \qquad \text{. . . . . (145b}$$

which give true scale at $\Phi \approx \pm 40°30\_'$. Representative plotting co-ordinates for the unit globe, with longitudes relative to the chosen central meridian are immediately below.

	$\dot{x}$							$\dot{y}$
	0°	±30°	±60°	±90°	±120°	±150°	±180°	all
±90°	0.000	0.221	0.442	0.663	0.884	1.105	1.327	1.414
±60°	0.000	0.345	0.691	1.036	1.382	1.727	1.169	1.169
±30°	0.000	0.418	0.836	1.254	1.672	2.090	2.508	0.643
0°	0.000	0.442	0.884	1.327	1.769	2.211	2.653	0.000

each co-ordinate to have the sign respectively of the relative longitude and of the latitude.

Unit plotting co-ordinates for the Eckert IV projection
to 3 decimal places of approximation.

The Eckert IV is illustrated earlier on page 142, the complete semi-ellipses of its meridians contrasting with the straight meridians of Eckert II above it and with the partial semi-ellipses of the P_2' on page 162.

The **Eckert V** and **Eckert VI** projections use full sinusoidal curves for meridians, identical in shape to those of the Sinusoidal projection.

The plotting equations for the Eckert V, with equally spaced parallels, are:

$$\dot{x} = \frac{R}{\sqrt{2+\pi}}(1+\cos\phi)\lambda \quad = 0.44101\_R\,(1+\cos\phi)\lambda$$
$$\dot{y} = \frac{2R}{\sqrt{2+\pi}}\phi \quad = 0.88203\_\,R\phi \qquad \text{. . . . . (145c}$$

which has true scale at $\Phi = \pm 37°55..'$. The map co-ordinates of any point are identically the mean of those of the Sinusoidal and a Plate-Carrée of identical equatorial scale.

The plotting equations for the equal-area Eckert VI are:

$$\dot{x} = \frac{1}{\sqrt{2+\pi}}\left(1+\cos\psi\right)R\lambda \qquad = 0.441\,01_{\sim}R\left(1+\cos\psi\right)\lambda$$
$$\dot{y} = \frac{2}{\sqrt{2+\pi}}R\psi \qquad = 0.882\,03_{\sim}\,R\psi \qquad \text{. . . . . (146a}$$

with ψ the solution of $\psi + \sin\psi = \left(1+\frac{\pi}{2}\right)\sin\phi$. This gives true scale at $\Phi \approx \pm 49°16..'$.

The first of the projections introduced by Otto Winkel of Germany [1921b] has co-ordinates of any point identically the mean of those of the Sinusoidal and an Equirectangular. Since the latter is a secantal projection, there is unlimited choice for the parallels of true scale in it, hence in the combination; the shape changes correspondingly. The plotting equations for the **Winkel I** projection with equally spaced parallels are:

$$\dot{x} = R\frac{\left(\cos\phi_0 + \cos\phi\right)}{2}\lambda$$
$$\dot{y} = R\phi \qquad \text{. . . . . (146b}$$

where ϕ_0 is the chosen standard parallel for the Equirectangular, hence the latitude of true scale in the Winkel. If $\Phi_0 = \pm 50°28..'$ as Winkel originally chose, the map is overall authalic; his projection is illustrated opposite. If $\Phi_0 = 0°$, the Equirectangular becomes the Plate-Carrée, the Winkel I becomes the Eckert V.

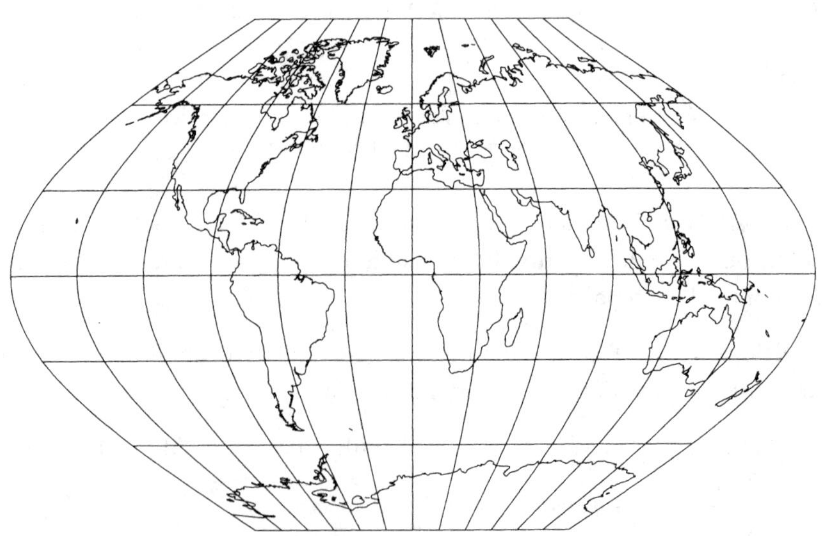

Exhibit 6–9: The Winkel I projection at Simple aspect.
Standard parallels ±50°28..'.

The flat-polar Putnins projections

Putnins' flat-polar versions of his six point-polar Pseudocylindricals [1934a, see also Endnote A of this chapter] have the central meridian and polar parallels half the length of the equator. The meridians are consequentially spaced out further.

The **Putnins P_1'** and **Putnins P_2'** projections have meridians of the same elliptical shape as the P_1 and P_2, fitting the formula of Equation 135a.

In the plotting equations for equally spaced parallels we must have $\dot{y} = 2s\phi$. Substitution in that equation gives

$$\frac{\dot{x}^2}{(2\lambda s)^2} = 1 - \frac{3(2s\phi)^2}{(2\pi s)^2}$$

$$\dot{x}^2 = 4s^2\lambda^2\left(1 - 3\frac{\phi^2}{\pi^2}\right)$$

Using the specific factor s_1' evaluated in Endnote A at Equation 165c, the plotting equations for the P_1', with its equally spaced parallels, are:

$$\dot{x} = 2s_1'R\sqrt{1-3\left(\frac{\phi}{\pi}\right)^2}\ \lambda \qquad = 0.86310_{\sim}R\sqrt{1-3\left(\frac{\phi}{\pi}\right)^2}\ \lambda \qquad \text{. . . . . (147a}$$

$$\dot{y} = 2s_1'R\phi \qquad = 0.86310_{\sim}R\phi$$

The equator is $5.423\,02_{\sim}R$ in length, just modestly less than for a Mollweide map, reflecting the squaring affect of the long polar parallels. True scale occurs at $\Phi \approx \pm 35°55_{\sim}'$.

The equal-area P_2' and the identical **Wagner IV** [1932b] projection have the same gratiĉule as P_1' except for the spacing of parallels. The plotting equations for the P_2', with A_1' defined by Equations 165b, s_1' by Equation 165c, are:

$$\dot{x} = 2s_1'R(\cos\psi)\ \lambda \qquad = 0.863\,095_{\sim}R(\cos\psi)\ \lambda$$

$$\dot{y} = \frac{2}{A_1's_1'}R\sin\psi \qquad = 1.565\,481_{\sim}R\sin\psi \qquad \text{. . . . . (147b}$$

with ψ the solution of $2\psi + \sin 2\psi = A_1'\sin\phi$. This gives true scale at $\Phi \approx \pm 42°59_{\sim}'$. The P_2' **is** illustrated on page 162 facing the endnote detailing Putnins' approach. Its partial ellipses can be contrasted with the full semi-ellipses of the Eckert IV on page 142.

If enlarged in each direction by $1.158\,62_{\sim}$ to produce true scale along the equator, the P_2' becomes the **Werenskiold III** projection [1944a].

The **Putnins P_3'** and **Putnins P_4'** projections have meridians of the same parabolic shape as the P_3 and P_4. Again both are authalic pole-to-pole between meridians, with parallels equally spaced in the former and spaced for equal-area in the latter.

The plotting equations for the P_3' projection with equally spaced parallels are:

$$\dot{x} = 2\sqrt{\frac{3}{5\pi}}\, R\left\{1-2\left(\frac{\phi}{\pi}\right)^2\right\}\lambda \quad = 0.874\,039_{\sim}R\left\{1-2\left(\frac{\phi}{\pi}\right)^2\right\}\lambda$$

$$\dot{y} = 2\sqrt{\frac{3}{5\pi}}\, R\,\phi \quad = 0.874\,039_{\sim}R\,\phi \qquad \ldots\ldots\ldots (148a$$

giving true scale at $\Phi \approx \pm 36°46._{\sim}'$.

The plotting equations for the equal-area P$_4'$ projection are:

$$\dot{x} = 2\sqrt{\frac{3}{5\pi}}\, R\,\frac{\cos\psi}{\cos\frac{\psi}{3}}\,\lambda \quad = 0.874\,039_{\sim}R\left(\sec\frac{\psi}{3}\right)(\cos\psi)\,\lambda$$

$$\dot{y} = 2\sqrt{\frac{6\pi}{5}}\, R\sin\frac{\psi}{3} \quad = 3.883\,252_{\sim}R\sin\frac{\psi}{3} \qquad \ldots\ldots\ldots (148b$$

where $\psi = \arcsin\left\{\frac{5\sqrt{2}}{8}\sin\phi\right\}$. The P$_4'$, illustrated below, has true scale at $\Phi \approx \pm 46°20._{\sim}'$.

The **Putnins P$_5'$** and **Putnins P$_6'$** projections have meridians of the same hyperbolic shape as the P$_5$ (Exhibit 6–5 on page 136) and P$_6$. The choice of hyperbolic curves used is similar to that for the partial ellipses in P$_1'$ etc., with the outside meridians (i.e., $\lambda = \pm\pi$) having the hyperbolic formula $\dot{x}^2 - 3\dot{y}^2 = (2\pi\, s)^2$ in place of the elliptical formula of Equation 135b.

Again, both are authalic pole-to-pole between meridians, with parallels equally spaced in the P$_5'$, spaced for equal-area in P$_6'$. Using factor s_5' as defined in Equations

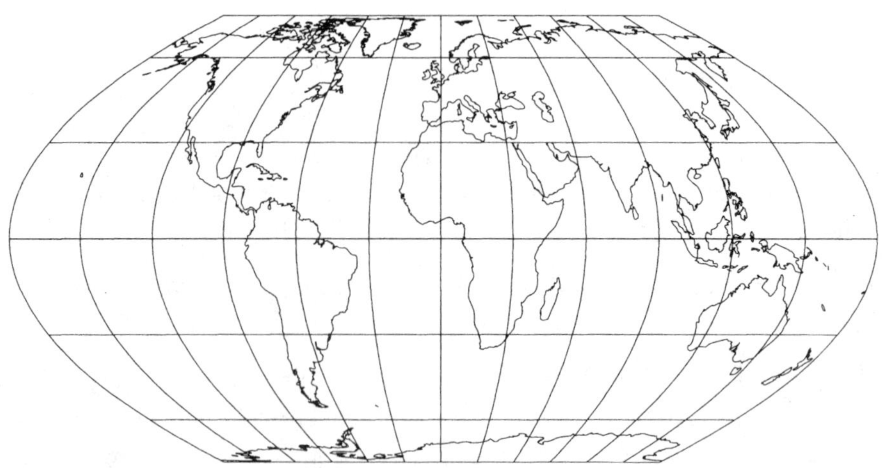

Exhibit 6–10: The P4′ projection at Simple aspect.

166t of Endnote A, the plotting equations for the P_5' projection with equally spaced parallels are:

$$\dot{x} = 2s_5'R\left(3 - \sqrt{1 + 3\left(\frac{2\phi}{\pi}\right)^2}\right)\lambda \quad = 0.443\,293\_R\left(3 - \sqrt{1 + 3\left(\frac{2\phi}{\pi}\right)^2}\right)\lambda \qquad \text{(149a}$$

$$\dot{y} = 4s_5'R\phi \quad = 0.886\,586\_R\phi$$

giving true scale at $\Phi \approx \pm 39°16\_'$.

The plotting equations for the equal-area P_6' projection are:

$$\dot{x} = 2s_5'R\left(3 - \sqrt{1 + \psi^2}\right)\lambda \quad = 0.443\,293\_\ R\left(3 - \sqrt{1 + \psi^2}\right)\lambda \qquad \text{(149b}$$

$$\dot{y} = \frac{2\pi}{\sqrt{3}}s_5'R\psi \quad = 0.804\,057\_R\psi$$

with $\psi = \sqrt{3}$ for $\phi = \pi/2$ otherwise the solution of

$$\left(6 - \sqrt{1 + \psi^2}\right)\psi - \ln\left(\psi + \sqrt{1 + \psi^2}\right) = A_5'\sin\phi$$

and A_5' defined in Equations 166t. The P_6', illustrated below, has true scale at $\Phi \approx \pm 50°40\_'$.

In each of the three Putnins flat-polar maps with equispaced parallels, i.e., P_1', P_3', P_5', the co-ordinates of any point are identically the mean of those in the corresponding point-polar version and a Plate-Carrée (i.e., Equidistant Cylindrical) projection of identical equatorial scale.

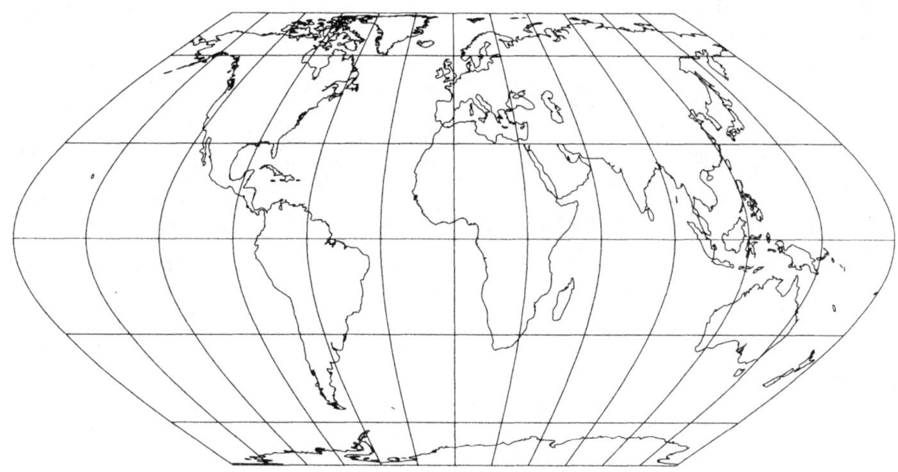

Exhibit 6–11: The P6′ projection at Simple aspect.

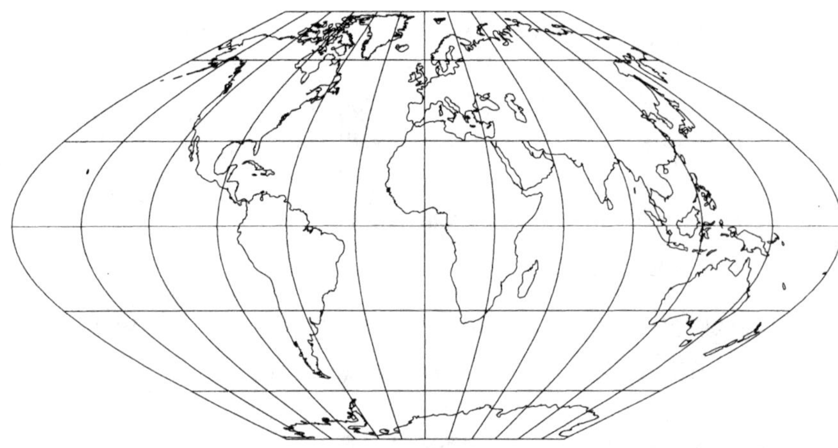

Exhibit 6–12: The McBryde-Thomas III or Flat-polar Sinusoidal projection at Simple aspect.

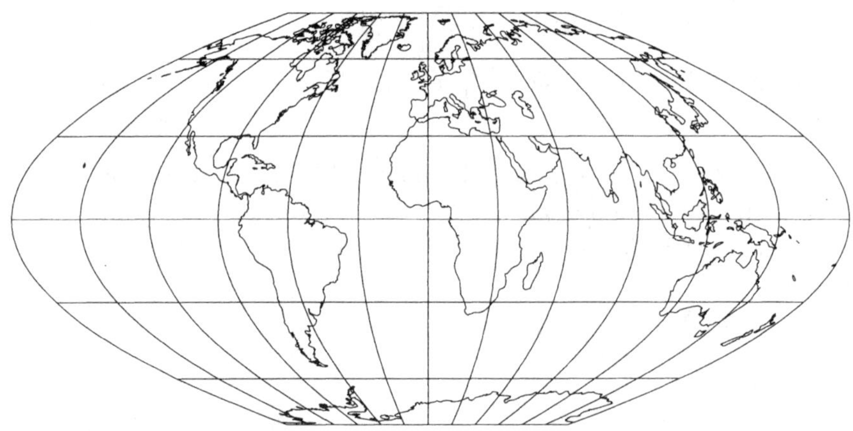

Exhibit 6–13: The McBryde-Thomas IV or Flat-polar Parabolic projection at Simple aspect.

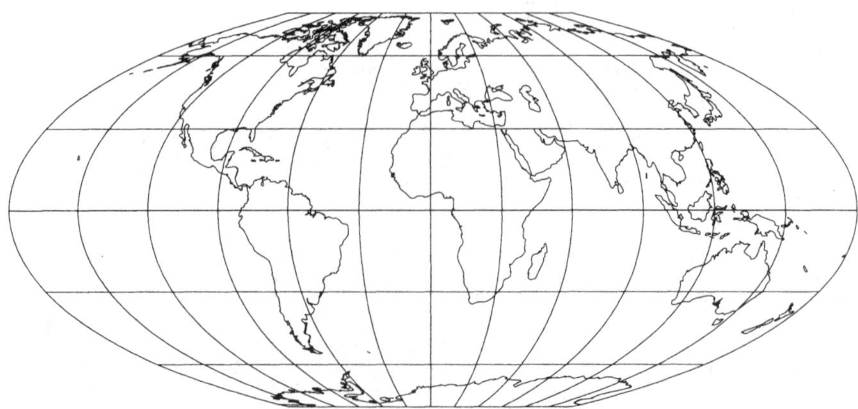

Exhibit 6–14: The McBryde-Thomas V or Flat-polar Quartic projection at Simple aspect.

The flat-polar McBryde-Thomas projections

A 1:3 ratio for length of polar parallel to length of equator was adopted by F. W. McBryde and P. D. Thomas of the U.S.A. [1949b] in three flat-polar equal-area projections using the meridian shapes of the Sinusoidal, Craster Parabolic, and Quartic Authalic projections. They are shown together opposite (at slightly smaller scale than were preceding maps).[†]

The **McBryde-Thomas III** or **Flat-polar Sinusoidal** projection, which retains the correct 1:2 ratio of central meridian to equator, has plotting equations:

$$\dot{x} = \frac{2}{3}\sqrt{\frac{6}{4+\pi}}\,(0.5+\cos\psi)\,R\,\lambda \qquad = 0.611\,06_{\smile}R(0.5+\cos\psi)\,\lambda$$
$$\dot{y} = \sqrt{\frac{6}{4+\pi}}\,R\,\psi \qquad = 0.916\,60_{\smile}R\,\psi \qquad \text{. . . . . (151a}$$

with ψ the solution of $\psi + 2\sin\psi = 2\left(1+\frac{\pi}{4}\right)\sin\phi$, using Newton-Raphson approximation for each specific value of ϕ. True scale occurs at $\Phi \approx \pm 55°51_{\smile}'$.

McBryde and Thomas elaborated their work on the sinusoids to produce general plotting equations. These are discussed later along with their associated work with ellipses.

The **McBryde-Thomas IV** or **Flat-polar Parabolic** projection, which, for computational convenience, has the central meridian just $3/2\pi = 0.477_{\smile}$ of the equator, has plotting equations:

$$\dot{x} = \sqrt{\frac{6}{7}}\left\{2\left(\cos\frac{2\psi}{3}\right)-1\right\}R\,\lambda \qquad = 0.925\,82_{\smile}\,R\left(2\cos\frac{2\psi}{3}-1\right)\lambda$$
$$\dot{y} = \frac{9}{\sqrt{7}}\left(\sin\frac{\psi}{3}\right)R \qquad = 3.401\,68_{\smile}\,R\left(\sin\frac{\psi}{3}\right) \qquad \text{. . . . . (151b}$$

with $\psi = \arcsin\left\{\frac{7\sin\phi}{3\sqrt{6}}\right\} = \arcsin(0.952\,58_{\smile}\sin\phi)$. True scale occurs at $\Phi \approx \pm 45°30_{\smile}'$.

The **McBryde-Thomas V** or **Flat-polar Quartic** projection, which, for computational convenience, has the central meridian just $0.450\,16_{\smile}$ of the equator (as with the Quartic above), has plotting equations:

[†] The **Denoyer Semi-elliptical** projection, which is not-quite semi-elliptical and is not pseudocylindrical because its meridians are not quite equally spaced, has a similar appearance to the McBryde-Thomas Flat-polar projections. It has the 1:2 ratio of central meridian to equator and has polar parallels nearly one-third of the equator in length. Its parallels are straight and equally spaced. The Denoyer is one of many projections hand-crafted for in-house use by a commercial corporation; like most of these, the exact formulation used is not public knowledge.

Φ	$\dot{x}$	$\dot{y}$ taking sign of ϕ	
90	0.5322	1.0000	= 0.9761 + 0.0239
85	0.5722	0.9761	= 0.3720 + 0.0367
80	0.6213	0.9394	= 0.3720 + 0.0458
75	0.6732	0.8936	= 0.3720 + 0.0501
70	0.7186	0.8435	= 0.3720 + 0.0532
65	0.7597	0.7903	= 0.3720 + 0.0557
60	0.7986	0.7346	= 0.3720 + 0.0577
55	0.8359	0.6769	= 0.3720 + 0.0693
50	0.8679	0.6176	= 0.3720 + 0.0605
45	0.8962	0.5571	= 0.3720 + 0.0613
40	0.9216	0.4958	= 0.3720 + 0.0618
35	0.9427	0.4340	= 0.3720 + 0.0620
30	0.9600	0.3720	= 0.3100 + 0.0620
25	0.9730	0.3100	= 0.2480 + 0.0620
20	0.9822	0.2480	= 0.1860 + 0.0620
15	0.9900	0.1860	= 0.1240 + 0.0620
10	0.9954	0.1240	= 0.6200 + 0.0620
5	0.9986	0.0620	= 0.0000 + 0.0620
0	1.0000	0.0000	

Table of unit grid values for Robinson projection.

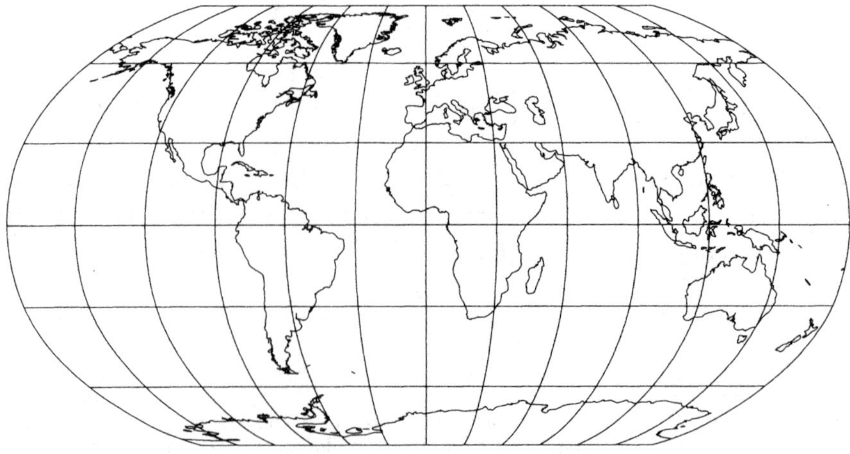

Exhibit 6–15: The Robinson projection at Simple aspect.

$$\dot{x} = \frac{\left(1 + 2\cos\psi \sec\frac{\psi}{2}\right)}{\sqrt{3}\sqrt{2+\sqrt{2}}} R\lambda \qquad = 0.31246_{\sim}\, R\left(1 + 2\cos\psi \sec\frac{\psi}{2}\right)\lambda \qquad \text{. . . . . (153a}$$

$$\dot{y} = \frac{2\sqrt{3}}{\sqrt{2+\sqrt{2}}} R \sin\frac{\psi}{2} \qquad = 1.87476_{\sim}\, R \sin\frac{\psi}{2}$$

where ψ is the solution of

$$\sin\psi + \sin\frac{\psi}{2} = \left(\frac{2+\sqrt{2}}{2}\right)\sin\phi.$$

True scale occurs at $\Phi \approx \pm 33°45._{\sim}'$.

Seeking good shape

The **Robinson** projection [1974m] resulted from efforts to create, within the pseudocylindrical rules, a map of flat-polar style with a desired shapeliness. It was developed expressly for atlases of the Rand McNally Company and is illustrated on the opposite page. It is markedly different mathematically from all the projections met to this point; it uses a numeric grid, rather than plotting equations. The grid values are tabulated above its illustration. Plotting within the grid is by interpolation. Including a multiplier to make the map overall authalic, the plotting equations are:

$$\dot{x} = 0.8487_{\sim} R\, X\, \lambda \qquad \text{. . . . . (153b}$$

$$\dot{y} = 1.3523_{\sim} R\, Y\, \phi$$

where X and Y are linearly interpolated according to latitude from the table opposite. This gives a central meridian 0.51 the length of the equator and polar parallels 0.53 the length of the equator. The spacing between parallels for successive intervals of 5° is uniform out to 35°, then slowly declines. The spacing does not provide equal-area [1989e].

Both the *Oxford Atlas* from Oxford University Press and *The Times Atlas* from John Bartholomew & Co. have used pseudocylindrical projections of believed better appearance developed expressly for the respective publications — by Guy Bomford and John Moir respectively. Both have parallels spaced as in the Gall version of the secantal Stereographic Cylindrical (which had true scale at ±45°) and both have meridians slightly curved inwards towards polar parallels having lengths of over ⅔ the equator — but both stop short of including the poles. Such slight curving is achieved by including in u^* of Equation 127b a small subtractive factor dependent on a power of ϕ — e.g., setting

$$u^* = 1 - \frac{1}{25\phi^4}$$

János Baranyi [1968c] developed maps similar to the Robinson that were used in the Hungarian national atlases. Various hybrids of one Pseudocylindrical with either another else a 'true' cylindrical have also been put forth (which produces a Pseudocylindrical). Some are composite maps, discussed in the next chapter, but others are individual projections that are a mathematical blending of two distinct projections. J. P. Snyder [1977a] covers some of these in his extensive survey of Pseudocylindricals, citing Waldo Tobler, for instance, as suggesting a blending of Sinusoidal and Mollweide using a weighting function based on latitude.

The Loximuthal projection

Accidentally pseudocylindrical, the **Loximuthal** projection was originated by Siemon [1935c]. It was, as with his Quartic, recreated and named by another [Tobler, 1966a]. For a chosen central point, all points on the loxodrome from there of bearing β are placed on a radiating straight line at azimuthal angle β and correct loxodromic distance. Thus, all loxodromes passing through the central point are straight, while in a radial sense the projection is equidistant from there. (The Mercator also had straight loxodromes, but that applied to everywhere on that conformal map — such loxodromes, however, are not true in scale. The Azimuthal Equidistant shows true distances from its central point, but the radial lines follow great-circle paths — i.e., are *orthodromes*, not loxodromes.)

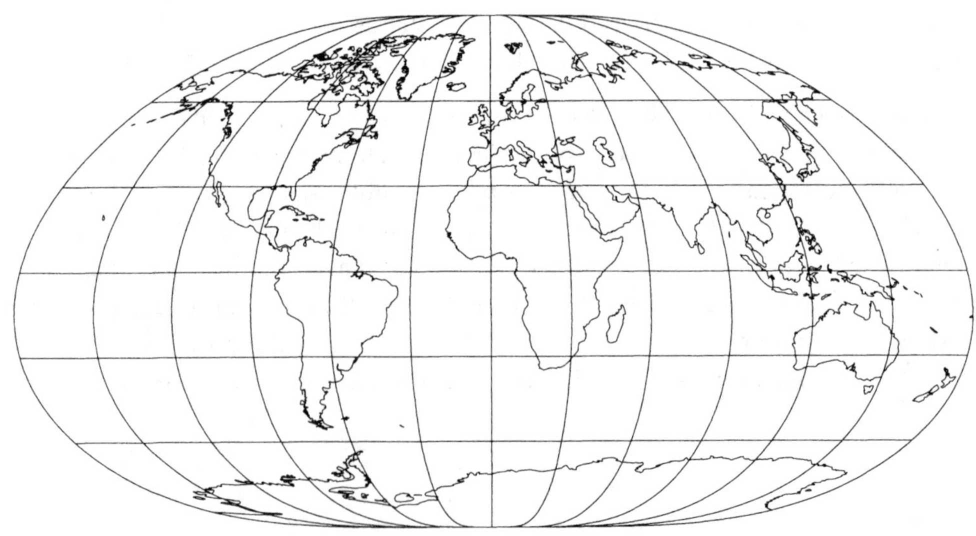

Exhibit 6–16: The Loximuthal projection
Focused for Greenwich, i.e., 0°, 51°30′N

Let $P_0 = (\phi_0, \lambda_0)$ be the central point and $P = (\phi, \lambda)$ the general point, and denote by β the bearing of the loxodrome from P_0 to P and by s the geodesic length of that line. Make P_0 the origin for the map. If the two points are on the same parallel, then that parallel is the loxodrome and its length s is merely the cosine-corrected longitudinal difference. The parallel maps onto the y axis and the plotting equations are:

$$\dot{x} = (\cos\phi)\,\lambda \qquad \dot{y} = 0 \qquad \text{. . . . . (155a}$$

Otherwise, using Equations 88a and 88b, the plotting equations are:

$$\dot{x} = s\sin\beta = s\cos\beta\tan\beta = \frac{R(\phi-\phi_0)\lambda}{\ln\tan\left(\dfrac{\pi}{4}+\dfrac{\phi}{2}\right)-\ln\tan\left(\dfrac{\pi}{4}+\dfrac{\phi_0}{2}\right)} \qquad \text{. . . . . (155b}$$

$$\dot{y} = s\cos\beta = R(\phi-\phi_0)$$

Because the latter co-ordinate is independent of longitude, all latitudinal parallels on the map are straight lines parallel with the equator. Hence, the map classes as a Pseudocylindrical. It is illustrated, centred for Greenwich, on the opposite page. Symmetry about the central parallel occurs when that is the equator. The projection is not conformal; indeed, as is visible from the illustrative map, the meridians are not generally orthogonal to the parallels.

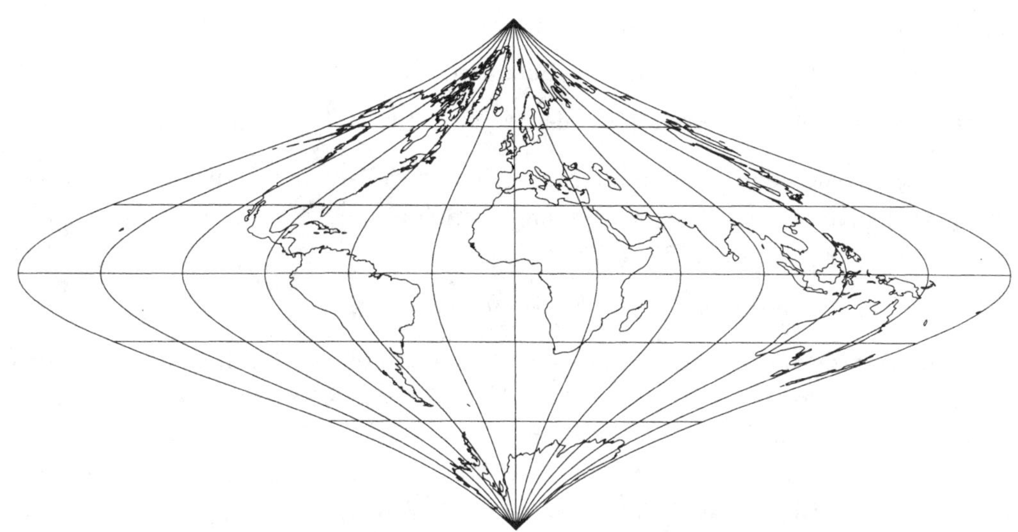

Exhibit 6–17: The Foucaut Stereographic Equivalent projection at Simple aspect.
See page 140 for description.

PSEUDOCYLINDRICAL PROJECTIONS GENERALIZED

General sinusoidal meridians

McBryde and Thomas [1949b] developed general formulae for projections having equispaced parallels and sinusoidal meridians — covering point- and flat-polar forms, and accommodating partial as well as full sinusoids (meaning the curve of $\cos\alpha$ for α ranging from $-\pi/2$ to $+\pi/2$). These are:

$$\dot{x} = \frac{M(m + \cos\psi)}{m+1} R\lambda \qquad \dot{y} = MR\psi \qquad \text{. . . . . (156a}$$

where $M = \sqrt{\frac{m+1}{n}}$ for arbitrary constants m and n, and ψ is the solution for a given ϕ of

$$f(\psi) = m\psi + \sin\psi - n\sin\phi = 0 \qquad \text{. . . . . (156b}$$

If $f(\psi)$ is such that $\psi(\phi) = \phi$ for $\phi = \pm\pi/2$, then the last becomes

$$m\tfrac{\pi}{2} + \sin\tfrac{\pi}{2} - n\sin\tfrac{\pi}{2} = 0 \quad \text{or} \quad 2n = \pi m + 2 \qquad \text{. . . . . (156c}$$

and, from Equation 156a, the length of a polar parallel is

$$2\frac{M(m+0)}{m+1} R\pi \qquad \text{. . . . . (156d}$$

which is zero, indicating point-polar form, if and only if $m = 0$. Then $n = 1$ by Equation 156c and $M = 1$ by Equation 156d, making Equations 156a into Equations 130a (i.e., the plotting equations of the Sinusoidal projection). If $m \neq 0$ the meridians are partial sinusoids and the projection is flat-polar; from Equations 156d & 156a the polar parallels have length $m/(m+1)$ times that of the equator. Examples then include:

if $m = 1$, $n = (1 + \pi/2)$ give the Eckert VI (Equations 146a).

if $m = 1/2$, $n = (1 + \pi/4)$ give the McBryde-Thomas Flat-polar Sinusoidal (Equations 151a).

If $f(\psi)$ is such that $\psi(\phi) < \phi$ for $\phi = \pm\pi/2$, the meridians are partial sinusoids. The value $\psi(\phi) = 2\phi/3$ for $\phi = \pm\pi/2$ (i.e., auxiliary angle ranging from $-\pi/3$ to $+\pi/3$) has been used in several projections. Equation 156b becomes

$$m\frac{\pi}{3} + \sin\frac{\pi}{3} - n\sin\frac{\pi}{2} = 0 \quad \text{or} \quad 6n = 2\pi m + 3\sqrt{3}$$

and the polar parallels have length ratio $(m+1/2):(m+1)$ to that of the equator. Choosing to make this ratio 1:2 gives $m = 0$, hence $n = \sqrt{3}/2$. From Equations 156a this produces a ratio of 1:3 for central meridian to equator. Multiplication/division of the respective parameters by $\sqrt{(2/3)}$ alters this to the true ratio of 1:2 without changing areas, giving the **Wagner I** [1932b] and identical **Kavraiskiy VI** [1934a] projections with plotting equations:

$$\dot{x} = \sqrt{\frac{2}{3}}\sqrt{\frac{2}{\sqrt{3}}}\frac{\cos\psi}{1}R\lambda \quad = \frac{2\sqrt[4]{3}}{3}\cos\psi\, R\lambda \quad = 0.8774_{\sim} R\cos\psi\,\lambda$$
$$\dot{y} = \sqrt{\frac{3}{2}}\sqrt{\frac{2}{\sqrt{3}}}\,R\psi \quad = \sqrt[4]{3}\,R\psi \quad = 1.3161_{\sim} R\psi \qquad \text{(157a}$$

where, from Equation 156b,

$$\sin\psi = \frac{\sqrt{3}}{2}\sin\phi \quad \text{so} \quad \cos\psi = \sqrt{\left(1 - \tfrac{3}{4}\sin^2\phi\right)} \qquad \text{(157b}$$

From the first of Equations 157a, the length of the equator, where $\psi = \phi = 0$ hence $\cos\psi = 1$, is the stated fraction of its true length. The **Werenskiold II** projection [1944a] has both co-ordinates divided by this factor to give the equator true scale without changing the shape of the map, yielding, with ψ again provided by Equation 157b, the plotting equations:

$$\dot{x} = \frac{3}{2\sqrt[4]{3}}\frac{2\sqrt[4]{3}}{3}\cos\psi\, R\lambda \quad = R(\cos\psi)\lambda$$
$$\dot{y} = \frac{3}{2\sqrt[4]{3}}\sqrt[4]{3}\,R\phi \quad = \frac{3}{2}R\phi \quad = 1.5\,R\phi \qquad \text{(157c}$$

Karlheinz Wagner produced two variants of his Wagner I projection. One variation — applied also to the Wagner IV to produce the **Wagner V** projection — has the non-polar parallels respaced to produce a progressive increase in areal scale as latitudes increase, and the overall map enlarged slightly to take the areal scale from true at the equator to 1.2 times true at the 60th parallels [1949c]. The resulting plotting equations for the **Wagner II** projection are:

$$\dot{x} = 0.924\,83_{\sim} R(\cos\psi)\lambda$$
$$\dot{y} = 1.387\,25_{\sim} R\,\phi \qquad \text{(157d}$$

where $\sin\psi = 0.880\,22_{\sim} \sin(0.8855_{\sim}\,\phi)$.

The other variant has the y co-ordinate reduced to that of the Sinusoidal — making the central meridian of true scale — then a fractional multiplier applied to the x co-ordinate to give true scale along a specified pair of parallels. Using ϕ_0 to denote the latter, the plotting equations for the variable **Wagner III** [1932b] are:

$$\dot{x} = \frac{\cos\phi_0\, R\cos\left(\frac{2\phi}{3}\right)}{\cos\left(\frac{2\phi_0}{3}\right)}\lambda$$
$$\dot{y} = R\,\phi \qquad \text{(157e}$$

N. A. Urmayev of the U.S.S.R., in 1950 [see 1960a], set m = 0 and let n range from 0 to 1 for a series of equal-area flat-polar projections, producing partial-sinusoid meridians and a varying relative length and spacing of polar parallels. The x co-ordinates are multiplied by one third of $2\sqrt{(n\sqrt{3})}$. Then:

$$\text{length of Polar parallels} = \frac{2\sqrt{n\sqrt{3}}}{3} \text{ times length of Equator}$$

$$\text{length of Equator} = \frac{\left(4\sqrt{3}\right)\pi n}{8 \arcsin n} \text{ times length of central meridian}$$

The plotting equations for the **Urmayev IV** projection are:

$$\dot{x} = \frac{2\sqrt[4]{3}}{3} R(\cos\psi)\lambda$$
$$\dot{y} = \frac{3}{2n\sqrt[4]{3}} R\psi \qquad \text{. . . . . (158a}$$

where $\psi = \arcsin\{n\sin\phi\}$. The meridians are cosine curves between the angles $\pm\arcsin n$; the latitudes of true scale are

$$\pm\arcsin\left\{\frac{\sqrt{9-4\sqrt{3}}}{9-4n^2\sqrt{3}}\right\}$$

With n = $\sqrt{3}/2$, this becomes Wagner I.

General elliptical meridians

General formulae have also been developed for projections having elliptical meridians — covering point- and flat-polar forms, and accommodating partial as well as full semi-ellipses. If ξ is such that $\cos\xi$ is the length ratio of polar parallel to equator (hence $\pi/2$ for a Mollweide and other point-polar projections, $\pi/3$ for those having the polar parallel half the length of the equator, and so on), then:

$$\dot{x} = \frac{2r}{\pi}(\cos\psi)\lambda$$
$$\dot{y} = \frac{r}{\sin\xi}\sin\psi \qquad \text{. . . . . (158b}$$

with ψ the solution of

$$2\psi + \sin 2\psi = (2\xi + \sin 2\xi)\sin\phi \qquad \text{. . . . . (158c}$$

and

$$r = R\sqrt{\frac{2\pi\sin\xi}{(2\xi + \sin 2\xi)}} \qquad \text{. . . . . (158d}$$

If $\xi = \pi/2$, then $\cos\xi = 0$ and the full Mollweide is created. Otherwise, the meridians are truncated semi-ellipses. If $\xi = \pi/3$, then $\cos\xi = 1/2$ and the Wagner IV is created and, after the pertinent scale changes, the Werenskiold III.

Polar parallels 2/3 and 3/4 the length of the equator were adopted by Masataka Hatano of Japan [1972c] for Pseudocylindricals with partial elliptical arcs (taking the maps back towards the width of the 'true' Cylindricals). Typically used together for different hemispheres in one composite map, these are discussed in the Chapter 7 (see page 172).

General formulae for particular spacing of parallels

McBryde and Thomas also developed, from earlier work by E. J. Baar [1947a], general formulae for equal-area projections having the spacing of parallels set by sine values of some angle derived from latitude, and by tangent values similarly.

The general plotting equations for sine relationships are:

$$\dot{x} = \frac{p}{q} R \frac{\cos\phi}{\cos\frac{\phi}{2}} \lambda$$

$$\dot{y} = q R \sin\frac{\phi}{p}$$

where p and q are arbitrary constants.

If $p=2$ and $q=2$ these become Equations 140a for the Quartic. The **McBryde-Thomas I** is a flat-polar equal-area projection with the polar parallels ⅓ the length of the equator. The defining feature is that the equatorial x co-ordinate for the 20° meridian is 0.85 times the y co-ordinate for the 20° parallel. If $\alpha = \arccos(3\cos 80°)$ then the parameters become:

$$p = \frac{4\pi}{9\alpha} \qquad = 1.365\ 09_{\sim}$$

$$q = \sqrt{\frac{80\pi}{9\left\{0.85\alpha \sin\frac{\alpha}{4}\right\}}} \qquad = 1.488\,75_{\sim}$$

The **Kavraiskiy V** projection, used for maps of the Pacific Ocean, has:

$$q = \frac{35}{\left(°\arccos 0.9\right)} \qquad = 1.354\ 39_{\sim}$$

$$p = \frac{q}{0.9} \qquad = 1.504\ 88_{\sim} \qquad \text{. . . . . (159a}$$

to obtain an equal-area map with scale true at ±35° latitude. The scale is 0.9 of true at the equator.

The general plotting equations for tangent relationships are:

$$\dot{x} = \frac{q}{p} R \left\{\cos\phi \cos^2\frac{\phi}{q}\right\} \lambda$$

$$\dot{y} = p R \tan\frac{\phi}{q}$$

where p and q are again arbitrary constants. Stereographic projecting has $q=2$.

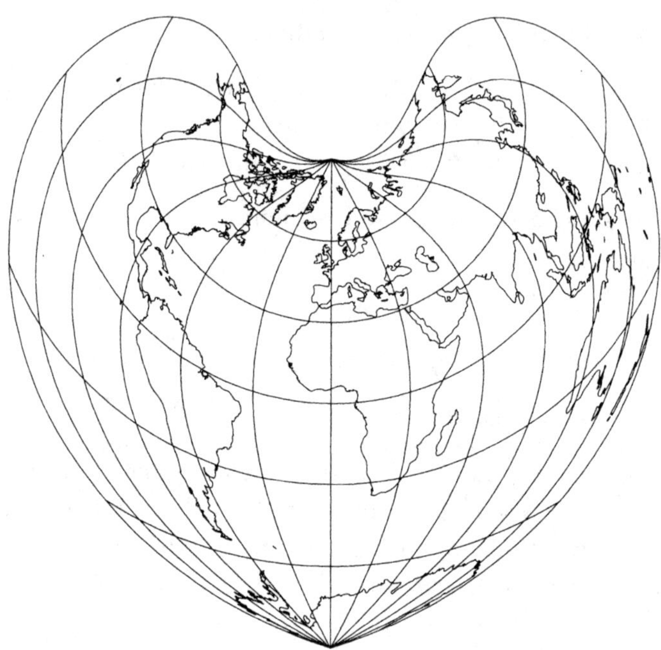

Exhibit 6–18: The Bonne projection at Simple aspect.
Standard parallel set as 45°N.

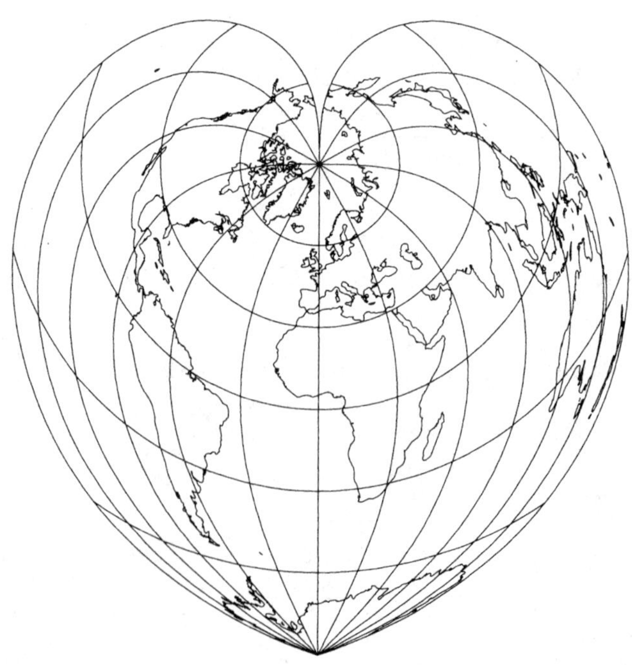

Exhibit 6–19: The Werner projection —
the Bonne projection with standard parallel set as 32°42..'N, i.e., polar distance equal to 1 rad.

PSEUDOCONICAL PROJECTIONS

Basis

Pseudoconical projections bend meridians while keeping the parallels as concentric circular arcs; expressly, they keep the central meridian straight and bend others in mirror image about the central line. The only form of any note is the equivalent of the Sinusoidal — i.e., an equal-area adaptation of the Equidistant Conical. We find again that these two options define the exact shape of the meridians, hence the lengths of the parallels, with mathematical development initiated by making the spacing of the meridians along each parallel equal to the corresponding spacing on the globe.

The basic pseudoconical projection dates back to Ptolemy around 100CE and was further developed in the cartographically busy 16th Century. It became widely used only in the 18th Century. It was promoted in France (where it continues to be well used) by Rigobert Bonne, after whom it is usually named.

We use the tangential Conical, for which we again denote the standard parallel by ϕ_0. For the equidistant form, the radius $\dot{\rho}$ of the arc for parallel ϕ is shown in Equation 69c to be

$$\dot{\rho} = R\{(\phi_0 + \cot\phi_0) - \phi\}$$

If the distance along parallel ϕ between longitude λ and the central meridian on the map is to be the same as that between those longitudes on the globe, then it must be $R(\cos\phi)\lambda$. Hence, the circular arc covering this longitudinal span subtends at the centre of the corresponding circle the angle

$$\begin{aligned}\dot{\theta} &= \frac{\cos\phi}{\rho} R\,\lambda \;=\; \frac{\cos\phi}{R\{\phi_0 + \cot\phi_0 - \phi\}} R\,\lambda \\ &= \frac{\cos\phi}{\phi_0 + \cot\phi_0 - \phi}\lambda\end{aligned} \qquad \text{. . . . . (161a}$$

relative to the straight line of the central meridian. For $\phi = \phi_0$, this becomes $\sin\phi_0$, the familiar constant of the cone (which no longer applies beyond its own latitude because of the adoption of bent meridians). The plotting trigonometry remains unchanged. So we have (Equations 59b):

$$\dot{x} = \dot{\rho}\sin\dot{\theta}$$

$$\dot{y} = \dot{\rho}_0 - \dot{\rho}\cos\dot{\theta} = R\cot\phi_0 - \dot{\rho}\cos\dot{\theta}$$

Substitution from the above equations creates the plotting equations for the **Bonne** projection as:

$$\begin{aligned} \dot{x} &= R\{(\phi_0 + \cot\phi_0) - \phi\}\sin\left\{\frac{\cos\phi}{\phi_0 + \cot\phi_0 - \phi}\lambda\right\} \\ \dot{y} &= R\left[\cot\phi_0 - \{(\phi_0 + \cot\phi_0) - \phi\}\cos\left\{\frac{\cos\phi}{\phi_0 + \cot\phi_0 - \phi}\lambda\right\}\right] \end{aligned} \quad \ldots\ldots\ldots (162a$$

However, computation is better served by evaluating the intermediate parameters then substituting arithmetically.

The plotting table and picture depend on the choice of ϕ_0 as well as the area to be covered. The pseudocylindrical Sinusoidal is just the limiting case as the constant of the cone approaches zero. The Sinusoidal is used mostly for World maps; the pseudoconical Bonne is used mostly for regions of relatively modest size.

As noted earlier for the Equidistant Conical (at Simple aspect), the pole on the map is the centre of the circular arcs if the standard parallel has polar distance equal to one radian, i.e., the standard latitude is 32°42′15.‿″. With this particular choice, the projection can be plotted continuously around the nearer pole. This was done about 1500CE by Johannes Stabius or Stab. It was promoted by Johannes Werner of Nuremberg from 1514 on [see 1993a], and consequently usually called the **Werner** projection. It is also referred to descriptively as cordiform (meaning heart shaped).

The Bonne and Werner are illustrated together at Simple aspect on page 160.

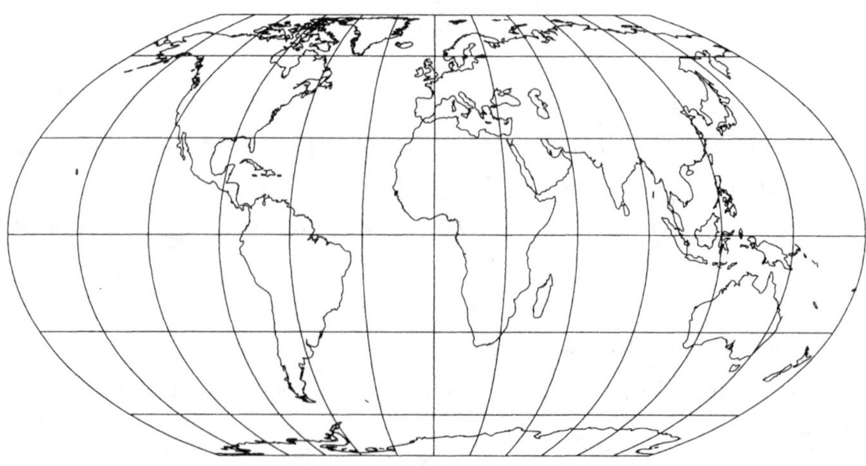

Exhibit 6–20: The Putnins P2′ or Wagner IV projection at Simple aspect.

ENDNOTE A: PUTNINS' APPROACH

The idea

Within the large repertoire of equal-area pseudocylindrical projections the Sinusoidal is unique in having equally spaced parallels, the Mollweide distinctive by having a full ellipse forming its outer meridians. Putnins appears to have set out to create hybrids of these using elliptical curves slightly less than the full semi-ellipse, then acted similarly with the parabola and the hyperbola. By not using a sinusoidal curve for meridians, he made equally spaced parallels and being equal-area mutually incompatible. So he created pairs of projections on a common meridional graticule dimensioned to be overall authalic. The first in each pair has equally spaced parallels, the second is equal-area. These were labelled P_1 and P_2 for the elliptical, P_3 and P_4 for the parabolic and P_5 and P_6 for the hyperbolic. The P_4 and P_5 are illustrated on page 136; comparison with the Sinusoidal on page 126 shows the close similarity to that eminent predecessor achieved by Putnins.

Putnins also created corresponding flat-polar projections for each of the six, with the polar parallels equal in length to the central meridian (and half the equator). These are labelled P_1' and P_2' for the elliptical, P_3' and P_4' for the parabolic and P_5' and P_6' for the hyperbolic. Again the two members of a pair share a common meridional graticule.

For each choice of curve style, the curved arcs used in the flat-polar form occur identically in the point-polar form, although the latter has extended versions too. Establishing defining factors for these arcs is done more clearly through the flat-polar form, which we now address for the elliptical case, illustrated opposite.

The method applied with elliptical curves for flat-polar projections

Let $4\pi s$ be the length of the equator, hence $2\pi s$ the length of the central meridian and of each polar parallel. Then, any ellipse providing the $\pm\lambda$ meridians intersects the x axis at $(\pm 2\lambda s, 0)$ and has the equation

$$\frac{x^2}{(2\lambda s)^2} + \frac{y^2}{b^2} = 1 \qquad \text{........ (163a}$$

Having uniformity of scale along each parallel (standard for Pseudocylindricals) implies that the adopted ratio between polar parallel and equator applies likewise to any interval of longitude. So the ellipse must pass through the points $(\pm\lambda s, \pm\pi s)$, giving

$$\frac{(\lambda s)^2}{(2\lambda s)^2} + \frac{(\pi s)^2}{b^2} = 1 \quad \text{so} \quad b = \frac{2\pi s}{\sqrt{3}} \qquad \text{........ (163b}$$

Thus, the length of the vertical axis of the ellipse defining any meridian depends purely on the choice of the length of the central meridian, is independent of its longitude. That is, the ellipses for all meridians pass through the common points $(0, \pm^{2\pi s}/_{\sqrt{3}})$. Equation 163a for the general ellipse becomes

$$\frac{\dot{x}^2}{(2\lambda s)^2} + \frac{3\dot{y}^2}{(2\pi s)^2} = 1 \qquad \text{. . . . . (164a}$$

which for the outside meridians (i.e., $\lambda = \pm\pi$) is

$$\dot{x}^2 + 3\dot{y}^2 = (2\pi s)^2 \qquad \text{. . . . . (164b}$$

which intersects the x axis at $(\pm 2\pi s, 0)$.

Two meridians must lie on a circle, which must be of radius b to pass through $(0, \pm^{2\pi s}/_{\sqrt{3}}) = (0, \pm b)$. That circle intersects the polar parallels at $(\pm^{\pi s}/_{\sqrt{3}}, \pm\pi s)$, which mark the end points of the meridians. The diagram in Exhibit 6–21 below illustrates the scene, with the pair that fit the circle continued beyond the polar parallels along with the outer perimeter meridians as well as its minor axis ends at $(0, \pm b)$.

Because the length of the section of either polar parallel straddling the circle is $^{2\pi s}/_{\sqrt{3}}$, identically the radius b of the circle, the angle it subtends at the centre as a chord of the circle is $^{\pi}/_{3}$. Looking at the sector either chord defines and the triangle between the chord and the centre of the circle we have

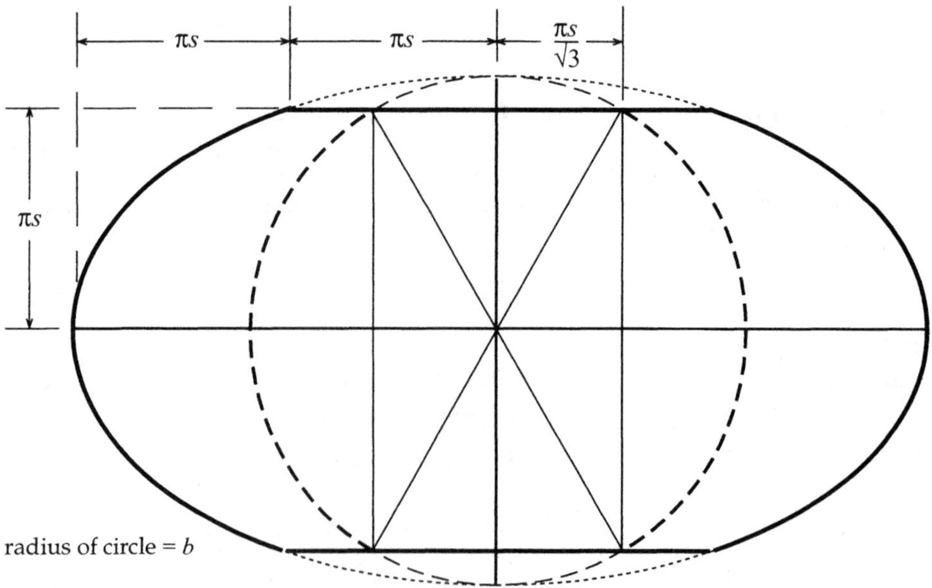

Exhibit 6–21: Example geometry of the Putnins projections.

$$\text{area of sector } = \frac{1}{2}\frac{\pi}{3}b^2 = \frac{\pi}{6}b^2$$

$$\text{area of sectorial triangle } = \frac{1}{2}\pi s b = \frac{\sqrt{3}}{4}\left(\frac{2\pi s}{\sqrt{3}}\right)b = \frac{\sqrt{3}}{4}b^2$$

The difference between these is the area of either segment of the circle beyond a polar parallel. From that, we can obtain the area of the circle clipped by those parallels, as follows:

$$\text{area of clipped circle } = \left(\pi - 2\left(\frac{\pi}{6} - \frac{\sqrt{3}}{4}\right)\right)b^2 = \left(\frac{2\pi}{3} + \frac{\sqrt{3}}{2}\right)b^2 = \frac{4\pi + 3\sqrt{3}}{6}b^2 \quad \text{. . . . (165a}$$

The quotient in the final expression is effectively an areal proportionality factor. We denote this by A_1', so

$$A_1' = \left\{\frac{4\pi + 3\sqrt{3}}{6}\right\} = 2.960\,421_{\sim} = 0.942_{\sim}\pi \quad \text{. . . . . (165b}$$

indicating that the area of the clipped circle is about 94% of the whole. The meridians on the circle pass through the points $(\pm{}^{2\pi s}/_{\sqrt{3}}, 0)$, hence occur for $\lambda = \pm{}^{\pi}/_{\sqrt{3}}$ (i.e., $\Lambda = \pm 103°55._{\sim}'$); they span ${}^{2\pi}/_{\sqrt{3}}$ of longitude. Equating this with the area on the unit globe for such a span with the area given by Equations 165a, substituting for b from Equations 163b, we get

$$A_1'\left(\frac{2\pi s}{\sqrt{3}}\right)^2 = 2\left(\frac{2\pi}{\sqrt{3}}\right) \quad \text{or} \quad A_1'\pi s^2 = \sqrt{3}$$

Adding distinguishing features to factor s as derived for this situation, we then get

$$\left(s_1'\right)^2 = \left\{\frac{\sqrt{3}}{\pi A_1'}\right\} = \left\{\frac{6\sqrt{3}}{\pi\left(4\pi + 3\sqrt{3}\right)}\right\} = 0.186\,233_{\sim} = \left(0.431\,548_{\sim}\right)^2 \quad \text{. . . . . . (165c}$$

The method applied with elliptical curves for point-polar projections

For the corresponding point-polar projections, the elliptical arc of every meridian must contact the y axis at the polar points. The design uses the same curves as those fitted between the polar parallels in the flat-polar situation, with the resulting map having to be enlarged compared with the flat-polar version to compensate for the reduction of contained area. Those that are circular arcs in the point-polar map are identically the curves on the circle of the flat-polar version. So, Exhibit 6–21 is applicable again, as the area between each circular arc and the central meridian is that of a lateral segment. The sum of two such equals the difference between that of the clipped circle and the central rectangle. Hence,

$$\frac{4\pi + 3\sqrt{3}}{6} b^2 - 2\pi s b = \left(\frac{4\pi + 3\sqrt{3}}{6} - \sqrt{3} \right) b^2 = \frac{4\pi - 3\sqrt{3}}{6} b^2$$

giving an areal proportionality factor

$$A_1 = \left\{ \frac{4\pi - 3\sqrt{3}}{6} \right\} = 1.228\,370_{\sim} = 0.391_{\sim}\pi \qquad \text{(166a}$$

indicating that the area is now less than 40% of the whole circle. Because the point-polar projections are still designed to make the equator twice the length of the central meridian, the meridians that are circular arcs must be for longitudes exactly half that for those in the flat-polar projections (i.e., $\lambda = \pm^{\pi}/_{2\sqrt{3}}$, spanning $^{\pi}/_{\sqrt{3}}$ of longitude. Equating areas with the corresponding area on the unit globe, we get

$$A_1 \left(\frac{2\pi s}{\sqrt{3}} \right)^2 = \left(\frac{2\pi}{\sqrt{3}} \right) \quad \text{or} \quad A_1 \pi s^2 = \frac{\sqrt{3}}{2}$$

Adding a distinguishing suffix to factor s as derived for this situation, we get

$$(s_1)^2 = \left\{ \frac{\sqrt{3}}{2\pi A_1} \right\} = \left\{ \frac{6\sqrt{3}}{2\pi\left(4\pi - 3\sqrt{3}\right)} \right\} = 0.224\,414\,9_{\sim} = \left(0.473\,724\,5_{\sim}\right)^2 \qquad \text{(166b}$$

The factors for the non-elliptical projections

Corresponding factors can be developed geometrically for the other eight projections. However, those involving parabolic curves are addressed more directly. The others use the hyperbola

$$\frac{\dot{x}^2}{(2\lambda s)^2} - \frac{3\dot{y}^2}{(2\pi s)^2} = 1$$

which, for the outside meridians (i.e., $\lambda = \pm\pi$), is

$$\dot{x}^2 - 3\dot{y}^2 = (2\pi s)^2$$

The resulting factors are

$$A_5{}' = 4\sqrt{3} - \ln\left(2 + \sqrt{3}\right) = 5.611\,245_{\sim} \quad \text{and} \quad \left(s_5{}'\right)^2 = \frac{\sqrt{3}}{2\pi A_6{}'} = \left(0.221\,646\,4_{\sim}\right)^2 \qquad \text{(166c}$$

for the flat-polar $P_5{}'$ and $P_6{}'$ projections. For the point-polar P_5 and P_6 projections the factors are

$$A_5 = 2\sqrt{3} - \ln\left(2 + \sqrt{3}\right) = 2.147\,144_{\sim} \quad \text{and} \quad \left(s_5\right)^2 = \frac{\sqrt{3}}{4\pi A_5} = \left(0.253\,364\,0_{\sim}\right)^2 \qquad \text{(166d}$$

CHAPTER 7

Cut and Paste: Interrupted and composite maps

Preamble

Pseudocylindrical projections avoid increasing lateral exaggeration but at the price of great angular distortion in outer zones. The Sinusoidal and others can serve well the whole Equatorial zone and the zone astride the full length of the central meridian. Whole-world maps deteriorate away from those zones, however, impeding greatly usage that their other qualities favour. Splitting is an alternative means to stretching the map, effectively creating multiple central meridians. If our level of interest varies over the map — e.g., when addressing terrestrial features that do not apply to oceans — then cutting the map discerningly from pole to Equator into pieces can relegate major distortions to insignificant zones. Such action is tantamount to creating separate regional maps and juxtaposing them. The pieces need to be compatibly scaled and shaped at their junctions, but do not have to be of one projection; they could be a composite of two or more. We now look into this useful technique, for the evolution of which Dahlberg [1962a] provided a lengthy discussion.

This chapter has little mathematics — and none that is unfamiliar.

The varied maps have no common theme; some illustrate fragmentation just by their outline while the later ones illustrate construction from infinitely many parts. Their graticules are correspondingly variable. A common scale is used for the interrupted world maps.

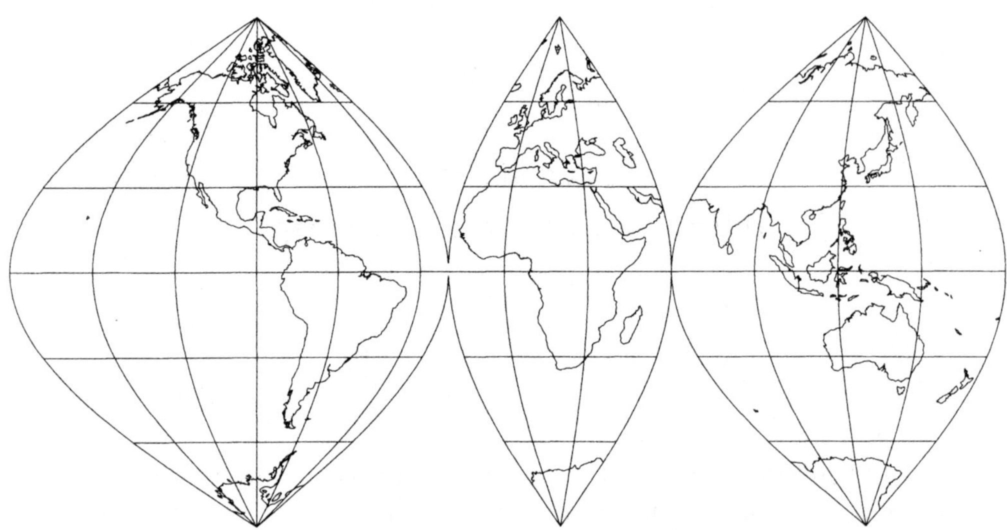

Exhibit 7–1: Three maps as one — an interrupted world Sinusoidal map.
i) all latitudes; longitudes 180°W to 20°W, centred on 90°W
ii) all latitudes; longitudes 20°W to 60°E, centred on 20°E
iii) all latitudes; longitudes 60°E to 180°E, centred on 130°E.

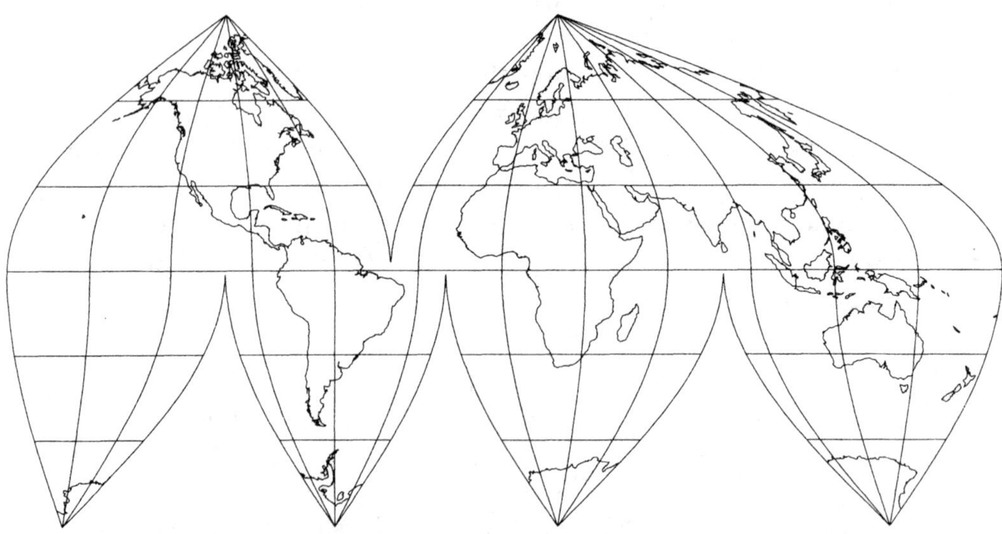

Exhibit 7–2: Six maps as one — an interrupted world Sinusoidal map.
i) N latitudes; longitudes 180°W to 40°W, centred on 100°W
ii) N latitudes; longitudes 40°W to 180°E, centred on 20°E
iii) S latitudes; longitudes 180°W to 100°W, centred on 160°W,
iv) S latitudes; longitudes 100°W to 20°W, centred on 60°W
v) S latitudes; longitudes 20°W to 80°E, centred on 20°E
vi) S latitudes; longitudes 80°E to 180°E, centred on 140°E.

THE APPROACH

Joining maps

Any two maps can be joined satisfactorily into one if the shape is identical along the intended joining line and the scale along that line is consistent at every point. It is preferable that adjacent features transverse to the line are consistent, too, but it is certainly not necessary. (See the Eckert II in Exhibit 6–7 on page 142 for an example of inconsistency across the equator of meridional alignment within a single projection.) Meeting these requirements is easy if the joining line is merely a point, but extremely difficult if it is a proper curve. Having them meet at a point serves little purpose except where the constituent maps are of the same projection, including scale. Joining along a straight line offers many opportunities to enhance our repertoire of projections. Joining along straight meridians offers little except the facility to contrast projections, but the straight parallels of the 'true' cylindrical and pseudocylindrical projections offer interesting potential for the creation of maps that are the composite of two projections (and of even more).

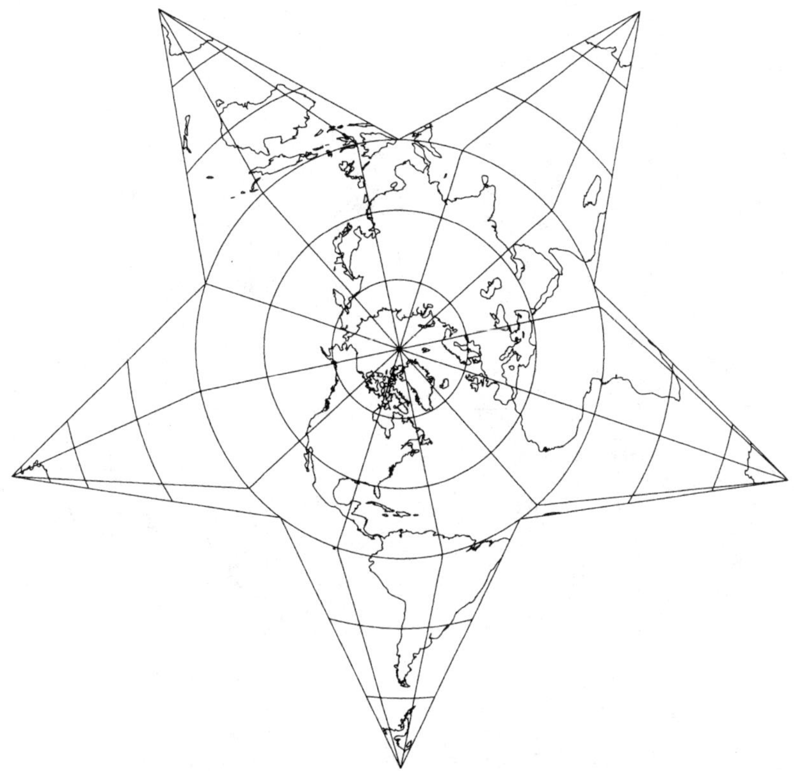

Exhibit 7–3: The Berghaus Star world map.
Straight meridians: 126°E, 54°E, 18°W, 90°W, 62°W.

INTERRUPTED MAPS

Basis

Although the pseudocylindrical projections were illustrated in the previous chapter with whole-world maps, all of them can be used for regional coverage too. Indeed the Sinusoidal is an excellent projection for maps that run along the Equator, as with Indonesia, else along a meridian, as with Chile. It can serve well the presentation of whole continents, notably Africa and the Americas, and regions such as Southeast Asia. Exhibit 7–1 on page 168 shows this with three Sinusoidal maps that collectively cover the whole world. They are presented touching at their Equatorial limit points in a manner that provides a continuous map of the Equator; hence, they can be seen as one map rather than three — a map whose coverage at any non-zero latitude, although complete of the globe, suffers breaks on the map. (All flat maps suffer breaks from outer edge to outer edge; here we mean intermediate breaks.) The general qualifier for any map with intermediate breaks is *interrupted;* where this involves repositioning central meridians they are also termed *re-centred.* R. E. Dahlberg [1962a] surveyed the evolution of this useful technique.

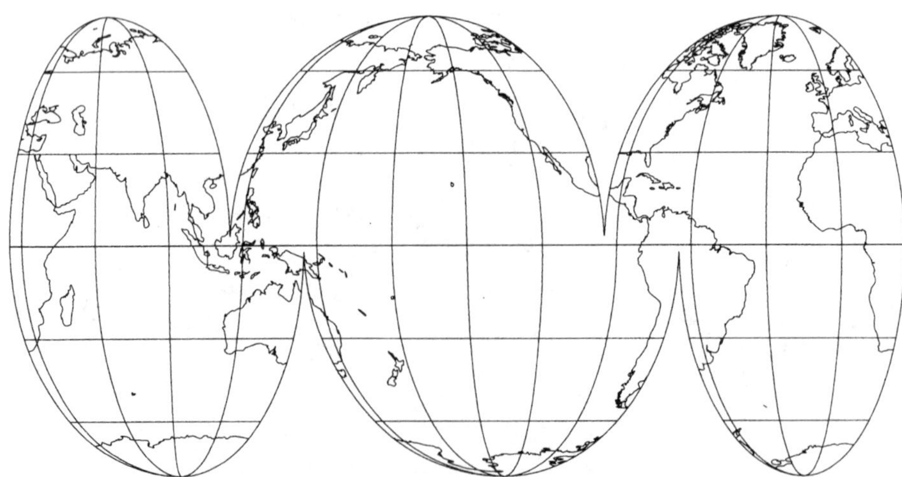

Exhibit 7–4: An interrupted world Mollweide map, with emphasis on oceans.
i) N latitudes; longitudes 25°E to 115°E, centred on 65°E
ii) N latitudes; longitudes 115°E to 95°W, centred on 165°W
iii) N latitudes; longitudes 95°W to 25°E, centred on 25°W
iv) S latitudes; longitudes 25°E to 145°E, centred on 95°E
v) S latitudes; longitudes 145°E to 65°W, centred on 135°W
v) S latitudes; longitudes 65°W to 25°E, centred on15°W.

In Exhibit 7–1 on page 168, each piece has been projected with its central meridian at the central longitude of its span. It is not necessary to be so balanced about any central meridian or the Equator. Interruption lines can be different between North and South hemispheres — hence half meridians rather than full meridians. Then there are separate constituent maps meeting across the Equator. And the central meridian does not have to be literally central to each piece; if the zone of prime interest lies off-centre, the central longitude can be likewise. Such licence is taken in Exhibit 7–2 on that same page, giving South America a central meridian different from the 100°W that is ideal for North America, so moving the split of the North Atlantic ocean westwards to include Iceland wholly with Europe. Eurasia stays in one piece but a split south of that, between Africa and southern Asia, remains.

Those two maps could serve various land-oriented interests well, particular those of little relevance in polar regions, for instance vegetation cover, population density, language groups, concentrations of religions. It could satisfactorily show major trade routes within the respective regions, e.g., within the Americas, within/between Europe and Africa, and within the greater Southeast Asia. It would be poor, however, for showing Pacific-rim and trans-Atlantic trade, because of the interruptions. If oceans provide the preponderant interest, the interruptions can be made in the land just as easily; the map opposite illustrates this using the Mollweide projection.

An interrupted map can repeat fragments of global surface across its interruptions, e.g., to avoid the splitting of Greenland and the isolation of eastern Siberia from its adjoining land, while still having it placed, along with the Aleutian islands, with North America. Any repetition across inner boundaries must not overlap or confusingly adjoin another piece, so it is effectively limited to higher latitudes.

The star-like map on page 169 illustrates a different form of interruption, but one that involves modification of the basic projection, so technically composite. Its essence is the Equidistant Azimuthal projection but, beyond the Equator, meridians are reoriented to converge to five selected points, each representing the South Pole (with all parallels still equidistant circular arcs centred on the North Pole). Called the **Berghaus Star**, this composite projection is of novelty use only, but has one notable use — despite being of German origin [1845a], this American-style star has since 1911 been the logo of the *Association of American Geographers*. The five evenly spaced straight meridians cannot be placed to consistent purpose; here Australia is integral while South America is riven apart.

The idea of star projections appears to have been initiated with the **Jäger Star** [1865a], with eight uneven triangles attached to an irregular octagon, all parallels being sectionally straight and equally spaced. The **Petermann Star** was created by the eponymous journal publisher as an adaptation having the parallels concentric arcs and its eight points evenly spaced — arbitrarily breaking the southern land masses. See 1935a.

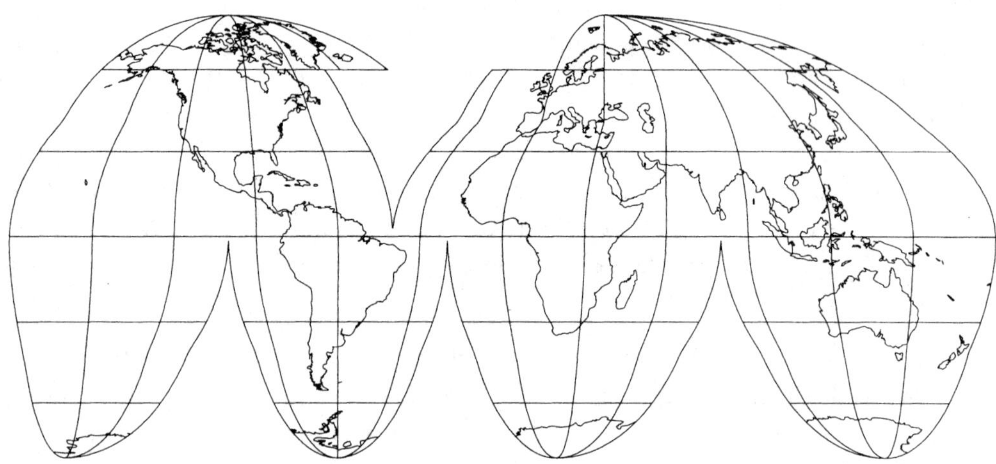

Exhibit 7–5: An interrupted world Homolosine map, with emphasis on land masses.
i) N latitudes; longitudes 180°W to 40°W, to 0° above 60°N, centred on 100°W
ii) N latitudes; longitudes 40°W except 0° above 60°N to 180°E, centred on 20°E
iii) S latitudes; longitudes 180°W to 100°W, centred on 160°W
iv) S latitudes; longitudes 100°W to 20°W, centred on 60°W
v) S latitudes; longitudes 20°W to 80°E, centred on 20°E
vi) S latitudes; longitudes 80°E to 180°E, centred on 140°E.

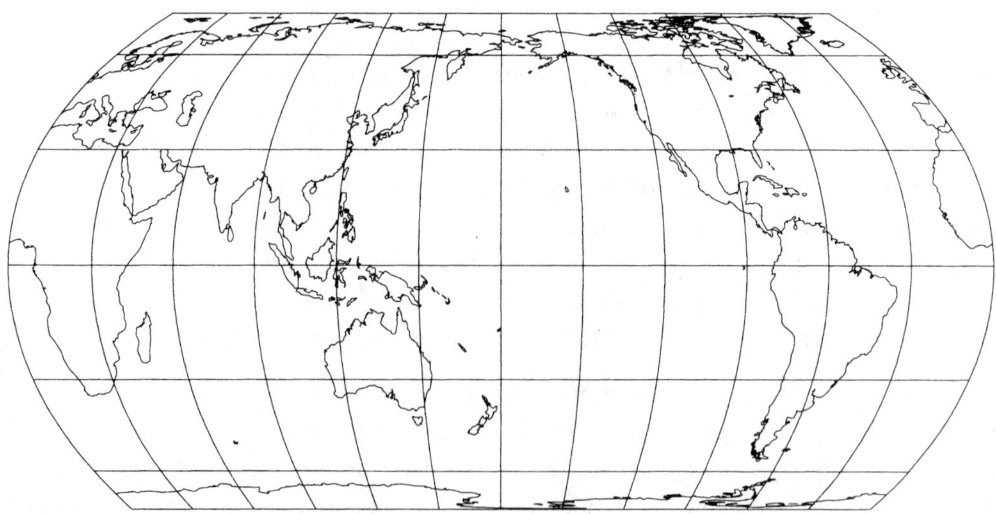

Exhibit 7–6: The Hatano Asymmetrical Equal-area projection.

MIXING PROJECTIONS

Composite maps

Except for the Berghaus Star, each of the interrupted maps we have shown was uniformly of one projection and consistent between the pieces in scale. The exception maintained the equidistant Azimuthal nature throughout, merely refocusing meridians at their further reaches. All could be seen as composites of multiple regional maps — three just touching at points for the simplistic Sinusoidal used as the initial illustration, but a composite of maps sharing lines along the straight Equator for the Mollweide on page 170. It had the half meridians differently centred in the two hemispheres, sharing lines along another, curved, parallel for the Star. We could have interrupted the Sinusoidal else any other of our Pseudocylindricals as we did for Mollweide as many times as we wished, as long as we maintained a common scale at their junctions. Given a common shape and scale for the boundary to be shared, we can also construct composite maps employing different projections, beyond what was done for the Berghaus Star. This is easily achieved between different Azimuthals, with their circular parallels, and between different Cylindricals and Pseudocylindricals, which all have straight parallels. Conicals, with their variably curved parallels, present only one, albeit notable, opportunity for composite use.

Any two Azimuthal maps of different projections having the same scale along a specific parallel — e.g., having true scale by being the Orthographic else being constructed secantally at that parallel — can be cut along the said parallel and put one disc inside the other outer rim to produce a composite Azimuthal map. However, such maps serve little purpose, and are rare. The like holds for 'true' Cylindricals but not for Pseudo-cylindricals, where one composite in particular, of two projections, is widely used; usually referred to as a projection, it is described in the next subsection. For Conicals, as already mentioned, there is also one notable composite; described further below, it is a composite of an infinite number of maps employing one projection differently, and is, appropriately, itself called a projection.

The centrally projected Gnomonic offers a special incentive for piece-wise use as a composite map by its having all great circles as straight lines. However, this requires the oblique aspect — the subject of the following chapter — so discussion is delayed accordingly.

COMPOSITE CYLINDRICAL MAPS

Composites of 'true' Cylindricals

Any two 'true' Cylindricals can be juxtaposed along any parallel if they are of the same width, but neither conformality nor equal-area can be achieved for both. There is scope to maintain either in the main band of the map while surrendering it in higher latitudes — thus, for instance, having the poles appear on a map that is essentially a Mercator — but little interest has been shown on any such composite constructions.

Sinusoidal plus other Pseudocylindrical

The Sinusoidal projection was expressly developed to have true scale along all parallels. Other pseudocylindrical projections and the 'true' Cylindricals have true scale along only one parallel in a hemisphere. Because all parallels are straight lines, we can cut along the one parallel of true scale for any of the latter, and along the same parallel on a Sinusoidal map, then adjoin the contrasting parts to produce an acceptable map that maintains the equal-area attribute of the Sinusoidal and of the other if it is equal-area. If that other projection is symmetric about the Equator too, we could use the same method in both hemispheres and produce another symmetric map. We could, however, do otherwise. If

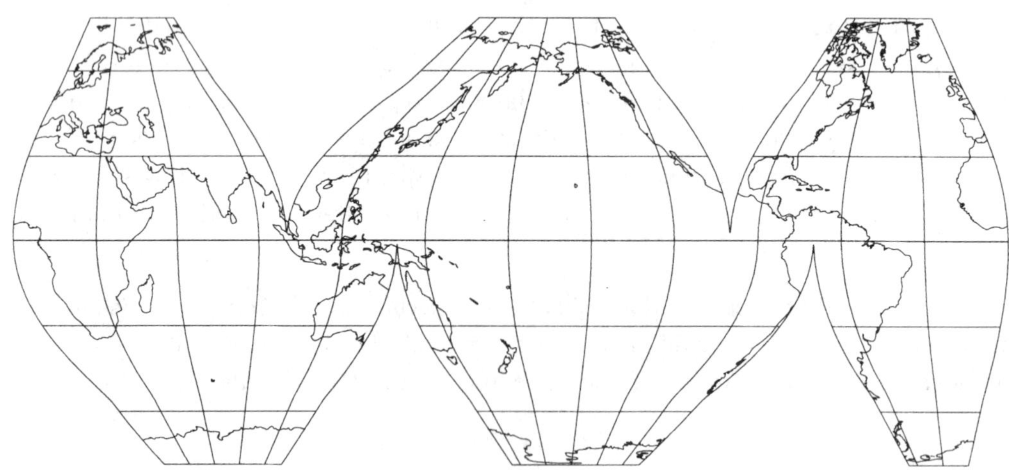

Exhibit 7–7: An interrupted McBryde S3 projection with emphasis on oceans.
i) N latitudes; longitudes 0° to 100°E, centred on 40°E
ii) N latitudes; longitudes 100°E to 100°W, centred on 160°W
iii) N latitudes; longitudes 100°W to 0°, centred on 40°W
iv) S latitudes; longitudes 0° to 140°E, centred on 80°E
v) S latitudes; longitudes 140°E to 70°W, centred on 160°W
vi) S latitudes; longitudes 70°W to 0°, centred on 25°W.

we chose to use the Equatorial section of the Sinusoidal, we could opt to use a projection other than the Sinusoidal for the outer sections of each hemisphere — even different ones for each. Examples of the latitude of true scale for equal-area Pseudocylindricals are:

±25°42.‿′ for P6
±33°45.‿′ for McBryde-Thomas V or Flat-polar Quartic
±36°46.‿′ for the Parabolic and P4
±37°28.‿′ for P2
±40°15.‿′ for the Boggs Eumorphic
±40°30.‿′ for Eckert IV
±40°44.‿′ for the Mollweide
±42°59.‿′ for P2′
±45°30.‿′ for McBryde-Thomas IV or Flat-polar Parabolic
±49°16.‿′ for Eckert VI
±50°40.‿′ for P6′
±55°10.‿′ for Eckert II
±55°51.‿′ for McBryde-Thomas III or Flat-polar Sinusoidal.

As an example of the mixed form, we could take a Sinusoidal map from 36°46.‿′S to 40°44.‿′N and append a Mollweide map from 40°40.‿′N to the North Pole and a Parabolic map from 36°46.‿′S to the South Pole. This three-part example and most two-part examples are rarely, if ever, seen. The symmetric composite of Sinusoidal for the Equatorial zone and Mollweide for latitudes beyond ±40°44.‿′ is, however, widely used. Called the **Homolosine** projection — from conjoining Homolographic (the generic name for the Mollweide) and Sinusoidal — the projection is usually prefixed possessively with the surname of the cartographer that introduced it, Paul Goode, and whose atlases have promoted it [1925a]. It is typically used for whole-world purpose, and interrupted. The characteristic change of line of meridians at the critical latitudes can be easily seen in Exhibit 7–5 (page 172). The merits of this equal-area composite are seen as a retention of exact equi-spacing of and true scale along parallels almost half way to the Pole (i.e., to ±40°44.‿′) without the sharp pointedness of the Poles of the full Sinusoidal.

The **Philbrick** or **Sinu-Mollweide** projection [1953b] uses the Mollweide north of +40°44.‿′, but uses the Sinusoidal elsewhere; as put forward by its originator, it was used obliquely and subject to multiple interruptions in its further regions. György Érdi-Krauss [1968b] also made a composite of Mollweide and a Pseudocylindrical with sinusoidal meridians, but using a flat-polar derivative of the Sinusoidal. The Mollweide, necessarily enlarged, was used beyond the 60th parallels. While retaining equal-area within each component, the proportions were markedly different between the two projections.

After teaming with Thomas to produce their many Pseudocylindricals described in Chapter 6, F. W. McBryde adapted three of them, along with the Eckert VI, to produce composite maps retaining the equal-area property. Two used the Sinusoidal and two used other point-polar projections as the second projection. With the Sinusoidal cases, the fusion lines are those of true scale; with the others, it is otherwise. Using McBryde's labels (and initials for the McBryde-Thomas projections), these were:

McBryde S2 = Eckert VI	+ Sinusoidal	fused at ±49°16..′	factor 0.084 398
McBryde S3 = M-T III	+ Sinusoidal	fused at ±55°51..′	factor 0.069 065
McBryde Q3 = M-T IV	+ Quartic Authalic	fused at ±52°09..′	factor 0.042 686
McBryde P3 = M-T V	+ Craster Parabolic	fused at ±49°20..′	factor 0.082 818

In each, the poles appear as lines a little shorter than in the component flat-polar projection, and the y co-ordinate reduced by the indicated factor times R. All are usually used in interrupted form. The **McBryde S3** projection is illustrated on page 174; it employs the Sinusoidal further toward the Poles than any other composite using that projection. In 1982, all four new projections were covered by U.S. patent 4,315,747 (following a pioneer mentioned in Chapter 10).

Further composites of Pseudocylindricals

As multiplying the x co-ordinates by some number and dividing the $\dot{y}$ co-ordinates by the same does not impair equal-area status, maps of any equal-area Pseudocylindrical can be joined along any parallels to produce an equal-area composite.

Hatano [1972c] used related projections for the respective hemispheres, namely those generically specified by Equations 158b, but with different lengths of Polar parallel — specifically the North being ⅔ the Equator, the South ¾ the Equator — conditioning the projecting via Equations 158c & 158d. For the northern hemisphere the plotting equations are:

$$\dot{x} = 0.85\,R(\cos\psi)(\lambda - \lambda_0)$$
$$\dot{y} = 1.758\,59_{\sim}R\sin\psi \qquad \text{(176a}$$

with ψ the solution of

$$2\psi + \sin 2\psi - 2.675\,95_{\sim}\sin\phi = 0$$

For the southern hemisphere the plotting equations are:

$$\dot{x} = 0.85\,R(\cos\psi)(\lambda - \lambda_0)$$
$$\dot{y} = 1.930\,52_{\sim}R\sin\psi \qquad \text{(176b}$$

with ψ the solution of

$$2\psi + \sin 2\psi - 2.437\,63_{\sim}\sin\phi = 0$$

A starting value of $\psi = \phi/2$ suffices for the necessary Newton-Raphson approximations for both. The projection is illustrated on page 172.

Érdi-Krauss [1968b] and János Baranyi [1968c], both of Hungary, produced numerous Pseudocylindricals of a mixed nature — some clearly composites, others hand-crafted in various ways. The former author favoured equal-area, at least for a central band of the map, while Baranyi addressed shape without demanding conformality. Projections from both authors have been used in national atlases of their country.

Pseudocylindrical plus 'true' Cylindrical

Scope exists to juxtapose the Sinusoidal and other Pseudocylindricals with 'true' Cylindricals, including retaining equal-area. The verticality of the 'true' form is visually at odds with the curvature of the other, however, making the results inferior to various 'true' Cylindricals alone.

Blended composites

Whereas each of the above composites involves discrete juxtapositions of sectional maps each derived from one projection, the Boggs Eumorphic described in the Chapter 6 is everywhere a composite of the Sinusoidal and Mollweide, with all spacing of parallels set at the arithmetic mean of the spacing in those two (then the lengths of the parallels set to maintain equal-area). The Boggs is thus a composite map, but of a different style that we can call *blended*. It has the simple blending of a 50-50 nature, but Waldo Tobler [1973a] explores blending on a more general basis. Taking the Equal-area Cylindrical projection and Equidistant Cylindrical projection, with respective weights w and $(1-w)$, the **Tobler blended composite** projection has the spacing of parallels set by (see page 97 for source equations)

$$\dot{y} = R\{(1-w)\phi + w\sin\phi\}$$

Equal-area can then be established by ensuring orthogonal scales are inverse everywhere within the map. The scale along the meridians is

$$\frac{\mathrm{d}\dot{y}}{\mathrm{d}\phi} = R\{(1-w) + w\cos\phi\}$$

Allowing for the cosine factor, the scale along the parallels must then be

$$\frac{\mathrm{d}\dot{x}}{\mathrm{d}\lambda} = R\frac{\cos\phi}{(1-w) + w\cos\phi}$$

Hence,

$$\dot{x} = R\frac{\cos\phi}{(1-w) + w\cos\phi}\lambda$$

To make the non-point pole a specific proportion of the Equator, the equations can be subjected to reciprocal transformations, making the plotting equations

$$\dot{x} = \sigma R \frac{\cos\phi}{(1-w) + w\sin\phi}\lambda$$
$$\dot{y} = \sigma^{-1} R\{(1-w)\phi + w\sin\phi\} \qquad \text{. . . . . (178a}$$

Besides his Stereographic Equivalent (i.e., equal-area) projection (a Pseudocylindrical described on page 140), Foucault [1862a] adopted the blending technique to the Equal-area Cylindrical and Sinusoidal projections using a weighted average of the y co-ordinates then setting the x co-ordinate to achieve equal-area again. The plotting equations for the **Foucault blended composite** projection are of the form:

$$\dot{x} = R \frac{\cos\phi}{(1-w) + w\sin\phi}\lambda$$
$$\dot{y} = R\{(1-w)\phi + w\sin\phi\} \qquad \text{. . . . . (178b}$$

Alternately, the blending can be applied to the spacing of meridians, followed by derivation of spacing for parallels, e.g., by A. M. Neil [1890a] and Hammer [1900a].

Tobler also considered using the geometric mean rather than the arithmetic mean. Choosing again to start with spacing the parallels, we have:

$$\dot{y} = R \sqrt[(\alpha+\beta)]{\phi^{\beta}\sin^{\alpha}\phi} = R\left(\phi^{\beta}\sin^{\alpha}\phi\right)^{\frac{1}{\alpha+\beta}}$$
$$\dot{x} = R\cos\phi\left(\frac{\mathrm{d}\dot{y}}{\mathrm{d}\phi}\right)^{-1}(\lambda - \lambda_0)$$

Alternately, starting with spacing the meridians, we have:

$$\dot{x} = R \sqrt[(\alpha+\beta)]{\lambda^{\alpha}\lambda^{\beta}\cos^{\beta}\phi} = R\left(\lambda^{\alpha}\lambda^{\beta}\cos^{\beta}\phi\right)^{\frac{1}{\alpha+\beta}}$$
$$\dot{y} = R\int_0^{\phi}\cos\phi\left(\frac{\mathrm{d}\dot{y}}{\mathrm{d}\phi}\right)^{-1}\mathrm{d}\phi$$

In conclusion

There is limitless scope for fitting together any two projections from among the Cylindricals and Pseudocylindricals, indeed infinite scope to do so within the confines of producing an equal-area map. Except for Goode's Homolosine — developed for and used by Rand McNally in many editions of their school atlas — few have received significant acceptance. J. P. Snyder, following his coverage of Pseudocylindricals with some attention to composites [1977a], has a wide-ranging discussion of the scene in his *Flattening the Earth* [1993a]. There is also scope to fit together an infinite number of mappings using a single projection. This technique, invented for the Polyconic projection described in the next section, is discussed as the Polycylindrical projection following the presentation of that classic American projection and its derivatives.

COMPOSITE CONICAL MAPS

The challenge

Conical projections offer no equivalent to the cylindrical type projections for making composite maps, because the curvature of the parallel for any given latitude is different for each different single standard parallel, and for any pair of standard parallels when equal-area or other such feature is adopted. One can join two different projections along a common standard parallel, which has been done in several atlases (notably by the famous Bartholomew cartographic publishing family of Edinburgh). Otherwise, any two conical maps are inherently incompatible. The Bipolar Oblique Conformal Conical projection, described in Chapter 8, is an example of a composite of two that have been handcrafted to deal with the incompatibilities.

John Bartholomew joined the Equidistant Conical projection with the Bonne projection, using the latter, interrupted, for higher latitudes. The arcs forming parallels are concentric for the two projections. The Equidistant Conical projection is limited to one hemisphere, with standard parallels at 22°30′ and 67°30′. The **Kite** projection [1942a] has the Equidistant in the northern hemisphere, solely between those latitudes and Bonne both northwards and southwards; the latter projection has three lobes reaching each Pole. The **Regional** projection varies this, switching to four lobes in one atlas [1948a] then [1958b], extending the Equidistant to 80°N to embrace more northerly habitations, discarding mapping beyond that and reverting to three lobes to the south. The **Lotus** projection, designed to address oceans, inverts the process. The Equidistant covers 22°30′N and 67°30′S, the latter value being the southern limit of the map. Three lobes of the Bonne reach to the North Pole, with interruptions within the land masses (at 75°W, 15°E and 105°E).

Mixing an infinite variety

Incompatibility in the Conics can be avoided by the aggregation of an infinite collection of infinitesimally narrow Conical maps, each with its own latitude as its (single) standard parallel. Thus, every parallel is a circular arc of true scale, but concentricity of parallels, which was retained for the Pseudoconicals, is forfeited. Called *Polyconic* as a class, the unlimited range of feasible projections form a special type of composite map. In each the central meridian remains straight, but open licence exists, as in the Pseudoconicals, to bend all others in mirror image about the central meridian. Because all parallels are standard parallels (hence of true scale), the only licence is the spacing of the parallels. As a class, however, the label Polyconic is applied to any projection having non-concentric circular arcs, regardless of scale along the parallels, so the licence is much wider.

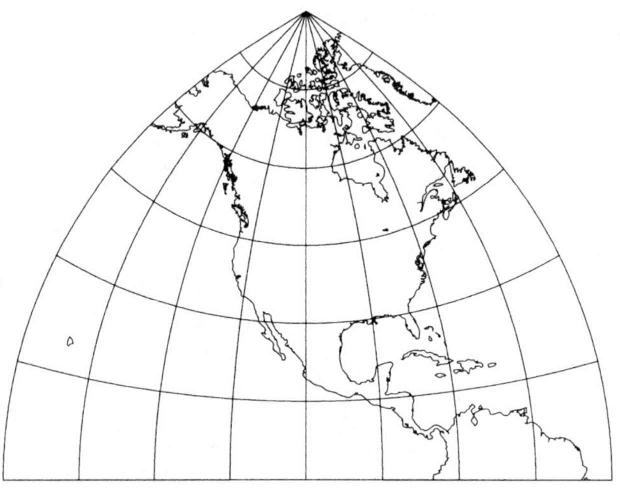

Exhibit 7–8: A regional Polyconic projection at Simple aspect.
Centred on 105°W

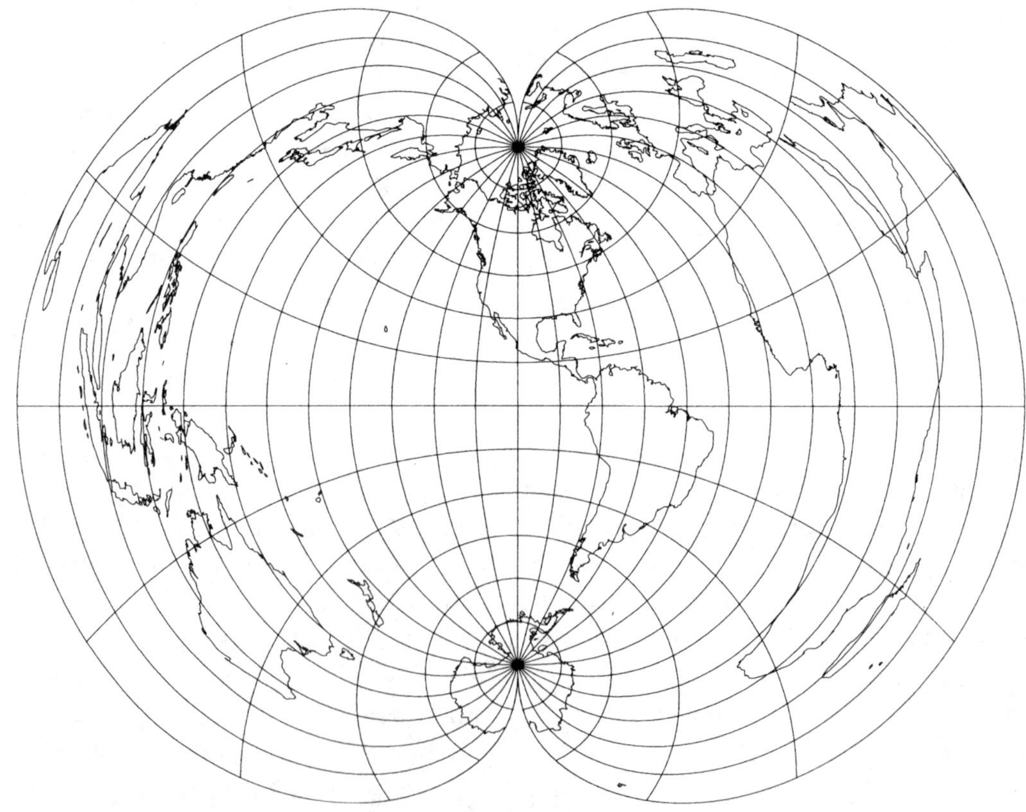

Exhibit 79: A whole-world Polyconic projection at Simple aspect.
Centred on 105°W

The Polyconic projection

One specific projection of this class is named likewise. It is based on the Equidistant, thus has true scale along the central meridian, as well as along the arcs of all parallels. This makes all other meridians curved and not equidistant along that curve. Substituting the general ϕ for the particular ϕ_0 in Equation 57d, we find that the arc for a single standard parallel at latitude ϕ has radius

$$\dot{\rho} = R\frac{\cos\phi}{c} = R\frac{\cos\phi}{\sin\phi} = R\cot\phi$$

with $c = \sin\phi$ being the constant of the corresponding cone (now not applicable uniformly over the whole map as it was previously).

There is now no obvious latitude to choose to have passing through the origin on our plot. Let ϕ_x be the choice. Then, with true scale along the central meridian (assumed to lie along the y axis) the parallel for general latitude ϕ must cut the y axis at $R(\phi - \phi_x)$. Hence, the basic plotting equations become:

$$\begin{aligned} \dot{x} &= \dot{\rho}\sin\dot{\theta} &&= R\cot\phi\sin\dot{\theta} \\ \dot{y} &= R(\phi-\phi_x) + \dot{\rho} - \dot{\rho}\cos\dot{\theta} &&= R\left\{(\phi-\phi_x) + \cot\phi\left(1-\cos\dot{\theta}\right)\right\} \end{aligned} \qquad \text{. . . . . (181a}$$

Because all parallels have true scale, we have

$$\dot{\theta} = c(\lambda - \lambda_0) = (\lambda - \lambda_0)\sin\phi$$

for every point (ϕ, λ) that is mapped. Hence, the plotting equations for the **Polyconic** projection are

$$\begin{aligned} &\text{if } \phi = 0 \quad \dot{x} = R(\lambda-\lambda_0), \quad \dot{y} = -R\phi_x \\ &\text{else} \qquad \dot{x} = R\cot\phi\sin\{(\lambda-\lambda_0)\sin\phi\} \\ &\qquad\qquad \dot{y} = R\left\{\phi - \phi_x + \cot\phi\left(1-\cos\{(\lambda-\lambda_0)\sin\phi\}\right)\right\} \end{aligned} \qquad \text{. . . . . (181b}$$

The Polyconic (called also the **American Polyconic** projection, the **Ordinary Polyconic** projection and the **American** projection) was invented by F. R. Hassler (the Swiss-born founding head of the U.S. Coast and Geodetic Survey) specifically to meet the needs for high-quality survey maps of the country [1825b]. Illustrated opposite, in the upper exhibit regionally and in the lower for the whole world, the Polyconic is neither equal-area nor conformal, but it was the standard projection for survey use in the U.S.A. until supplanted by conformal projections around 1950. It was used piecewise, exploiting the fact that distortion is trivial close to the central meridian. Used in pieces as small as 15′ longitude and latitude, the curvature of parallels and meridians is indistinguishable from straight lines at a scale of 1:50 000. Over larger expanses the curvatures become noticeable, and the distortion significant. Despite this, the U.S. military cartographic grid was based on multiple Polyconic projections until around 1960, at spans of 8° longitude between central meridians. Survey use needed incorporation of the ellipsoidal shape of

Earth, so the plotting equations used were more elaborate than shown above. The fuller versions are shown in Chapter 17. C. H. Deetz [1918c] gives an extended comparison of the Polyconic and rival projections.

The briefly used graphically adapted **Equidistant Polyconic** projection was different from the ordinary Polyconic by having the meridians of true scale [1854a].

Other polyconic projections

Equal-area projections of the polyconic style have been put forward [e.g., 1935a], as have a lot of projections using non-concentric circular arcs for parallels without them generally being standard parallels in terms of curvature and scale — projections that can be classed as polyconic in a looser sense. In the Bonne projection of Chapter 6 (and its special form, the Werner), all parallels are circular arcs of true scale, but their curvatures are not consistent with being standard parallels. The Nicolosi Globular and van der Grinten projections of Chapter 9 similarly meet the looser definition.

Also in this looser class is the **Rectangular Polyconic** projection illustrated opposite covering the Americas. It maintains the true scale along the central meridian and the curvature of the standard parallel for every latitude (showing that is true for both hemispheres on one map), but forfeits having true scale along all parallels in favour of orthogonality of meridians and parallels. While being consistent with conformality in that particular, the projection is not generally conformal. Developed and used by the British military (so sometimes called the **War Office** projection), the resulting map is barely distinguishable from the Polyconic. Originally created as an adaptation of the American Polyconic for maps of larger regions [1854a], it is covered, including for the ellipsoidal world, in a wider discussion by O. S. Adams in his *General Theory of Polyconic Map Projections* [1934c].

The mapping equations depend on the choice of latitude having true scale. If not zero, then it applies identically in both hemispheres. Whatever the choice, the Equator, being of the curvature it has as a standard parallel, is straight. Labelling the choice ϕ_0 as usual, we define an auxiliary function as

$$l = \tan\left\{\frac{(\lambda-\lambda_0)\sin\phi_0}{2}\right\}\sin^{-1}\phi_0 \quad \text{if } \phi_0 \neq 0$$

$$= \frac{\lambda-\lambda_0}{2} \quad \text{if } \phi_0 = 0 \qquad \text{. . . . . (182a}$$

which, for a specific choice of ϕ_0, is a function of relative longitude alone. Putting ϕ_x again as the latitude at which the x axis intersects the central meridian, the plotting equations for points on the straight Equator become:

$$\dot{x} = 2R\,l$$
$$\dot{y} = R\,(\phi - \phi_x) \qquad \text{(183a}$$

If $\phi > 0$ the intermediate parameters become

$$\dot{\rho} = R \cot\phi$$
$$\dot{\theta} = 2\arctan(l \sin\phi)$$

which substitute into Equation 181a to give the plotting equations of the Rectangular Polyconic projection for non-zero latitudes as:

$$\dot{x} = R \cot\phi \sin\{2\arctan(l \sin\phi)\}$$
$$\dot{y} = R\{(\phi - \phi_x) + \cot\phi(1 - \cos\{2\arctan(l \sin\phi)\})\} \qquad \text{(183b}$$

with l specified in Equations 182a.

If $\phi_0 = 0$ (i.e., the Equator is the parallel of true scale), the plotting equations for non-zero latitudes simplify to:

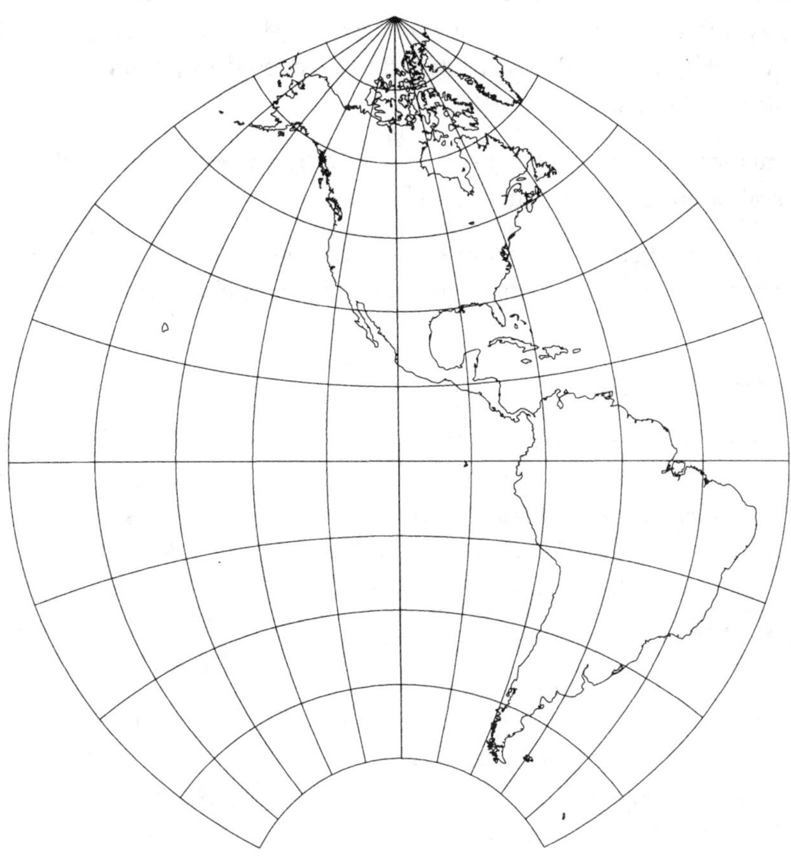

Exhibit 7–10: The Rectangular Polyconic projection at Simple aspect.
Centred on 105°W.

$$\dot{x} = R\cot\phi$$
$$\dot{y} = 2\arctan\left\{\frac{\sin\phi_0}{2}(\lambda - \lambda_0)\right\}$$

George McGaw [1921h] offered a further variant of the Rectangular Polyconic, and various other variants have been proposed and used, notably in Russia [1960a, 1995a].

The **Modified Polyconic for the International Map of the World** projection was created by Charles Lallemand [1911a] specifically for a project aimed at producing compatible low-distortion sectional maps covering much of the world at 1:1 000 000. A sectional span of six degrees of longitude and four degrees of latitude was the standard up to latitudes ±60°. All meridians are straight lines, with those one sixth in from each edge (i.e., 1° in from the edge for the prime regions) being of true scale. Each parallel had the curvature as though a standard parallel, but only the ones at the upper and lower boundaries were of true scale. This arrangement meant that any two maps adjoining laterally or vertically had a compatible common boundary. Between latitudes 60° and 76°, the longitudinal span of sections and likewise between their true-scale meridians was doubled. Beyond ±76° they were doubled again. The basic concept can be seen as continuing with the Mercator-based UTM scheme (covered in Chapter 18) that displaced this Polyconic-based scheme in the 1960s.

An equal-area Polyconic has been sought, but can be achieved only at the cost of loss of true scale along the central meridian [1927c].

Polyconics are discussed, generically, toward the close of Chapter 10 and again at the start of Chapter 15, relative to the Lambert and Lagrange projections.

From polyconic to polycylindric

Inspired by over 100 years of use of the Polyconic projection for his country, Tobler [1986a] took the relatively simple step of compounding from an infinite number of Cylindricals to create **Polycylindric** or **Polycylindrical** projections. Each projection in Tobler's novel set has every parallel straight and of true length, but their spacing a matter of choice. Thus the plotting equations for every Polycylindrical projection are of the form:

$$\dot{x} = \lambda\cos\phi \qquad \dot{y} = f(\phi) \qquad \text{. . . . . (184a}$$

for some differentiable, strictly monotonic function f of ϕ. If that function is simply $f(\phi) = \phi$ the result is the Sinusoidal projection — the one Pseudocylindrical with true scale along all parallels. Otherwise, such equations produce new projections. The function f can be that of any cylindrical projection (e.g., the $\sin\phi$ of the orthographic projecting) else any other that meets the loose constraints of the specifications (e.g., a power of ϕ).

COMPOSITE AND INTERRUPTED

Poly-centring

B. D. Dent [1987c] devised a world map by applying the (azimuthal) Orthographic projection fragmentarily to the globe, with foci in each of the two Americas, Australia, Africa (embracing Europe and southwestern Asia), and Asia generally. Various subjective approaches were made in the selection of appropriate foci, plus objective application of the centre-of-gravity method of Maling [1973a]. The motivation was the production of a world map with fair shapes for all the continental masses, but even so the areal distortion was modest due to the repeated use of the central zone of the underlying projection. Efforts were made to harmonize the adjoining regions, but the resulting map is necessarily interrupted.

Side-by-side compounding

The interrupted forms discussed at the start of this chapter have their interruptions along meridians (and hence are also referred to as re-centred); each of the various composites presented has its constituent projection serving all longitudes along a latitudinal band. D. G. Watts [1970c] turned this around using a variety of projections side-by-side, putting the interruptions along parallels. The motivation was to reduce the conspicuous angular distortion that depreciates most Pseudocylindricals, without interrupting areas of interest. The large longitudinal span of Eurasia at higher latitudes is the obvious case, but even North America presents something of a problem.

It was noted in Chapter 6 that the pseudocylindrical Sinusoidal is the limiting case of the pseudoconical Bonne projection. Both are equal-area and both have true scale along all parallels and the central meridian. Because that meridian is straight, one has full consistency of line and scale across the join line when appropriate sections of the two are juxtaposed laterally. Watts applied this technique to several matched pairs of projections, including point-polar projections. He also applied it equivalently to over a dozen pseudocylindrical projections by curving their parallels beyond the central meridian into concentric circular arcs, maintaining spacing of parallels and meridians. Using Eckert I and II, the curved sections are identically the Simple Conic and the Equal-area Conical projections. Otherwise they represent unnamed Pseudoconicals.

An interruption can be made along any parallel, and such interruptions can be used in maps that also have the familiar interruptions along meridians. While of value for world maps, this new technique has clear potential for regional maps, notably for Eurasia.

TUTORIAL 19 SPHERICAL-TRIANGLE FORMULAE

A triangle on the surface of a sphere is mathematically much more complicated than a planar triangle. For instance, the surface angles do not add up to a consistent figure, as can readily be seen by considering a triangle with one side on the Equator and the other point at a pole. It has two right angles plus the angle equal to the longitudinal span, giving a total ranging from barely over 360° to virtually 540°. There are numerous formulae relating the surface angles to the length of the sides, however, so to the angles subtended by those sides at the centre of the sphere.

Let α, β and γ be the surface angles of a triangle on the surface of a sphere of unit radius, and let *a*, *b*, and *g* be the lengths of the respective opposite sides — which are sections of great circles and thus subtend angles *a*, *b*, and *g* at the centre of the sphere, as illustrated.

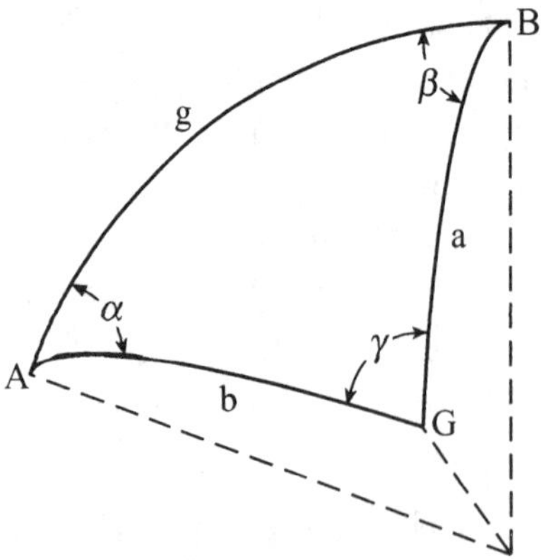

The following formulae can then be derived (the word *side* being synonymous with central angle).

Law of sines — relating all sides and surface angles:

$$\frac{\sin a}{\sin\alpha} = \frac{\sin b}{\sin\beta} = \frac{\sin g}{\sin\gamma}$$

Law of cosines for sides — each relating all sides and one surface angle:

$$\cos a = \cos\alpha \sin b \sin g + \cos b \cos g$$
$$\cos b = \cos\beta \sin c \sin a + \cos g \cos a$$
$$\cos g = \cos\gamma \sin a \sin b + \cos a \cos b$$

Law of cosines for angles — each relating all surface angles and one side:

$$\cos\alpha = \cos a \sin\beta \sin\gamma - \cos\beta\cos\gamma$$
$$\cos\beta = \cos b \sin\gamma \sin\alpha - \cos\gamma\cos\alpha$$
$$\cos\gamma = \cos g \sin\alpha \sin\beta - \cos\alpha\cos\beta$$

Law of tangents — each relating two sides and their subtended surface angles:

$$\frac{\tan\frac{1}{2}(\beta-\gamma)}{\tan\frac{1}{2}(\beta+\gamma)} = \frac{\tan\frac{1}{2}(b-g)}{\tan\frac{1}{2}(b+g)}$$
$$\frac{\tan\frac{1}{2}(\gamma-\alpha)}{\tan\frac{1}{2}(\gamma+\alpha)} = \frac{\tan\frac{1}{2}(g-a)}{\tan\frac{1}{2}(g+a)}$$
$$\frac{\tan\frac{1}{2}(\alpha-\beta)}{\tan\frac{1}{2}(\alpha+\beta)} = \frac{\tan\frac{1}{2}(a-b)}{\tan\frac{1}{2}(a+b)}$$

Four-part formulae — each relating two sides and two surface angles. Either:

$$\sin\gamma\cot\beta = \sin a \cot b - \cos a \cos\gamma$$
$$\sin\alpha\cot\gamma = \sin b \cot g - \cos b \cos\alpha$$
$$\sin\beta\cot\alpha = \sin g \cot a - \cos g \cos\beta$$

else the mirror images of these asymmetric formulae:

$$\sin\beta\cot\gamma = \sin\alpha \cot g - \cos a \cos\beta$$
$$\sin\gamma\cot\alpha = \sin b \cot a - \cos b \cos\gamma$$
$$\sin\alpha\cot\beta = \sin g \cot b - \cos g \cos\alpha$$

CHAPTER 8

Skewing the Simple: At oblique aspect

Preamble

To this point, all mapping has been done at Simple aspect — i.e., with the mapping plane symmetrical about the axis of the globe (the axial line of the globe corresponding to the rotational axis of Earth). This arrangement has made both mathematics and perception easier, but it severely limits us. For instance, because it projects all great circles through its focal point as straight lines of consistent proportional length, the Equidistant Azimuthal is of major potential value for illustrating airline routes from any airport — the literal Poles being of trivial interest in such a context. Likewise, any cylindrical projection of value in a tropical context has comparable value for any region extending similarly some distance along any other great circle. We now look into applying all our projections at any aspect, allowing the potential of each to be fully realized for any place on Earth. Any aspect other than Simple can be called *oblique,* though the term can be relegated in favour of a more particular name in some circumstances.

Using familiar algebra and geometry more deeply, this chapter turns to the complications of the spherical triangle, i.e., the triangle on the surface of a sphere.

No new projections are developed in this chapter, but there are many maps using earlier projections reoriented. The Greenwich meridian is central to the maps, with 0°E, 51°N (nominally London) the focus for maps at general Oblique aspect.

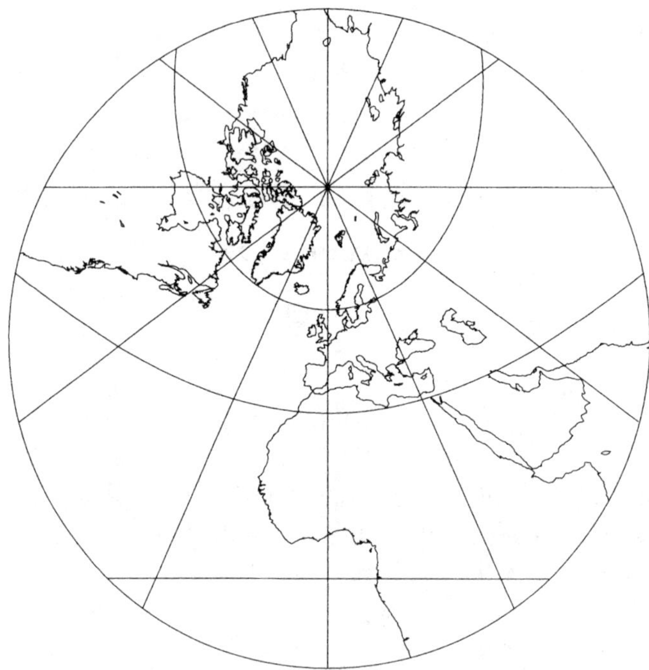

Exhibit 8–1: The Gnomonic projection at Oblique aspect.

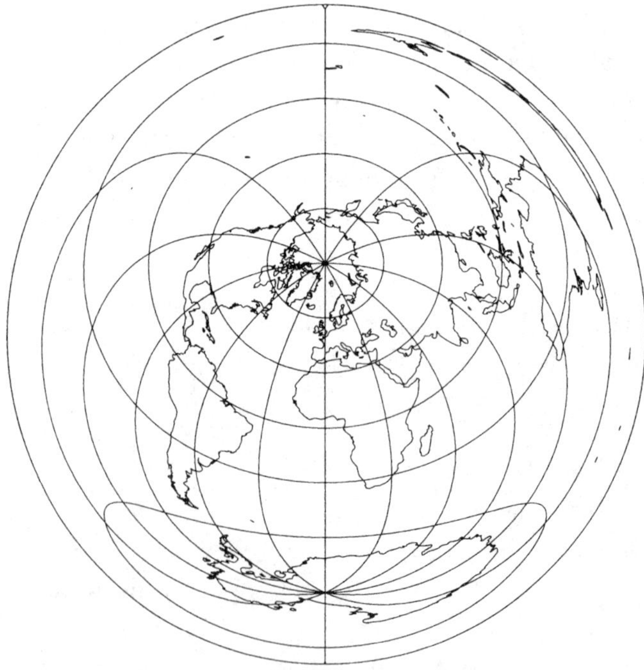

Exhibit 8–2: The Equidistant Azimuthal projection at Oblique aspect.

THE APPROACH

Customizing longitude and latitude

Simple aspect made the mathematics and perception of our preceding developments easier not directly because of Earth's rotation, but because the standard longitude/latitude scheme is based, naturally and understandably, on that rotation.

Were we addressing a non-spherical globe, as we do later, any diameter line of the globe we use as axis would be pinned by the shape of the globe. With a spherical globe, however, the axis could with impunity be any diameter; the mathematics would be unchanged — assuming longitude and latitude are correspondingly defined. Meridians and parallels are not etched on Earth; they are arbitrary human creations, on paper. So, any of our projections could be applied to any point as pole, and to any appropriate circles on the globe's surface as parallels. However, of course, only in rare instances would the resulting arbitrary parallels and meridians be wanted on the map. For the Equidistant Azimuthal they would be valuable in the sense of such parallels being points equidistant from the centre and the meridians direct routes, but not to be confused as latitude and longitude. A map using a magnetic pole as the focus of projection could, perhaps, have its own parallels and meridians. Normal maps need notation and representation of longitude and latitude to accord with conventional practice, and generally can be produced only from conventional longitude and latitude figures.

While plotting equations can be devised in terms of conventional longitude and latitude for projections at any aspect, it is more straightforward and efficient, in development and application, to see the general aspect addressed as a two-stage process — a transformation from conventional to customized longitude and latitude then plotting by the established means. That is, we adopt our own longitude and latitude, which we call *auxiliary longitude* and *auxiliary latitude* (also *meta-longitude* and *meta-latitude* in more academic circles), putting our particular point of interest at the auxiliary pole etc., and applying suitable transformational formulae to obtain these auxiliary values, for both our geographical details and for the conventional parallels and meridians. We then apply the established plotting equations to our auxiliary values. We need to establish those formulae, universally true (for a sphere) regardless of projection. The rest has already been done, in the form of our various different projections (with more to come).

The term *oblique* applies to this shifting from Simple aspect, but is usually replaced by Equatorial or Transverse for an Azimuthal projection that has a point on the Equator as its central point, and by Transverse or Polar for a Cylindrical at right angles to Simple aspect (i.e., having its axis in the Equatorial Plane rather than along the natural axis of the globe — hence its centre line a bimeridian). The new term is used here capitalized for the aspect exclusive of those particular circumstances.

There is no special geographical advantage for the Equatorial over other oblique Azimuthals, but the Transverse Mercator, used piecewise, has become the cornerstone for worldwide mapping, as is discussed in Chapter 18. Conicals, too, can be set at Oblique aspect, but they are rarely so. The advantageous semi-realism of their curved parallels for mid latitudes is lost for little offsetting gain if the Simple aspect is discarded. However, the transformational mathematics applies regardless of projection, so covers all three modes equally.

Whether an Azimuthal, Cylindrical else Conical, any oblique aspect for a projection involves the relocation of the poles. For an Azimuthal focussed on a given point, the pole is relocated to that point. For a Cylindrical and likewise a Conical, the relocation must be to an indirect point implied by our focus requirements. We deal with this particular later, assuming at this stage that the point of relocation is known.

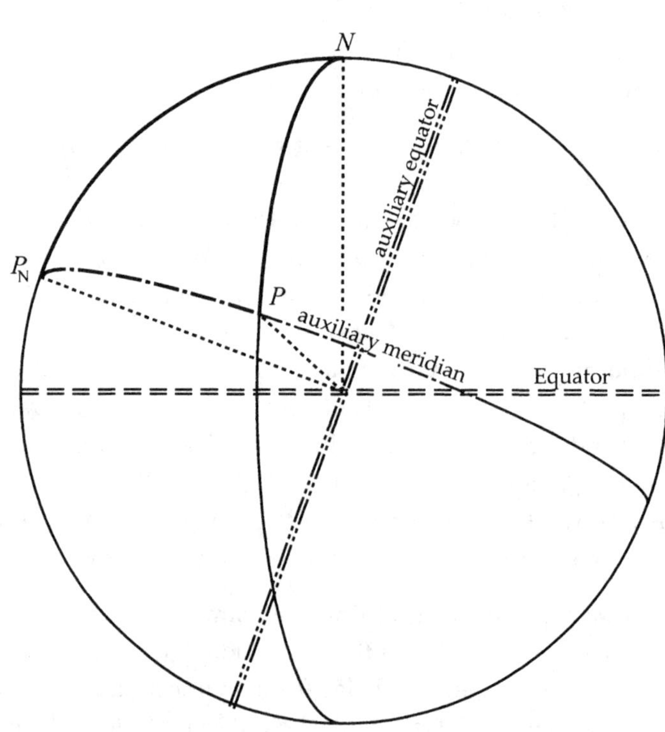

Exhibit 8–3: The Greenwich and auxiliary graticules.

RELOCATING THE POLES

Methodology

The essence of our methodology is to apply established projections as though our focal point is a Pole, but then to connect the resulting equations to the standard scheme via the Greenwich longitude and latitude of the relocation point.

Let $P_N = (\Lambda_N, \Phi_N)$, expressed in conventional longitude and latitude terms, be our relocation point which, for initial convenience, we assume is in the Northern hemisphere. We consider two graticules — the standard one, which we can refer to as Greenwich regardless of the central meridian, and an *auxiliary* graticule with P_N as its northern pole. The diagram opposite illustrates the scene.

We require the algebra connecting the two graticules, which can be developed by studying the moving from one to the other. Moving the standard graticule to make P_N the pole of the auxiliary graticule can be achieved by rotating the graticule, by the angle equal to the polar distance of P_N, about an axis transverse to the meridional plane of that point. This can be effected easily, by sliding the true Pole down the meridian containing P_N by the amount subtending $(90-\Phi_N)°$ at the centre of the globe. This has the simple result of making the auxiliary latitude of our specified point 90°, but has a complicated impact on other points. In the diagram we choose to put this crucial standard meridian on the extreme periphery. Labelling our auxiliary longitude and latitude by the familiar symbols augmented by the prefixed suffix $_a$, our challenge is to establish for general point $P = (\Lambda, \Phi)$ its auxiliary co-ordinates $_a\Lambda$ and $_a\Phi$ in terms of Λ, Φ, Λ_N and Φ_N (actually the equivalent in radians).

Development of our transformational formulae depends on three standard formulae of spherical geometry shown in Tutorial 19 (page 186) — the Law of cosines, the Law of sines, and the Four-part formula. These deal essentially with angles, but involve the length of the sides of the triangle on the spherical surface; hence, they involve the radius of the sphere it is on. Since the geometry is independent of the size of the sphere, one can choose a unit sphere, making the length of a side identically the radians of angle that it subtends at the centre of the sphere. (As explained in Tutorial 19, the three angles of a spherical triangle add up to a variable amount — a minimum of 180°, a maximum of 540°, as can be seen if we consider any triangle based on the Equator with its apex at a Pole. So, unlike with the planar triangle, knowing two angles does not give us the third. The excess over the minimum 180° for any spherical triangle is the same proportion of the maximal excess of 360° as is the area of the triangle to that of a hemisphere.)

The diagram opposite illustrates the geometry for the globe, with the true North Pole at N, the auxiliary pole at P_N, and a general point P. The auxiliary pole is deliberately placed on the bimeridian forming the circular perimeter, which, because it then passes through all four poles, is called the *panmeridian*. The Greenwich meridian through P is shown as a continuous line and the true Equator as a double dashed line. Lines having dots and dashes show the auxiliary equator and the auxiliary meridian through P. With both poles positioned on it, the circular perimeter represents the great circle through them and carries the meridian of each in the scheme based on the other. The spherical triangle with corners at N, P_N, and P, illustrated magnified and in isolation below, is key to our development. Looking to use those formulae of Tutorial 19, we establish the lengths of the sides and the angles between them in the two schemes — Greenwich and auxiliary.

From the Greenwich scheme with points N, $P=(\lambda,\phi)$ and $P_N=(\lambda_N,\phi_N)$, we have:

$$\text{length of side } NP = \tfrac{\pi}{2} - \phi$$

$$\text{length of side } NP_N = \tfrac{\pi}{2} - \phi_N$$

with $P=({}_a\lambda, {}_a\phi)$ within the auxiliary scheme, which has the pole at P_N,

$$\text{length of side } PP_N = \tfrac{\pi}{2} - {}_a\phi$$

Since the surface angle between sides NP and NP_N is the difference in Greenwich longitude from P_N to P, i.e., equals $(\lambda-\lambda_N)$, the Law of cosines for the sides about this angle gives

$$\cos\left(\tfrac{\pi}{2} - {}_a\phi\right) = \cos\left(\tfrac{\pi}{2} - \phi_N\right)\cos\left(\tfrac{\pi}{2} - \phi\right) + \sin\left(\tfrac{\pi}{2} - \phi_N\right)\sin\left(\tfrac{\pi}{2} - \phi\right)\cos\left(\lambda - \lambda_N\right)$$

or, swapping sines of angles for cosines of their co-angles and vice versa,

$$\begin{aligned} \sin {}_a\phi &= \sin\phi_N \sin\phi + \cos\phi_N \cos\phi \cos(\lambda - \lambda_N) \\ &= \cos\phi_N \cos\phi\{\tan\phi_N \tan\phi + \cos(\lambda - \lambda_N)\} \end{aligned} \quad \text{. . . . . (192a}$$

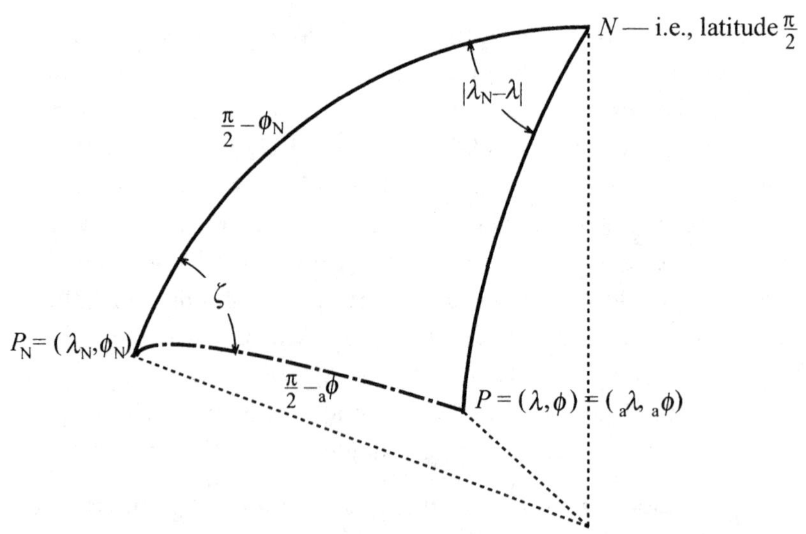

Exhibit 8–4: The spherical geometry of the longitude and latitude scheme.

The surface angle between sides NP_N and PP_N is the azimuth of P at P_N. If we denote this by ζ, then likewise we get

$$\cos\left(\tfrac{\pi}{2}-\phi\right)=\cos\left(\tfrac{\pi}{2}-\phi_N\right)\cos\left(\tfrac{\pi}{2}-{}_a\phi\right)+\sin\left(\tfrac{\pi}{2}-\phi_N\right)\sin\left(\tfrac{\pi}{2}-{}_a\phi\right)\cos\zeta$$

so

$$\sin\phi=\sin\phi_N\sin{}_a\phi+\cos\phi_N\cos{}_a\phi\cos\zeta \qquad \text{(193a}$$

Were we, within our auxiliary scheme, to adopt the meridian through N as zero longitude, ζ would equal the negative of auxiliary longitude. We opt instead, for convenience in developing the trigonometric formulae, to have the zero meridian pass a multiple of 90° offset from that other, but do it differently for azimuthal versus cylindrical and conical projections. We pursue these details under the respective headings.

Applying the Law of sines to the two sides for which we know their subtended surface angles gives

$$\frac{\sin\zeta}{\sin\left(\frac{\pi}{2}-\phi\right)}=\frac{\sin(\lambda-\lambda_N)}{\sin\left(\frac{\pi}{2}-{}_a\phi\right)}$$

or

$$\cos{}_a\phi\sin\zeta=\cos\phi\sin(\lambda-\lambda_N) \qquad \text{(193b}$$

The more complicated Four-part formula, which relates two surface angles to two sides, asymmetrically, provides a means for expressing longitude without embedding the associated latitude. Applying it in mirror-image forms to our triangle gives:

$$\cot\zeta\sin(\lambda-\lambda_N)=\sin\left(\tfrac{\pi}{2}-\phi_N\right)\cot\left(\tfrac{\pi}{2}-\phi\right)-\cos\left(\tfrac{\pi}{2}-\phi_N\right)\cos(\lambda-\lambda_N)$$

$$\cot(\lambda-\lambda_N)\sin\zeta=\sin\left(\tfrac{\pi}{2}-\phi_N\right)\cot\left(\tfrac{\pi}{2}-{}_a\phi\right)-\cos\left(\tfrac{\pi}{2}-\phi_N\right)\cos\zeta$$

Hence:

$$\cot\zeta=\frac{\cos\phi_N\tan\phi-\sin\phi_N\cos(\lambda-\lambda_N)}{\sin(\lambda-\lambda_N)}=\frac{\cos\phi_N\sin\phi-\sin\phi_N\cos\phi\cos(\lambda-\lambda_N)}{\cos\phi\sin(\lambda-\lambda_N)} \qquad \text{(193c}$$

$$\cot(\lambda-\lambda_N)=\frac{\cos\phi_N\tan{}_a\phi-\sin\phi_N\cos\zeta}{\sin\zeta}=\frac{\cos\phi_N\sin{}_a\phi-\sin\phi_N\cos{}_a\phi\cos\zeta}{\cos{}_a\phi\sin\zeta} \qquad \text{(193d}$$

providing the longitude in each system without involving its associated latitude — with ζ being a proxy for ${}_a\lambda$. Inversion of these gives:

$$\tan\zeta=\frac{\sin(\lambda-\lambda_N)}{\cos\phi_N\tan\phi-\sin\phi_N\cos(\lambda-\lambda_N)}=\frac{\cos\phi\sin(\lambda-\lambda_N)}{\cos\phi_N\sin\phi-\sin\phi_N\cos\phi\cos(\lambda-\lambda_N)}$$

$$\tan(\lambda-\lambda_N)=\frac{\sin\zeta}{\cos\phi_N\tan{}_a\phi-\sin\phi_N\cos\zeta}=\frac{\cos{}_a\phi\sin\zeta}{\cos\phi_N\sin{}_a\phi-\sin\phi_N\cos{}_a\phi\cos\zeta}$$

Using Equation 193b to extract the sine component of the left-hand tan terms gives:

$$\cos\zeta = \cot\zeta\,\sin\zeta \;=\; \frac{\cos\phi_N \sin\phi - \sin\phi_N \cos\phi \cos(\lambda-\lambda_N)}{\cos_a\phi} \quad \ldots \text{(194a}$$

$$\cos(\lambda-\lambda_N) = \cot(\lambda-\lambda_N)\sin(\lambda-\lambda_N) = \frac{\cos\phi_N \sin_a\phi - \sin\phi_N \cos_a\phi \cos_a\lambda}{\cos\phi} \quad \ldots \text{(194b}$$

completing the full repertoire of basic trigonometric functions for ζ and $(\lambda-\lambda_N)$.

Allowing that ζ is just a fixed offset from ${}_a\lambda$, we thus have the equations to transform either way between our two schemes of longitude and latitude. For conventional to auxiliary, we use Equation 192a and get

$${}_a\phi = \arcsin\{\sin\phi_N \sin\phi + \cos\phi_N \cos\phi \cos(\lambda-\lambda_N)\} \quad \ldots\ldots\ldots \text{(194c}$$

(which provides great-circle distance — see later in this chapter), while from Equation 194a we get

$$\begin{aligned}\zeta &= \operatorname{arccot}\left\{\frac{\cos\phi_N \sin\phi - \sin\phi_N \cos\phi \cos(\lambda-\lambda_N)}{\cos\phi \sin(\lambda-\lambda_N)}\right\} \\ &= \operatorname{arccot}\left\{\frac{\cos\phi_N \tan\phi - \sin\phi_N \cos(\lambda-\lambda_N)}{\sin(\lambda-\lambda_N)}\right\} \\ &= \arctan\left\{\frac{\cos\phi \sin(\lambda-\lambda_N)}{\cos\phi_N \sin\phi - \sin\phi_N \cos\phi \cos(\lambda-\lambda_N)}\right\} \\ &= \arctan\left\{\frac{\sin(\lambda-\lambda_N)}{\cos\phi_N \tan\phi - \sin\phi_N \cos(\lambda-\lambda_N)}\right\}\end{aligned} \quad \ldots\ldots\ldots \text{(194d}$$

The transformation from auxiliary to conventional can be obtained likewise from Equations 193d & 194b.

The trigonometric formulae apply wherever on the globe the general point is positioned. The only problem is selecting the specific angular quadrants in converting from the trigonometric solutions to actual angles. Equation 194c always yields a unique result because $-\pi/2 < {}_a\phi < +\pi/2$. In general, Equation 194d will give two possible values for ζ, such as $+\pi/4$ and $-3\pi/4$. This can be resolved by determining $\cos\zeta$ from Equation 194a. If $\cos\zeta$ has the same sign as $\tan\zeta$ then $\cos\zeta \geq 0$ else $\cos\zeta < 0$. (In a computing context, using the arctan function can give inaccurate results when $\zeta \approx \pm\pi/2$, but Equation 194a provides an alternative route to the proper value of ζ.)

For the *transverse form of oblique projecting,* putting the auxiliary pole on the Equator making $\phi_N = 0$ hence $\sin\phi_N = 0$, $\cos\phi_N = 1$, the transformation equations simplify to:

$${}_a\phi = \arcsin\{\cos\phi \cos(\lambda-\lambda_N)\} \quad \ldots\ldots\ldots \text{(194e}$$

$$\zeta = \operatorname{arccot}\left\{\frac{\sin\phi}{\cos\phi \sin(\lambda-\lambda_N)}\right\} = \arctan\{\cot\phi \sin(\lambda-\lambda_N)\} \quad \ldots\ldots\ldots \text{(194f}$$

APPLYING THE TRANSFORMATION

Methodology

Our auxiliary scheme provides the means to regard any Oblique aspect as effectively Simple aspect. We must first establish the point $P_N = (\Lambda_N, \Phi_N)$ that is to be the auxiliary pole — which is easy for an Azimuthal map, but not so easy for the other modes (which are discussed further below). Conditional on choosing a position for the auxiliary zero meridian, hence the offset of ζ from $_a\lambda$, Equations 194c & 194d provide for transforming conventional into auxiliary longitude and latitude, for all points of concern. The plotting equations for the desired projection at Simple aspect are then interpreted as being in auxiliary longitude and latitude and applied to the transformed values.

This can be a two-stage process, with the numeric auxiliary longitude and latitude values derived for every point. Alternatively, we can incorporate the algebraic transformation formulae directly into the Simple plotting equations. While minimizing arithmetic steps, this complicates, sometimes greatly, the plotting equations. Consolidation produces plotting equations that serve the Simple aspect too, so nothing is lost. As with the original plotting equations, where complexity sometimes contra-indicated consolidation of intermediate variables into generic plotting equations, we find cases favouring each approach regarding these transformations. In some cases, the general Oblique is too complicated for consolidation but transverse versions are not.

Applying the transformation to Azimuthals

The formulae for oblique transformation were used in Chapter 2 to obtain the general plotting equations for the inherently oblique Aerial projection. While there is a use for most Azimuthals at Simple aspect, and such merits as the universal straight-line depiction of great circles on the Gnomonic apply to the limits of the map, the escalating distortion away favours oblique focusing generally. For the Gnomonic, Stereographic and Equal-area Azimuthals, such focussing can be regional, but the benefits of correct azimuths and radial distances from the focus prompt specific focusing for the Equidistant. The Equidistant map in Exhibit 8–2 on page 188, centred on London, shows well the gross distortion and discouraging unfamiliarity of shapes in further parts that must be suffered to achieve those benefits. The azimuthal Stereographic, illustrated at Oblique and Equatorial aspects overleaf, is the preferred projection for much aerial navigation, its radial pattern of distortion suiting even square maps more than its conformal cylindrical alternative — the Mercator — which suffers distortion orthogonally to a central line.

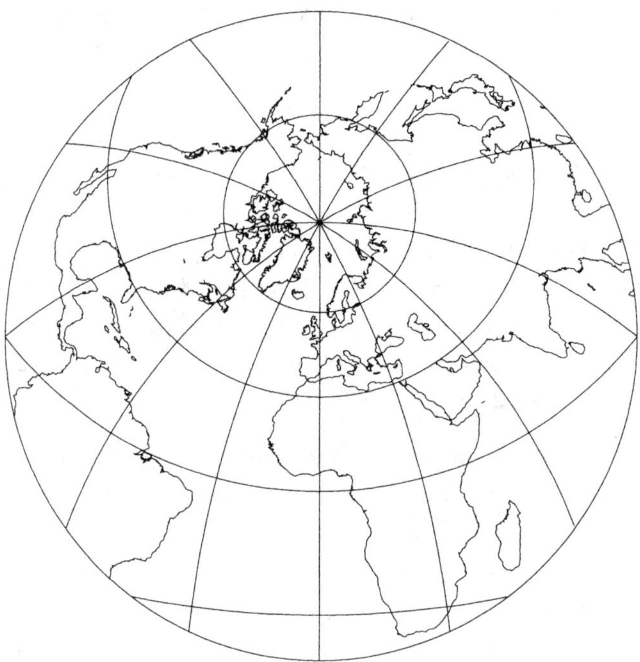

Exhibit 8–5: The Stereographic projection at Oblique aspect.

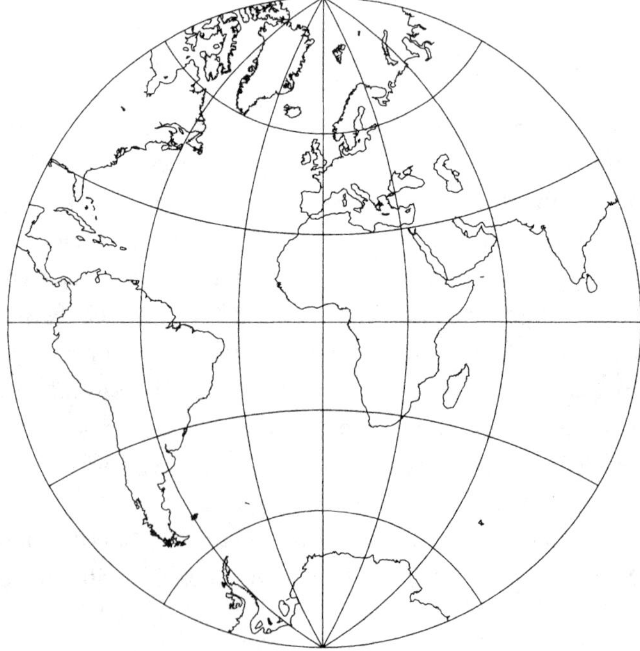

Exhibit 8–6: The Stereographic projection at Equatorial aspect.

As the point $P_N = (\lambda_N, \phi_N)$ that is our auxiliary pole is inherently the central point of our intended oblique azimuthal map, it is reasonable to have the true meridian of that point along the *y* axis. For the map to have true North upwards we require the auxiliary meridian through *N* to be on the +*y* half-axis. That meridian is, of course, the meridian from which ζ was measured (clockwise), so our transformation equations apply with $\zeta = -_a\lambda$ to measure longitude counter-clockwise. Hence (see Tutorial 10, page 38), we get:

$$\sin\zeta = -\sin_a\lambda, \quad \cos\zeta = \cos_a\lambda, \quad \tan\zeta = -\tan_a\lambda, \quad \cot\zeta = -\cot_a\lambda$$

so Equations 193b, 193c & 194a become:

$$-\sin_a\lambda = \frac{\cos\phi\,\sin(\lambda-\lambda_N)}{\cos_a\phi} \qquad \text{(197a}$$

$$-\cot_a\lambda = \frac{\cos\phi_N\,\sin\phi - \sin\phi_N\,\cos\phi\,\cos(\lambda-\lambda_N)}{\cos\phi\,\sin(\lambda-\lambda_N)} \qquad \text{(197b}$$

$$\cos_a\lambda = \frac{\cos\phi_N\,\sin\phi - \sin\phi_N\,\cos\phi\,\cos(\lambda-\lambda_N)}{\cos_a\phi}$$

respectively.

Adapting Equations 30b (for the North Pole) to accommodate $_a\lambda$ being relative to the +*y* half-axis, then substituting from the three equations immediately above, we get the following as basic generic plotting equations for all Azimuthals:

$$\dot{x} = \dot{\rho}\cos\theta = \dot{\rho}\cos\left({}_a\lambda + \tfrac{\pi}{2}\right) = -\dot{\rho}\sin_a\lambda = \dot{\rho}\frac{\cos\phi\,\sin(\lambda-\lambda_N)}{\cos_a\phi}$$

$$\dot{y} = \dot{\rho}\sin\theta = \dot{\rho}\sin\left({}_a\lambda + \tfrac{\pi}{2}\right) = +\dot{\rho}\cos_a\lambda = \dot{\rho}\frac{\cos\phi_N\,\sin\phi - \sin\phi_N\,\cos\phi\,\cos(\lambda-\lambda_N)}{\cos_a\phi} \qquad \text{(197c}$$

At Transverse or Equatorial aspect, with $\sin\phi_N = 0$, $\cos\phi_N = 1$, these become:

$$\dot{x} = \dot{\rho}\frac{\cos\phi\,\sin(\lambda-\lambda_N)}{\cos_a\phi}$$

$$\dot{y} = \dot{\rho}\frac{\sin\phi}{\cos_a\phi} \qquad \text{(197d}$$

The linear intermediate parameter is specific to each projection, but in all Azimuthals at Simple aspect it involves the latitude solely, via the polar distance. Previously we converted from the latter to the former to express our formulae for this parameter; here we can find it convenient in places to retain the original angle. From Equation 194c, we get

$$\tfrac{\pi}{2} - {}_a\phi = \arccos\{\sin\phi_N\,\sin\phi + \cos\phi_N\,\cos\phi\,\cos(\lambda-\lambda_N)\} \qquad \text{(197e}$$

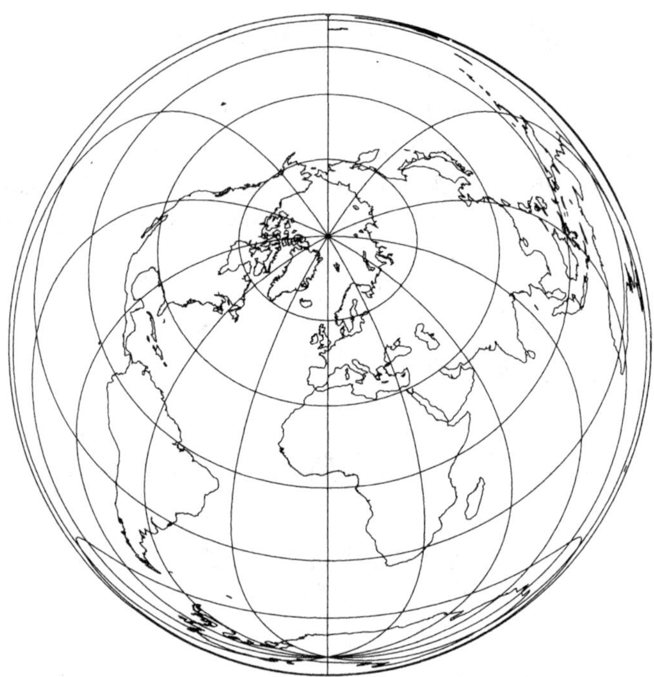

Exhibit 8–7: The (Lambert) Equal-area Azimuthal projection at Oblique aspect.

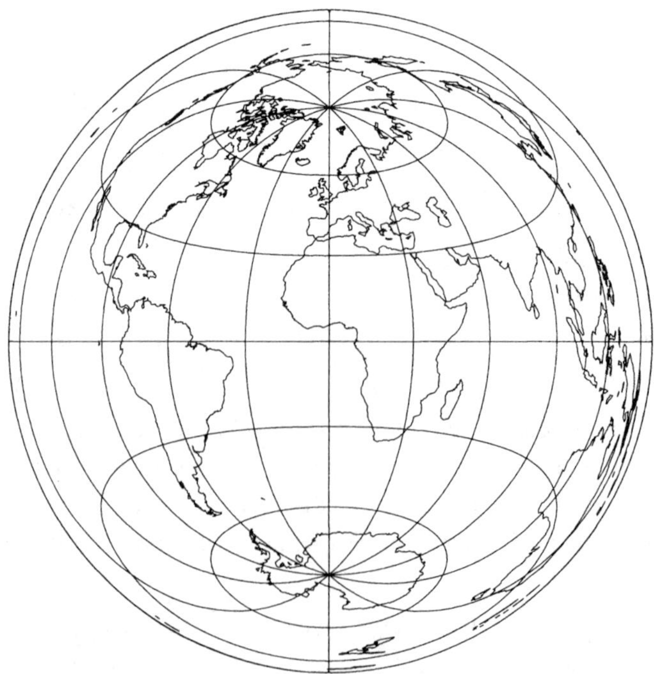

Exhibit 8–8: The (Lambert) Equal-area Azimuthal projection at Equatorial aspect.

For the Gnomonic or Central (Azimuthal) projection (which is inherently restricted to less than a hemisphere), at Simple aspect in the auxiliary scheme we have

$$\dot{\rho} = \sigma R \cot{}_{a}\phi \quad = \sigma R \frac{\cos_{a}\phi}{\sin_{a}\phi}$$

from Equation 45a. Substituting this for ρ in the penultimate forms of Equations 197c gives:

$$\dot{x} = -\sigma R \frac{\cos_{a}\phi}{\sin_{a}\phi} \sin{}_{a}\lambda \quad = -\sigma R \frac{\cos_{a}\phi \sin_{a}\lambda}{\sin_{a}\phi}$$

$$\dot{y} = +\sigma R \frac{\cos_{a}\phi}{\sin_{a}\phi} \cos{}_{a}\lambda \quad = +\sigma R \frac{\cos_{a}\phi \sin{}_{a}\lambda \cot{}_{a}\lambda}{\sin_{a}\phi}$$

for the general oblique situation. Translating these into conventional longitude and latitude using Equation 197a for the initial pair of terms in the numerator, Equation 197b for the cot term, and Equation 192a for the denominator, gives plotting equations for the Gnomonic projection at Oblique aspect:

$$\dot{x} = \sigma R \frac{\cos\phi \sin(\lambda - \lambda_{N})}{\sin\phi_{N} \sin\phi + \cos\phi_{N} \cos\phi \cos(\lambda - \lambda_{N})}$$

$$\dot{y} = \sigma R \frac{\cos\phi_{N} \sin\phi - \sin\phi_{N} \cos\phi \cos(\lambda - \lambda_{N})}{\sin\phi_{N} \sin\phi + \cos\phi_{N} \cos\phi \cos(\lambda - \lambda_{N})} \qquad \text{(199a}$$

assuming the denominator is strictly positive — i.e., that ${}_{a}\phi$ is greater than zero.

For the General Perspective Azimuthal projection from Equation 50b we have, at Simple aspect in the auxiliary scheme,

$$\dot{\rho} = \frac{\sigma R \cos_{a}\phi}{1 - h^{-1}(1 - \sin_{a}\phi)}$$

Substituting in the final forms of Equations 197c gives plotting equations for the General Perspective Azimuthal at Oblique aspect:

$$\dot{x} = H \cos\phi \sin(\lambda - \lambda_{N})$$

$$\dot{y} = H \{\cos\phi_{N} \sin\phi - \sin\phi_{N} \cos\phi \cos(\lambda - \lambda_{N})\} \qquad \text{(199b}$$

where $$H = \frac{R}{1 - h^{-1}(1 - \sin_{a}\phi)}$$

with the $\sin_{a}\phi$ term, expressed in Equation 192a, being retained for economy.

For the Stereographic projection (with $h = 2$), using Equation 192a gives

$$H = \frac{\sigma R}{1 - \frac{1}{2}(1 - \sin_{a}\phi)} = \frac{2\sigma R}{1 + \sin_{a}\phi} = \frac{2\sigma R}{1 + \cos\phi_{N} \cos\phi\{\tan\phi_{N} \tan\phi + \cos(\lambda - \lambda_{N})\}}$$

$$= \frac{2\sigma R \sec\phi_{N} \sec\phi}{\sec\phi_{N} \sec\phi + \tan\phi_{N} \tan\phi + \cos(\lambda - \lambda_{N}))}$$

TUTORIAL 20 MATRIX NOTATION

In mathematics, a *matrix* (plural: *matrices*) is just a rectangular array of elements. Using bold upright capital Roman letters for denoting them, examples of matrices are:

$$\mathbf{G} = \begin{bmatrix} a & b \\ c & d \end{bmatrix}, \quad \mathbf{H} = \begin{bmatrix} a_{11} & a_{12} & a_{13} \\ a_{21} & a_{22} & a_{23} \end{bmatrix}$$

$$\mathbf{L} = \begin{bmatrix} 1 & 7 & 3 \\ 2 & 2 & 5 \\ 5 & 0 & 1 \end{bmatrix}, \quad \mathbf{V_1} = \begin{bmatrix} x & y & z \end{bmatrix}, \quad \mathbf{V_2} = \begin{bmatrix} x \\ y \\ z \end{bmatrix}$$

The number of rows followed by the number of columns represents the dimension of each; thus the matrix **H** is referred to as a 2x3 matrix. As with vectors, the dimensions often fit the 2- and 3-dimensional world but can be of any size. The style of **H** is typical, with a subscript that shows row then column number, without an intervening comma. Matrices are generically of the form

$$\mathbf{M} = \begin{bmatrix} a_{11} & a_{12} & . & . & a_{1n} \\ a_{21} & a_{22} & . & . & a_{2n} \\ . & . & . & . & . \\ . & . & . & . & . \\ a_{m1} & a_{m2} & . & . & a_{mn} \end{bmatrix}$$

or, written compactly,

$$\mathbf{M} = \left[\left\{a_{ij}\right\}\right] \quad \text{for} \quad i = 1,2,\ldots m; j = 1,2\ldots n$$

The matrix that has its columns identical with the rows of another is known as the *transpose* of that other; reciprocally, that other is the transpose of the first. The transpose of matrix **M** is denoted by $\mathbf{M}^{\mathrm{T}}$.

The *determinant* of a matrix, denoted by vertical bars about the array or its symbol, is a simple derived scalar quantity. The rigorous definition of determinants is more demanding than required for this work. For 2x2 arrays the definition is

$$\begin{vmatrix} a_{11} & a_{12} \\ a_{21} & a_{22} \end{vmatrix} = a_{11}a_{22} - a_{21}a_{12}$$

The definition for 3x3 arrays is shown at the foot of this page. If a determinant equals zero, the matrix is termed *singular*; otherwise it is *non-singular*.

A matrix with its two dimensions equal is described as a *square matrix*. A square matrix with zeros everywhere except on the prime diagonal is called a *diagonal matrix*; if every element on the diagonal is 1, it is the *unit matrix*.

Multiplication of a matrix by a scalar yields a matrix with the same dimensions, its elements the scalar multiple of the original elements. Multiplication of two matrices is covered in Tutorial 24.

$$\begin{vmatrix} a_{11} & a_{12} & a_{13} \\ a_{21} & a_{22} & a_{23} \\ a_{31} & a_{32} & a_{33} \end{vmatrix} = \left(a_{11}a_{22}a_{33} + a_{12}a_{23}a_{31} + a_{13}a_{21}a_{32}\right) - \left(a_{11}a_{23}a_{32} + a_{12}a_{21}a_{33} + a_{13}a_{22}a_{31}\right)$$

so the plotting equations for the Stereographic projection at Oblique aspect become:

$$\dot{x} = \frac{2\sigma R \sec\phi_N \sin(\lambda - \lambda_N)}{\sec\phi_N \sec\phi + \tan\phi_N \tan\phi + \cos(\lambda - \lambda_N)}$$
$$\dot{y} = \frac{2\sigma R \{\tan\phi - \tan\phi_N \cos(\lambda - \lambda_N)\}}{\sec\phi_N \sec\phi + \tan\phi_N \tan\phi + \cos(\lambda - \lambda_N)} \qquad \text{. . . . . (201a}$$

At Equatorial aspect, with $\phi_N = 0$, the $\sin_a\phi$ term simplifies to

$$\sin_a\phi = \cos\phi \cos(\lambda - \lambda_N)$$

and the plotting equations for the General Perspective Azimuthal at Equatorial aspect become:

$$\dot{x} = H \cos\phi \sin(\lambda - \lambda_N)$$
$$\dot{y} = H \sin\phi \qquad \text{. . . . . (201b}$$

where $$H = \frac{\sigma R}{1 - h^{-1}\{1 - \cos\phi \cos(\lambda - \lambda_N)\}} \qquad \text{. . . . . (201c}$$

Substituting appropriate values for h into Equation 201c then for H in Equations 199b, we get:

for the Orthographic projection at Equatorial aspect, with $h = \infty$

$$H = \frac{\sigma R}{1 - 0} = \sigma R$$

for the Stereographic projection at Equatorial aspect, with $h = 2$

$$H = \frac{2\sigma R}{2 - 1 + \sin_a\phi} = \frac{2\sigma R}{1 + \cos\phi \cos(\lambda - \lambda_N)} \qquad \text{. . . . . (201d}$$

for the Central or Gnomonic projection at Equatorial aspect, with $h = 1$

$$H = \frac{\sigma R}{1 - (1 - \sin_a\phi)} = \frac{\sigma R}{1 - 1 - \cos\phi \cos(\lambda - \lambda_N)} = \frac{-\sigma R}{\cos\phi \cos(\lambda - \lambda_N))}$$

Being the conformal azimuthal projection, and maintaining that conformality at every aspect, the Stereographic projection is widely used in oblique form. (This oblique form was attributed as the **Werner IV** projection in 1514, although used long before that.) Following its inclusion in the Mercator Atlas of 1587 for separate maps of the eastern and western hemispheres (compiled by the famous man's son Rumold), this Stereographic was habitually used in such manner for 200 years. The Stereographic is also often used for maps recording the path of eclipses and, at Equatorial aspect, to show Earth as two round hemispheres — one is shown as Exhibit 8–6 on page 196 Though the modesty of its escalation of radial spacing is of obvious value in such usage, it does not have the uniformity of scale along the Equator possessed by the Equidistant Azimuthal projection, and required for the globular projections of Chapter 9. It does, however, have the

TUTORIAL 21 VECTOR MULTIPLICATION

Vectors were defined and their multiplication introduced in Tutorial 4.

If $\mathbf{p}_1 = (X_1, Y_1, Z_1)$ and $\mathbf{p}_2 = (X_2, Y_2, Z_2)$ are two 3-dimensional vectors the *dot-product* of the two was defined as

$$\mathbf{p}_1 \cdot \mathbf{p}_2 = X_1X_2 + Y_1Y_2 + Z_1Z_2 = |\mathbf{p}_1||\mathbf{p}_1| \cos\delta$$

where δ is the (lesser) angle between the two in their common plane. It follows from this that the dot-product of two orthogonal vectors is zero, and the dot-product of a vector with itself equals the square of the length of that vector, i.e.,

$$\mathbf{p}_1 \cdot \mathbf{p}_1 = X_1X_1 + Y_1Y_1 + Z_1Z_1 = |\mathbf{p}_1|^2$$

The ordinary arithmetic properties hold for dot-products, namely,

$$\mathbf{p}_1 \cdot \mathbf{p}_2 = \mathbf{p}_2 \cdot \mathbf{p}_1$$

and, if $\mathbf{p}_3 = (X_3, Y_3, Z_3)$ is a third vector:

$$\mathbf{p}_1 \cdot (\mathbf{p}_2 + \mathbf{p}_3) = \mathbf{p}_1 \cdot \mathbf{p}_2 + \mathbf{p}_1 \cdot \mathbf{p}_3$$

$$(\mathbf{p}_1 + \mathbf{p}_2) \cdot \mathbf{p}_3 = \mathbf{p}_1 \cdot \mathbf{p}_3 + \mathbf{p}_2 \cdot \mathbf{p}_3$$

However, since the result of the dot-product is not a vector, the associative laws are not applicable.

The above formulae apply likewise to 2-dimensional vectors. However, the *cross-product*, which produces another vector, is restricted to the 3-dimensional scene. Also called the *vector product*, it is defined for vectors $\mathbf{p}_1$ and $\mathbf{p}_2$ as

$$\mathbf{p}_1 \times \mathbf{p}_2 = (Y_1Z_2 - Z_1Y_2, Z_1X_2 - X_1Z_2, X_1Y_2 - Y_1X_2)$$

If $\breve{\mathbf{i}}, \breve{\mathbf{j}}$ and $\breve{\mathbf{k}}$ are unit vectors respectively lying along the positive X, Y and Z half-axes, then, using determinants (see Tutorial 20) we equivalently have

$$\mathbf{p}_1 \times \mathbf{p}_2 = \begin{vmatrix} Y_1 & Y_2 \\ Z_1 & Z_2 \end{vmatrix} \breve{\mathbf{i}} + \begin{vmatrix} Z_1 & Z_2 \\ X_1 & X_2 \end{vmatrix} \breve{\mathbf{j}} + \begin{vmatrix} X_1 & X_2 \\ Y_1 & Y_2 \end{vmatrix} \breve{\mathbf{k}}$$

Taking the dot-product of either of its two constituent vectors with the product vector, simple multiplication gives

$$\mathbf{p}_1 \cdot (\mathbf{p}_1 \times \mathbf{p}_2) = \mathbf{p}_2 \cdot (\mathbf{p}_1 \times \mathbf{p}_2) = 0$$

hence, from Tutorial 4, the resulting vector must be orthogonal to both its constituent vectors. It follows that the set of all vectors $\mathbf{p} = (X, Y, Z)$ such that

$$\mathbf{p}_1 \times \mathbf{p} = 0$$

form a plane orthogonal to $\mathbf{p}_1$, so any one vector can be seen as defining a plane.

Clearly,

$$\mathbf{p}_2 \times \mathbf{p}_1 = -(\mathbf{p}_1 \times \mathbf{p}_2)$$

while if s is an ordinary scalar number and $\mathbf{0}$ the null vector (0, 0, 0):

$$s\mathbf{p}_1 \times \mathbf{p}_2 = \mathbf{p}_1 \times s\mathbf{p}_2 = s(\mathbf{p}_1 \times \mathbf{p}_2)$$

$$\mathbf{p}_1 \times \mathbf{0} = \mathbf{0} \times \mathbf{p}_1 = \mathbf{0}$$

$$\mathbf{p}_1 \times \mathbf{p}_1 = \mathbf{0}$$

$$\mathbf{p}_1 \times (\mathbf{p}_2 + \mathbf{p}_3) = \mathbf{p}_1 \times \mathbf{p}_2 + \mathbf{p}_1 \times \mathbf{p}_3$$

$$(\mathbf{p}_1 + \mathbf{p}_2) \times \mathbf{p}_3 = \mathbf{p}_1 \times \mathbf{p}_3 + \mathbf{p}_2 \times \mathbf{p}_3$$

Since any non-zero cross-product vector is orthogonal to both its constituent vectors, the cross-product of such product with another vector must lie in the plane of the original pair. Interestingly, this brings in the dot-product, with

$$\mathbf{p}_1 \times (\mathbf{p}_2 \times \mathbf{p}_3) = (\mathbf{p}_1 \cdot \mathbf{p}_3)\mathbf{p}_2 - (\mathbf{p}_1 \cdot \mathbf{p}_2)\mathbf{p}_3$$

Applying a dot-product to a cross-product gives

$$\mathbf{p}_1 \cdot (\mathbf{p}_2 \times \mathbf{p}_3) = (\mathbf{p}_1 \times \mathbf{p}_2) \cdot \mathbf{p}_3$$

The length of a cross-product vector can be related to the lengths of its constituent vectors via the angle δ between them, as

$$|\mathbf{p}_1 \times \mathbf{p}_2| = |\mathbf{p}_1||\mathbf{p}_2| \sin\delta$$

advantage of both meridians and parallels being circular arcs, albeit non-concentric arcs (the two being orthogonal for conformality) — indeed, with the Stereographic, all circles map into circles. The precise geometry of circular arcs for meridians and parallels is considered with the Lambert-Lagrange projection late in Chapter 10, for which the Stereographic at Equatorial aspect forms the basis. At general Oblique aspect, it is the basis for the Miller Oblated Stereographic projection (developed in Chapter 14).

For the Equidistant Azimuthal projection Equations 69a & 197e give

$$\dot{\rho} = \sigma R\left(\tfrac{\pi}{2} - {}_a\phi\right) = \sigma R \arccos\left\{\sin\phi_N \sin\phi + \cos\phi_N \cos\phi \cos\left(\lambda - \lambda_N\right)\right\} \quad \text{. . . . (203a}$$

which, substituted in Equations 197c, provides the plotting equations for the projection at Oblique aspect. Exhibit 8–2 (page 196) illustrates the result with a map focused on London, shown immediately below the Gnomonic focussed identically. While great circles through London appear as straight lines on both maps, such lines are of a uniform scale — hence are proportional to distance — on the Equidistant but not on the Gnomonic map. The projection at Equatorial aspect is illustrated in whole-world form on page 240, and in hemispheric form on page 222. In the former it is with the Aitoff projection, which is derived from it, and in the latter it is associated with globular projections. Putting $\phi_N = 0$ in Equation 203a, then substituting that into Equations 197d, gives the plotting equations for Equidistant Azimuthal projection at Equatorial aspect as:

$$\begin{aligned} \dot{x} &= \sigma R \arccos\left\{\cos\phi \cos\left(\lambda - \lambda_N\right)\right\} \cos\phi \sin\left(\lambda - \lambda_N\right) \\ \dot{y} &= \sigma R \arccos\left\{\cos\phi \cos\left(\lambda - \lambda_N\right)\right\} \cos\phi_N \sin\phi \end{aligned} \quad \text{. . . . . (203b}$$

For the (Lambert) Equal-area Azimuthal Equations 73c & 192a give

$$\dot{\rho} = \sigma R\sqrt{2(1 - \sin{}_a\phi)} = \sigma R\sqrt{2\left\{1 - \sin\phi_N \sin\phi - \cos\phi_N \cos\phi \cos\left(\lambda - \lambda_N\right)\right\}} \quad \text{. . . (203c}$$

which, substituted in Equations 197c, provides the plotting equations for the projection at Oblique aspect. Putting $\phi_N = 0$ in Equation 203c then substituting into Equations 197d gives the plotting equations for the Equal-area Azimuthal projection at Equatorial aspect as:

$$\begin{aligned} \dot{x} &= \sigma R\sqrt{2\left\{1 - \cos\phi \cos\left(\lambda - \lambda_N\right)\right\}} \cos\phi \sin\left(\lambda - \lambda_N\right) \\ \dot{y} &= \sigma R\sqrt{2\left\{1 - \cos\phi \cos\left(\lambda - \lambda_N\right)\right\}} \cos\phi_N \sin\phi \end{aligned} \quad \text{. . . . . (203d}$$

The projection is shown at Oblique and Equatorial aspects on page 198. At Equatorial aspect it is shown also on page 242 above two projections — the Hammer and the Eckert-Greiffendorf — that are derived from it. Like the Stereographic, the Equal-area Azimuthal is similar to the globular projections but lacks their uniformity of scale along the Equator. Another oblique version is shown on page 244 above the Briesemeister projection, which derives from it via the Hammer projection.

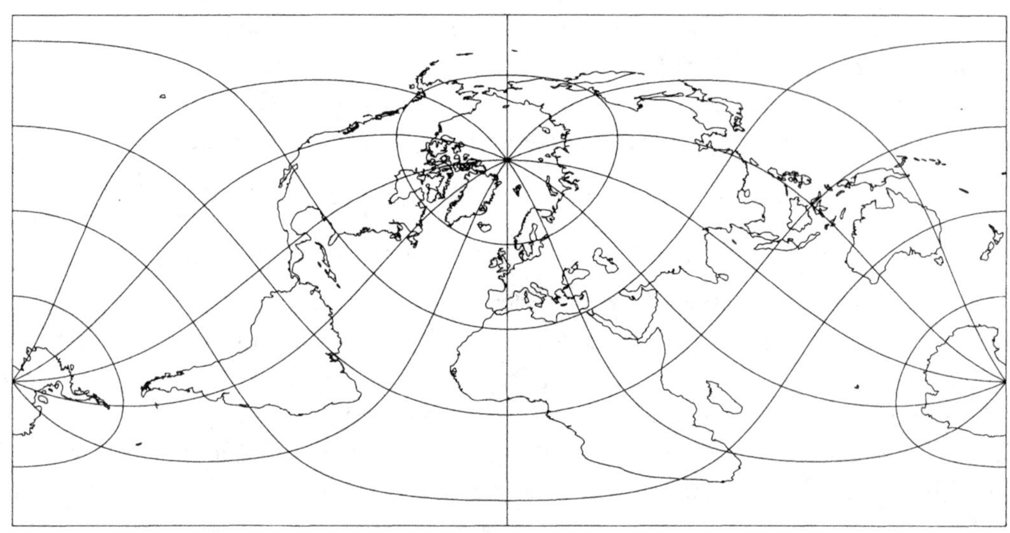

Exhibit 8–9: The Equidistant Cylindrical (or Plate-Carrée) projection at Oblique aspect.

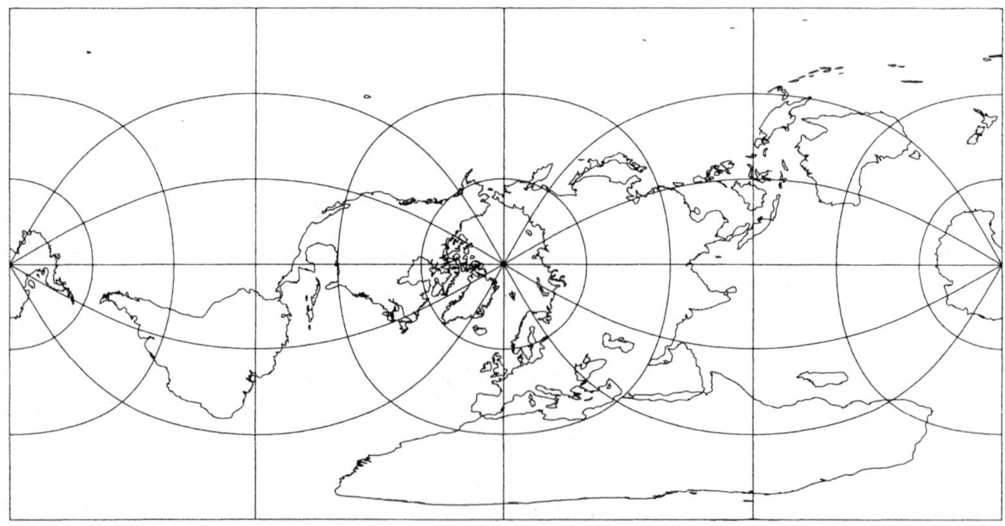

Exhibit 8–10: Equidistant Cylindrical (or Plate-Carrée) projection at Transverse aspect – the Cassini projection

Applying the transformation to Cylindricals

The general Oblique aspect makes the conformal Mercator projection and the Equal-area Cylindrical projection of notable value for mapping elongated entities such as Sumatra, Madagascar and New Zealand. The country of Chile is even more elongated and narrow, but its orientation is so close to north-south as to make the Transverse aspect preferable, because being centred on a section of any great circle passing through the Poles makes the map clearer (and the plotting easier). We will find in Chapter 17 that the Transverse Mercator provides the basis for the modern universal co-ordinate system for most of Earth.

With Cylindricals at Simple aspect, the Poles are peripheral if not wholly absent. In turning them to an oblique position it is not obvious what longitude and latitude to adopt as auxiliary pole. The answer is provided by reference to vector cross-products (see Tutorial 21, page 202). Let $P_1 = (\lambda_1, \phi_1)$ and $P_2 = (\lambda_2, \phi_2)$ be two discrete points that we wish to have on the auxiliary Equator, with (X_1, Y_1, Z_1) and (X_2, Y_2, Z_2) their respective expression in Cartesian co-ordinates (Tutorial 3). Let $\mathbf{p}_1 = (X_1, Y_1, Z_1)$ and $\mathbf{p}_2 = (X_2, Y_2, Z_2)$ be the corresponding vectors (Tutorial 4). Then the cross-product $\mathbf{p}_1 \times \mathbf{p}_2$, being orthogonal to each individual vector, is orthogonal to the auxiliary Equatorial plane. Hence, the new vector lies along the auxiliary axis so effectively making it represent one auxiliary pole, its negative the other.

We assume that point $P_N = (\lambda_N, \phi_N)$ is established as the auxiliary north pole. In developing our transformation formulae above, we chose the great circle through that point and the junction of auxiliary and true Equator as zero longitude. We will assume that meridian is to be our central meridian.

If ${}_a\lambda_0$ is the chosen central meridian in the auxiliary scheme, the intermediate parameters for Cylindricals, adapted to the auxiliary longitude/latitude scheme, are:

$$\dot{u} = \sigma R\left({}_a\lambda - {}_a\lambda_0\right)$$
$$\dot{v} = v\left({}_a\phi\right),$$

the latter being a function solely of ${}_a\phi$ and particular to the projection.

These intermediate parameters apply respectively along the auxiliary Equator and orthogonally thereto, along the chosen central meridian. We choose it such that it crosses the auxiliary Equator at its intersection with the true Equator. That puts it offset by 90° from the Greenwich bimeridian forming the circular perimeter in Exhibit 8–3. Choosing of the two intersections the one on the face of our drawing makes $\zeta = -{}_a\lambda + {}^{\pi}\!/_2$. Hence:

$$\sin\zeta = \cos_a\lambda, \quad \cos\zeta = \sin_a\lambda, \quad \tan\zeta = \cot_a\lambda, \quad \cot\zeta = \tan_a\lambda$$

so Equations 193b, 193c and 194a become

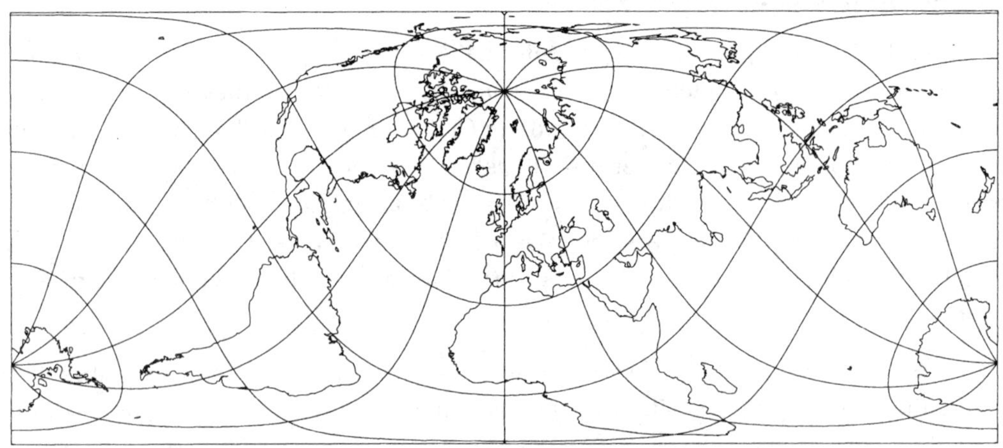

Exhibit 8–11: The (Lambert) Equal-area Cylindrical projection at Oblique aspect.

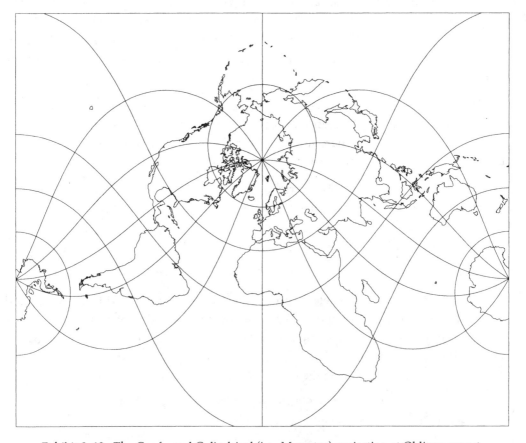

Exhibit 8–12: The Conformal Cylindrical (i.e., Mercator) projection at Oblique aspect.

$$\cos_{\mathrm{a}}\phi \cos_{\mathrm{a}}\lambda = \cos\phi \sin(\lambda - \lambda_{\mathrm{N}})$$

$$\begin{aligned}\tan_{\mathrm{a}}\lambda &= \frac{\cos\phi_{\mathrm{N}} \sin\phi - \sin\phi_{\mathrm{N}} \cos\phi \cos(\lambda - \lambda_{\mathrm{N}})}{\cos\phi \sin(\lambda - \lambda_{\mathrm{N}})} \\ &= \frac{\cos\phi_{\mathrm{N}} \tan\phi_{\mathrm{N}} - \sin\phi_{\mathrm{N}} \cos(\lambda - \lambda_{\mathrm{N}})}{\sin(\lambda - \lambda_{\mathrm{N}})}\end{aligned} \qquad \text{. . . . . (207a}$$

$$\sin_{\mathrm{a}}\lambda = \tan_{\mathrm{a}}\lambda \cos_{\mathrm{a}}\lambda = \frac{\cos\phi_{\mathrm{N}} \sin\phi - \sin\phi_{\mathrm{N}} \cos\phi \cos(\lambda - \lambda_{\mathrm{N}})}{\cos_{\mathrm{a}}\phi}$$

respectively, while choice of the correct spherical quadrant can be assured by ensuring that

$$\cos_{\mathrm{a}}\phi \sin_{\mathrm{a}}\lambda = \sin\phi \cos\phi_{\mathrm{N}} - \cos\phi \sin\phi_{*} \cos(\lambda - \lambda_{\mathrm{N}}) \qquad \text{. . . . . (207b}$$

Substituting from Equation 207a for ${}_{\mathrm{a}}\lambda$, we get the equation, common to all Cylindricals,

$$\dot{u} = \sigma R\,{}_{\mathrm{a}}\lambda = \sigma R \arctan\frac{\cos\phi_{\mathrm{N}} \sin\phi - \sin\phi_{\mathrm{N}} \cos\phi \cos(\lambda - \lambda_{\mathrm{N}})}{\cos\phi \sin(\lambda - \lambda_{\mathrm{N}})} \qquad \text{. . . . . . (207c}$$

At Transverse aspect, with $\phi_{\mathrm{N}}=0$, this becomes

$$\dot{u} = \sigma R \arctan\frac{\sin\phi}{\cos\phi \sin(\lambda - \lambda_{\mathrm{N}})} \quad = R \arctan\frac{\tan\phi}{\sin(\lambda - \lambda_{\mathrm{N}})} \qquad \text{. . . . . (207d}$$

Up to now the plotting equations for Cylindricals have had the plotting variables identically the intermediate parameters. If used here, the result is to have the auxiliary Equator along the x axis, the natural Equator elsewhere. Depending on the specific map, a rotation may be desirable. Using Equations 86a to provide for counter-clockwise rotation by angle χ, we get:

$$\begin{aligned}\dot{x} &= \dot{u}\cos\chi - \dot{v}\sin\chi \\ \dot{y} &= \dot{u}\sin\chi + \dot{v}\cos\chi\end{aligned} \qquad \text{. . . . . (207e}$$

A rotation of 90° would put North at the top, put the map at Transverse aspect.

For the Lambert Equal-area Cylindrical projection, from Equations 114a & 192a

$$\dot{v} = \sigma^{-1} R \sin_{\mathrm{a}}\phi \quad = \sigma^{-1} R\{\sin\phi_{\mathrm{N}} \sin\phi + \cos\phi_{\mathrm{N}} \cos\phi \cos(\lambda - \lambda_{\mathrm{N}})\} \qquad \text{. . . . . (207f}$$

This is substituted, along with $\dot{u}$ from Equation 207c, in Equations 207e with the chosen χ. At Transverse aspect, with $\phi_{\mathrm{N}}=0$ and choosing $\chi={}^{\pi}\!/_{2}$, the plotting equations, noting Equation 207d, become:

$$\begin{aligned}\dot{x} &= -\dot{v} = -\sigma^{-1} R \cos\phi \cos(\lambda - \lambda_{\mathrm{N}}) \\ \dot{y} &= +\dot{u} = \sigma R \arctan\frac{\tan\phi}{\sin(\lambda - \lambda_{\mathrm{N}})}\end{aligned} \qquad \text{. . . . . (207g}$$

The Oblique aspect is illustrated in the upper map opposite, the Transverse in the upper map overleaf.

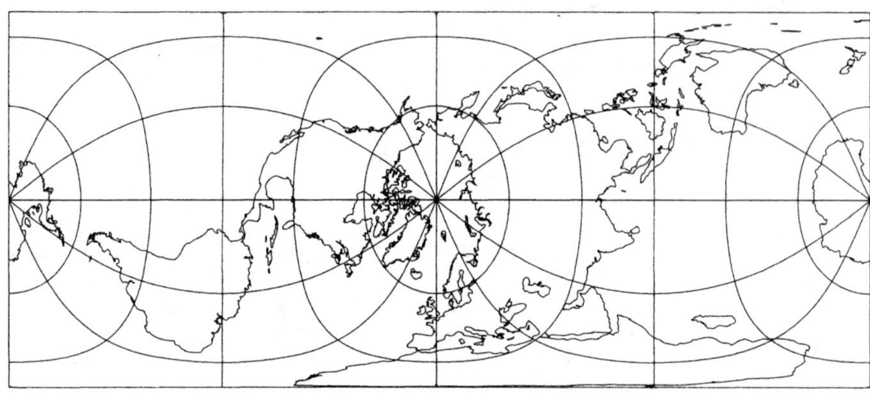

Exhibit 8–13: The (Lambert) Equal-area Cylindrical projection at Transverse aspect.

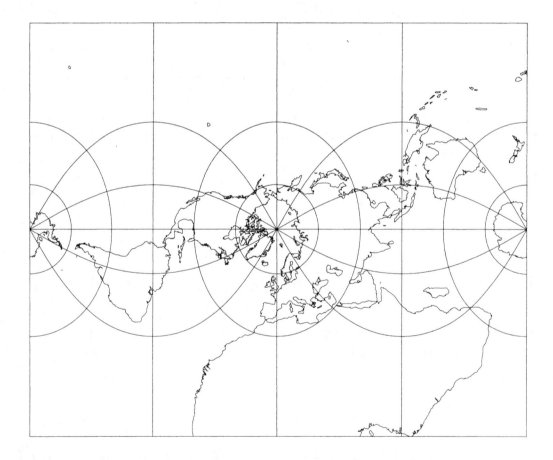

Exhibit 8–14: The Conformal Cylindrical (i.e., Mercator) projection at Transverse aspect.

For the General Perspective Cylindrical projection, from Equation 56a

$$\dot{v} = \frac{R \sin_a \phi}{1 - h^{-1}(1 - \cos_a \phi)} \qquad \text{(209a}$$

to be substituted (along with $\dot{u}$ from Equation 207c) in Equations 207e with the chosen rotational angle χ, and with $\sin_a\phi$, hence ${}_a\phi$ and $\cos_a\phi$, defined by Equation 192a.

For the Plate-Carrée or Equidistant Cylindrical projection from Equation 69b

$$\dot{v} = \sigma R\, {}_a\phi = \sigma R \arcsin\{\sin\phi_N \sin\phi + \cos\phi_N \cos\phi \cos(\lambda - \lambda_N)\} \qquad \text{(209b}$$

to be substituted (along with $\dot{u}$ from Equation 207c) in Equations 207e with the chosen χ. At Transverse aspect, choosing $\chi = \pi/2$ and noting Equation 207d, the plotting equations become:

$$\dot{x} = -v = -\sigma R \arcsin\{\cos\phi \cos(\lambda - \lambda_N)\}$$
$$\dot{y} = +u = +\sigma R \arctan \frac{\sin\phi}{\cos\phi \sin(\lambda - \lambda_N)} = R \arctan \frac{\tan\phi}{\sin(\lambda - \lambda_N))} \qquad \text{(209c}$$

Both Oblique and Transverse aspects are illustrated on page 204. At Transverse aspect, the projection is distinctively named the **Cassini** projection, commemorating its usage by César François Cassini for a survey of France in the 1780s, when he was Director of the Paris Observatory. This projection at Oblique aspect is shown also in Exhibit 8–19 on page 217 for a special use described on its facing page.

For the Mercator or Conformal Cylindrical projection, Equation 119a gives

$$\dot{v} = \frac{\sigma R}{2} \ln\left\{\frac{1 + \sin_a\phi}{1 - \sin_a\phi}\right\} \qquad \text{(209d}$$

which becomes by substitution for $\sin_a\phi$ from Equation 192a

$$\dot{v} = \frac{\sigma R}{2} \ln\left\{\frac{1 + \sin\phi_N \sin\phi + \cos\phi_N \cos\phi \cos(\lambda - \lambda_N)}{1 - \sin\phi_N \sin\phi - \cos\phi_N \cos\phi \cos(\lambda - \lambda_N)}\right\} \qquad \text{(209e}$$

but Equation 209d is computationally more efficient. At Transverse aspect, the latter becomes

$$\dot{v} = \frac{\sigma R}{2} \ln\left\{\frac{1 + \cos\phi\,(\lambda - \lambda_N)}{1 - \cos\phi \cos(\lambda - \lambda_N)}\right\} \qquad \text{(209f}$$

Again these values are to be substituted (along with $\dot{u}$ from Equation 207c, else for the Transverse aspect from Equation 207d) in Equations 207e with the chosen χ.

The Oblique aspect is illustrated in the lower map on page 206, the Transverse aspect in the lower map on page 208. The Transverse version is discussed further in Chapter 18, for its heavily interrupted use in the UTM scheme.

An example of the transformation

To illustrate the process, let us assume we want to have the pole at Vancouver — at (conventional) longitude 122°W, latitude 49°20′N. Our auxiliary reference meridian becomes the section of the conventional meridian through Vancouver that lies between it and the true North Pole, plus its continuation beyond that Pole for a further 49°20′ down the corresponding meridian, i.e., 58°E. Hence the point 58°E, 40°40′N in the conventional scheme will be on the auxiliary Equator, with auxiliary longitude and latitude zero. Substituting $\Lambda_N = +122°$ and $\Phi_N = +49°20'$ in Equations 194c & 194d gives transformations:

$$_{a}\lambda = \arctan\left\{\frac{\cos\Phi\sin(122-\Lambda)}{0.652\,89_{\sim}\sin\Phi - 0.757\,45_{\sim}\cos\Phi\cos(122-\Lambda)}\right\}$$

$$_{a}\phi = \arcsin\left\{0.757\,45_{\sim}\sin\Phi + 0.652\,89_{\sim}\cos\Phi\cos(122-\Lambda)\right\}$$

applicable to any point (Φ, Λ), the results being in radians — as appropriate for input to the specific projection formulae. The specific quadrant is identified by the equality

$$\cos_{a}\phi\cos_{a}\lambda = 0.65289_{\sim}\sin\Phi - 0.757\,45_{\sim}\cos\Phi\cos(122-\Lambda)$$

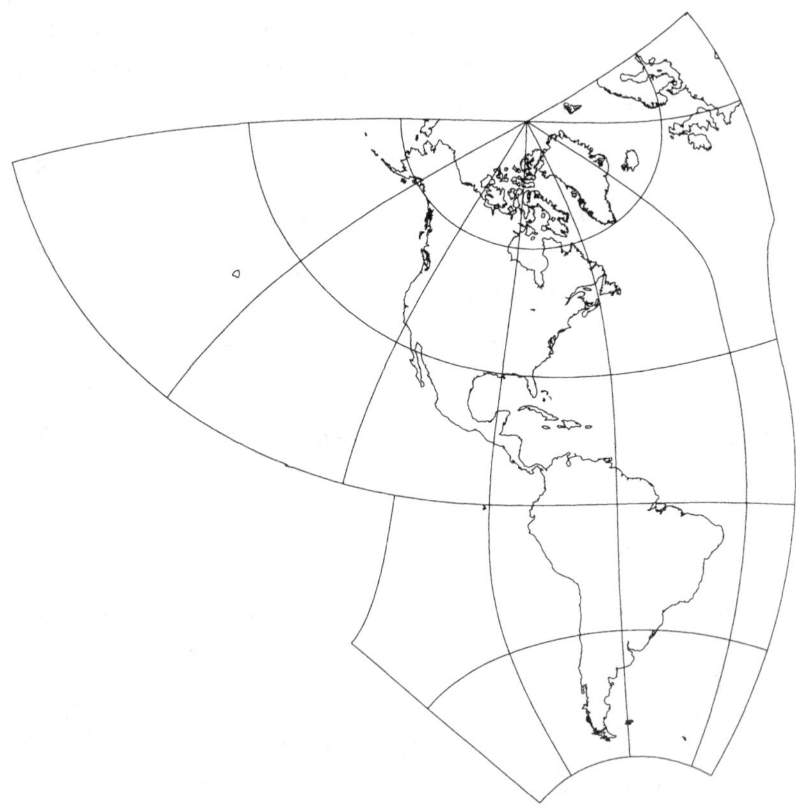

Exhibit 8–15: Bipolar Oblique Conical projection.

Applying the transformation to Conicals

The elaborateness of the plotting formulae of Conicals at Simple aspect precludes any advantageous consolidation with formulae for transformation of aspect, even at Transverse aspect. Plotting is carried out in two stages. Equations 194c & 194d else, if Transverse, Equations 194e & 194f, are applied as the first step, followed by application of the plotting equations for Simple aspect of the chosen projection. As with the Cylindricals, however, the resulting map may need to be rotated to have North at the top. This can be achieved by adding a fixed amount to the angular intermediate parameter.

As with the other modes, the Conical loses the visual features of its Simple graticule when turned from that mode, but Conicals tend to lose more. However, the fact that the band of lower distortion follows an arched shape means that the Oblique aspect can prove beneficial for selected regions. North America has a degree of eastward-facing concavity. C. H. Deetz [1919c] applied the Polyconic transversely to that continent, putting an auxiliary pole on the Equator at 70°W and making the 160°W meridian the auxiliary equator, with latitude 45°N thereon being on the central auxiliary meridian. New Zealand [1938a], Japan [its *National Atlas*] and the Chinese coast have more obvious curvature, preferably addressed with two standard parallels to maximize the band of low distortion.

An interesting use of conical projecting at Oblique aspect occurs with the **Bipolar Oblique Conical** projection developed by O. M. Miller and William Briesemeister [1941d]. Addressing the eastward concavity of North America along with the westward concavity of South America, they juxtaposed in an opposed sense two conformal maps. The work originated in the pursuit of a single map for South America on a scale no larger than 1:4 000 000, for which the Conformal Conical projection with auxiliary pole in the Pacific Ocean at 20°S, 110°W proved effective. Standard parallels at 17° and 59° auxiliary latitude kept the linear distortion within reasonable bounds across that continent, but barely as the map spread northward into central America and certainly not beyond. They then applied the same projection to North America, with the auxiliary pole swapped into the Atlantic ocean, at 45°N, 19°59′36.‿″W (chosen to be exactly 104° of arc from the other pole), using the same standard latitudes.

The adjective *Bipolar* comes from the juxtaposition of the two, separated by the line between their respective poles. As the scaling does not match along the line except at the points of intersection of their respective standard parallels, some handcrafting is applied along a narrow strip straddling the line. This destroys conformality there. Otherwise the composite map is conformal, and provides a useful option for depicting the paired continents. It is illustrated opposite.

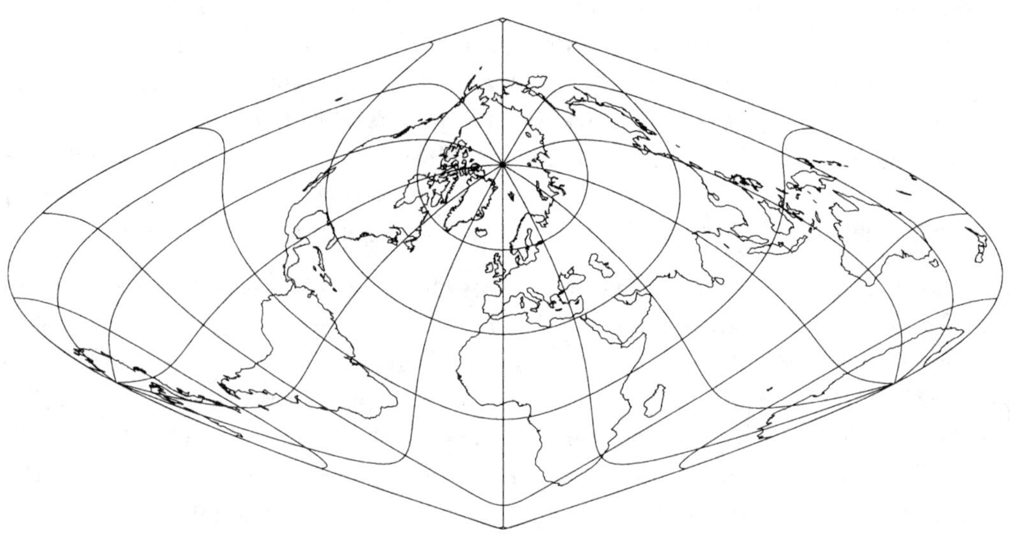

Exhibit 8–16: The Sinusoidal projection at Oblique aspect.

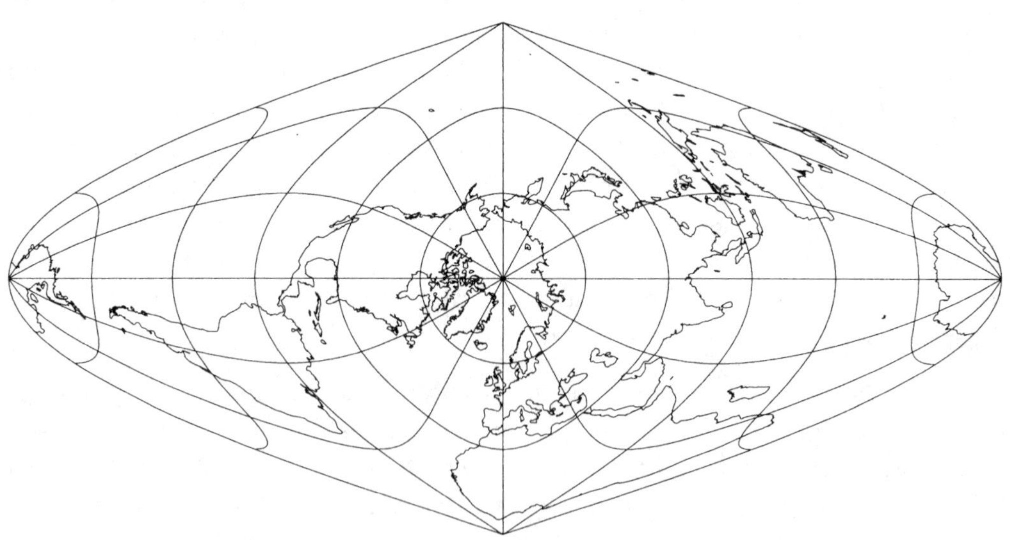

Exhibit 8–17: The Sinusoidal projection at Transverse aspect.

Applying the transformation to Pseudocylindricals

The Sinusoidal is the most used obliquely of the Pseudocylindricals. It is illustrated at Oblique aspect directly opposite, then twice at Transverse aspect — below opposite then directly below after rotation by –90°. There is little purpose in developing consolidated formulae for any. With Mollweide and many others there is no straightforward mathematical expression to consolidate. C. F. Close [1927b] published a transverse

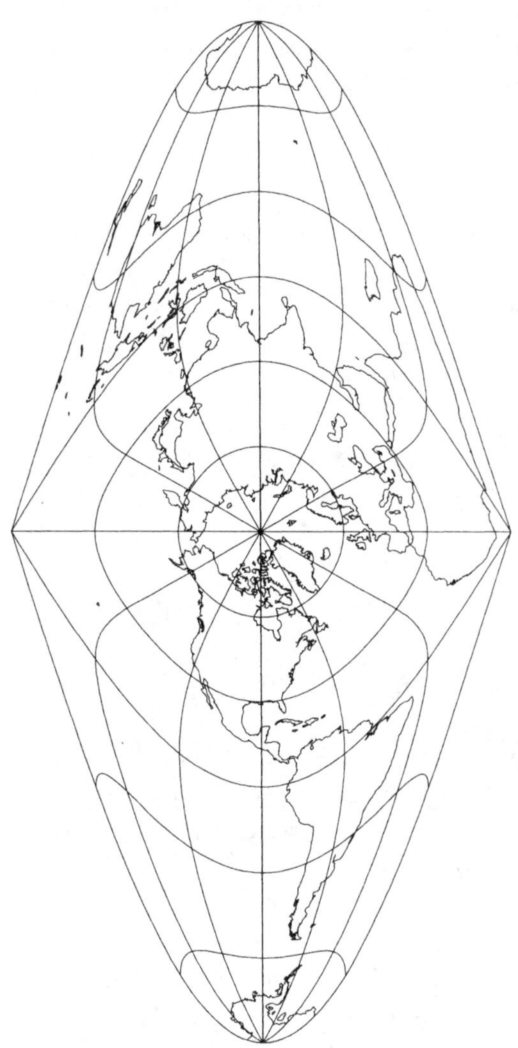

Exhibit 8–18: The Sinusoidal projection at Transverse aspect after rotation.

version of Mollweide as **The Transverse Elliptical Equal-Area Projection of the Sphere**. Close then [1929d], prompted by another's "new projection" [1928a], transformed it by 45°, which maintains something of its normal appearance, developing for it a table of numeric plotting values.

Applying the transformation to other projections

There is little of note regarding the transformation of other projections, and nothing to add mathematically. Any projection can be transformed to better fit a given region, and both conformal projections (further discussed in Chapter 15) and equal-area projections can be of significantly greater value when transformed. The Briesemeister projection in Chapter 10, for instance, is an equal-area world map centred on 45°N, 10°W. obtained by obliquely transforming, then re-proportioning another projection.

In conclusion

With the universal availability of computers having sufficient capacity and speed, the transformation of any projection to any auxiliary pole has become a routine two-step process. For world and regional maps, the above transformation and the spherical world suffice. For maps of surveyor quality the ellipsoidal world is necessary — a subject that we discuss in Chapter 17. In the past, this applied to several projections, notably the Conical and Polyconic in the U.S.A. and the Airy in Britain. Now it occurs most notably for the Transverse Mercator used in UTM and its finer relatives, along with the conformal Stereographic of UPS — both discussed in Chapter 18.

TUTORIAL 22 COMPUTING SURFACE DISTANCE – 2

Tutorial 5 (page 14) provided one means for calculating great-circle distance, applying it to get the distance between

Calgary International Airport (YYC, location 114°01′W, 51°07′N) and
London Heathrow Airport (LHR, location 0°27′W, 51°28′N).

We now use Equation 215d for the same purpose. Adopting its notation with YYC as point P_0 and LHR as point P_i we have:

$\sin\phi_0 = \sin\Phi_0 = +0.778\,426_\sim$, $\cos\phi_0 = \cos\Phi_0 = +\,0.627\,737_\sim$

$\sin\phi_i = \sin\Phi_i = +0.782\,246_\sim$, $\cos\phi_i = \cos\Phi_i = +0.622\,970_\sim$

$\tan(\lambda_0 - \lambda_i) = \tan(\Lambda_0 - \Lambda_i) = \tan(114.0167_\sim° - 0.45°) = \tan 113.567_\sim° = -0.399\,816_\sim$

$\cos\delta_{0,i} = (0.778\,426_\sim)(0.782\,246_\sim) + (0.627\,737_\sim)(0.622\,970_\sim)(-0.399\,816_\sim) = 0.452\,57_\sim$

Surface distance between YYC and LHR $= R \arccos\delta_{0,i} = R \arccos 0.452\,57_\sim$

(taking $R = 6371$ km) $= 1.1012_\sim \times 6371$ km $= 7015._\sim$ km $= 3788._\sim$ INM

AZIMUTHS AND GREAT-CIRCLE DISTANCES

Spherical triangles and great circles

The sides of spherical triangles are by definition arcs of great circles, and the various formulae relating sides and angles of a spherical triangle allow us to establish general formulae for azimuthal bearings and surface distances.

Azimuths

Azimuth refers to the angle of bearing relative to North, measured clockwise (hence the reverse of the mathematics practice, as discussed at the end of Chapter 1). Because the spherical triangle NPP_N of Exhibit 8–4 on page 192 connects a general point $P=(\lambda, \phi)$ with a specific point $P_N=(\lambda_N, \phi_N)$ and the North Pole (denoted N), the angle NP_NP is the azimuth of P at P_N. Denoted by ζ — the standard symbol for azimuths in this work — in the exhibit, its measure is provided by Equation 193c. Changing the notation to have $P_0=(\lambda_0, \phi_0)$ in the place of P_N and $P_i=(\lambda_i, \phi_i)$ in the place of P, and then denoting the azimuth more specifically as $\zeta_{0,i}$ (being the azimuth at P_0 of P_i), the equation gives the following parallel expressions:

$$\begin{aligned}\cot\zeta_{0,i} &= \frac{\cos\phi_0\tan\phi_i - \sin\phi_0\cos(\lambda_i - \lambda_0)}{\sin(\lambda_i - \lambda_0)} \\ &= \frac{\cos\phi_0\sin\phi_i - \sin\phi_0\cos\phi_i\cos(\lambda_i - \lambda_0)}{\cos\phi_i\sin(\lambda_i - \lambda_0)} \qquad \text{(215a)} \\ &= \cos\phi_0\sin\phi_i\csc(\lambda_i - \lambda_0) - \sin\phi_0\cos\phi_i\cot(\lambda_i - \lambda_0)\end{aligned}$$

which simplifies when P_0 and P_i are a right-angle apart longitudinally to

$$\cot\zeta_{0,i} = \cos\phi_0\sin\phi_i \qquad \text{(215b)}$$

Great-circle distance

Noting that the angle subtended by the arc between $P_N=(\lambda_N, \phi_N)$ and $P=(\lambda, \phi)$ in Exhibit 8–4 is the complement of angle ${}_a\phi$, Equation 194c gives the arc length as

$$R\left(\frac{\pi}{2} - {}_a\phi\right) = R\arccos\{\sin\phi_N\sin\phi + \cos\phi_N\cos\phi\cos(\lambda - \lambda_N)\} \qquad \text{(215c)}$$

which provides a new formula for calculating the shortest distance between points on the sphere (see Tutorial 22). Again replacing P_N and P by $P_0=(\lambda_0, \phi_0)$ and $P_i=(\lambda_i, \phi_i)$, and denoting the angle subtended at the centre of the sphere by the arc spanning these points by $\delta_{0,i}$ we have

$$\delta_{0,i} = \arccos\{\sin\phi_0\sin\phi_i + \cos\phi_0\cos\phi_i\cos(\lambda_i - \lambda_0)\} = \delta_{i,0} \qquad \text{(215d)}$$

which simplifies when P_0 and P_i are a right angle apart longitudinally to

$$\delta_{0,\hat{i}} = \arccos\{\sin\phi_0 \sin\phi_{\hat{i}}\} \qquad \text{(216a}$$

For any point $P_i = (\lambda_i, \phi_i)$ on the boundary of the hemisphere centred at $P_0 = (\lambda_0, \phi_0)$

$$R \arccos\{\sin\phi_0 \sin\phi_i + \cos\phi_0 \cos\phi_i \cos(\lambda_i - \lambda_0)\} = \tfrac{\pi}{2} R$$

i.e., $$\sin\phi_0 \sin\phi_i + \cos\phi_0 \cos\phi_i \cos(\lambda_i - \lambda_0) = \cos\tfrac{\pi}{2} = 0$$

So the boundary is the set of points $P_i = (\lambda_i, \phi_i)$ fulfilling the equation

$$\cot\phi_i \cos(\lambda_i - \lambda_0) = -\tan\phi_0$$

The point at given azimuth and distance

Equations 215a & 215d give us — relative to point $P_0 = (\lambda_0, \phi_0)$ — the azimuth and the angular distance away of point $P_i = (\lambda_i, \phi_i)$. We now look into the reverse situation — finding the point $P_i = (\lambda_i, \phi_i)$ that is a given angular distance $\delta_{0,i}$ from point $P_0 = (\lambda_0, \phi_0)$ at a given azimuthal bearing $\zeta_{0,i}$. Such a point is, of course, on the great circle having that azimuth. Identifying point P_N with $P_0 = (\lambda_0, \phi_0)$ and point P with $P_i = (\lambda_i, \phi_i)$, Exhibit 8–4 sets the scene if we have $\delta_{0,i}$ for $\pi/2 - \phi$. Adapting Equations 193a & 193d then yields

$$\sin\phi_i = \sin\phi_0 \cos\delta_{0,i} + \cos\phi_0 \sin\delta_{0,i} \cos\zeta_{0,i} \qquad \text{(216b}$$

$$\cot(\lambda_i - \lambda_0) = \frac{\cos\phi_0 \cot\delta_{0,i} - \sin\phi_0 \cos\zeta_{0,i}}{\sin\zeta_{0,i}} = \frac{\cos\phi_0 \cos\delta_{0,i} - \sin\phi_0 \sin\delta_{0,i} \cos\zeta_{0,i}}{\sin\delta_{0,i} \sin\zeta_{0,i}} \qquad \text{(216c}$$

giving us the longitude and latitude of point $P_i = (\lambda_i, \phi_i)$.

The Equidistant Azimuthal projection provides a ready means for plotting the point a given azimuth and distance from a starting point — provided the map is focussed at that starting point. This projection provides a very appropriate map for use in travel offices in the region of that starting point, though its deformities make it grossly unfamiliar. It is notably advantageous for long-distance travel at higher latitudes.

F. V. Botley [1951a] promoted the Plate-Carrée (or Equidistant Cylindrical) for such plotting. At Simple aspect the projection provides true scale from both Poles to any point along a meridian (see page 19), while the equally spaced meridians provide measure of the rotational angle. Hence, at Oblique aspect the projection does likewise relative to any pair of auxiliary poles, which are the upper and lower bounding lines of the rectangle. Whatever points are chosen for the auxiliary poles, a standard graduated outline rectangle can provide both azimuthal bearing and distance from the starting point. Exhibit 8–19 opposite mimics Botley's original, with London as the upper boundary and a rectilinear grid superimposed on the curved graticule. The degree readings below give the azimuth from London for all points on each vertical grid line. The figures at the sides provide the measure of distance from London for all points on each horizontal grid line, with linear distance in kilometres at the left and angular distance in radians at the right.

The points along a given great circle

Any two distinct non-antipodean points on the globe define a great circle. Let $P_0 = (\lambda_0, \phi_0)$ and $P_2 = (\lambda_2, \phi_2)$ be two such points. Equation 215a then shows the azimuth $\zeta_{0,2}$ of P_2 at P_0 as given by

$$\begin{aligned}\cot \zeta_{0,2} &= \frac{\cos\phi_0 \tan\phi_2 - \sin\phi_0 \cos(\lambda_2 - \lambda_0)}{\sin(\lambda_2 - \lambda_0)} \\ &= \frac{\cos\phi_0 \sin\phi_2 - \sin\phi_0 \cos\phi_2 \cos(\lambda_2 - \lambda_0)}{\sin(\lambda_2 - \lambda_0)}\end{aligned} \quad \text{. . . . . (217a}$$

Because the azimuth at P_0 is identical for all points along that great circle in the same direction until the antipodean point is reached, substitution of the value of $\zeta_{0,2}$ for $\zeta_{0,i}$ in Equations 216b & 216c gives the longitude and latitude for the point distance $\delta_{0,i}$ from P_0.

From Equation 215d, the distance between points P_0 and P_2 is

$$\delta_{0,2} = \arccos\{\sin\phi_0 \sin\phi_2 + \cos\phi_0 \cos\phi_2 \cos(\lambda_2 - \lambda_0)\} \quad \text{. . . . . (217b}$$

Substitution of any fractional multiple of this value for $\delta_{0,i}$ in Equations 216b & 216c gives the longitude and latitude for the point that is that proportion along the circle from P_0 to P_2. Being along the pertinent great circle, this readily provides intermediate points between which the less direct but more convenient lines of constant compass-bearing can be deployed for navigation from P_0 to P_2 (see Loxodromes on page 87).

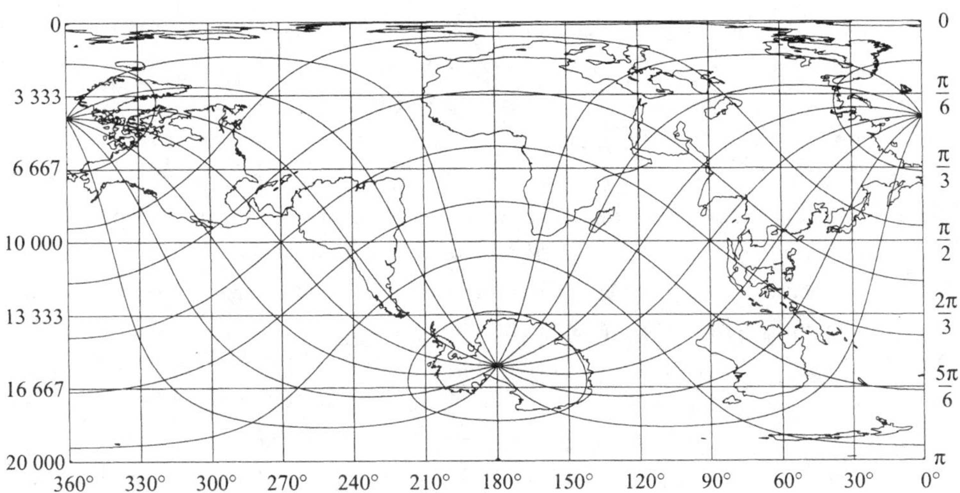

Exhibit 8–19: Bearings and distance from one point to any other point via a Cylindrical after F. V. Botley [1951a].
Showing relative to London (centre of top boundary)
a) the distance in kilometres (left-hand labels)
b) the polar distance in radians (right-hand labels)
c) the azimuthal bearing in degrees (bottom labels)
the labels being applicable to all points on the associated line.

OBLIQUITY AND ASPECTS CLASSIFIED

Three aspects, else seven?

Clearly, there are infinitely many choices even for the auxiliary pole used in the transformation — indeed, even when restricted to Transverse aspect. Thomas Wray [1974b] saw seven distinguishable classes of choice or seven classes of aspect. Using the prefix *meta-* for the auxiliary framework, and the term *Direct* for what has usually been *Simple* aspect in the preceding pages here, Wray defines:

node = an intersection of Equator and meta-equator

ascending node = nodal intersection where eastwards along meta-equator is into northern hemisphere

panmeridian = the one great circle passing through the Poles and meta-poles

Starting at his Direct aspect, Wray progressively moves the original graticule to the meta position via three rotations, as follows:

eastwards about the Polar axis to have the prime meridian include the south meta-pole

counter-clockwise (as seen from ascending node) about the internodal axis to have the North Pole coincident with north meta-pole

clockwise about the meta-polar axis to have the prime meridian coincident with the chosen prime meta-meridian

We again use the left suffix to denote auxiliary elements but retain Wray's more academic terminology. So ${}_a\Lambda$ denotes meta-longitude and ${}_a\Phi$ denotes meta-latitude. We define the two functions $\theta_x = \arccos(\cos_a\Phi \cos_a\Lambda)$ and $\theta_y = \arccos(\cos_a\Phi \sin_a\Lambda)$. The results from the various rotations can be classified according to the positions of the Equator and the bimeridian in relation to the meta-equator, meta-prime-meridian and the meta-meridians at ±90°, which we call the meta-offset-meridian from the prime. The results can also be expressed according to the position of the meta-polar-axis within the standard 3-dimensional co-ordinate scheme (X, Y, Z), in both angular and Cartesian terms. Using *p*, *q* and *r* as variable values in the latter, the seven classes are:

	meta-equator	*meta-prime-meridian*	*meta-offset-meridian*	- - *Polar* X	*axis* Y	- - Z	*p q r*
Direct	Equator	bimeridian	bimeridian	90	90	0	0 0 1
Simple Oblique	—	bimeridian	—	${}_a\Phi$	90	$90-{}_a\Phi$	*p* 0 *r*
First Transverse	bimeridian	bimeridian	Equator	0	90	90	1 0 0
Transverse Oblique	bimeridian	—	—	${}_a\Lambda$	$90-{}_a\Lambda$	90	*p q* 0
Second Transverse	bimeridian	Equator	bimeridian	90	0	90	0 1 0
Skew	—	—	bimeridian	90	0	$90-{}_a\Phi$	0 *q* r
Plagal/Scalene	—	—	—	θ_x	θ_y	$90-{}_a\Phi$	*p q* r

Wray illustrated the particulars using the Mollweide projection, starting with its characteristic pattern at Simple aspect of a central circle formed by the meridians for ±90° covering the inner hemisphere, and a 2:1 ellipse straddling it to envelope the whole world. The central meridian forms a central vertical line and the equator forms a central horizontal line. The results are:

Direct	Prime meridian along central vertical. North Pole at top of central vertical.
Simple Oblique	Prime meridian along central vertical. North Pole thereby on central vertical, but specifically not on ±90° circle, and not on central horizontal.
First Transverse	Prime meridian along central vertical. North Pole thereby on central vertical, specifically at intersection with central horizontal.
Transverse Oblique	Prime meridian along central horizontal. North Pole thereby on central horizontal, but specifically not on ±90° circle, and not on central vetical.
Second Transverse	Prime meridian along central horizontal. North Pole thereby on central horizontal, specifically at intersection with central vertical.
Skew	Prime meridian not along central vertical, and not along ±90° circle, and not along central horizontal North Pole thereby not on central vertical, and specifically not on ±90° circle, and not on central horizontal.
Plagal/Scalene	Prime meridian along ±90° circle North Pole thereby on ±90° circle, but specifically not on central vertical, and not on central horizontal.

TUTORIAL 23 COMPLEX NUMBERS

If $i = \sqrt{-1}$ and x and y are real numbers,

$$z = x + iy$$

is called a *complex number*. Because i is not a real entity, the component y is known as the *imaginary part* of the number, leaving x as the *real part*. If y is zero, z is effectively an ordinary real number; if not and x is zero, $z = iy$ is a *pure imaginary number*.

The two parts are independent of each other, and akin to the two variables in 2-dimensional space. Given an appropriate context, the complex number can be written accordingly as the ordered tuple (x, y), and can be plotted like a 2-dimensional vector. Many operations with complex numbers look identical to those of the vectors — e.g., there is a *modulus* function defined by

$$|z| = +\sqrt{x^2 + y^2}$$

which can equal zero only for $x = y = 0$.

If $z' = x' + iy'$ or (x', y'), then

$z = z'$ if and only if $x = x'$ and $y = y'$.

Further:

$$z + z' = (x + x', y + y')$$
$$-z' = (-x', -y')$$
$$z - z' = z + (-z') = (x - x', y - y')$$

Since

$$(x + iy)\,(x' + iy') = xx' + xiy' + iyx' + iyiy'$$
$$= xx' + i(yx' + xy') - yy'$$

multiplication is defined by

$$z{\cdot}z' = (xx' - yy', yx' + xy')$$

The commutative and associative laws apply to addition — and to this multiplication, with the distributive law

$$(z + z')z'' = zz'' + z'z''$$

for any number $z'' = x'' + iy''$ or (x'', y''). If we make $z'' = z{\cdot}z'$, then:

$$x'' = xx' - yy', \quad y'' = yx' + xy'$$

Solving for x' and y' gives

$$x' = \frac{yy'' + xx''}{x^2 + y^2}, \quad y' = \frac{xy'' - x''y}{x^2 + y^2}$$

If the numerator is not zero — i.e., $z \neq (0,0)$ — we have a complex number $z' = (x', y')$ that is the quotient of z'' divided by z, thereby defining division other than by zero. If $z'' = (1,0)$ — i.e., the real unit value — then:

$$x' = \frac{x}{x^2 + y^2}, \quad y' = \frac{-y}{x^2 + y^2}$$

giving the reciprocal for any non-zero complex number $z = (x, y)$ as

$$z^{-1} = z' = (x', y') = |z|^{-1}(x, -y)$$

If

$$\rho = |z| = +\sqrt{x^2 + y^2}$$

there must, if $\rho > 0$, be an angle θ such that

$$z = \rho(\cos\theta + i\sin\theta) = \rho e^{i\theta}$$

This is called *polar form*. The bracketed term can be abbreviated to cisθ. While

$$\theta = \arctan\frac{y}{x}$$

only one of its two values in any 2π span satisfies signage requirements.

Because their components are mutually independent, complex numbers have played a major part in many sectors of mathematics involving two variables — see Tutorial 28 and Chapter 15.

CHAPTER 9

World in the Round: Globular projections

Preamble

The Aerial projection of Chapter 2 delivered a map that pictured Earth as it could be seen in proxy form via the globe — and can now be seen by astronauts directly — as a circular disc. That disc could show at most slightly less than half the whole world. Extended to an infinite viewpoint to become the Orthographic, however, it can show a full hemisphere — as can other Azimuthal projections. At Equatorial aspect, these projections show the Equator as a horizontal straight line across the centre of the circle, the Poles as points at top and bottom, and mirror-image symmetry about both Equator and the straight central meridian. There is something inherently appealing in presenting a hemisphere in such manner, but in all of those projections the scaling is very distorted toward the ends of the Equator. Various efforts have been made to create similar projections with uniform scaling along the Equator. The Mollweide applied at Simple aspect to a hemisphere meets the criteria by accident, but various so-called globular projections were created explicitly far earlier — centuries before the majority of the projections described so far were introduced.

There is no new mathematics in this chapter.

A 15° graticule is used for showing the varying formulations in this genre. A hemisphere is the usual content of the maps, so 90°E is chosen as the central longitude to provide a full spread of land mass across the map.

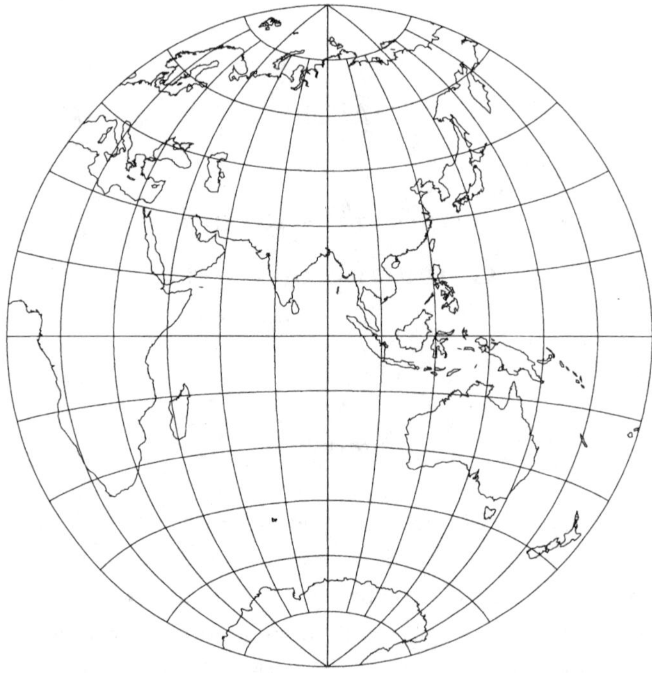

Exhibit 9–1: A natural globular projection — Equidistant Azimuthal at Equatorial aspect.

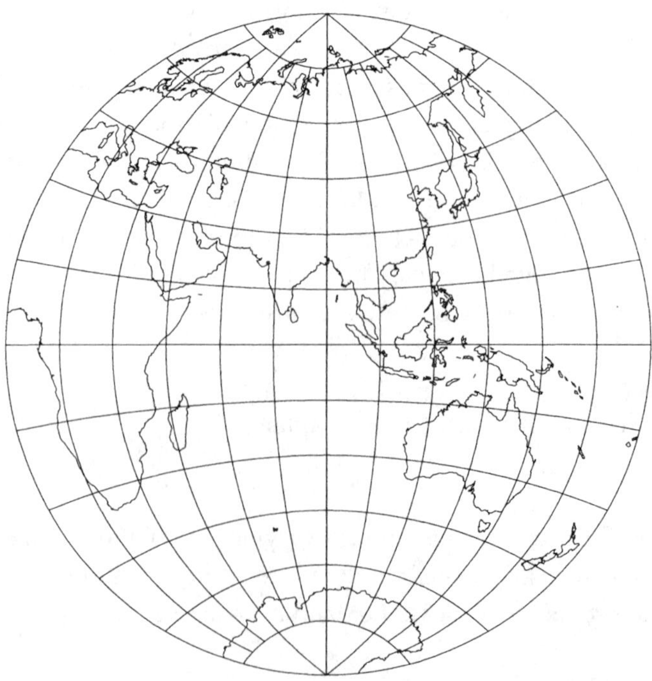

Exhibit 9–2: A constructed globular projection — the archetypical Nicolosi.

METHODOLOGY

Fixed perimeter, variable graticule

Beyond the basic criteria of a circular outline for a hemisphere, Poles as points at top and bottom of the circle and uniform scale along the Equator, globular projections can have any shape and spacing for the meridians and parallels. The Mollweide projection, described and illustrated (in whole-world form) on page 126 within the more important class of Pseudocylindricals, has semi-ellipses for meridians and straight lines for parallels (spaced to achieve equal-area). While explicitly globular projections occur with straight parallels, many have proper curves for both meridians and parallels, with each parallel an arc concave toward its nearer Pole. The Azimuthal Equidistant meets these criteria when applied to a hemisphere at Equatorial aspect, with complex curves for parallels. An example is shown in Exhibit 9–1. More commonly in globular projections, the curves for parallels are circular arcs, as in Exhibit 9–2. This brings any such projection within the loose Polyconic class.

While this last facet may suggest conical-type intermediate parameters for developing the projection, most globular projections are developed directly into orthogonal rather than polar co-ordinates. The plotting axes are routinely centred with the disc of the map, placing the Equator along the x axis and the Poles on the y axis.

Equispacing along the Equator implies a uniform scale thereon, but not necessarily true scale. Though such is often the case, other options are consistent with being classed as Globular, notably making the map overall authalic — it is solely a matter of what scale is stated for the map. Using σ again to denote the relative scale along the Equator, then:

$\sigma = 1$ to have true scale along the Equator

$\sigma = \frac{2\sqrt{2}}{\pi}$ to have the circular hemispheric map authalic

making the semi-Equator presented of length $\pi\,\sigma R$, the central meridian the same.

Using λ to denote relative longitude, for all projections:

if $\phi = 0$ then $\dot{x} = \sigma R \lambda,\ \dot{y} = 0$ by equispacing along Equator

if $\phi = \pm\frac{\pi}{2}$ then $\dot{x} = 0,\ \dot{y} = \sigma R \phi$ to have the Poles at top and bottom . . . . . (223a

if $\lambda = 0$ then $\dot{x} = 0$ being on the straight central meridian

Together these fix, for any specific longitude, three points on its meridian. If a well defined shape is chosen for the meridians — e.g., circular arcs, semi-ellipses — the three define the complete curve.

PROJECTIONS

Nicolosi Globular projection

Believed to have been created by the eminent Arabic-Persian scholar Mohammad ibn Ahmad al-Biruni a thousand years ago, but recreated by Giambattista Nicolosi of Rome [1660a] — and named after him — the archetypical Globular projection has both parallels and meridians curved. Both are actually circular arcs, putting the projection within the loosely defined Polyconic class (Chapter 7).

The circular arc that represents any meridian is precisely defined by the conditions of Equations 223a. Tutorial 17 provides formulae for such meridians. Using the notation of that tutorial modified as map variables we have for longitude λ:

$$\dot{x}_a = 0, \quad \dot{x}_b = \sigma R\lambda, \quad \dot{a} = \tfrac{\pi}{2}\sigma R, \quad \dot{b} = |\sigma R\lambda| = \sigma R|\lambda|$$

Substituting in the formula of the tutorial gives the radius of the circular arc as

$$r_{\rm m} = r = \frac{\tfrac{\pi}{2}\sigma R}{2}\left\{\frac{\tfrac{\pi}{2}\sigma R}{\sigma R|\lambda|} + \frac{\sigma R|\lambda|}{\tfrac{\pi}{2}\sigma R}\right\}$$

$$= \frac{\pi\sigma R}{4}\left\{\frac{\pi}{2|\lambda|} + \frac{2|\lambda|}{\pi}\right\} \qquad \text{. . . . . (224a}$$

$$= \sigma R l_1$$

where $l_1 = \dfrac{\pi}{4}\left\{\dfrac{\pi}{2|\lambda|} + \dfrac{2|\lambda|}{\pi}\right\}$ . . . . (224b

While these formulae for $r_{\rm m}$ and l_1 are infinite when $\lambda = 0$, that circumstance is already covered by Equations 223a. The equations for a meridian as a circular arc are thus:

if $\lambda = 0 \quad \dot{x} = 0$

if $\lambda \neq 0 \quad \dot{x} = (\sigma R\lambda \pm \sigma R l_1) \mp \sqrt{(\sigma R l_1)^2 - y^2}$ . . . (224c

$$= \frac{\lambda}{|\lambda|}\sigma R\left\{|\lambda| - l_1 + \sqrt{l_1^2 - \left(\frac{y}{\sigma R}\right)^2}\right\}$$

Being deduced from the mere assumption of circular arcs along with the standard features of the Globular projections, this applies to any projection with meridians of such shape.

Equispacing along the central meridian implies uniform scaling (of relative scale σ) along it, hence the parallel for latitude ϕ must pass through the point

$$(0, \sigma R\,\phi)$$

Equispacing along the bounding meridians implies that it must pass through the points

$$\left(\pm\tfrac{\pi}{2}\sigma R\cos\phi, \tfrac{\pi}{2}\sigma R\sin\phi\right)$$

The circular arc that represents any parallel is precisely defined by these three points. Using formulae from Tutorial 17 and its notation, modified as map variables, gives for latitude ϕ:

$$\dot{y}_a = \tfrac{\pi}{2}\sigma R \sin\phi, \quad \dot{a} = \left|\tfrac{\pi}{2}\sigma R \cos\phi\right| = \tfrac{\pi}{2}\sigma R \cos\phi,$$

$$\dot{y}_b = \sigma R \phi, \quad \dot{b} = \left|\tfrac{\pi}{2}\sigma R \sin\phi - \sigma R \phi\right| = \tfrac{1}{2}\sigma R\,|2\phi - \pi\sin\phi|$$

Substituting in the formula of the tutorial gives the radius r_p of this circular arc as

$$r_p = r = \frac{\tfrac{\pi}{2}\sigma R\cos\phi}{2}\left\{\frac{\tfrac{\pi}{2}\sigma R\cos\phi}{\tfrac{1}{2}\sigma R\,|2\phi-\pi\sin\phi|} + \frac{\tfrac{1}{2}\sigma R\,|2\phi-\pi\sin\phi|}{\tfrac{\pi}{2}\sigma R\cos\phi}\right\}$$

$$= \frac{\pi\sigma R\cos\phi}{4}\left\{\frac{\pi\cos\phi}{|2\phi-\pi\sin\phi|} + \frac{|2\phi-\pi\sin\phi|}{\pi\cos\phi}\right\} \quad \text{. . . . . (225a}$$

$$= \sigma R f_1$$

$$\text{where } f_1 = \frac{\pi\cos\phi}{4}\left\{\frac{\pi\cos\phi}{|2\phi-\pi\sin\phi|} + \frac{|2\phi-\pi\sin\phi|}{\pi\cos\phi}\right\} \quad \text{. . . . . (225b}$$

While these formulae for both r_p and f_1 are infinite when $|\phi| = \pi/2$, that circumstance is already covered by Equation 223a. The equations for a parallel as a circular arc are thus:

$$\text{if } \phi = \pm\tfrac{\pi}{2} \quad \dot{y} = \sigma R \phi$$

$$\text{if } \phi \neq \pm\tfrac{\pi}{2} \quad \dot{y} = \left(\sigma R\phi + \frac{\phi}{|\phi|}\sigma R f_1\right) - \frac{\phi}{|\phi|}\sqrt{r_p^{\,2} - x^2} \quad \text{. . . . . (225c}$$

$$= \frac{\phi}{|\phi|}\sigma R\left\{|\phi| + f_1 - \sqrt{f_1^{\,2} - \left(\frac{x}{\sigma R}\right)^2}\right\}$$

Because Equations 225c & 225b are mathematically independent, the two can be solved to provide separate formulae for the two variables. Incorporating the exceptions covered by Equations 223a, the general plotting equations for the **Nicolosi Globular** projection become:

$$\text{if } \phi = 0 \quad \dot{x} = \sigma R\lambda, \quad \dot{y} = 0$$

$$\text{if } \phi = \pm\tfrac{\pi}{2} \quad \dot{x} = 0, \quad \dot{y} = \sigma R\phi$$

$$\text{if } \lambda = 0 \quad \dot{x} = 0, \quad \dot{y} = \sigma R\phi$$

$$\text{if } \lambda = \pm\tfrac{\pi}{2} \quad \dot{x} = \sigma R\lambda\cos\phi, \quad \dot{y} = \sigma R|\lambda|\sin\phi \quad \text{. . . . . (225d}$$

$$\text{else} \quad \dot{x} = \frac{\pi\sigma R}{2}\left\{D + \frac{\lambda}{|\lambda|}\sqrt{D^2 + C\cos^2\phi}\right\}$$

$$\dot{y} = \frac{\pi\sigma R}{2}\left\{E - \frac{\phi}{|\phi|}\sqrt{E^2 - B\left(\frac{F_1^{\,2}\sin^2\phi}{L_1^{\,2}} + F_1\sin\phi - 1\right)}\right\}$$

where $f_2 = \dfrac{2\phi}{\pi}$ a function purely of ϕ,

$$F_1 = \frac{1 - f_2^{\,2}}{\sin\phi - f_2} \quad \text{a function purely of } f_2 \text{ and } \phi \text{, hence of } \phi$$

$$L_1 = \frac{\pi}{2\lambda} - \frac{2\lambda}{\pi} \quad \text{a function purely of } \lambda,$$

$$B = \left\{ 1 + \left(\frac{F_1}{L_1} \right)^2 \right\}^{-1},$$

$$C = \left\{ 1 + \left(\frac{L_1}{F_1} \right)^2 \right\}^{-1},$$

$$D = C\left(\frac{L_1 \sin\phi}{F_1} - \frac{L_1}{2} \right),$$

and $$E = B\left(\frac{F_1^{\,2} \sin\phi}{L_1^{\,2}} + \frac{F_1}{2} \right).$$

Fournier Globular projection

Georges Fournier of France [1643a] kept circular arcs for parallels, identically with Nicolosi, but adopted semi-ellipses for meridians. Equispaced along the Equator, the meridional graticule is the same as that of the later but better known Mollweide. The result, illustrated in Exhibit 9–3, is barely different from the Nicolosi on page 222.

Because the meridian for longitude λ passes through the Equatorial point $(\sigma R\lambda, 0)$ and the Poles at $(0, \pm\sigma R^{\pi}\!/_2)$, the formula for the complete ellipse is

$$\left(\frac{\dot{x}}{\sigma R \lambda} \right)^2 + \left(\frac{\dot{y}}{\sigma R \frac{\pi}{2}} \right)^2 = 1 \qquad \text{or} \qquad (\pi \dot{x})^2 + (2\dot{y})^2 = (\pi \sigma R \lambda)^2$$

Adopting abbreviations

$$F_2 = \frac{\left(\frac{\pi}{2}\right)^2 - \phi^2}{\pi|\sin\phi| - 2|\phi|} \quad \text{a function purely of } \phi \qquad \text{. . . . . (226a}$$

and $$L_2 = \left(\frac{2\lambda}{\pi} \right)^2 - 1 \quad \text{a function purely of } \lambda. \qquad \text{. . . . . (226b}$$

and incorporating the exceptions covered by Equations 223a the general plotting equations for the **Fournier Globular** projection are:

$$\begin{aligned}
&\text{if } \phi = 0 && \dot{x} = \sigma R \lambda, \quad \dot{y} = 0 \\
&\text{if } \lambda = 0 && \dot{x} = 0, \quad \dot{y} = \sigma R \phi \\
&\text{if } \lambda = \pm\tfrac{\pi}{2} && \dot{x} = \sigma R \lambda \cos\phi, \quad \dot{y} = \sigma R \lambda \sin\phi \qquad \text{(227a} \\
&\text{else} && \dot{y} = \frac{\phi}{|\phi|}\left\{\sqrt{F^2 - L\left(\frac{\pi^2}{4} - F\pi|\sin\phi| - \lambda^2\right)} - F\right\}\frac{\sigma R}{L} \\
& && \dot{x} = \left\{\sqrt{1 - \left(\frac{2\dot{y}}{\pi\sigma R}\right)^2}\right\}\sigma R \lambda
\end{aligned}$$

where $F = F_2$ of Equation 226a and $L = L_2$ of Equation 226b.

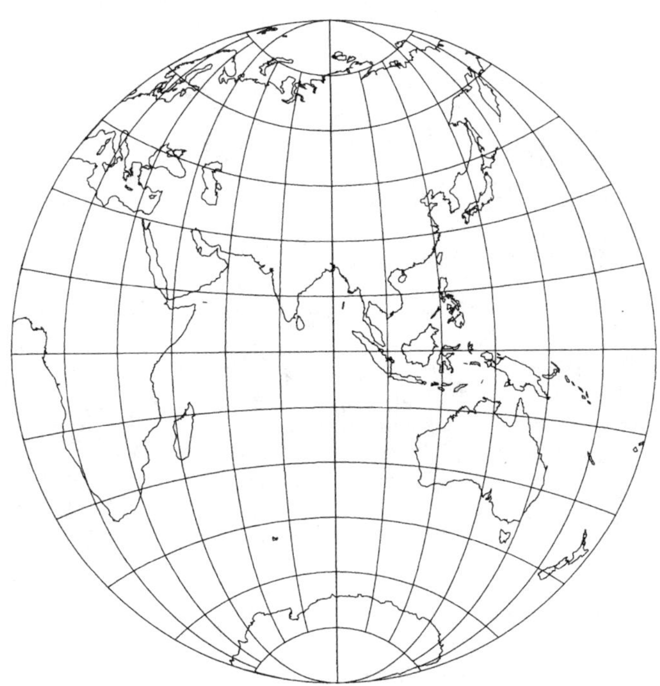

Exhibit 9–3: The Fournier Globular I projection.

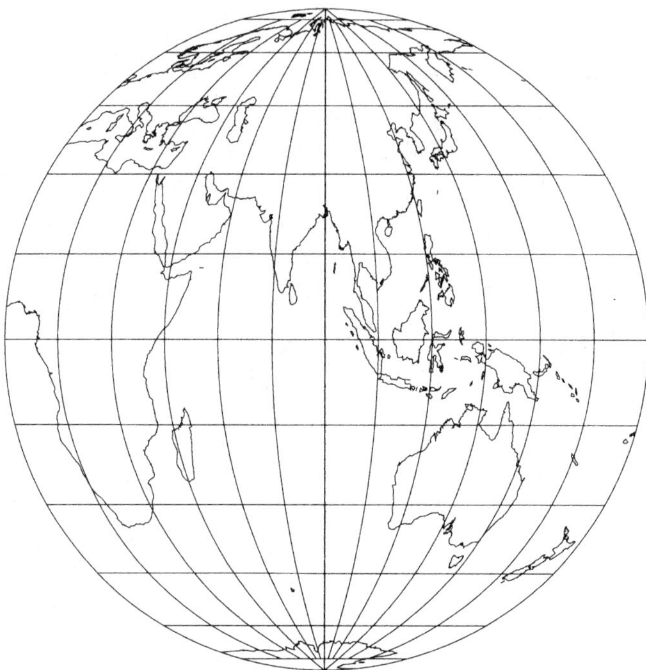

Exhibit 9–4: The Bacon projection of the 13th Century.

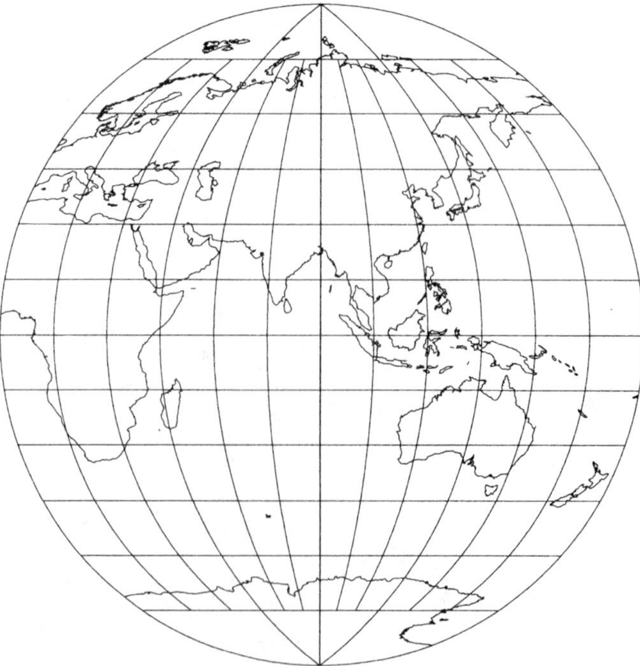

Exhibit 9–5: The Apian I projection of the 16th Century.

Bacon and Apian I Globular projections

Introduced by Roger Bacon of England [1266a], the **Bacon Globular** projection has straight parallels equally spaced along the circular perimeter meridians, so

$$\dot{y} = \tfrac{\pi}{2}\sigma R \sin\phi \qquad \text{. . . . . (229a}$$

for all longitudes and latitudes, providing one plotting equation. The meridians are circular arcs, so Equation 224c applies; substituting this formula into that earlier equation, the remaining plotting equations for the Bacon Globular (or **Glareanus**) projection are:

$$\text{if } \lambda = 0 \quad \dot{x} = 0$$

$$\text{if } \lambda \neq 0 \quad \dot{x} = \frac{\lambda}{|\lambda|}\sigma R\left\{|\lambda| - l_1 + \sqrt{l_1^{\,2} - \left(\frac{\tfrac{\pi}{2}\sigma R\sin\phi}{\sigma R}\right)^2}\right\} \qquad \text{. . . . . (229b}$$

$$= \frac{\lambda}{|\lambda|}\sigma R\left\{|\lambda| - l_1 + \sqrt{l_1^{\,2} - \left(\frac{\pi\sin\phi}{2}\right)^2}\right\}$$

where l_1 is given by Equation 224b.

The projection is shown in Exhibit 9–4. (While equispaced along the Equator the meridians are not equispaced along other parallels; Mollweide shows that, with straight parallels, equispacing of meridians carries over from Equator to all parallels only with semi-elliptical meridians.)

Peter Apian of Saxony [1524a] also produced a projection with circular-arc meridians and straight parallels, but with the latter equispaced along the central meridian. With those latitudinal parallels geometrically parallel to the Equator, their general formula is hence

$$\dot{y} = R\phi \qquad \text{. . . . . (229c}$$

Substituting in Equations 224c, the remaining plotting equations for the **Apian Globular I** projection are:

$$\text{if } \lambda = 0 \quad \dot{x} = 0$$

$$\text{if } \lambda \neq 0 \quad \dot{x} = \frac{\lambda}{|\lambda|}\sigma R\left\{|\lambda| - l_1 + \sqrt{l_1^{\,2} - \left(\frac{\sigma R\,\phi}{\sigma R}\right)^2}\right\} \qquad \text{. . . . . (229d}$$

$$= \frac{\lambda}{|\lambda|}\sigma R\left\{|\lambda| - l_1 + \sqrt{l_1^{\,2} - \phi^2}\right\}$$

where l_1 is given by Equation 224b.

The projection, shown in Exhibit 9–5, has a further life as the basis of the Ortelius Oval described in Chapter 10.

The equispaced Globular projection

Except for the Mollweide, all these Globular projections have, by their selection of meridional shape, precluded equispacing of meridians along parallels other than the Equator. The Nicolosi comes close, and can be modified to have such equispacing with little alteration to its appearance — and significant simplification of the derivation of the plotting equations. We merely take Equation 225c (with its supporting Equation 224a) and define the location of points along each parallel proportionally by longitude. Since the parallels are circular arcs, we resort to intermediate parameters of the Conical style, as with the Polyconic projection. The radius of curvature for latitude ϕ is the r_{p} of Equation 225a, which we now equate with $\dot{\rho}$. Its centre of curvature is at $(0, R\phi + \dot{\rho})$ for northern latitudes, at $(0, R\phi - \dot{\rho})$ for southern. The complete hemispheric parallel between the bounding meridians subtends at the centre of curvature twice the angle, to get

$$\arccos\left\{\frac{\rho - R\left|\phi - \frac{\pi}{2}\sin\phi\right|}{\rho}\right\} = \arccos\left\{1 - \frac{R}{\rho}\left|\phi - \frac{\pi}{2}\sin\phi\right|\right\} = \arccos\left\{\frac{\cos^2\phi - \left(\sin\phi - \dfrac{2\phi}{\pi}\right)^2}{\cos^2\phi + \left(\sin\phi - \dfrac{2\phi}{\pi}\right)^2}\right\}$$

Dividing proportionally to longitude gives the second parameter for longitude λ as

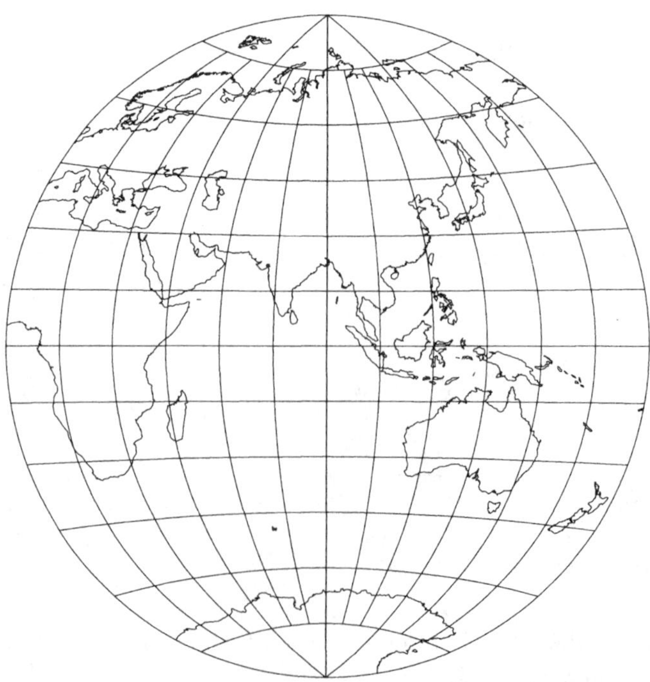

Exhibit 9–6: The van der Grinten IV projection - globular.

$$\dot{\theta} = \frac{2\lambda}{\pi}\arccos\left\{1 - \frac{R}{\rho}\left|\phi - \frac{\pi}{2}\sin\phi\right|\right\} = \frac{2\lambda}{\pi}\arccos\left\{\frac{\cos^2\phi - \left(\sin\phi - \frac{2\phi}{\pi}\right)^2}{\cos^2\phi + \left(\sin\phi - \frac{2\phi}{\pi}\right)^2}\right\} \quad \text{. . . (231a}$$

The Cartesian plotting equations for the **Equispaced Globular** projection are:

$$\dot{x} = \dot{\rho}\sin\dot{\theta}$$
$$\dot{y} = \sigma R\phi + \frac{\phi}{|\phi|}\dot{\rho}\left(1 - \cos\dot{\theta}\right) \quad \text{. . . . . (231b}$$

with $\dot{\rho} = r_P$ from Equation 225a, $\dot{\theta}$ from Equation 231a.

The van der Grinten Globular

Centuries after the fashion for globular maps had had its heyday, Alphons van der Grinten of the U.S.A. presented [1904a] a Globular projection with circular arcs for both meridians and parallels, with each set equispaced along the central specimen of the other set. His presentation was actually of a whole-world map that included its central hemisphere in a circle. Patented in the U.S.A. [1905a], the map of the whole world has a butterfly-like appearance (but the name *Apple-shaped*) illustrated on page 237. As can be seen there, the outer meridians are major parts of circles with diameters greater than the length of the central meridian, resulting in the higher latitudes on the map being further than the Poles from the Equator. Now known as the van der Grinten IV projection (see next section for others), its formulae developed for the Globular hemispheric form extend to the more-used whole-world coverage. As all the parallels are circular arcs, even with such extension, the projection falls within the loosely defined Polyconic class.

The characteristic that distinguishes this van der Grinten from the Nicolosi is that the circular arcs forming parallels are not set to make the circular meridian of the Globular form equispaced. As with the other Globular projections, the scale is usually set to be true along the central meridian and the Equator. Hence, Equations 223a hold, but the equispacing along the central meridian means that, if $\lambda = 0$, then, for all latitudes:

$$\dot{x} = 0$$
$$\dot{y} = \sigma R\phi$$

Let f_3 be the ratio of a general latitude ϕ to its nearer Polar latitude — i.e.,

$$f_3 = \frac{|\phi|}{\frac{\pi}{2}} = \frac{2|\phi|}{\pi} = \left|\frac{2\phi}{\pi}\right| \quad \text{a function purely of } \phi \quad \text{. . . . . (231c}$$

then define

$$G = \frac{8f_3 - f_3^{\,4} - 2f_3^{\,2} - 5}{2f_3^{\,3} - 2f_3^{\,2}} \quad \text{a function purely of } f_3 \text{ so purely of } \phi,$$

$$L_3 = \frac{2\lambda}{\pi} + \frac{\pi}{2\lambda} \quad \text{a function purely of } \lambda$$

and $$H = \frac{|\lambda| - \frac{\pi}{2}}{\left||\lambda| - \frac{\pi}{2}\right|}\sqrt{L_3^{\,2} - 4} \quad \text{a function purely of } L_3 \text{ so purely of } \lambda.$$

and finally the mixed functions

$$M = (f_3 + G)^2\left(f_3^{\,2} + G^2H^2 - 1\right) + \left(1 - f_3^{\,2}\right)\left(f_3^{\,2}\left\{(f_3 + 3G)^2 + 4G^2\right\} + 12f_3G^3 + 4G^4\right)$$

and $$N = \frac{H\left\{(f_3 + G)^2 + G^2 - 1\right\} + 2\sqrt{M}}{4(f_3 + G)^2 + H^2}$$

Including the general scaling factor, the plotting equations for the **van der Grinten IV** projection then become:

$$\dot{x} = \frac{\lambda}{|\lambda|}\frac{\pi\sigma R}{2}N$$

$$\dot{y} = \frac{\phi}{|\phi|}\frac{\pi\sigma R}{2}\sqrt{1 + H\cdot|N| - N^2} \qquad \text{. . . . . (232a}$$

where, again, the initial term in each equation merely accords the sign of longitude else latitude to the co-ordinate value. It is shown in globular form, with a hemisphere in a circle, on page 230.

Hans Maurer [1921c, 1935a] also offered globular projections equally scaled along Equator and central meridian and with all cardinal lines circular arcs, at least two being extensible to the whole world.

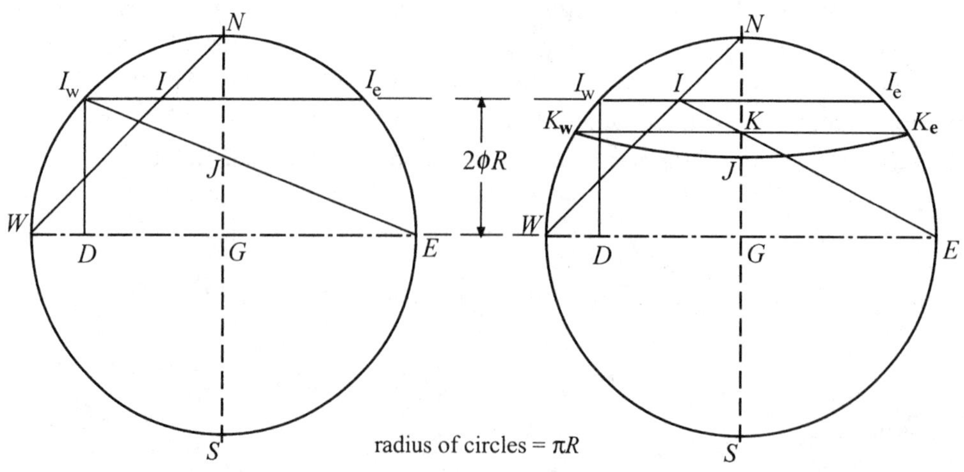

Exhibit 9–7: The contriving geometry for van der Grinten I projection

FURTHER GLOBULAR-STYLE PROJECTIONS

The original van der Grinten

A projection akin to the Globular style but designed to have the whole world within a circular disc was also produced by van der Grinten (and granted a U.S. patent) [1905a]. Apparently intended to do that in a superior manner to the century-old Lambert projection (developed in Chapter 10) while retaining that other's circular arcs for both meridians and parallels, the new projection necessarily lost the conformality that prompted the old.

Labelled subsequently the **van der Grinten I** projection, it has the Poles on its perimeter, making the length of the central meridian that of the full Equator. Assuming the co-ordinates are centred with the map, then:

if $\phi = 0$ then $\dot{x} = \sigma R \lambda,\ \dot{y} = 0$ by equi-spacing along Equator

if $\lambda = 0$ then $\dot{x} = 0$ being on the straight central meridian . . . (233a

if $\phi = \pm\frac{\pi}{2}$ then $\dot{x} = 0,\ \dot{y} = 2\sigma R\phi$ to have the Poles at top and bottom

with the whole map being bounded by a circle of radius $\pi\sigma R$.

Both the spacing and the radii of curvature of parallels are geometrically contrived, the geometry being shown in Exhibit 9–7. To help clarity, we assume true scale along the Equator — i.e., relative scaling factor $\sigma = 1$, allowing σ to be omitted. For latitude ϕ, we first draw the line parallel to and distance $2R|\phi|$ from the Equator (hence equispaced along the focal meridian). This is line $I_w I_e$. We then draw the line from the nearer Pole to an Equatorial end point (line WN), and label as I the point of intersection of our two lines.

To establish the new spacing along the central meridian, we draw the line $I_w E$. Its intersection with the line SN (the focal meridian), labelled J, is to be on the parallel for latitude ϕ. To gain a formula for its distance from the Equator (i.e., the length of GJ), we drop the vertical from I_w to the line WE and label its foot as D, then from the left-hand diagram opposite, we can see that:

$$\text{length of } I_W D = 2R|\phi| \text{ and length of } DG = \sqrt{(\pi R)^2 - (2R\phi)^2} = R\sqrt{\pi^2 - (2\phi)^2}$$

$$\text{length of } GJ = \frac{\text{length of } EG}{\text{length of } ED}\text{length of } I_W D = \frac{\pi R}{\pi R + R\sqrt{\pi^2 - (2\phi)^2}} 2\pi R|\phi|$$

Adopting Cartesian co-ordinates with origin at G, the co-ordinates for J are thus:

$$\dot{x} = 0, \quad \dot{y} = \frac{2R\phi}{\left(1 + \sqrt{1 - \left(\frac{2\phi}{\pi}\right)^2}\right)} = \frac{2R\phi}{\left(1 + \sqrt{1 - f_3{}^2}\right)} \quad \text{. . . . . (233b}$$

where f_3 is given by Equation 231c.

The right-hand diagram in Exhibit 9–7 shows the method for defining the specific arc. We now draw the line *IE* and label its intersection with the line *SN* as *K*. We then draw the line through *K* parallel to the Equator and label its intersections with the bounding circle as K_w and K_e. We make the parallel for latitude ϕ the circular arc through K_w, *J*, and K_e.

To reduce the complexity of the plotting equations we introduce the abbreviations

$$F_3 = \sqrt{1 - f_3^{\,2}}\,, \qquad \text{. . . . . (234a}$$

$$S = \frac{F_3}{f_3 + F_3 - 1} = \frac{\sqrt{1 - f_3^{\,2}}}{f_3 - 1 + \sqrt{1 - f_3^{\,2}}} \qquad \text{. . . . . (234b}$$

and

$$T = S\left(\frac{2}{f_3} - 1\right) \qquad \text{. . . . . (234c}$$

all being functions of ϕ alone, via f_3 as given by Equation 231c.

The circular arcs forming the meridians are precisely defined by the three conditions of Equations 233a. We again define a compound function purely of longitude, now

$$L_4 = \frac{1}{2}\left|\frac{\pi}{\lambda} - \frac{\lambda}{\pi}\right| \qquad \text{. . . . . (234d}$$

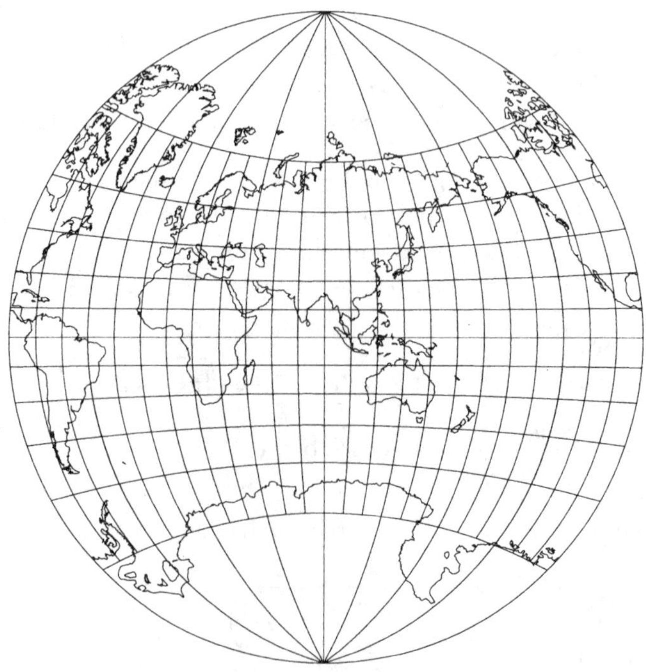

Exhibit 9–8: The van der Grinten I projection.

Writing F for F_3 of Equation 234a and L for L_4 of Equation 234d, and reinstating the relative scaling factor (which has to be $2/\pi$ for an authalic map), the general plotting equations for the van der Grinten I (or just **van der Grinten**) projection then become:

$$\text{If } \lambda = 0 \quad \dot{x} = 0, \quad \dot{y} = \frac{2\sigma R}{1+F}$$

$$\text{else} \quad \dot{x} = \frac{\lambda}{|\lambda|}\frac{\pi\sigma R}{T^2+L^2}\left(L\left(S-T^2\right)+\sqrt{L^2\left(S-T^2\right)^2-\left(T^2+L^2\right)\left(S^2-T^2\right)}\right) \quad \text{(235a}$$

$$\dot{y} = \frac{\phi}{|\phi|}\frac{\pi\sigma R}{T^2+L^2}\left(T\left(L^2+S\right)-L\sqrt{\left(L^2+1\right)\left(T^2+L^2\right)-\left(L^2+S\right)^2}\right)$$

where S is defined in Equation 234b, T in Equation 234c.

Although a given parallel crosses the central meridian further from the Equator than it does with the Lambert (Exhibit 10–13, page 254), the spacing of parallels still escalates rapidly as distance from the Equator increases, giving major areal exaggeration in the high latitudes.

Variants of the van der Grinten I

Two variations of the van der Grinten I were produced by that author, with unchanged meridians and parallels identically spaced along the central meridian, but with the parallels differing in their shape. They are illustrated overleaf.

In the first of these variants — the **van der Grinten II** projection — the parallels remain circular arcs but are orthogonal to the meridians — so are more sharply curved. (The intersection with the bounding circle of parallels for latitudes 30° and 45° in projection II are virtually the same respectively as those for 45° and 60° in projection I.) Writing f for f_3 of Equation 231c, F for F_3 of Equation 234a, L for L_4 of Equation 234d and

$$V = \frac{F\sqrt{1+L^2}-LF^2}{1+L^2f^2} = \frac{\sqrt{\left(1-f^2\right)\left(1+L^2\right)}-L\left(1-f^2\right)}{1+L^2f^2}$$

the plotting equations for the van der Grinten II projection become:

$$\text{If } \lambda = 0 \quad \dot{x} = 0, \quad \dot{y} = \frac{2\sigma R}{1+F}$$

$$\text{else} \quad \dot{x} = \frac{\lambda}{|\lambda|}\pi\sigma R\, V \quad \text{(235b}$$

$$\dot{y} = \frac{\phi}{|\phi|}\pi\sigma R\sqrt{1-V^2-2LV}$$

The **van der Grinten III** projection has straight parallels. It is usually classed as a modified Pseudocylindrical (it lacks uniform scale along all parallels). Its essence is that of the preceding two projections. Setting

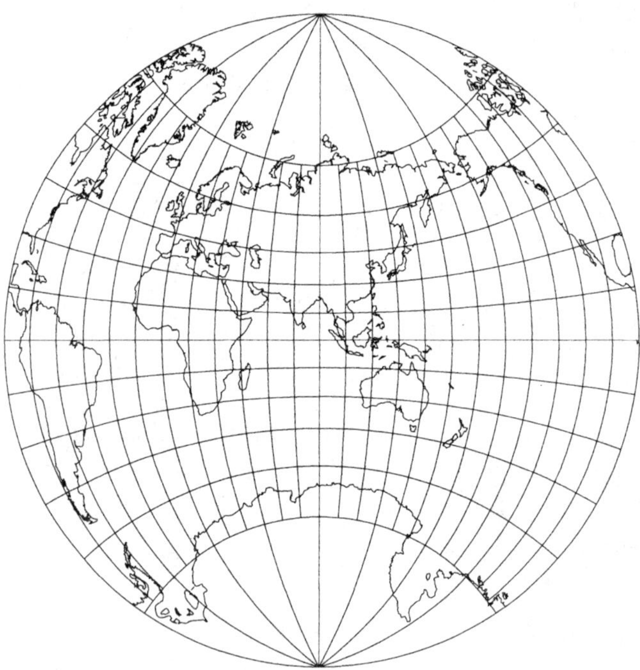

Exhibit 9–9: The world van der Grinten II projection.

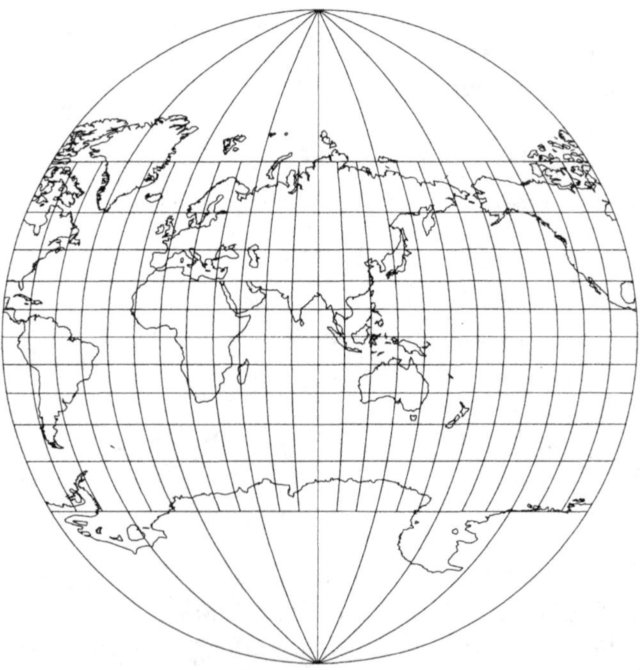

Exhibit 9–10: The world van der Grinten III projection.

$$W = \frac{f}{1 + \sqrt{\left(1 - f^2\right)}}$$

where $f = f_3$ of Equation 231c, making W a function purely of ϕ, the plotting equations for the van der Grinten III projection become:

$$\dot{x} = \frac{\lambda}{|\lambda|}\sigma\pi R\left\{\sqrt{L^2 + 1 - W^2} - L\right\}$$

$$\dot{y} = \frac{\phi}{|\phi|}\sigma\pi RW \qquad \ldots\ldots\ldots (237a$$

where $L = L_4$ of Equation 234d.

The projection is illustrated in Exhibit 9–10. [See 1977c for some later notes on this projection, 1980c for corrections thereto. See also 1981b for related calculations from grid values to longitude and latitude.]

Concerned with the angular distortion of the van der Grinten I when used for whole-world maps, W. E. Brooks and C. E. Roberts [1976c] developed another variant with straight parallels. Called a **Modified van der Grinten Equatorial Arbitrary Oval** projection, it has the parallel for latitude ϕ crossing the central meridian at the same point as with the van der Grinten, i.e., point J in our diagram. Its advantages seem marginal.

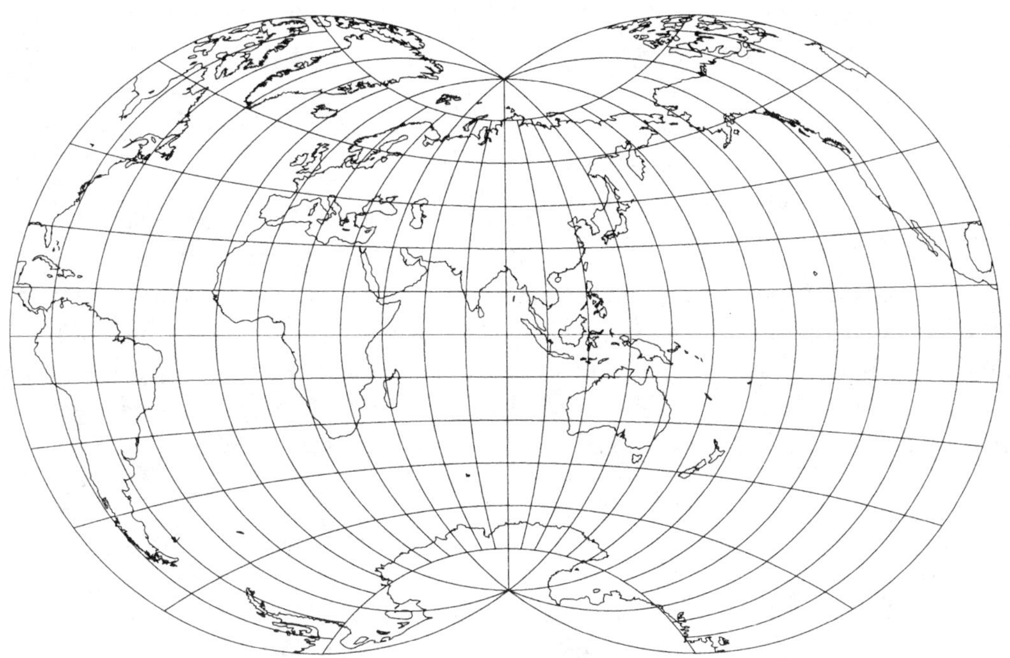

Exhibit 9–11: The van der Grinten IV projection – whole world.

TUTORIAL 24 MATRIX MULTIPLICATION

Transposition, addition and scalar multiplication of matrices along with the determinants are covered in Tutorial 20, which presented the matrix as merely an array. The value of matrices in mathematics comes, however, from their representation of functions representing transformations. Multiplication of matrices then equates with compounding of transformations.

Any two matrices can be multiplied if their adjacent dimensions are identical — e.g., a 2×3 followed by a 3×4. Such multiplication is seen as the transformation by the first matrix of the consequence of the second. The multiplication produces a matrix with dimensions identical with the non-adjacent dimensions of the incoming matrices (i.e., 2×4 for the cited example). The individual elements equal the dot-product of the corresponding row of the first matrix with the corresponding column of the second matrix, effectively regarding the sequence of elements in rows and columns as vectors.

It follows from the above that two matrices with identical dimensions can be directly multiplied only if they are square, but if either is transposed they can be multiplied whatever their dimensions.

While matrices of any dimensions can facilitate various steps in the more advanced mathematics of cartography, 2×2 and 3×3 matrices are of particular value because they represent the essential transformations of map projecting. The 2×2 matrix relates readily to the transformation from longitude and latitude to a pair of intermediate parameters then to x, y co-ordinates. The 3×3 matrix provides similar facility in a 3-dimensional context. For example, if P is the general point (X,Y,Z), then

$$\begin{bmatrix} a_{11} & a_{12} & a_{13} \\ a_{21} & a_{22} & a_{23} \\ a_{31} & a_{32} & a_{33} \end{bmatrix} \begin{bmatrix} X \\ Y \\ Z \end{bmatrix} = \begin{bmatrix} a_{11}X + a_{12}Y + a_{13}Z \\ a_{21}X + a_{22}Y + a_{23}Z \\ a_{31}X + a_{32}Y + a_{33}Z \end{bmatrix}$$

produces a single-column matrix representing the transformed point.

If **A** is a non-singular $m \times m$ matrix, then values of scalar k that render the determinant $|\mathbf{A} - k\mathbf{I}|$ equal to zero are called *eigenvalues* of **A**. There are m such values, although they are not necessarily distinct. If m is odd, there is at least one solution that is a real number; any others can be complex numbers.

If **V** is a non-zero $m \times 1$ matrix such that $\mathbf{A} \cdot \mathbf{V} = k\,\mathbf{V}$, then **V** is called an *eigenvector* of **A**. Under transformation **A**, every such vector is merely multiplied in length by k. Likewise all points on the line of **V** have their co-ordinates multiplied by k. Clearly, any scalar multiple of an eigenvector must be an eigenvector too.

If **A** is a 3×3 matrix it must have at least one real eigenvalue. If it has just one, then it is a simple rotation, with the corresponding eigenvector defining the line of the rotational axis.

CHAPTER 10

Divide and Conquer: Extension by arithmetic tricks

Preamble

The Gnomonic and Mercator projections extend infinitely. At Simple aspect, the former cannot reach the Equator, the latter cannot reach the Poles. The Equidistant and Equal-area Azimuthal projections at Equatorial aspect present interesting maps, but the parallels are overly curved, and the Poles are buried within the map if the whole world is covered. The Apian Globular has better curves, but does not go beyond one hemisphere. All of these perceived failings have been addressed by simple arithmetical manipulation to enhance the maps, creating revised projections that are accorded distinct names.

Again there is no new mathematics in this chapter, just a revised cartographic approach.

The maps are generally centred again on the Greenwich meridian to have the land masses spanning them. One projection created explicitly centred on 10°E, 45°N and (along with a matched general projection) is presented so. The mimic of Lambert's original map (on page 254) is centred on Greenwich, but his original was centred somewhat east of the modern 0° meridian.

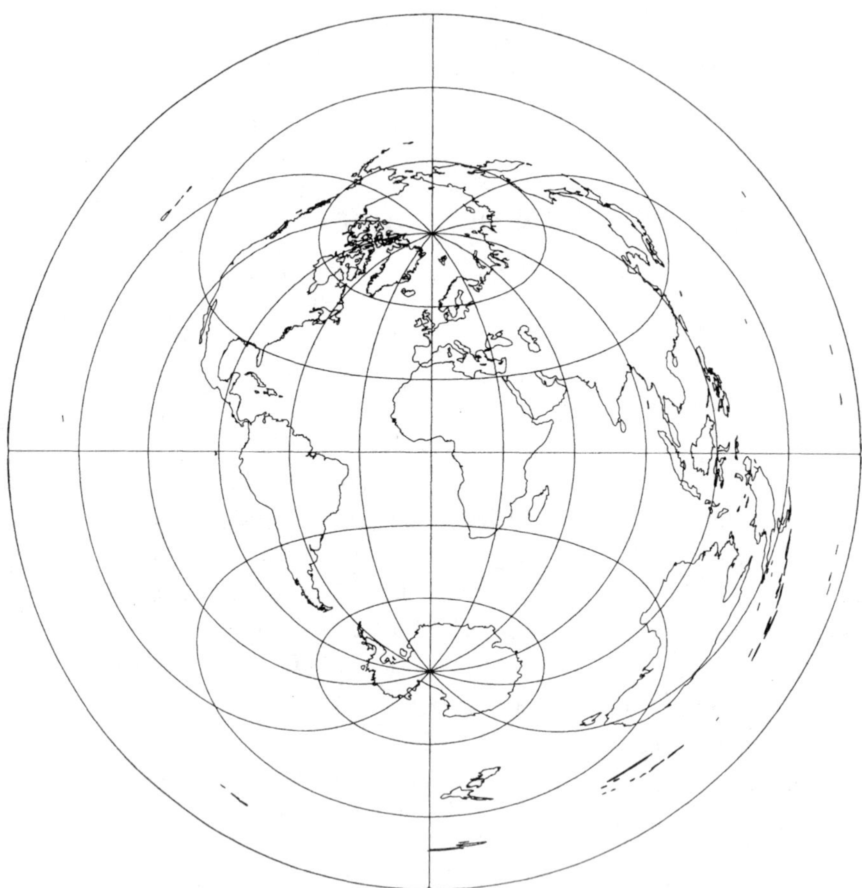

Exhibit 10–1: The Equidistant Azimuthal projection at Equatorial aspect.

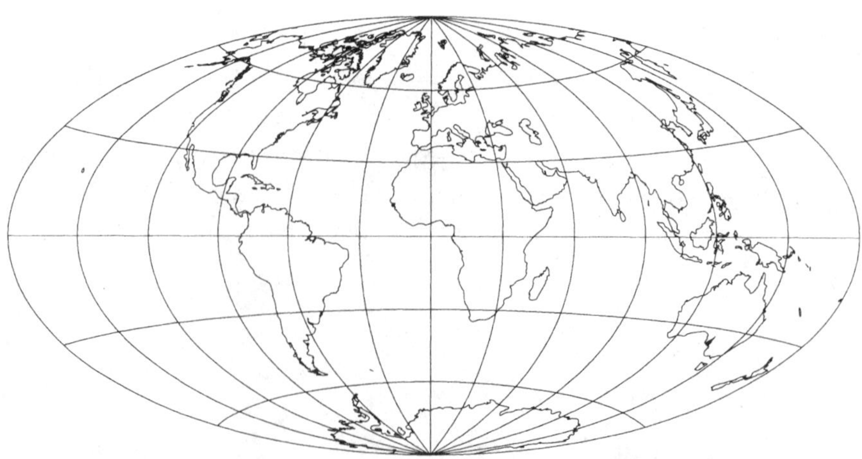

Exhibit 10–2: The Aitoff projection at its Simple aspect.

SHRINKING LONGITUDES

Equidistant Azimuthal

The Equidistant Azimuthal is illustrated in Exhibit 10–1 at Equatorial aspect in whole-world form and in Exhibit 9–1 on page 222 in hemispheric form as a globular map. With the focus lying on the Equator, both maps provide consistent scale along the full length of the Equator and between the Poles. The latter, however, has the disadvantage (shared by all whole-world Azimuthals at Equatorial aspect) of the heavy distortion beyond the Poles, while the globular version has its hemispheric restriction.

An interesting and attractive compromise was created by David Aitoff [1889a], deploying the simple expedient of halving the true longitudes to bring everywhere on Earth within the range ±90°, then plotting as in the Equidistant Azimuthal in globular style, except that he stretched the map laterally by doubling the x co-ordinate. The result is illustrated in Exhibit 10–2. The stretching reduces the curvature of the parallels and returns the Equator to a length twice that of the central meridian. The other parallels and meridians are complex curves, but the bounding meridians form an ellipse, hence have linear continuity at the Poles. The equidistant feature of the base projection means that the meridians are equispaced along the central meridian and the parallels are equispaced along the central meridian. Overall the map is akin to the Pseudocylindricals, particularly the Mollweide, but the curved parallels exclude it from that class. The Aitoff is neither equal-area nor conformal nor equidistant, but its moderated curvature and smooth point-Poles make it an attractive map for general whole-world maps.

Using Equations 203b for the basic projection, the plotting equations for the **Aitoff** projection are, assuming relative scale σ along the Equator and hence along the central meridian too:

$$\text{if } \phi = 0 \quad \dot{x} = \sigma R \lambda$$

$$\dot{y} = 0$$

$$\text{else} \quad \dot{x} = \frac{\lambda}{|\lambda|} 2\sigma R\, C \sqrt{1 - \left(\frac{\sin\phi}{\sin C}\right)^2} \qquad \text{(241a)}$$

$$\dot{y} = \sigma R\, C \frac{\sin\phi}{\sin C}$$

where

$$C = \arccos\left\{\cos\phi \cos\left(\frac{\lambda}{2}\right)\right\}$$

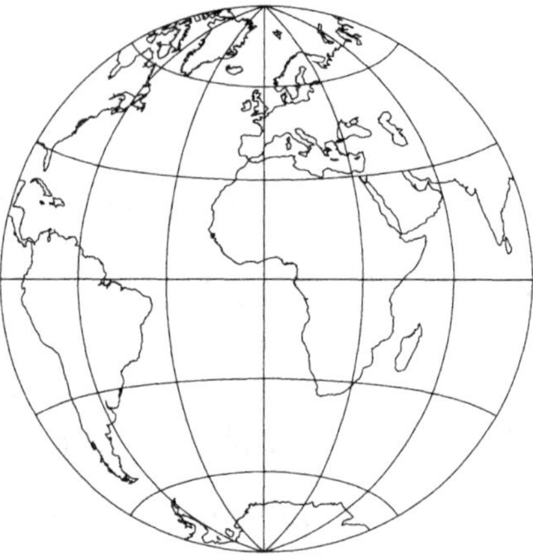

Exhibit 10–3: A hemispheric Equal-area Azimuthal projection at Equatorial aspect.

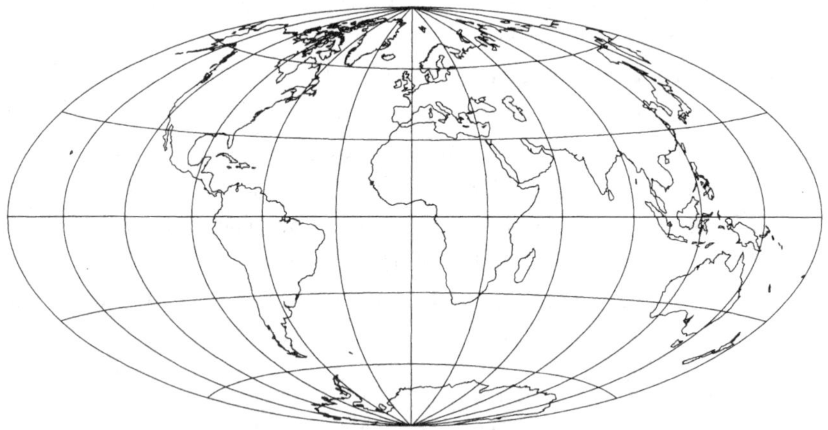

Exhibit 10–4: The whole-world Hammer projection (at its Simple aspect).

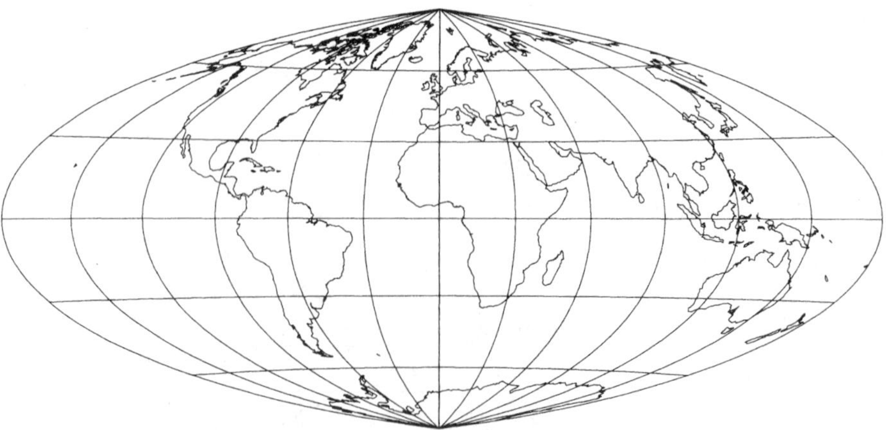

Exhibit 10–5: The whole-world Eckert-Greifendorff projection (at its Simple aspect).

Equal-area Azimuthal

Aitoff's ingenuity quickly prompted Ernst von Hammer of Germany to do the same with Lambert's Equal-area Azimuthal. The **Hammer** projection [1892a] again has straight lines of correct proportion for the Equator and central meridian, but the scale along each, inherited from the Azimuthal, necessarily varies. Other parallels and meridians are again complex curves, but the bounding meridians form an ellipse, producing linear continuity at the Poles. The equal-area attribute carries through from the base projection to Hammer's new projection, according it merit over the Aitoff for many uses.

The Hammer is shown in Exhibit 10–4 with the Equal-area Azimuthal at Equatorial aspect in hemispheric form immediately above it.

The reduction of the curvature of the parallels produced by the halving-then-doubling process is clearly visible from the juxtaposed maps. The **Eckert-Greifendorff**[†] projection [1935b], shown in Exhibit 10–7 immediately below those others, enhances that reduction of curvature by repeating the process. That is, the longitude is halved before feeding into the Hammer formulae, then its x co-ordinate doubled — or, equivalently, longitude is quartered, fed into the formulae for the Equatorial Equal-area Azimuthal, then its resulting x co-ordinate quadrupled. The projection retains the equal-area attribute, but loses the smoothness at the Poles. Its appearance is very close to that of the point-polar Quartic Pseudocylindrical (discussed in Chapter 6 but not itself illustrated). The notable difference is the straight parallels of that other — a modest difference given the considerable attenuation of their curvature in the Eckert-Greifendorff. Clearly, any other divide-multiply factor could be used, with value 3 being an obvious choice that lies unclaimed, along with π and others.

From Equations 203d, using F for the selected fraction term (i.e., 0.5 for the Hammer and 0.25 for the Eckert-Greifendorff), the plotting equations for the Hammer (also known as **Hammer-Aitoff** and, distinguished by its outline shape as the **Hammer Elliptical**) projection, Eckert-Greifendorff and similar projections, are:

$$\begin{aligned} \dot{x} &= \sigma R \frac{\sqrt{D}}{F} \cos\phi \sin(F\,\lambda) \\ \dot{y} &= \sigma R \sqrt{D} \sin\phi \end{aligned} \qquad \text{(243a}$$

where

$$D = \frac{2}{1 + \cos\phi \cos F\,\lambda} .$$

[†] reflecting the revised personal name of the Max Eckert that produced several Pseudocylindricals.

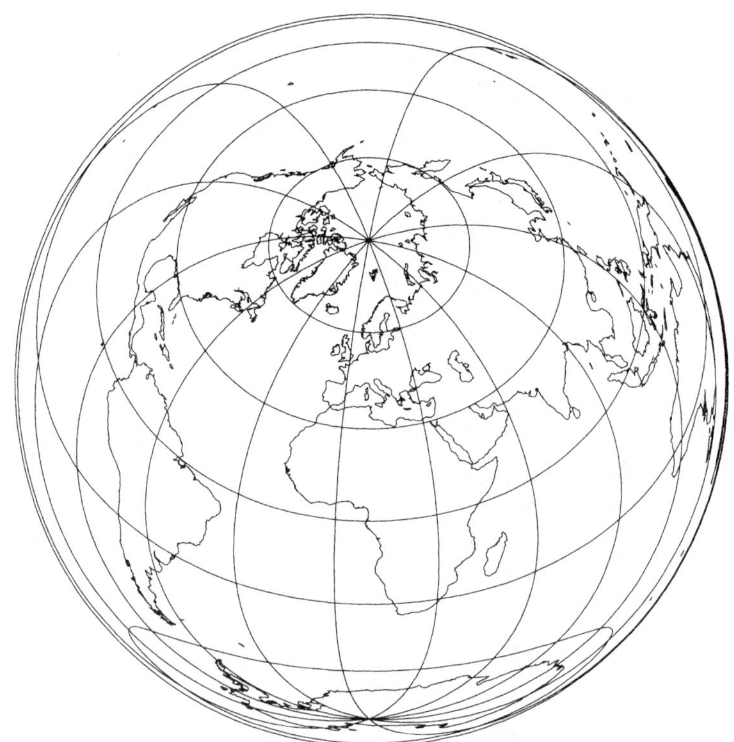

Exhibit 10–6: The (Lambert) Equal-area Azimuthal projection at Oblique aspect.
Focused at 10°E, 45°N.

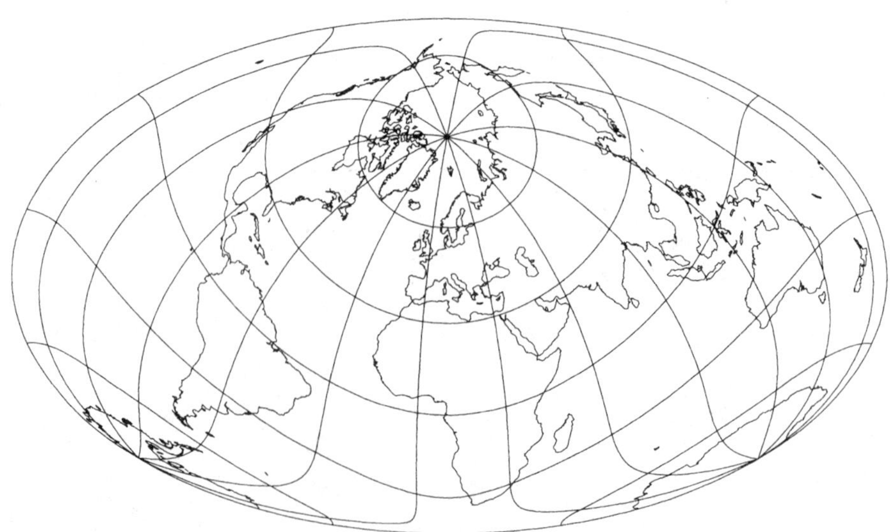

Exhibit 10–7: The Briesemeister (equal-area) projection.
At standard focus of 10°E, 45°N.

As with the Globular projections, these modifications to the Azimuthal projections are aimed at maps at what can be called Simple aspect for them, i.e., Equator along the x axis, Poles on the y axis, as with the Pseudocylindricals. Oblique versions, however, have been used to show the world's land masses. The so-called **Bartholomew Nordic** projection is just an Oblique Hammer centred horizontally on the Greenwich meridian but vertically on the point at 45°N latitude thereon.

William Briesemeister [1953c] used an oblique version of Hammer too, focussed on 10°E, 45°N. To enhance the image of the world's land masses, however, the height-to-width ratio of the map's elliptical boundary is reduced to 4:7 from the natural 1:2 or 4:8 that the Hammer. This is achieved by multiplying/dividing the plotting co-ordinates by $\sqrt{(7/8)}$. The resulting plotting equations for the **Briesemeister** projection are thus:

$$\begin{aligned} \dot{x} &= \sigma R\, 2\sqrt{\tfrac{7}{8}D}\cos {}_a\phi \sin\frac{{}_a\lambda}{2} &&= \sigma R\sqrt{\tfrac{7}{2}D}\cos {}_a\phi \sin\frac{{}_a\lambda}{2} \\ \dot{y} &= \sigma R\sqrt{\tfrac{8}{7}D}\sin {}_a\phi &&= \sigma R\, 2\sqrt{\tfrac{2}{7}D}\sin {}_a\phi \end{aligned} \qquad \text{. . . . (245a}$$

where

$${}_a\phi = \arcsin\left\{\frac{\sin\Phi - \cos\Phi\cos(\Lambda - 10)}{\sqrt{2}}\right\},$$

$${}_a\lambda = \arccos\left\{\frac{\sin\Phi + \cos\Phi\cos(\Lambda - 10)}{\sqrt{2}\cos {}_a\phi}\right\} \text{ with sign of } (\Lambda - 10)$$

and

$$D = \frac{2}{1 + \cos {}_a\phi \cos\frac{{}_a\lambda}{2}}.$$

The Briesemeister is shown in Exhibit 10–7, immediately below the (Lambert) Equal-area Azimuthal — its progenitor — at the same focus.

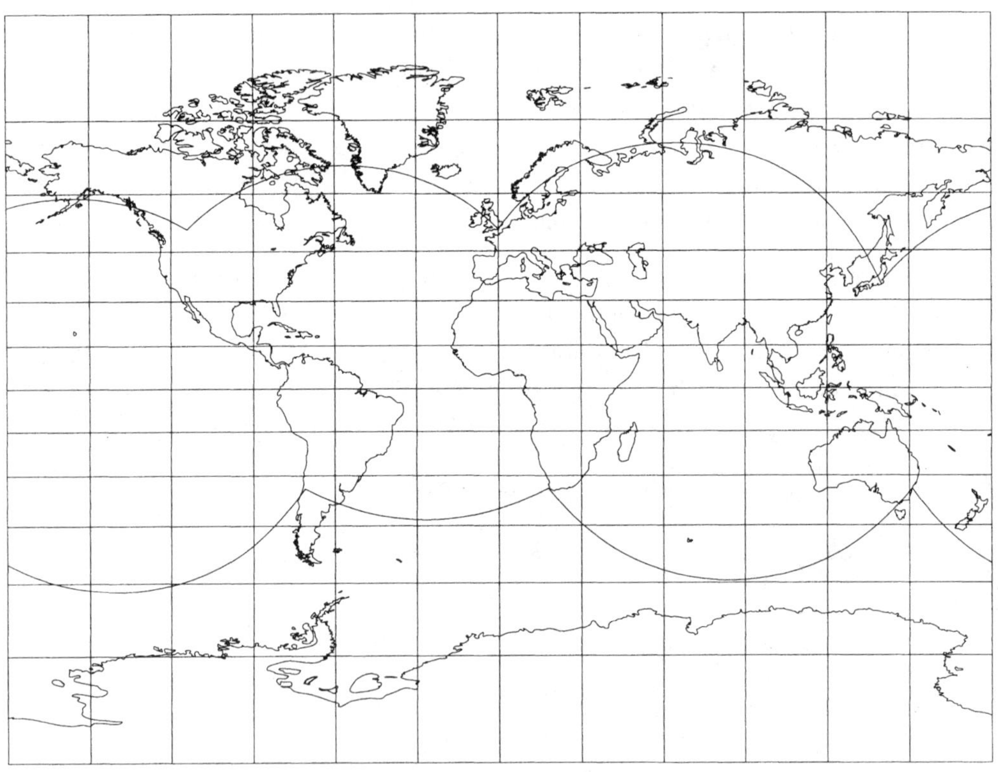

Exhibit 10–8: The Miller Cylindrical projection.
Graticule 15° for parallels, 30° for meridians.
With great-circle routes as detailed for Exhibit 4–3 on page 82.

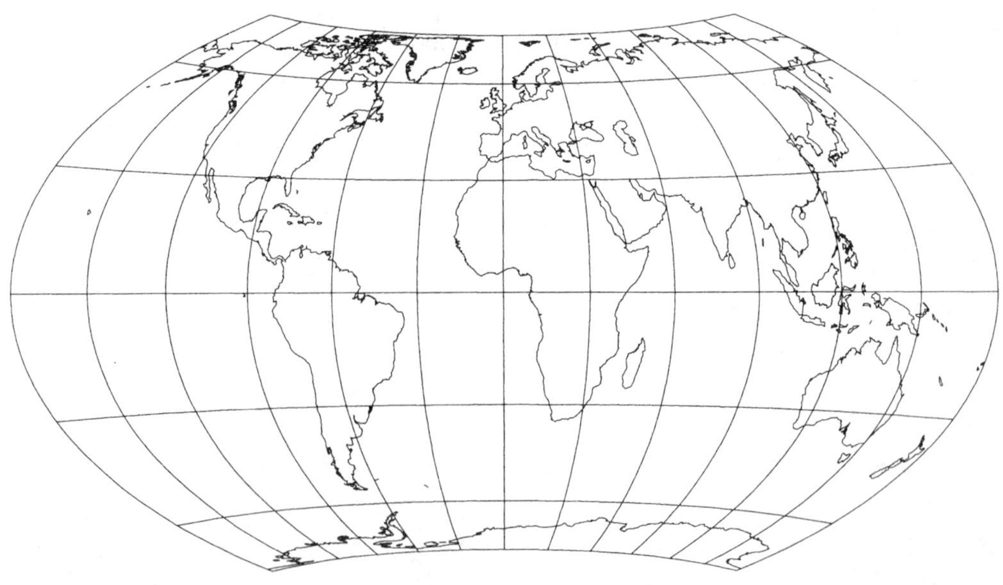

Exhibit 10–9: The Wagner VII projection.

SHRINKING LATITUDES

The Cylindricals

Maps produced by the Mercator projection appear everywhere, but no single such map can show everywhere. At Simple aspect, the Poles are an infinite distance from the Equator. On any such map the (incomplete) central meridian is more than half the length of the Equator by 67° latitude, and longer than the Equator should the map reach 85°. This is inherent to the substance of the development of this conformal cylindrical projection, and is expressed in Equation 81a covering the intermediate parameter $\dot{v}$.

Many people seem to see a Mercator map as correctly representing the world in general appearance, so efforts have been made to tame the projection so as to have it encompass the whole world. The clear first step is to apply some fractional multiplier to the latitude, so that the Polar latitudes of ±90° produce arithmetically finite values for insertion in the equation. O. M. Miller developed the notable success. The **Miller Cylindrical** projection {1942b], which is illustrated in Exhibit 10–8 with selective great-circle routes shown, uses 0.8 as the factor (so feeding a value of 72° at most into the crucial equation). It was one of four developed by Miller in pursuit of a balance between distortion of shape and of area. In it, Miller chose to divide the resulting formula value for latitudinal parameter by the same amount, thus expanding the resulting co-ordinate (and height of the map) by 25%. The revised equation is

$$\dot{v} = \frac{\sigma R}{0.8} \ln\left\{\tan\left(\frac{\pi}{4} + \frac{0.8\,\phi}{2}\right)\right\} = 1.25\,\sigma R \ln\left\{\tan\left(\frac{\pi}{4} + 0.4\,\phi\right)\right\} \quad \text{. . . . . (247a}$$

The alteration of the input latitude value alone destroys the conformality that led to the original projection (and along with it the unique feature of straight loxodromes). The subsequent division by the same 0.8 that multiplied the real latitude approximately restores the proportions of the map to that of the typical nearly whole-world Mercator map, but appears to have no specific purpose (unlike with the Aitoff etc., where such multiplication restored the 1:2 ratio of central meridian to Equator). The spacing between parallels is shown graphically in Exhibit 10–8, numerically in the table below. Comparison with the Mercator equivalents on page 83 shows that the whole world in the Miller fits clearly within the dimensions of the world to ±80° in that other.

lat	0°		10°		20°		30°		40°		50°		60°		70°		80°		90°
distce	0.0000		0.1751		0.3537		0.5396		0.7375		0.9536		1.1968		1.4813		1.8324		2.3034
diffce		0.1751		0.1786		0.1859		0.1979		0.2161		0.2432		0.2845		0.3511		0.4710	

Miller Cylindrical projection: spacing of parallels for unit globe.
(Tangential form at Simple aspect.)

The Gall Stereographic projection, shown in Exhibit 5–1 on page 102, produces a similar map to the Miller, with distortion nil along the ±45° parallels rather than the Equator.

The Azimuthals

Any such manipulation of the Gnomonic would destroy its unique feature — having straight lines identically great circles — so avoiding infinity with it has found no favour. Arithmetic manipulation of Azimuthals is oriented, as described above, to reducing the curves of the parallels for a whole-world map with the Poles on the periphery of the map.

A similar divide-multiply process to that used for the Aitoff can be applied to latitude. The **Wagner VII** projection does this to the Hammer, shifting the Poles from their pointed positions to the curved lines originally representing latitude 65°. The resulting plotting equations are:

$$\dot{x} = 2.667\,23_{\sim}\sigma R\, C\, f\, \sin\frac{\lambda}{3}$$

$$\dot{y} = 1.241\,04_{\sim}\sigma^{-1} R\, C \sin 65^\circ \sin\phi$$

where

$$f = \sqrt{1 - (\sin 65^\circ \sin\phi)^2},$$

and

$$C = \sqrt{\frac{2}{1 + f\cos\frac{\lambda}{3}}}.$$

Using double-angle formulae from Tutorial 10, the plotting equations can be re-expressed compactly as:

$$\dot{x} = 2.667\,23_{\sim}\sigma R \sec\frac{\alpha}{2} \cos\beta \sin\frac{\lambda}{3}$$

$$\dot{y} = 1.241\,04_{\sim}\sigma^{-1} R \sec\frac{\alpha}{2} \sin\beta \qquad \text{. . . . . (248a}$$

where $\sin\beta = \sin 65^\circ \sin\phi$

and $\cos\alpha = \cos\beta \cos\frac{\lambda}{3}$

The multiplier numbers in the plotting equations ensure the retention of equal-area status. If factor σ equals 1, the Equator is twice the length of the central meridian.

The Wagner VII projection, shown in Exhibit 10–9 on page 246, is unusual outside of the Conicals by having curved lines for Poles.

These derivatives of Azimuthal projections are, with certain others, classed as Modified Azimuthals. Like those of the Pseudocylindrical class — a likeness to which that has already been noted — they can be used interrupted where appropriate.

ADDING LONGITUDES

The Ortelius Oval

Battista Agnese of Italy in 1544 expanded the coverage of the globular Apian I projection (on page 228) from hemispheric to whole-world just by repeating the semicircular arcs of the outermost meridians (±90° relative longitude) at the same uniform scale along the Equator — until double the longitude was reached. Named misleadingly after an early promoter and user (and qualified by an ambiguous geometric descriptor), the **Ortelius Oval** projection has Equation 229c applying throughout and Equation 229d applying untouched to the central hemisphere. Outside that (i.e., for $\pi/2 < |\lambda| \leq \pi$) the plotting equation for the longitude-dependent co-ordinate becomes

$$\dot{x} = \frac{\lambda}{|\lambda|} R \left\{ |\lambda| - \frac{\pi}{2} + \sqrt{\left(\frac{\pi}{2}\right)^2 - \phi^2} \right\} \qquad \text{(249a}$$

The projection is illustrated immediately below.

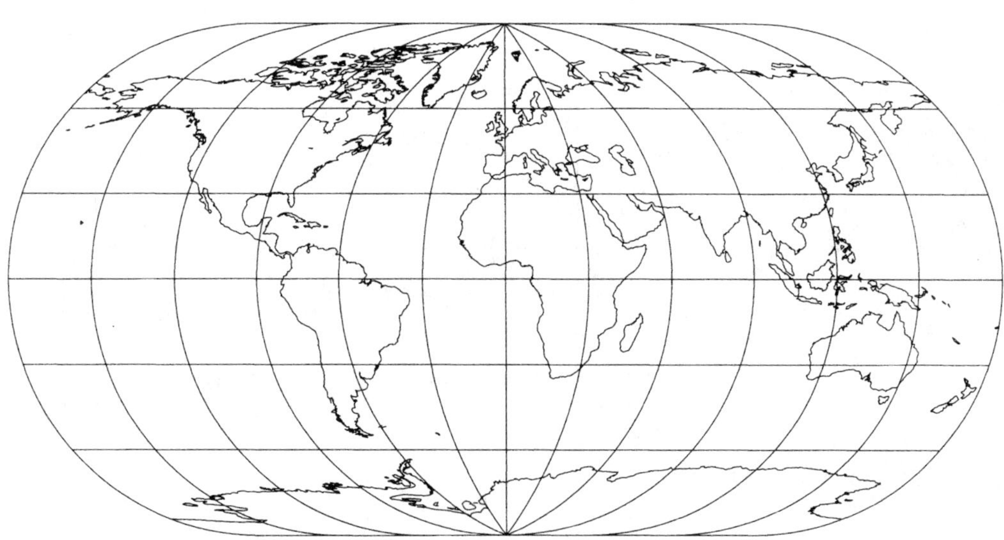

Exhibit 10–10: The Ortelius Oval projection.

SHRINKING LONGITUDES, CHANGING LATITUDES

Lambert and Lagrange

Long before others mentioned earlier in this chapter played such tricks, Lambert, in his monumental work [1772a], effectively adapted the Stereographic projection at Simple aspect by applying multipliers to both longitudes and latitudes. Just as Aitoff was to do a full hundred years later, Lambert merely halved the longitudes, bringing the whole world into the circle of the hemispheric Stereographic map (illustrated in Exhibit 8–6 on page 196). He then, however, developed a multiplier for latitudes that retained the conformality of the underlying projection.

The process pursued by Lambert was general to any fractional multiplier of longitude, and the resulting projection can be applied with other values than zero for the central latitude, with its straight line bisecting the map. The general form is usually referred to as the Lagrange projection; it is discussed in Chapter 15. The original form created by Lambert, covered by that development, is illustrated, after our own exposition, in Exhibit 10–13 on page 254.

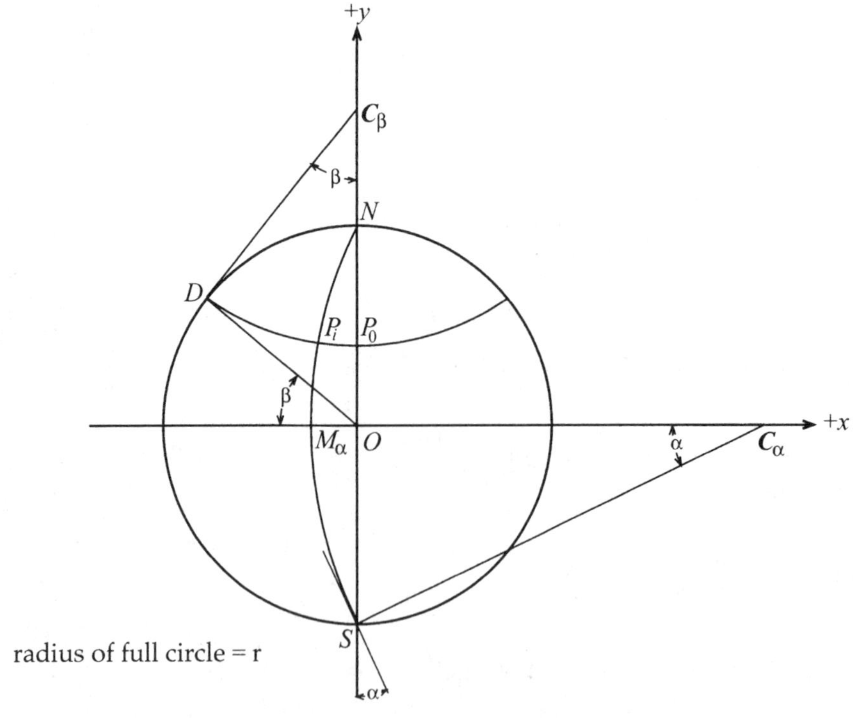

Exhibit 10–11: Geometry with circular arcs.

With the Stereographic at any aspect, circles on the globe become circles on the map, hence parallels and meridians are circular arcs — finite else infinite. At Equatorial aspect this means that for meridians arcs concave towards the central meridian and for parallels arcs concave towards the Poles, with a central straight one of each. That arrangement holds here, and we proceed to establish the representative equations for such arcs and the general point, and the condition for continuing conformality. We do this with factors as general as possible.

Applying conformality

Having the parallels orthogonal to the meridians is a necessary but not sufficient condition for conformality. Assuming a proper fractional multiplier of longitude (multiplying by 1 giving the Equatorial Stereographic itself), the meridians meet at the Poles at angles different from the true values, so conformality cannot apply at those two points. As was done to achieve the Mercator and the Lambert Conformal Conical projections (in Chapter 4), however, appropriate spacing of the parallels can provide conformality generally.

The challenge is to establish the spacing that produces identical scaling in all directions at any general point. It suffices to show equal scaling along meridian and parallel at such point. Further, since the shapes of parallels and meridians are set, it suffices to establish the spacing for identical scaling at points on the central meridian — i.e., to establish the necessary spacing along the central meridian. We calculate such spacing from the one straight parallel (which, as has been mentioned, is not necessarily the Equator).

We pursue Lambert's method [1772a], changing only various symbols to fit more comfortably with current practice. Of particular note is that Lambert worked entirely relative to northern polar distance rather than latitude; we denote it as usual by γ. Lambert also worked with the unit circle, but we use a circle of radius r.

Exhibit 10–11 illustrates the geometry for a general meridian striking (i.e., meeting) the central meridian of the map at angle α, at both Poles $N = (0, +r)$ and $S = (0, -r)$, and intersecting a general parallel at point P. With M_α as the midpoint of that meridian and C_α as the centre of its pertinent circle (necessarily on the x axis), the line from S to C_α, being a radius of the arc, must be orthogonal to the meridian. Then, from the diagram:

$$\text{angle } OC_\alpha N = \text{angle } OC_\alpha S = \alpha$$

$$\text{length of } ON = \text{length of } OS = r$$

$$\text{length of } NC_\alpha = \text{length of } SC_\alpha = r\csc\alpha = \text{length } M_\alpha C_\alpha$$

$$\text{length of } OC_\alpha = r\cot\alpha$$

Noting that for any angle θ

$$\frac{1-\cos 2\theta}{\sin 2\theta} = \frac{2\sin^2\theta}{2\sin\theta\cos\theta} = \tan\theta$$

we also have for the arc-radius of a meridian meeting with strike angle α

$$\text{length } M_\alpha C_\alpha = r\csc\alpha = r\left\{\frac{1}{\sin\alpha}\right\} = r\left\{\frac{1-\cos\alpha}{\sin\alpha} + \frac{\cos\alpha}{\sin\alpha}\right\} = r\left\{\tan\frac{\alpha}{2} + \cot\alpha\right\} \quad . . \text{(252a}$$

Addressing elements along the x axis, we have

$$\text{length } M_\alpha O = \text{length } M_\alpha C_\alpha - \text{length } OC_\alpha$$

$$= r\csc\alpha - r\cot\alpha \quad = r\frac{1-\cos\alpha}{\sin\alpha} \quad = r\tan\frac{\alpha}{2}$$

Let $P_0 = (0, y)$ and $P_\phi = (0, y\text{-}\Delta y)$ be at the intersections with the central meridian of the parallels of polar distance γ and $\gamma+\Delta\gamma$, respectively, as illustrated below. Let P_λ be the intersection of the extension of line $C_\alpha P_0$ with the meridional arc having strike angle α with the central meridian. Lambert then compared the lengths of and hence the scales along $P_0 P_\phi$ and $P_0 P_\lambda$ for an infinitesimal α, expressed as $\alpha = c\Delta\lambda$ with infinitesimal $\Delta\lambda$.

Exhibit 10–12 provides an exaggerated image of the scene. Since $P_0 P_\lambda$ becomes orthogonal to the central meridian as $\Delta\lambda$ goes to zero, its length must involve the cosine of latitude — i.e., the sine of polar distance. With $\Delta\lambda$ infinitesimal, its higher powers can be ignored, allowing its sine and tangent functions to be equated with the angle itself (likewise for half that angle). Applying the theorem of Pythagoras to triangle $C_\alpha P_0 O$ gives

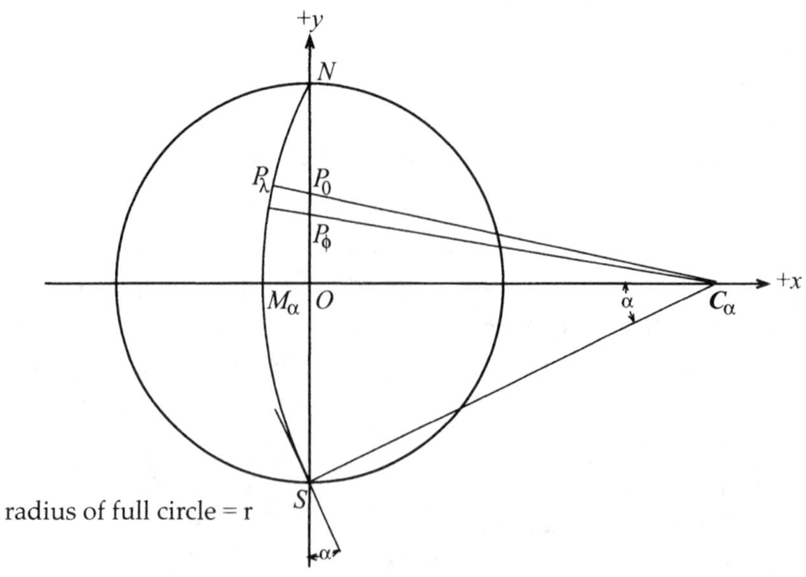

Exhibit 10–12: Geometry with circular arcs: infinitesimals.

$$\text{length } P_0C_\alpha = \sqrt{r^2\csc^2(c\,\Delta\lambda) + y^2} = \sqrt{\frac{r^2}{\sin^2(c\,\Delta\lambda)} + y^2} \approx \sqrt{\left(\frac{r}{c\,\Delta\lambda}\right)^2 + y^2}$$

$$\approx \frac{r}{c\,\Delta\lambda} + \frac{c\,\Delta\lambda}{2r}y^2 \quad \text{discarding higher terms in } \Delta\lambda$$

while Equation 252a gives

$$\text{length } P_\lambda C_\alpha = r\left(\frac{1}{\tan(c\,\Delta\lambda)} + \tan\frac{c\,\Delta\lambda}{2}\right) \approx r\left(\frac{1}{c\,\Delta\lambda} + \frac{c\,\Delta\lambda}{2}\right)$$

Differencing the latter length from the former gives

$$\text{length } P_\lambda P_0 \approx \left(r\left(\frac{1}{c\,\Delta\lambda} + \frac{c\,\Delta\lambda}{2r}\right) - \left(\frac{r}{c\,\Delta\lambda} + \frac{y^2c\,\Delta\lambda}{2r}\right)\right)$$

$$= r\left(\frac{c\,\Delta\lambda}{2} - \frac{y^2c\,\Delta\lambda}{2r^2}\right) = r\frac{c\,\Delta\lambda}{2}\left(1 - \frac{y^2}{r^2}\right) \quad \text{. . . . . (253a}$$

To have equal scale along these two lines, we must have

$$\frac{\text{length } P_\lambda P_0}{\text{length } P_\phi P_0} = \frac{(\sin\gamma)\Delta\lambda}{\Delta\gamma}$$

As the length of $P_\phi P_0$ equals $r\Delta y$, Equation 253a gives

$$-\frac{rc\,\Delta\lambda\left(1 - \frac{y^2}{r^2}\right)}{2r\Delta y} \approx \frac{(\sin\gamma)\Delta\lambda}{\Delta\gamma} \quad \text{. . . . . (253b}$$

Now, since $-r \le y \le +r$, there must be some angle υ such that $y = r\cos\upsilon$ (and consequently $dy = -r\sin\upsilon\, d\upsilon$). Substituting for y in Equation 253b, after taking the infinitesimals to their limits and obtaining differentials, gives

$$\frac{d\lambda\cos\gamma}{d\gamma} = \frac{rc\,d\lambda\,(1 - \cos^2\upsilon)}{-2r\sin\upsilon\, d\upsilon} = -\frac{c\,d\lambda\sin^2\upsilon}{2\sin\upsilon\, d\upsilon} = -\frac{c\sin\upsilon}{2}\frac{d\lambda}{d\upsilon}$$

or

$$\frac{2\,d\upsilon}{\sin\upsilon} = \frac{c\,d\gamma}{\cos\gamma}$$

Integrating yields, for some constant k,

$$\ln\tan^2\frac{\upsilon}{2} = \ln\tan^c\frac{\gamma}{2} + k \quad \text{. . . . . (253c}$$

If the Equator is chosen to be along the x axis (which was "incomparably more satisfying" to Lambert [1972b]), then $k = 0$ zero, hence

$$\tan^2\frac{\upsilon}{2} = \tan^c\frac{\gamma}{2} \quad \text{. . . . . (253d}$$

Evaluating y for $\lambda = 0$ using a double-angle formula from Tutorial 10, we get

$$y\Big]_{\lambda=0} = r\cos\upsilon = r\frac{1-\tan^2\frac{\upsilon}{2}}{1+\tan^2\frac{\upsilon}{2}} = r\frac{1-\tan^c\frac{\gamma}{2}}{1+\tan^c\frac{\gamma}{2}} \qquad \text{(254a}$$

Lambert, setting $c = \frac{1}{2}$ to bring the whole world exactly into the circle, calculated the spacings as follows:

distce	0.0000	0.043 53	0.088 88	0.136 48	0.188 44	0.247 46	0.317 83	0.408 56	0.543 46	1.
diffce		0.043 53	0.045 35	0.047 60	0.051 96	0.059 02	0.070 37	0.090 73	0.134 90	0.456 54

Lambert projection: spacing of parallels for unit circle.

As with the Mercator, the spacings escalate dramatically as they move away from the Equator — indeed initially more than with that familiar projection though the convergence of the meridians brings the Poles within a finite distance. The graphical result is well illustrated in the map below.

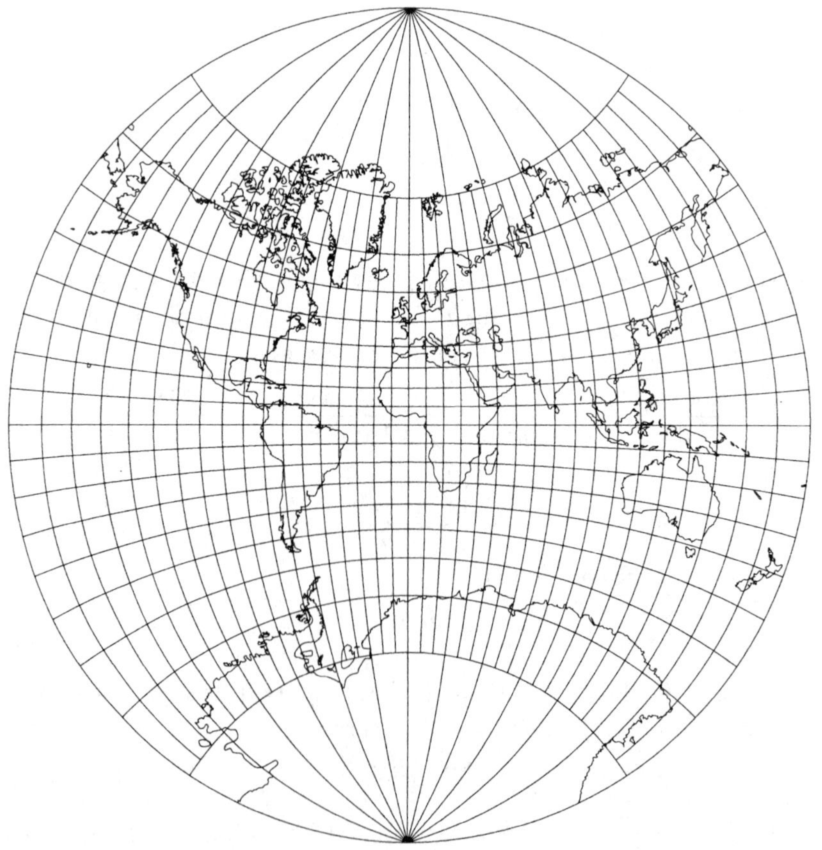

Exhibit 10–13: Lambert's original world-conformally-in-a-circle projection.

Lambert's calculations were laborious, and his illustration was based on geometric fitting of the circular arcs. We now look into establishing algebraic formulae for the parallels and meridians, and hence the plotting equations, keeping the various factors as general as possible.

Meridians as circular arcs

Returning to our general meridian striking the central meridian at angle α, Exhibit 10–11 on page 250 shows the centre of the pertinent circle to be $C_\alpha = (-r\csc\alpha, 0)$ and the equation for points on the meridian as

$$\{\dot{x} - (-r\cot\alpha)\}^2 + \{\dot{y}\}^2 = \{r\csc\alpha\}^2$$

Expanding and rearranging this becomes

$$\dot{x}^2 + (2r\cot\alpha)\dot{x} + r^2\cot^2\alpha + \dot{y}^2 = r^2\csc^2\alpha$$

$$\dot{x}^2 + (2r\cot\alpha)\dot{x} + \dot{y}^2 = r^2\csc^2\alpha - r^2\cot^2\alpha = r^2\frac{1-\cos^2\alpha}{\sin^2\alpha} = r^2\frac{\sin^2\alpha}{\sin^2\alpha} = r^2$$

or

$$\dot{x}^2 + (2r\cot\alpha)\dot{x} + \dot{y}^2 - r^2 = 0 \qquad \text{. . . . . (255a}$$

Parallels as circular arcs

Now let D be the intersection of the parallel with the circle of radius r and let β be the angle between line OD and the $-x$ half-axis — an angle which, being similar to that for latitude were this parallel straight, we can call *pseudo-latitude*. Because OD is a radius of that circle, it must be orthogonal to it, hence tangential to the coincident parallel (parallels and meridians being mutually orthogonal in the Stereographic). We draw the line passing through D orthogonal to OD, which in turn must be a radial line for the circular arc forming the parallel. The intersection of this new line with the y axis, which we label C_β, is the centre for that arc. Then, from the diagram:

angle $OC_\beta D = \beta$
length $OD = r$
length $OC_\beta = r\csc\beta$
length $C_\beta D = r\cot\beta$

Thus, $C_\beta = (0, r\csc\beta)$ and the equation for points on the parallel is

$$(\dot{x})^2 + (\dot{y} - r\csc\beta)^2 = (r\cot\beta)^2$$

Expanding and rearranging this becomes

$$\dot{x}^2 + \dot{y}^2 - 2(r\csc\beta)\dot{y} + r^2\csc^2\beta = r^2\cot^2\beta$$

so

$$\dot{x}^2 + \dot{y}^2 - 2(r\csc\beta)\dot{y} = r^2\cot^2\beta - r^2\csc^2\beta = r^2\frac{\cos^2\beta - 1}{\sin^2\beta} = r^2\frac{-\sin^2\beta}{\sin^2\beta} = -r^2$$

or

$$\dot{x}^2 + \dot{y}^2 - 2(r\csc\beta)\dot{y} + r^2 = 0 \qquad \text{. . . . . (256a}$$

The intersection of the circular arcs forming parallels and meridians

Subtracting Equation 256a from Equation 255a eliminates the squared variables, allowing development of formulae for the point $\dot{P}$, giving

$$2(r\cot\alpha)\dot{x} - r^2 + 2(r\csc\beta)\dot{y} - r^2 = 0$$

$$(\cot\alpha)\dot{x} + (\csc\beta)\dot{y} - r = 0 \qquad \text{. . . . . (256b}$$

which provides either $\dot{x}$ else $\dot{y}$ in terms of the other. Substitution for $\dot{x}$ in Equation 256a else for $\dot{y}$ in Equation 255a then provides a quadratic in the other variable. Pursuing the former alternative, multiplying Equation 256a through by $\sin^2\beta\cos^2\alpha$ gives

$$(\sin^2\beta\cos^2\alpha)\dot{x}^2 + (\sin^2\beta\cos^2\alpha)\dot{y}^2 - 2(r\sin\beta\cos^2\alpha)\dot{y} + r^2\sin^2\beta\cos^2\alpha = 0 \qquad \text{. . . (256c}$$

Equation 256b after multiplying through by $(\sin\beta\sin\alpha)$, gives

$$(\sin\beta\cos\alpha)\dot{x} = \sin\alpha\{r\sin\beta - \dot{y}\}$$

$$(\sin^2\beta\cos^2\alpha)\dot{x}^2 = r^2\sin^2\alpha\sin^2\beta - 2(r\sin^2\alpha\sin\beta)\dot{y} + (\sin^2\alpha)\dot{y}^2$$

Substituting for the left-hand side from Equation 256c gives

$$(\sin^2\alpha + \sin^2\beta\cos^2\alpha)\dot{y}^2 - 2r(\sin^2\alpha\sin\beta + \sin\beta\cos^2\alpha)\dot{y} + r^2(\sin^2\alpha\sin^2\beta + \sin^2\beta\cos^2\alpha) = 0$$

$$(\sin^2\alpha + \sin^2\beta\cos^2\alpha)\dot{y}^2 - 2r\sin\beta(\sin^2\alpha + \cos^2\alpha)\dot{y} + r^2\sin^2\beta(\sin^2\alpha + \cos^2\alpha) = 0$$

$$(1 - \cos^2\alpha + \sin^2\beta\cos^2\alpha)\dot{y}^2 - 2r(\sin\beta)\dot{y} + r^2\sin^2\beta = 0$$

$$\{1 - \cos^2\alpha(1 - \sin^2\beta)\}\dot{y}^2 - 2r(\sin\beta)\dot{y} + r^2\sin^2\beta = 0$$

$$(1 - \cos^2\alpha\cos^2\beta)\dot{y}^2 - 2(r\sin\beta)\dot{y} + (r^2\sin^2\beta) = 0$$

The routine solution to this quadratic equation (see Tutorial 18) is

$$\dot{y} = \frac{r\sin\beta \pm \sqrt{(r\sin\beta)^2 - (1 - \cos^2\alpha\cos^2\beta)(r^2\sin^2\beta)}}{1 - \cos^2\alpha\cos^2\beta}$$

$$= \frac{r\sin\beta \pm r\sqrt{\sin^2\beta - \sin^2\beta + \cos^2\alpha\sin^2\beta\cos^2\beta}}{1 - \cos^2\alpha\cos^2\beta}$$

$$= r\frac{\sin\beta \pm \sqrt{\cos^2\alpha\sin^2\beta\cos^2\beta}}{1 - \cos^2\alpha\cos^2\beta}$$

Taking the square root gives

$$\dot{y} = r\frac{\sin\beta \pm \cos\alpha \sin\beta\cos\beta}{1-\cos^2\alpha \cos^2\beta} = r\sin\beta\frac{1 \pm \cos\alpha \cos\beta}{(1-\cos\alpha \cos\beta)(1+\cos\alpha \cos\beta)}$$

the root with the + choice for the ± signs being within our circle. Hence,

$$\dot{y} = \frac{r\sin\beta}{(1+\cos\alpha \cos\beta)} \qquad \text{(257a}$$

Substituting in Equation 256b gives

$$(\cot\alpha)\dot{x} = r - \csc\beta\frac{r\sin\beta}{(1+\cos\alpha \cos\beta)}$$

$$= r\left\{1 - \frac{1}{(1+\cos\alpha \cos\beta)}\right\}$$

Hence,

$$\dot{x} = r\frac{\sin\alpha}{\cos\alpha}\left\{\frac{1+\cos\alpha \cos\beta - 1}{(1+\cos\alpha \cos\beta)}\right\} = r\frac{\sin\alpha}{\cos\alpha}\left\{\frac{\cos\alpha \cos\beta}{(1+\cos\alpha \cos\beta)}\right\}$$

or

$$\dot{x} = \frac{r\sin\alpha \cos\beta}{(1+\cos\alpha \cos\beta)}$$

completing, with Equation 257a, equations for the co-ordinates of the general map point in terms of strike angle a and pseudo-latitude β.

In the Lagrange and circular Polyconic projections generally, angle α is a simple multiple of longitude, hence the equations can be written as:

$$\dot{x} = \frac{r\sin c\lambda \cos\beta}{1+\cos c\lambda \cos\beta} = rW^{-1}\sin c\lambda$$

$$\dot{y} = \frac{r\sin\beta}{1+\cos c\lambda \cos\beta} = rW^{-1}\tan\beta \qquad \text{(257b}$$

where $W = \dfrac{1+\cos c\lambda \cos\beta}{\cos\beta} = \sec\beta + \cos c\lambda$

for a constant c akin to the constant of the cone. Setting $c = 1$, $\beta = \phi$ and $r = \sigma R$ in Equations 257b we get:

$$\dot{x} = \frac{2\sigma R\cos\phi \sin\lambda}{1+\cos\phi \cos\lambda}$$

$$\dot{y} = \frac{2\sigma R\sin\phi}{1+\cos\phi \cos\lambda}$$

i.e., the plotting equations for the Equatorial Stereographic given in Equations 201b with Equation 201d.

Converting from pseudo-latitude β to ordinary latitude ϕ

To make Equations 257b generally usable we need to obtain expressions for $\tan\beta$ and $\sec\beta$ in terms of latitude ϕ rather than pseudo-latitude β. We achieve this by ascertaining the length of line OP_0 expressed in each of the two variables. Equation 254a gives the length in terms of polar distance γ, thereby in ϕ. Subtracting length P_0C_β from length OC_β in the diagram of Exhibit 10–11 gives the length in terms of β, as $r(\csc\beta - \cot\beta)$. Hence

$$\csc\beta - \cot\beta = \frac{1 - \tan^c \frac{\gamma}{2}}{1 + \tan^c \frac{\gamma}{2}} = \frac{\cot^c \frac{\gamma}{2} - 1}{\cot^c \frac{\gamma}{2} + 1} \qquad \text{(258a)}$$

Using double-angle formulae of Tutorial 10, the left-hand side can be rationalized to a single trigonometric term, giving

$$\csc\beta - \cot\beta = \frac{1 - \cos\beta}{\sin\beta} = \frac{1 - \left(1 - 2\sin^2 \frac{\beta}{2}\right)}{2\sin\frac{\beta}{2}\cos\frac{\beta}{2}} = \frac{\sin\frac{\beta}{2}}{\cos\frac{\beta}{2}}$$

$$= \tan\frac{\beta}{2}$$

Writing t for this final term, we get from Equation 258a

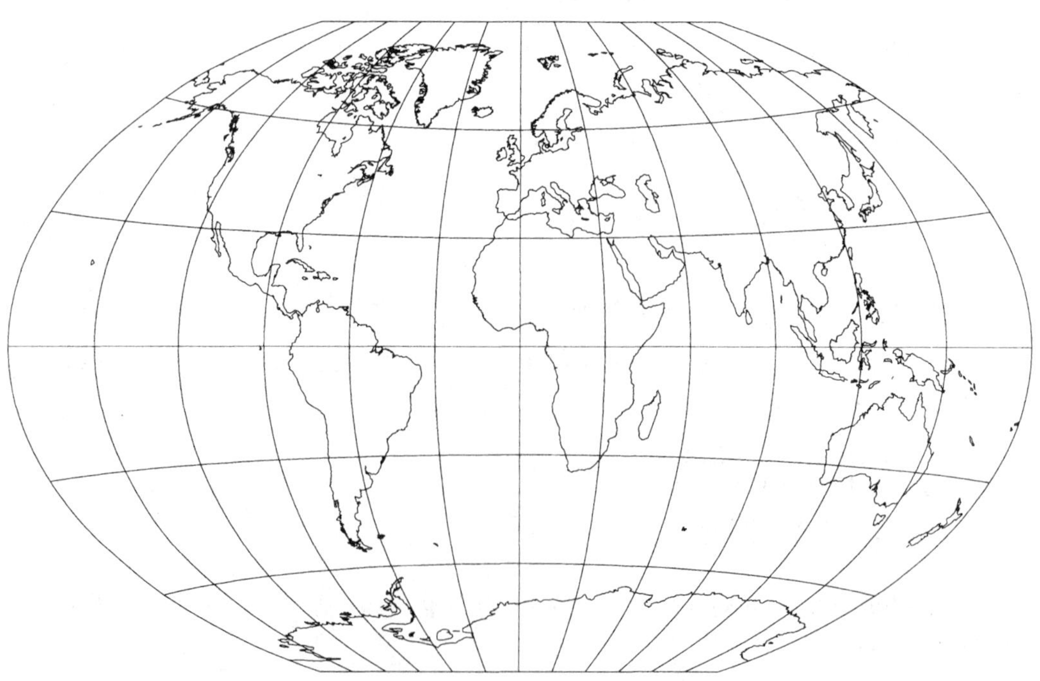

Exhibit 10–14: The Winkel Tripel projection.
[Standard parallels ±40° in underlying Equidistant Cylindrical projection.]

$$t = \tan\frac{\beta}{2} = \frac{V-1}{V+1} \qquad \text{(259a}$$

from which, by again using a double-angle formula, we get

$$V = \cot^c\frac{\gamma}{2} = \left\{\cot^2\frac{\gamma}{2}\right\}^{\frac{c}{2}} = \left\{\frac{1+\cos\gamma}{1-\cos\gamma}\right\}^{\frac{c}{2}} = \left\{\frac{1+\sin\phi}{1-\sin\phi}\right\}^{\frac{c}{2}} \qquad \text{(259b}$$

Further using double-angle formulae, we have for our target functions:

$$\tan\beta = \frac{2\tan\frac{\beta}{2}}{1-\tan^2\frac{\beta}{2}} = \frac{2t}{1-t^2}$$

$$\sec\beta = (\cos\beta)^{-1} = \left(\frac{1-\tan^2\frac{\beta}{2}}{1+\tan^2\frac{\beta}{2}}\right)^{-1} = \frac{1+t^2}{1-t^2}$$

Substituting from Equation 259a and multiplying generally by $(V+1)^2$, these become:

$$\tan\beta = \frac{2(V+1)(V-1)}{(V+1)^2-(V-1)^2} = \frac{2(V^2-1)}{4V} = \frac{V-V^{-1}}{2}$$

$$\sec\beta = \frac{(V+1)^2+(V-1)^2}{(V+1)^2-(V-1)^2} = \frac{2(V^2+1)}{4V} = \frac{V+V^{-1}}{2}$$

for $V\neq 0$ (i.e., $\gamma\neq 0$ or $\phi\neq \pi/2$).

We consider this general form further in Chapter 15. Here it suffices to consider the circumstances when $c = 1/2$, obtaining Lambert's original projection as shown on page 254.

Substituting for c, we get plotting equations for the **Lambert** projection as:

$$\text{If } \phi = \pm\frac{\pi}{2} \quad \dot{x} = 0, \quad \dot{y} = \frac{\phi}{|\phi|}2R$$

$$\text{else} \quad \dot{x} = 2RW^{-1}\sin\frac{\lambda}{2} \qquad \text{(259c}$$

$$\dot{y} = 2RW^{-1}\tan\beta$$

where $W = \sec\beta + \cos\frac{\lambda}{2}$,

$$\sec\beta = \frac{V+V^{-1}}{2},$$

$$\tan\beta = \frac{V-V^{-1}}{2}$$

and $V = \sqrt[4]{\dfrac{1+\sin\phi}{1-\sin\phi}}$.

AVERAGING

Approach

Various attempts have been made to produce a better map by combining the plotting formulae of two projections. The composite projections of Chapter 7 combined them by using each one exclusively in separate zones of the map, reaching the extreme in the Polyconic where the zones were infinitesimal in width and infinite in number. The Eumorphic was an exception in that all its parallels were spaced at the arithmetic mean of the values of two projections (Sinusoidal and Mollweide), with their lengths being computed to retain equal-area. More simply, projections can be combined by mere arithmetic averaging of the plotting formulae of two projections.

Winkel Tripel projection

Oswald Winkel used the Equidistant Cylindrical and Aitoff projections in 1921 to create a map reminiscent of the flat-polar Pseudocylindricals. Called the **Tripel** or triple projection, the Poles (like other non-zero parallels) are in fact slightly curved lines. The averaging is equal of the two component projections, but the Cylindrical component can be secantal, introducing a choice in that regard. Because the central meridian has true scale in both component projections, it also has true scale in the Tripel, but no other meridian or parallel has true scale in the derived projection. Although neither conformal nor equal-area, the projected map of the whole world has a pleasing appearance.

The plotting formulae are:

$$\ddot{x} = \tfrac{1}{2}(\dot{x}_a + \dot{x}_e)$$
$$\ddot{y} = \tfrac{1}{2}(\dot{y}_a + \dot{y}_e) \qquad \text{. . . . . (260a}$$

where the *a* subscript indicates co-ordinates of the Aitoff projection given by Equations 241a and *e* the like for the secantal Equidistant Cylindrical given by Equation 104b and the second of Equations 109a, i.e.,

$$\dot{x}_e = R(\cos\phi_1)\lambda$$
$$\dot{y}_e = R\phi$$

with $\pm\phi_1$ the latitudes of true scale for this component projection. Winkel chose $\phi_1 = 50.567°$.

Exhibit 10–14 on page 258 illustrates the projection with $\phi_1 = 40°$, as has been used for whole-world maps in the Bartholomew atlases.

CHAPTER 11

Double or Quits: Two-point and non-azimuthal Azimuthals

Preamble

The essential characteristic of Azimuthals, visibly displayed at Simple aspect, is of straight lines radiating from a single focus, each representing a great circle and all collectively spaced at exactly the angles between those circles on the globe. Hence, azimuths at this single point are true. In this chapter we look at having two such points on the one map, then we look at azimuthal-like maps that are not truly azimuthal.

Yet again there is no new mathematics in this chapter, just a new cartographic approach.

The maps in this chapter vary in appearance considerably. Some are highly unusual, even strange, in appearance. The centring of them varies too, responding to the typical application of each.

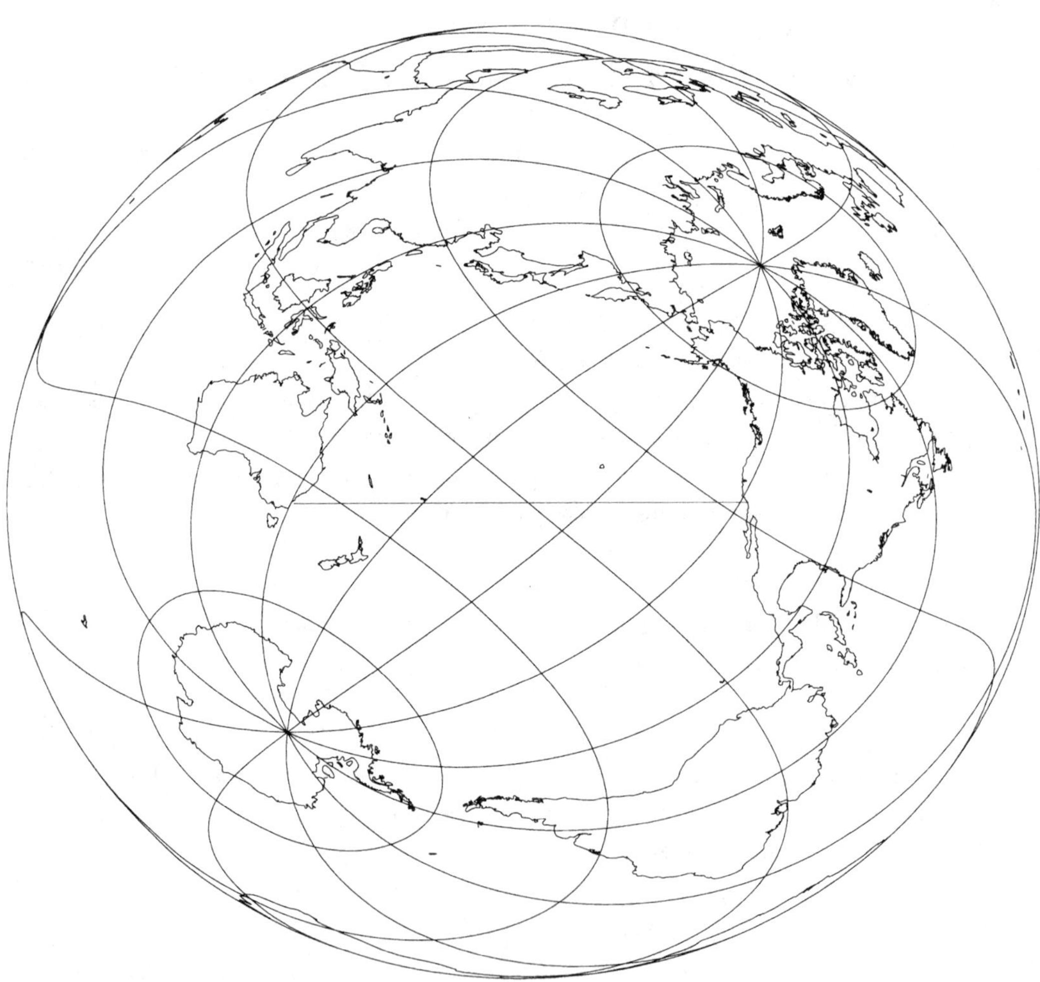

Exhibit 11–1: The Two-point Equidistant projection.
Focused at 151°E, 36°S (Sydney) and 118°W, 34°N (Los Angeles)
with great-circle route between shown.

BIFOCAL OR TWO-POINT PROJECTIONS

Approach

All the azimuthal projections covered in previous chapters focused on a single point, all other projections on a line. In positioning the general point on the map, the characteristic azimuthal map applied two criteria — the azimuthal bearing, which dictated a straight line upon which the mapped point must lie, and a distance deriving from the specific projection, which then dictated the exact position. Such criteria as compass bearing and equidistance can be applied, however, singly but simultaneously to two specific focal points relative to the general point. While these can be mixed on the one map, utility points to applying the same criterion to both points. We consider two examples, first true distance and second azimuthal bearing. The first results in a map that is no longer azimuthal in any bearing sense, but all projections using this two-point approach are usually classified as *modified azimuthal*. Hans Maurer of Germany and C. F. Close of Britain, working unknown to each other during World War I, were the dominant pioneers of this approach [1914a, 1919d, 1919f, 1921d, 1922a, 1927c, also 1935d for the mixed case by Close].

We label the two distinct focal points on the globe $P_a=(\lambda_a, \phi_a)$ and $P_b=(\lambda_b, \phi_b)$, their map points $\dot{P}_a = (\dot{x}_a, \dot{y}_a)$ and $\dot{P}_b = (\dot{x}_b, \dot{y}_b)$. To be useful these points must be a reasonable distance apart (they are typically within one region). Let that distance be $2d$. For ease of use of the map we generally choose to have the great circle through the foci lie along the x axis with the foci symmetrically placed about the origin, so $\dot{P}_a = (-d,0)$ and $\dot{P}_b = (+d,0)$.

Two-point Equidistant

The Two-point Equidistant map (see Exhibit 11–1) has the distance true from any point to each of the focal points. Clearly this is readily achievable; we merely have to place each map point on the intersection of appropriate circles centred on the two focal points, after those two points are placed their correct distance apart.

Using Equation 215c and putting

$$\delta_{a,b} = \arccos\{\sin\phi_a \sin\phi_b + \cos\phi_a \cos\phi_b \cos(\lambda_b - \lambda_a)\} = \delta_{b,a} \quad \ldots\ldots\ (263a$$

gives the great-circle distance between the two focal points as $R\delta_{a,b}$. If P_0 denotes the midpoint of this arc (which point maps onto the origin), then, using the same notation

$$\delta_{a,0} = \tfrac{1}{2}\delta_{a,b} = \tfrac{1}{2}\arccos\{\sin\phi_a \sin\phi_b + \cos\phi_a \cos\phi_b \cos(\lambda_b - \lambda_a)\} = \tfrac{1}{2}\delta_{b,a} = \delta_{0,b} \quad \ldots\ (263b$$

Having $\dot{P}_a$ and $\dot{P}_b$ their correct distance apart implies having $2d = R\delta_{a,b}$, hence:

$$\dot{P}_a = (-d,0) = (-R\delta_{a,0},0), \qquad \dot{P}_b = (+d,0) = (+R\delta_{b,0},0)$$

Similarly, using Equation 215c and putting:

$$\begin{aligned}\delta_{a,i} &= \arccos\left\{\sin\phi_a \sin\phi_i + \cos\phi_a \cos\phi_i \cos(\lambda_i - \lambda_a)\right\}\\ \delta_{b,i} &= \arccos\left\{\sin\phi_b \sin\phi_i + \cos\phi_b \cos\phi_i \cos(\lambda_i - \lambda_b)\right\}\end{aligned} \quad \text{. . . . . (264a}$$

the distances from points $\dot{P}_a$ and $\dot{P}_b$ to point $\dot{P}_i = (\dot{x}_i, \dot{y}_i)$ — the mapping of general global point $P_i = (\lambda_i, \phi_i)$ — must be $R\delta_{a,i}$ and $R\delta_{b,i}$ respectively.

Observing the geometry in the diagram below and using basic formulae, we then get:

$$(R\delta_{a,i})^2 = (\dot{x}_i + R\delta_{a,0})^2 + \dot{y}_i^2 \quad = \dot{x}_i^2 + 2\dot{x}_i R\delta_{a,0} + (R\delta_{a,0})^2 + \dot{y}_i^2 \quad \text{. . . . . (264b}$$

$$(R\delta_{a,i})^2 = (\dot{x}_i - R\delta_{b,0})^2 + \dot{y}_i^2 \quad = \dot{x}_i^2 - 2\dot{x}_i R\delta_{b,0} + (R\delta_{b,0})^2 + \dot{y}_i^2 \quad \text{. . . . . (264c}$$

Allowing that $\delta_{a,0} = \delta_{b,0}$ and subtracting the second from the first gives us

$$R^2\left(\delta_{a,i}^{\ 2} - \delta_{b,i}^{\ 2}\right) = 4\dot{x}_i R\delta_{a,0}$$

hence, after allowing that $2\delta_{a,0} = \delta_{a,b}$

$$\dot{x} = R\frac{\delta_{a,i}^{\ 2} - \delta_{b,i}^{\ 2}}{4\delta_{a,0}} \quad = R\frac{\delta_{a,i}^{\ 2} - \delta_{b,i}^{\ 2}}{2\delta_{a,b}} \quad \text{. . . . . (264d}$$

Adding Equation 264b and Equation 264c gives

$$R^2\left(\delta_{a,i}^{\ 2} + \delta_{b,i}^{\ 2}\right) = 2\dot{x}^2 + 2(R\delta_{a,0})^2 + 2\dot{y}^2$$

or

$$\dot{y}^2 = R^2\frac{\delta_{a,i}^{\ 2} + \delta_{b,i}^{\ 2}}{2} - (R\delta_{a,0})^2 - \dot{x}^2$$

Substituting from Equation 264d gives

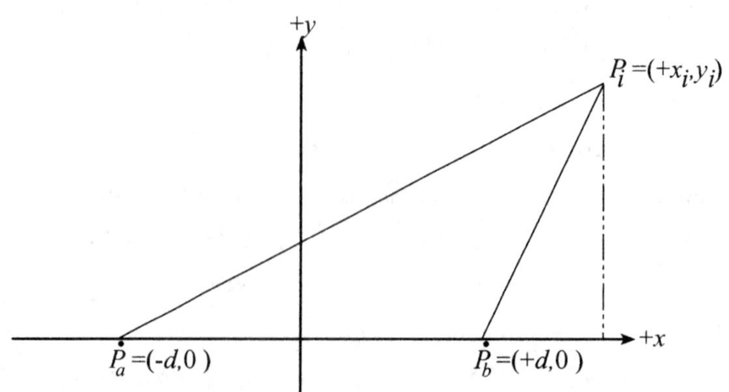

Exhibit 11–2: Two-point Equidistant projection: geometry.

$$\dot{y}^2 = R^2 \frac{\delta_{a,i}{}^2 + \delta_{b,i}{}^2}{2} - \left(R\delta_{a,0}\right)^2 - R^2 \frac{\left(\delta_{a,i}{}^2 - \delta_{b,i}{}^2\right)^2}{4\,\delta_{a,b}{}^2}$$

$$= R^2 \frac{2\delta_{a,b}{}^2\left(\delta_{a,i}{}^2 + \delta_{b,i}{}^2\right) - \left(2\delta_{a,b}\,\delta_{a,0}\right)^2 - \left(\delta_{a,i}{}^2 - \delta_{b,i}{}^2\right)^2}{4\,\delta_{a,b}{}^2}$$

Substituting $2\delta_{a,0} = \delta_{a,b}$ this resolves to

$$\dot{y}^2 = R^2 \frac{2\delta_{a,b}{}^2\left(\delta_{a,i}{}^2 + \delta_{b,i}{}^2\right) - \left(\delta_{a,b}\right)^4 - \left(\delta_{a,i}{}^2 - \delta_{b,i}{}^2\right)^2}{4\,\delta_{a,b}{}^2}$$

$$= R^2 \frac{2\left(\delta_{a,b}\,\delta_{a,i}\right)^2 + 2\left(\delta_{a,b}\,\delta_{b,i}\right)^2 - \left(\delta_{a,b}\right)^4 - \left(\delta_{a,i}\right)^4 + 2\left(\delta_{a,i}\,\delta_{b,i}\right)^2 - \left(\delta_{b,i}\right)^4}{4\,\delta_{a,b}{}^2}$$

$$= R^2 \frac{4\left(\delta_{a,b}\,\delta_{b,i}\right)^2 - \left\{\left(\delta_{a,b}\right)^2 - \left(\delta_{a,i}\right)^2 + \left(\delta_{b,i}\right)^2\right\}^2}{4\,\delta_{a,b}{}^2}$$

Thus, using definitions in Equations 263a & 264a, the plotting equations for the **Two-point Equidistant** or **Doubly Equidistant** projection become:

$$\dot{x} = R \frac{\left(\delta_{a,i}\right)^2 - \left(\delta_{b,i}\right)^2}{2\delta_{a,b}}$$

$$\dot{y} = \pm R \frac{\sqrt{4\left(\delta_{a,b}\,\delta_{b,i}\right)^2 - \left\{\left(\delta_{a,b}\right)^2 - \left(\delta_{a,i}\right)^2 + \left(\delta_{b,i}\right)^2\right\}^2}}{2\delta_{a,b}} \qquad \text{. . . (265a}$$

the latter taking the sign of the expression

$$\cos\delta_{a,i}\cos\delta_{b,i}\sin\phi\sin(\lambda_b - \lambda_a) - \cos\delta_{a,i}\sin\delta_{b,i}\cos\phi\sin(\lambda_i - \lambda_a) + \sin\delta_{a,i}\cos\delta_{b,i}\cos\phi\sin(\lambda_i - \lambda_b)$$

taking both signs when $\delta_{a,b} + \delta_{b,i} + \delta_{a,b} = 2\pi$, which produces the elliptical boundary.

The projection was first shown by Maurer [1919f] but soon after, independently, by Close [1921d]. Seen initially as serving relatively regional, often maritime purposes, the projection was later shown by Close for the whole world [1934d]. In his initial work, Close noted the special circumstances should a Pole be chosen as a focal point. The map then provides assistance with spherical geometry for any point having the latitude of the second focus. For example, given the declination and altitude of a star, its hour angle can be read off directly. A version of the Two-point Equidistant called the **Donald Elliptical** projection — surprisingly adjusted to reflect the ellipsoidal Earth — was used in North America to assess mileage for long-distance telephone calls [1957a]. A more appropriate usage would be to divide territory between stations providing air surveillance (e.g., for Search and Rescue). Although having true scale — representing distances measured along great circles — the line connecting the general point to a focal point is not itself part of a great circle except when the general point lies on the line through both focal points.

The whole world maps into an ellipse, with the focal points being its foci and their antipodean points lying at the reciprocal ends of the major axis.

Chapter 8 also provides the means for establishing the co-ordinates of the point $P_0=(\lambda_0, \phi_0)$ that maps into the origin. For azimuth $\zeta_{a,b}$ at P_a of P_b, Equation 215a shows

$$\tan\zeta_{a,b} = \frac{\sin(\lambda_b - \lambda_a)}{\cos\phi_a \tan\phi_b - \sin\phi_a \cos(\lambda_b - \lambda_a)}$$

which is identically the azimuth at P_a for all points on that great circle in the same general direction (the others being diametrically opposite). The angular distance from P_a to P_0 is $\delta_{a,0}$ for which Equation 263b provides a formula. Applying Equations 216b & 216c (identifying P_a here with P_0 there and P_0 here with P_i there), we get:

$$\phi_0 = \arcsin\{\sin\phi_a \cos\delta_{a,0} + \cos\phi_a \sin\delta_{a,0} \cos\zeta_{a,b}\} \quad \text{(266a}$$

$$\lambda_0 = \lambda_a + \arctan\left\{\frac{\sin\delta_{a,0} \sin\zeta_{a,b}}{\cos\phi_a \cos\delta_{a,0} - \sin\phi_a \sin\delta_{a,0} \cos\zeta_{a,b}}\right\} \quad \text{(266b}$$

Looking to P_0 from P_b instead of P_a, we likewise get the identically valued:

$$\phi_0 = \arcsin\{\sin\phi_b \cos\delta_{a,0} + \cos\phi_b \sin\delta_{a,0} \cos\zeta_{b,a}\}$$

$$\lambda_0 = \lambda_b + \arctan\left\{\frac{\sin\delta_{b,0} \sin\zeta_{b,a}}{\cos\phi_b \cos\delta_{b,0} - \sin\phi_b \sin\delta_{b,0} \cos\zeta_{b,a}}\right\}$$

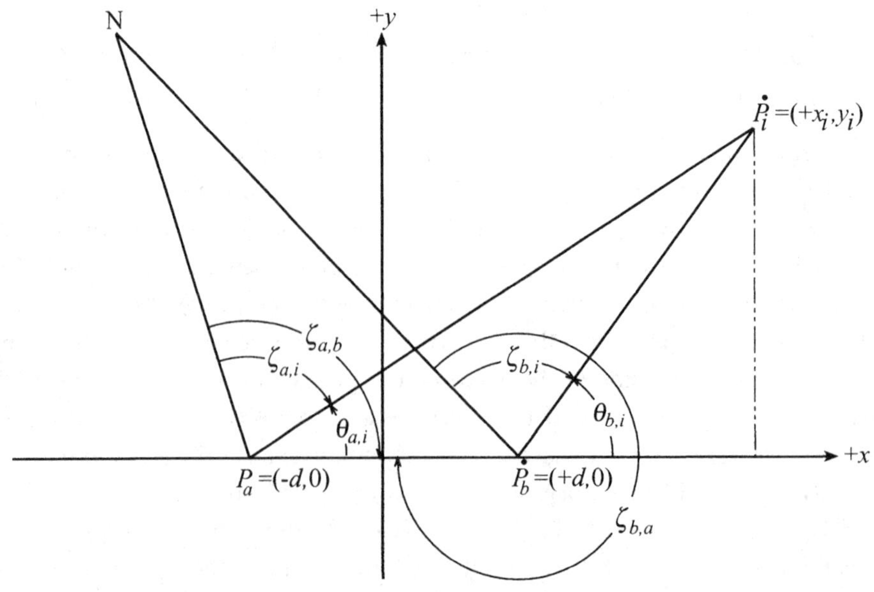

Exhibit 11–3: Two-point Azimuthal projection: geometry.

Two-point Azimuthal

Introduced by Maurer [1914a] and then, independently, by Close [1921d], the Two-point Azimuthal (or Doubly Azimuthal) map has the azimuth true from both of two focal points for any point within the map, and hence between the two. This implies that all great circles passing through the focal points must appear on the map as straight lines, and that they must intersect there at their correct angles. The Gnomonic projection has all great circles mapping into straight lines, but provides correct intersection angles only at the single focus. We find below that a simple modification can transfer this feature to the two foci, but first we explore the fundamentals. We choose again to have foci $P_a = (\lambda_a, \phi_a)$ and $P_b = (\lambda_b, \phi_b)$ map into points $(\pm d, 0)$ for some d, with P_a having a negative x co-ordinate. Equations 215a provide a choice of formulae measuring azimuths, which are denoted generically by ζ. We denote particular azimuths by subscripts as in the preceding paragraphs for distances.

Because the x axis of the map is the great circle through the two focal points, the meridians through each of the two must be straight lines intercepting the x axis at angles consistent with the pertinent azimuths (and intersecting each other unless the two foci are on the Equator). That is, the clockwise angle from the meridian through $\dot{P}_a$ to the $+x$ half-axis (i.e., the direction of $\dot{P}_b$) must be $\zeta_{a,b}$ and that from the meridian through $\dot{P}_b$ to the $-x$ half-axis (i.e., the direction of $\dot{P}_a$) must be $\zeta_{b,a}$. Exhibit 11–3 illustrates the scene.

The lines from $\dot{P}_a$ and $\dot{P}_b$ to $\dot{P}_i$ — the mapping of general global point $P_i = (\lambda_i, \phi_i)$ — must be respectively at (clockwise) azimuth angles $\zeta_{a,i}$ and $\zeta_{b,i}$ relative to the meridians. Writing $\theta_{a,i}$ and $\theta_{b,i}$ for the (counter-clockwise) mathematical angles at which these lines to $\dot{P}_i$ intercept the x axis (measured relative to the axis) we have:

$$\theta_{a,i} = \zeta_{a,b} - \zeta_{a,i}$$
$$\theta_{b,i} = \zeta_{b,a} - \pi - \zeta_{b,i}$$

From the diagram, we see

$$\frac{\dot{y}}{\dot{x}+d} = \tan\theta_{a,i} \quad \text{and} \quad \frac{\dot{y}}{\dot{x}-d} = \tan\theta_{b,i} \qquad \text{. . . . . (267a}$$

Manipulating these to give $\dot{y}$ from each, then equating the two forms gives

$$(\dot{x}+d)\tan\theta_{a,i} = \dot{y} = (\dot{x}-d)\tan\theta_{b,i}$$
$$(\dot{x}+d)\cot\theta_{b,i} = (\dot{x}-d)\cot\theta_{a,i}$$

Hence,

$$\dot{x}\left(\cot\theta_{b,i} - \cot\theta_{a,i}\right) = -d\left(\cot\theta_{b,i} + \cot\theta_{a,i}\right)$$

giving

$$\dot{x} = d\frac{\cot\theta_{a,i} + \cot\theta_{b,i}}{\cot\theta_{a,i} - \cot\theta_{b,i}} \qquad \text{. . . . . (267b}$$

Substitution for $\dot{x}$ in the first of Equations 267a gives

$$\begin{aligned}\dot{y} &= \frac{(\dot{x}+d)}{\cot\theta_{a,i}} = \frac{d}{\cot\theta_{a,i}}\left\{\frac{\dot{x}}{d}+1\right\} = \frac{d}{\cot\theta_{a,i}}\left\{\frac{\cot\theta_{a,i}+\cot\theta_{b,i}}{\cot\theta_{a,i}-\cot\theta_{b,i}}+1\right\} \\ &= \frac{d}{\cot\theta_{a,i}}\left\{\frac{2\cot\theta_{a,i}}{\cot\theta_{a,i}-\cot\theta_{b,i}}\right\} = \frac{2d}{\cot\theta_{a,i}-\cot\theta_{b,i}}\end{aligned} \qquad (268a$$

An angle-difference formula in Tutorial 10 allows these equations to be broken into the cotangents of the azimuthal angles. Then, any of the formulae of Equations 215a allow these, in turn, to be expressed in terms of the longitudes and latitudes of the three points $\dot{P}_a$, $\dot{P}_b$ and $\dot{P}_i$. Hence, we can obtain plotting equations for this projection. Substitution and manipulation of the successive lengthy expressions involved is inappropriate here, however. The interested reader is referred to Immler's lengthy paper [1919e].

As indicated above, we can alternatively adapt the Gnomonic projection. We first define an oblique Gnomonic mapping with $P_0 = (\lambda_0, \phi_0)$ as defined with Equations 266a & 266b at its centre; assuming correct scale at the centre, Equations 199a gives equations:

$$\begin{aligned}\dot{x} &= R\frac{\cos\phi\sin(\lambda-\lambda_0)}{\sin\phi_0\sin\phi+\cos\phi_0\cos\phi\cos(\lambda-\lambda_0)} \\ \dot{y} &= R\frac{\cos\phi_0\sin\phi-\sin\phi_0\cos\phi\cos(\lambda-\lambda_0)}{\sin\phi_0\sin\phi+\cos\phi_0\cos\phi\cos(\lambda-\lambda_0)}\end{aligned} \qquad (268b$$

Points $\dot{P}_a$ and $\dot{P}_b$ are symmetrically placed about the origin, on their great circle. From Equation 44a, they are distance $R\tan\delta_{a,0}$ from the origin, with $\delta_{a,0}$ given in Equation 263b.

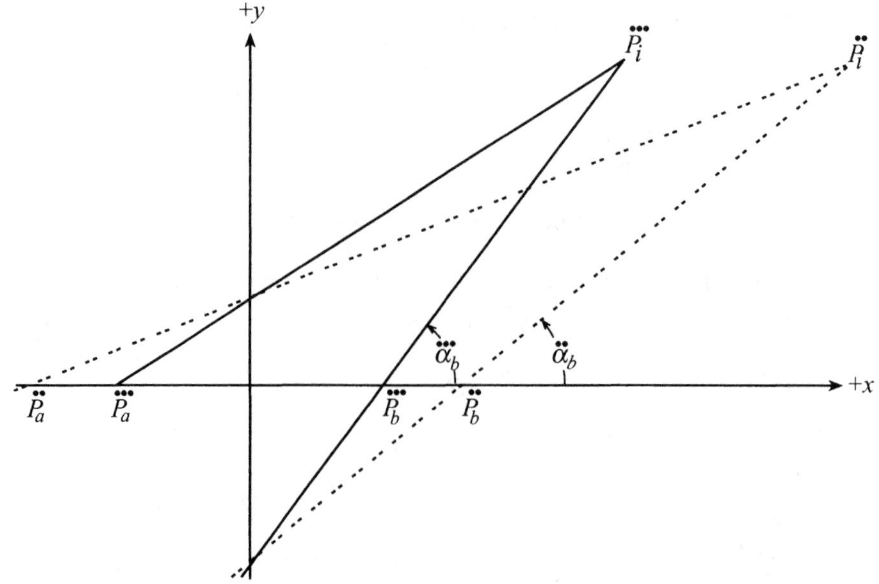

Exhibit 11–4: Modifying the Gnomonic projection to be doubly azimuthal.

If $\zeta_{0,a}$ is the azimuth at P_0 of P_a, then, from Equation 215a,

$$\begin{aligned}\zeta_{0,a} &= \operatorname{arccot}\left\{\frac{\cos\phi_0 \tan\phi_a - \sin\phi_0 \cos(\lambda_a - \lambda_0)}{\sin(\lambda_a - \lambda_0)}\right\} \\ &= \frac{3\pi}{2} - \arctan\left\{\frac{\cos\phi_0 \tan\phi_a - \sin\phi_0 \cos(\lambda_a - \lambda_0)}{\sin(\lambda_a - \lambda_0)}\right\}\end{aligned} \qquad \text{. . . . . (269a}$$

and, with azimuths being clockwise, the sequential rotational transformation (Equations 86a):

$$\begin{aligned}\ddot{x} &= -\dot{x}\sin\zeta_{0,a} - \dot{y}\cos\zeta_{0,a} \\ \ddot{y} &= +\dot{x}\cos\zeta_{0,a} - \dot{y}\sin\zeta_{0,a}\end{aligned} \qquad \text{. . . . . (269b}$$

will bring the focal points of the map onto the x axis symmetric about the origin at the mutual spacing they had with the original mapping, i.e.,

$$\ddot{P}_a = (-R\tan\delta_{a,0}, 0), \quad \ddot{P}_b = (+R\tan\delta_{a,0}, 0)$$

Exhibit 11–4 illustrates the situation using double-dotted lines for the great circles passing through each focal point and through the mapped general point, labelled $\ddot{P}_i$ at this stage. The angles of intersection between these and the great circle that is the x axis are distorted from those of the globe. From the analysis of the distortion in the pure Gnomonic projection on page 93, reverting from latitude to polar distance, we have at any point $(R\tan\gamma, 0)$ the distortion radially along the x axis equals $\cos^{-2}\gamma$ while the transverse distortion equals $\cos^{-1}\gamma$. Thus, if we compare the angle denoted $\ddot{\alpha}_b$ in the illustration with its original value α_b on the globe, we must have

$$\tan\ddot{\alpha}_b = \frac{\cos^2\gamma}{\cos\gamma}\tan\alpha_b \quad = \cos\gamma \tan\alpha_b$$

If we now apply the transformation:

$$\dddot{x} = \cos\delta_{a,0}\,\ddot{x}, \quad \dddot{y} = \ddot{y}$$

all great circles retain their intersections with the y axis unchanged and all remain straight lines, but all are brought closer to the verticality of the y axis, as shown by unbroken lines in the diagram. The angles of intersection with the x axis are increased, with

$$\tan\dddot{\alpha}_b \quad = \cos^{-1}(\delta_{a,0})\tan\ddot{\alpha}_b \quad = \frac{\cos\gamma}{\cos\delta_{a,0}}\tan\alpha_b$$

At the points $\dddot{P}_a = (-\sigma R\sin\delta_{a,0}, 0)$ and $\dddot{P}_b = (+\sigma R\sin\delta_{a,0}, 0)$ the quotient equals 1, so those angles become correct, as needed for the doubly azimuthal map.

The plotting equations for the **Two-point Azimuthal** or **Doubly Azimuthal** (or **Orthodromic** or **Close** or **McGaw** or **Immler**) projection, with optional scaling factor, are:

$$\begin{aligned}\dddot{x} &= \sigma\cos\delta_{a,0}\,\ddot{x} &&= \sigma\cos\delta_{a,0}\left(\dot{x}\sin\zeta_{0,a} + \dot{y}\cos\zeta_{0,a}\right) \\ \dddot{y} &= \sigma\ddot{y} &&= -\sigma\dot{x}\cos\zeta_{0,a} + \dot{y}\sin\zeta_{0,a}\end{aligned} \qquad \text{. . . . . (269c}$$

with the right-hand variables defined in Equations 263b, 268b & 269a.

THREE-POINT PROJECTIONS

Conflict

While there is no scope for mapping precisely in the above manner relative to three separate points, at least one projection has been developed that is referenced to three points. The **Chamberlin Trimetric** projection [1947b] is one attempt that comes close that is akin to the Two-point Equidistant. The three foci are placed their true distance apart. Then, for each map point, draw circles centred on the three foci of radii equal to the respective distances. The three pertinent pair-wise intersections define a small triangle. Wellman Chamberlin took the centre of that triangle, but modern computer methods favour a more elaborate weighted approach for fixing the point. So, except for scale along straight lines connecting foci, nothing is exactly true. The distortion is modest, however, if the foci are towards the outer limits of the map.

Clearly, the trimetric projection, illustrated covering Africa and adjacent areas in Exhibit 11–5, is usable only at a major regional level. The methodology can be adapted to extend the coverage beyond the basic triangle, but only very limitedly without significant change to effective scale.

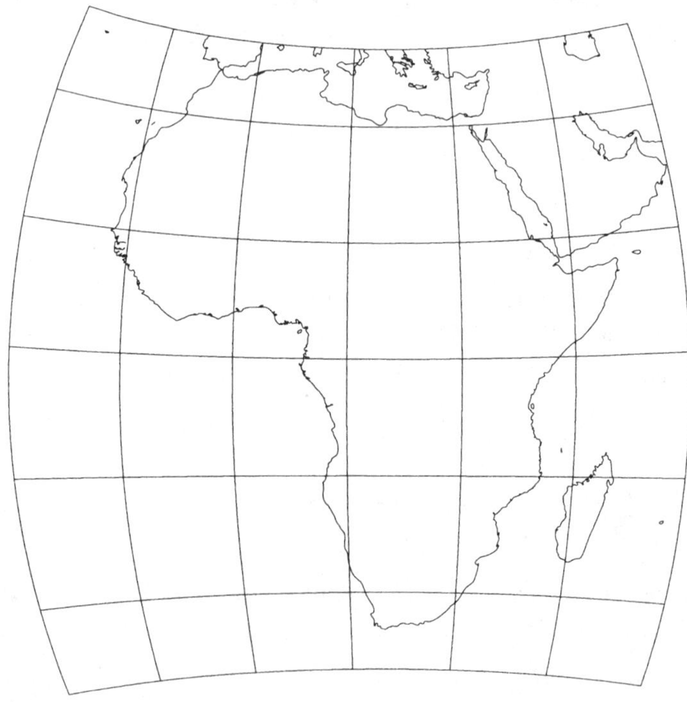

Exhibit 11–5: Chamberlin Trimetric projection.
Foci 19°W, 24°N; 59°E, 4°N; 20°E, 35°S.

RETRO-AZIMUTHAL PROJECTIONS

Purpose

Every azimuthal map provides an easy reading of the great-circle azimuth from the focus to any other point in the map. It also shows the reverse azimuth, but not with comparable ease of reading, even at Simple aspect. The Gnomonic goes further, providing azimuth between any two points, but only with much difficulty of reading for the general point at Oblique aspect — which is a necessity for widespread use because of the limited coverage of any one Gnomonic map. To avoid such difficulties, an open range of projections has been developed to provide easily read reverse azimuths to a preset focus. Pioneered by J. I. Craig of Britain [1927c], they are called *retro-azimuthal* projections.

Reversing the denotation of points, Equation 215a shows that the azimuth $\zeta_{i,0}$ at any point $P_i = (\lambda_i, \phi_i)$ of point $P_0 = (\lambda_0, \phi_0)$ is given by

$$\cot \zeta_{i,0} = \frac{\cos\phi_i \tan\phi_0 - \sin\phi_i \cos(\lambda_i - \lambda_0)}{\sin(\lambda_i - \lambda_0)} \qquad \text{. . . . . (271a}$$

Clearly, except for movement along the meridian of P_0 and, if P_0 has zero latitude, along the Equator, movement of P_i along any great else lesser circle produces a continual variation of the angular value. The variation is in fact continuous, except when traversing a Pole else the antipodean point — then it changes abruptly. The two Poles, therefore, mark a natural edge to any retro-azimuthal map, and, regardless of the latitude of the focal point, it is standard practice for zero to be the central latitude. Further, most such maps are kept to the hemisphere centred on the meridian of the focal point, since progress beyond produces an inverted replica of the prime hemisphere. If working from known longitude and latitude for point P_i, one only needs the left else right half of the prime hemisphere, because the arithmetic value of azimuth for any point outside this is the same as for the corresponding point within.

Straight parallel meridians for retro-azimuthal readings

Azimuths are angles relative to North, so ease of reading is best provided by a straight meridian. To apply this within the reality of a grid requires that all meridians be parallel, allowing the line through any point to be readily established. The even spacing of meridians is an obvious first choice, but there is no inherent requirement for any particular spacing.

Mapping our focal point P_0 onto the origin, and point P_i onto $\dot{P}_i = (\dot{x}_i, \dot{y}_i)$, the essence of a retro-azimuthal map with straight parallel meridians is to have

$$\dot{y}_i = \dot{x}_i \cot \zeta_{i,0} \qquad \text{. . . . . (271b}$$

If we choose evenly spaced meridians, then $x_i = c\,(\lambda_i - \lambda_0)$ for some constant c. This gives the **Craig Retro-azimuthal** projection, created by Craig [1910b] in a report for the Survey of Egypt. Illustrated in Exhibit 11–6, it is also known also as the **Mecca** (now Makkah) projection, that holiest of Islamic cities having been the focal point of the original. Allowing for the mirroring effect of Equation 264b and omitting the suffix i of the general point, the plotting equations for a focus in the northern hemisphere become:

$$\dot{x} = c\,(\lambda - \lambda_0)$$

$$\text{if } \lambda = \lambda_0 \quad \dot{y} = c\,\{\sin\phi - \cos\phi\,\tan\phi_0\} \qquad \text{(272a}$$

$$\text{else} \quad \dot{y} = c\,(\lambda - \lambda_0)\frac{\sin\phi\,\cos(\lambda - \lambda_0) - \cos\phi\,\tan\phi_0}{\sin(\lambda - \lambda_0)}$$

The projection has a practical longitudinal span of about 300°. A finer graticule is appropriate on such maps for locating points and facilitating the reading of azimuthal bearings, measured from the meridians by protractor.

Equation 271b suggests spacing proportional to sine of longitude. Thus:

$$\dot{x} = c\,\sin(\lambda - \lambda_0)$$

$$\dot{y} = c\,\sin(\lambda - \lambda_0)\cot\zeta = c\,\{\cos\phi\tan\phi_0 - \sin\phi\cos(\lambda - \lambda_0)\}$$

Squaring the latter and substituting for $c\cos(\lambda-\lambda_0)$ via $\sin(\lambda-\lambda_0)$ from the former equation of this pair shows the parallels to be the ellipses

$$\frac{\dot{x}^2}{c^2} + \frac{(c\cos\phi\,\tan\phi_0 - \dot{y})^2}{c^2\cos^2\phi} = 1$$

individually centred at $(0, c\cos\phi\,\tan\phi_0)$ with semi-axes of lengths c and $c\cos\phi$.

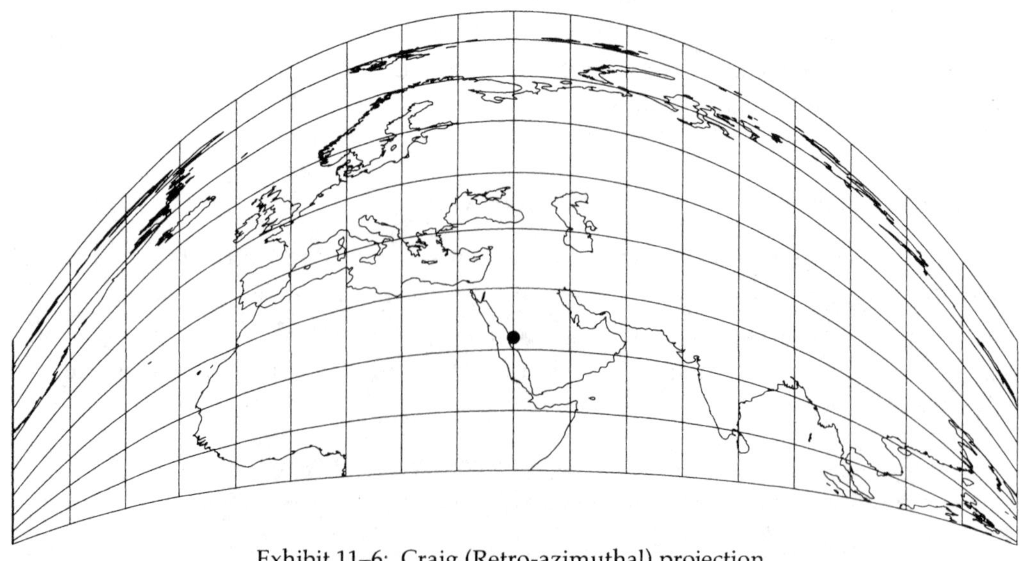

Exhibit 11–6: Craig (Retro-azimuthal) projection.
Focused at 39°49′E, 21°26′N (Makkah); 10° graticule.

Any function purely of longitude for the x co-ordinate produces straight parallel meridians, each with its own spacing of them and its own distinctive curvature for the parallels. Hyperbolic parallels are produced by $\dot{x} = c \tan\lambda$, parabolic if the longitude is halved. Setting $\dot{x} = \sin\lambda$ produces partial ellipses, with semi-axes of lengths c and $c \sin\phi_0$.

Curved meridians for retro-azimuthal readings

With straight parallel meridians, the azimuth is read by drawing the straight line from $\dot{P}$ to $\dot{P}_0$ then reading the angle relative to the meridian at the former. It could as well be read at the latter (where the meridian line could be ensured on the map), if a 180° correction is included else the line continued. Then, straight meridians other than one through the focal point are not necessary for our desired ease of reading.

Curved meridians result if the plotting equation for $\dot{x}$ involves the latitude. Using the compound denominator term of Equation 215a as the multiplier of longitude is an obvious idea that suggests simpler plotting equations. First presented [1910c] at about the same time as the Mecca, and eponymously named, the **Hammer Retro-azimuthal** projection (distinct from the Hammer projection in Chapter 10) has plotting equations:

$$\begin{aligned} \dot{x} &= +R\,A\cos\phi\,\sin(\lambda-\lambda_0) \\ \dot{y} &= -R\,A\{\sin\phi_0\cos\phi - \cos\phi_0\sin\phi\,\cos(\lambda-\lambda_0)\} \end{aligned} \qquad \text{(273a}$$

where, denoting by ${}_{a}\phi$ the auxiliary latitude of the general point under oblique transformation, here centred on P_0 rather than P_N, its value otherwise given via Equation 192a,

$$A = \frac{\frac{\pi}{2} - {}_{a}\phi}{\cos\left(\frac{\pi}{2} - {}_{a}\phi\right)} = \frac{\arccos\{\sin\phi_0\cos\phi + \cos\phi_0\sin\phi\,\cos(\lambda-\lambda_0)\}}{\sin\phi_0\cos\phi + \cos\phi_0\sin\phi\,\cos(\lambda-\lambda_0)}$$

The projection can cover the whole world, but the further hemisphere appears laterally inverted. The map in Exhibit 11–7 overleaf, which is restricted to a 90° span eastwards to avoid the overlap but spans 180° westwards so incurs the overlap in its left side, illustrates the considerable contortion that results for the meridians. While the map is focused as stated, it is centred at the intersection of the focal meridian with the Equator.

Prompted by E. A. Reeves drawing retro-azimuthal isolines on a Mercator map [1929e], A. R. Hinks created his own retro-azimuthal projection [1929f], differing from Hammer's work only by being extended to the whole world.

The most interesting and useful Retro-azimuthal is that introduced by J. J. Littrow of Austria [1833a]. A transverse form of the conformal Lagrange projection discussed later (in Chapter 15), the **Littrow** projection has the plotting equations:

$$\begin{aligned} \dot{x} &= c\,\sec\phi\,\sin(\lambda-\lambda_0) \\ \dot{y} &= c\,\tan\phi\,\cos(\lambda-\lambda_0) \end{aligned} \qquad \text{(273b}$$

The azimuth is read by drawing the straight line from starting point through destination on the focal meridian, then reading its ongoing angle relative to that meridian. The map in Exhibit 11–8 illustrates this for the direction to Accra from Ankara. The cotangent of the azimuth can be evaluated from the triangle completed by the dotted lines. By substitution from Equations 273b we then get

$$\cot\dot{\zeta} = \frac{\dot{y}_0 - \dot{y}}{-\dot{x}} = \frac{c\,\tan\phi_0 - c\,\tan\phi\,\cos(\lambda - \lambda_0)}{-c\,\sec\phi\,\sin(\lambda - \lambda_0)}$$

$$= \frac{\cos\phi\tan\phi_0 - \sin\phi\,\cos(\lambda - \lambda_0)}{-\sin(\lambda - \lambda_0)}$$

which means that the plotting equations meet the retro-azimuthal criterion Equation 271a. As this is for any general point P_0 of longitude λ_0, the projection is simultaneously retro-azimuthal to all points on the focal longitude.

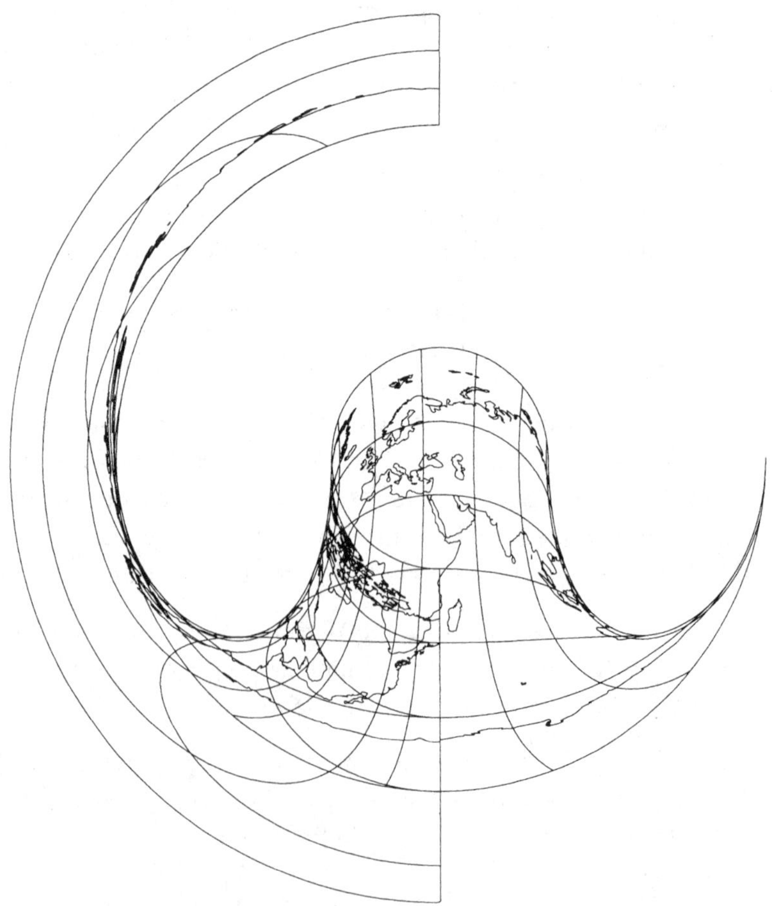

Exhibit 11–7: The Hammer Retro-azimuthal projection.

Rearranging, squaring, and adding the two Equations 273b to eliminate longitude, gives

$$\frac{\dot{x}^2}{c^2\sec^2\phi}+\frac{\dot{y}^2}{c^2\tan^2\phi}=\sin^2(\lambda-\lambda_0)+\cos^2(\lambda-\lambda_0)=1$$

so parallels are ellipses, of eccentricity

$$\varepsilon=\sqrt{\frac{c^2\sec^2\phi-c^2\tan^2\phi}{c^2\sec^2\phi}}=\sqrt{\frac{1-\sin^2\phi}{1}}=\sqrt{\cos^2\phi}=\cos\phi$$

Squaring, then differencing those two equations to eliminate latitude, gives the hyperbolae

$$\frac{\dot{x}^2}{c^2\sin^2(\lambda-\lambda_0)}-\frac{\dot{y}^2}{c^2\cos^2(\lambda-\lambda_0)}=1$$

While the Littrow projection cannot encompass the globe it covers a full hemisphere well, with the focal meridian spanning all latitudes within the hemisphere. Interestingly, it can thus be used generically for azimuths, and for angles of spherical triangles — which, as we saw in Chapter 8, all azimuths fundamentally are. C. P. Weir of Britain realized this general feature and published it in 1890 as the *Weir Azimuth Diagram* [see 1919d].

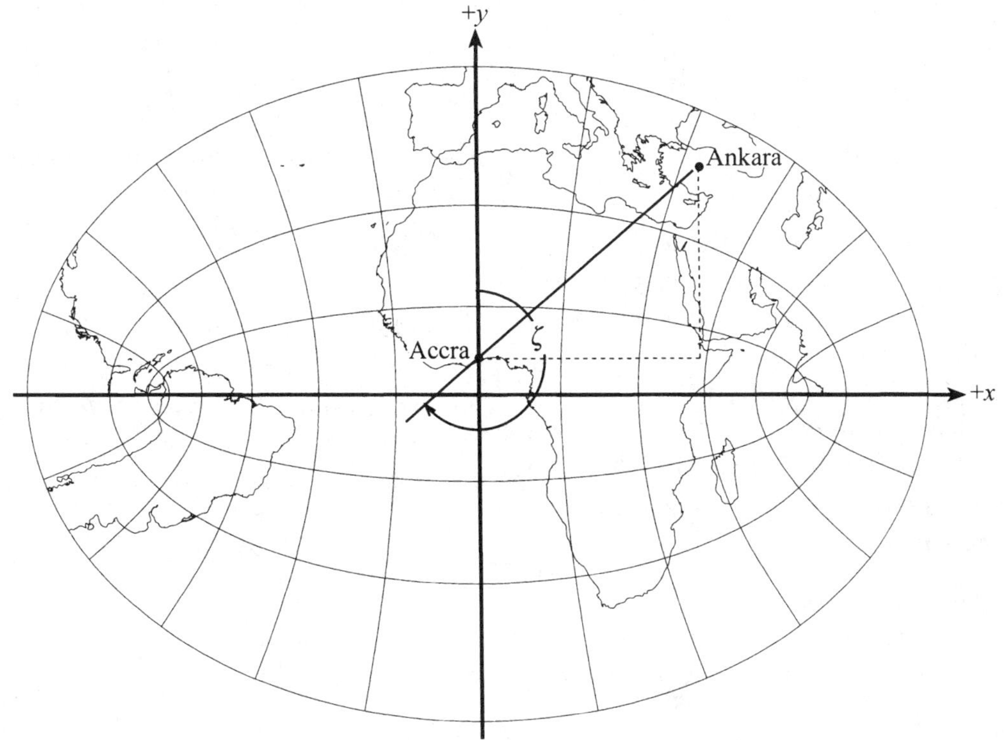

Exhibit 11–8: Littrow Retro-azimuthal projection.

TUTORIAL 25 PARTIAL DIFFERENTIATION

Differentiation as introduced in Tutorial 13 applied to functions that were of a single variable. The ordinary derivative of a function was effectively the slope function of a 1-dimensional line graph in a 2-dimensional plane.

For functions of two variables, corresponding to a 2-dimensional surface in a 3-dimensional setting, the slope is a complicated compound of the two underlying variables. Partial differentiation relates to functions of any number of variables, but our need is restricted to two.

Let $f(x, y)$ be such a function. Then the (first) *partial derivatives* of $f(x)$ with respect to x and y are denoted and defined as the limits

$$\frac{\partial f}{\partial x} = \lim_{\Delta x \to 0} \frac{f(x+\Delta x, y) - f(x,y)}{\Delta x}$$

and

$$\frac{\partial f}{\partial y} = \lim_{\Delta y \to 0} \frac{f(x, y+\Delta y) - f(x,y)}{\Delta y}$$

where such limits exist. Thus, each partial derivative is effectively an ordinary derivative with the other variable fixed. For the surface it represents the slope in the direction of its variable. Hence, a zero value for a partial derivative corresponds to a locally extreme value for the original function relative to that variable. The two partial derivatives define the tangent plane to the surface.

The full differential

The full differential of f, denoted by the familiar df, is the sum of these two partial derivatives multiplied by their respective ordinary differentials, i.e.,

$$\mathrm{d}f = \frac{\partial f}{\partial x}\mathrm{d}x + \frac{\partial f}{\partial y}\mathrm{d}y$$

The full derivative

If x and y are each differentiable functions of variable t, then the full derivative of f with respect to t is given by

$$\frac{\mathrm{d}f}{\mathrm{d}t} = \frac{\partial f}{\partial x}\frac{\mathrm{d}x}{\mathrm{d}t} + \frac{\partial f}{\partial y}\frac{\mathrm{d}x}{\mathrm{d}t}$$

Second and further partial derivatives

Assuming continuing differentiability, such differentiation can be continued, producing second, third and further partial derivates. Because each first derivative is typically a function of both variables, four potentially distinct derivatives must be recognized at the second level. These are

$$\frac{\partial}{\partial x}\left(\frac{\partial f}{\partial x}\right),\ \frac{\partial}{\partial x}\left(\frac{\partial f}{\partial y}\right),\ \frac{\partial}{\partial y}\left(\frac{\partial f}{\partial x}\right),\ \frac{\partial}{\partial y}\left(\frac{\partial f}{\partial y}\right)$$

The middle two, however, are identical if they are themselves continuous functions.

The Chain Rule

If $f(x, y)$ is a differentiable function of x and y, and each of those variables is a function of the two variables u and v (i.e., $x = g(u, v)$ and $y = h(u, v)$ for differentiable functions g and y), then:

$$\frac{\partial f}{\partial u} = \frac{\partial f}{\partial x}\frac{\partial x}{\partial u} + \frac{\partial f}{\partial y}\frac{\partial y}{\partial u}$$

$$\frac{\partial f}{\partial v} = \frac{\partial f}{\partial x}\frac{\partial x}{\partial v} + \frac{\partial f}{\partial y}\frac{\partial y}{\partial v}$$

The formula expressed by both is known as the *Chain Rule*.

CHAPTER 12

Shaping Mathematics: Differential geometry

Preamble

To this point we have used a variety of mathematical techniques, with tutorials to provide basic explanation. Having reached the stage where each of the intermediate parameters in a projection can involve both longitude and latitude, we need to delve into the much more complicated field of differential geometry. We do this within the main body of our text. This is appropriate, for it was cartographical needs that gave rise to this adaptation of differentiation to the curved surfaces of the 3-dimensional geometrical realm. That heritage is signalled by the term used for the curve along any surface that traces the shortest path between two points. It is *geodesic* — from the term geodesy that we have already met.

The calculus must now be expanded from ordinary to partial differentiation — a quintessential feature of differential geometry. The final pages of the chapter use matrices and determinants (linear algebra).

There are no maps in this chapter.

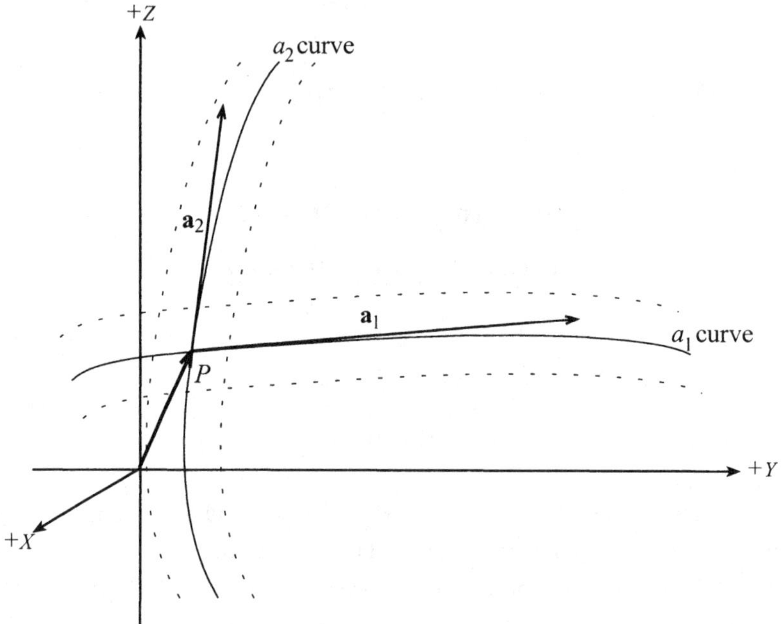

Exhibit 12–1: Parametric surfaces and curves.

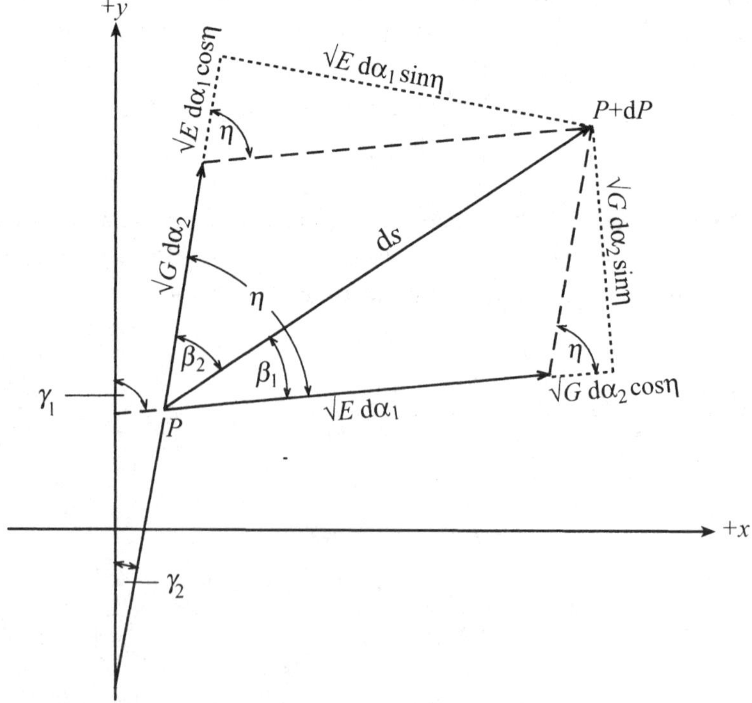

Exhibit 12–2: The parametric parallelogram.

GENERAL SURFACES

Parametric representation

For the global surface, all co-ordinates X, Y and Z are functions, for the given radius, of the two variables λ and ϕ (using equations from Tutorial 3). Although we never expressed them as 3-dimensional objects, the cylinder and cone have surfaces represented by the variable pairs $\dot{u},\dot{v}$ and $\dot{\rho},\dot{\theta}$. All surfaces are so representable, i.e., as the set of points

$$P = \left(X(a_1,a_2), Y(a_1,a_2), Z(a_1,a_2)\right)$$

for a pair of variables a_1 and a_2. These variables are known as *parameters*, and the expression of the surface in them is known as the *parametric form*.

If $\mathbf{p}$ is the vector to the point $P = (X, Y, Z)$ on the surface, then the scalar co-ordinates X, Y and Z and hence the vector $\mathbf{p}$ are functions of the parameters, i.e.,

$$\mathbf{p} = \mathbf{p}(a_1,a_2) = \left(X(a_1,a_2), Y(a_1,a_2), Z(a_1,a_2)\right)$$

If, at a given point, we develop the vectors with just one parameter varying from the value at the point, we obtain a curve passing through the point. If a_1 varies but a_2 is fixed, we call the resulting parametric curve the a_1 curve; if a_2 varies but a_1 is fixed, it is called the a_2 curve. Exhibit 12–1 illustrates the geometry, with $\mathbf{a}_1$ and $\mathbf{a}_2$ the tangent vectors at P along the a_1 and a_2 curves respectively. (The two parametric lines can intersect at any non-zero angle and their curvatures can be in any orientation.) The tangent vectors can be written as:

$$\mathbf{a}_1 = \frac{\partial \mathbf{p}}{\partial a_1} = \left(\frac{\partial X}{\partial a_1}, \frac{\partial Y}{\partial a_1}, \frac{\partial Z}{\partial a_1}\right), \quad \mathbf{a}_2 = \frac{\partial \mathbf{p}}{\partial a_2} = \left(\frac{\partial X}{\partial a_2}, \frac{\partial Y}{\partial a_2}, \frac{\partial Z}{\partial a_2}\right)$$

so, from Tutorial 25, the total differential of vector $\mathbf{p}$ is the *parametric incremental vector*

$$d\mathbf{p} = (dX, dY, dZ) = \frac{\partial \mathbf{p}}{\partial a_1} da_1 + \frac{\partial \mathbf{p}}{\partial a_2} da_2 \quad = \mathbf{a}_1\, da_1 + \mathbf{a}_2\, da_2$$

First fundamental quantities and form

If $ds = |d\mathbf{p}|$, i.e., the length of $d\mathbf{p}$,

$$(ds)^2 = (dX)^2 + (dY)^2 + (dZ)^2 \qquad \text{. . . . . (279a}$$

Alternatively, since the dot-product of a vector with itself equals the square of its length,

$$(ds)^2 = |d\mathbf{p}|^2 = d\mathbf{p}\cdot d\mathbf{p} = \{\mathbf{a}_1\, da_1 + \mathbf{a}_2\, da_2\}\cdot\{\mathbf{a}_1\, da_1 + \mathbf{a}_2\, da_2\}$$
$$= \mathbf{a}_1\cdot\mathbf{a}_1 (da_1)^2 + 2\mathbf{a}_1\cdot\mathbf{a}_2\, da_1\, da_2 + \mathbf{a}_2\cdot\mathbf{a}_2 (da_2)^2$$

So

$$ds^2 = E\, da_1{}^2 + 2F\, da_1\, da_2 + G\, da_2{}^2 \qquad \text{. . . . . . (279b}$$

where:

$$E = \mathbf{a}_1 \cdot \mathbf{a}_1 = |\mathbf{a}_1|^2 = \left\{ \left(\frac{\partial X}{\partial a_1}\right)^2 + \left(\frac{\partial Y}{\partial a_1}\right)^2 + \left(\frac{\partial Z}{\partial a_1}\right)^2 \right\}$$

$$F = \mathbf{a}_1 \cdot \mathbf{a}_2 = |\mathbf{a}_1||\mathbf{a}_2| \cos\eta = \left\{ \frac{\partial X}{\partial a_1}\frac{\partial X}{\partial a_2} + \frac{\partial Y}{\partial a_1}\frac{\partial Y}{\partial a_2} + \frac{\partial Z}{\partial a_1}\frac{\partial Z}{\partial a_2} \right\} \quad \ldots\ldots (280a$$

$$G = \mathbf{a}_2 \cdot \mathbf{a}_2 = |\mathbf{a}_2|^2 = \left\{ \left(\frac{\partial X}{\partial a_2}\right)^2 + \left(\frac{\partial Y}{\partial a_2}\right)^2 + \left(\frac{\partial Z}{\partial a_2}\right)^2 \right\}$$

angle η being the angle between vectors $\mathbf{a}_1$ and $\mathbf{a}_2$. The terms E, F, and G are called the *first fundamental quantities*. The composite expression in Equation 279b using them is called the *first fundamental form* for the surface.

The three fundamental quantities characterize a surface in many respects, whatever its shape. Related between the two surfaces involved, they also provide crucial indicators of the effect of any mapping process. Note, however, that, although labelled fundamental, their values depend on the choice of parameters. This is illustrated with the plane on page 286, where we begin qualifying the symbols for the fundamental quantities with subscripts to distinguish the parameter-pair chosen.

First fundamental quantities in two variables

Our interest in characterizing surfaces is to establish criteria for and evaluations of mapping transformations from one surface to another. While the 3-dimensional globe is the essential source of any map, the target of our mapping is a 2-dimensional plane. In later chapters, we have mappings that go from plane to plane, others that go from ellipsoid to sphere. These various surfaces — indeed every surface — can be represented by two variables. Then all can be treated mathematically as 2-dimensional. For the curved surfaces, we take the infinitesimal locality around a point as being planar, and use the tangent plane defined by $\mathbf{a}_1$ and $\mathbf{a}_2$ to represent that locality.

With some Cartesian axes x, y as reference, similar development to that producing Equations 280a gives for the 2-dimensional plane:

$$E = \left(\frac{\partial x}{\partial a_1}\right)^2 + \left(\frac{\partial y}{\partial a_1}\right)^2, \quad F = \frac{\partial x}{\partial a_1}\frac{\partial x}{\partial a_2} + \frac{\partial y}{\partial a_1}\frac{\partial y}{\partial a_2}, \quad G = \left(\frac{\partial x}{\partial a_2}\right)^2 + \left(\frac{\partial y}{\partial a_2}\right)^2 \quad \ldots\ldots (280b$$

In the remainder of this chapter, we use these quantities to establish various measurements for surfaces in general, then for the particular surfaces that concern us. Finally, we develop formulae relating, for any one surface, its fundamental quantities using different parameter pairs. The results form the basis for studying the transformation from one surface to another in the following chapter.

MEASUREMENTS ON THE SURFACE

Parametric axes

The tangent vectors $\mathbf{a}_1$ and $\mathbf{a}_2$ at a point for a chosen pair of parameters define the *parametric axes* specific to that point. The parallelogram defined by vectors $\mathbf{a}_1$ and $\mathbf{a}_2$ is referred to as the *parametric parallelogram*. Exhibit 12–2 shows this, with dashed lines completing the parallelogram. The vectors are labelled only with their lengths, as established in the next equations. We explore details of these axes, the parallelogram, and the incremental vector, viewing the infinitesimal differentials within the plane. We assume a pair of Cartesian axes within that plane, denoted by the usual x, y, and then use Equations 280b for the fundamental quantities.

Lengths

Since $\mathrm{d}a_2$ is zero for all points on the $\mathbf{a}_1$ curve, and likewise $\mathrm{d}a_1$ is zero for all points on the $\mathbf{a}_2$ curve, we get immediately from Equations 280b:

$$\begin{aligned}&\text{incremental distance along } \mathbf{a}_1 \text{ (the } a_1 \text{ curve)} = {}_{a_1}\mathrm{d}s = \sqrt{E\,\mathrm{d}a_1{}^2} = \sqrt{E}\,\mathrm{d}a_1\\ &\text{incremental distance along } \mathbf{a}_2 \text{ (the } a_2 \text{ curve)} = {}_{a_2}\mathrm{d}s = \sqrt{G\,\mathrm{d}a_2{}^2} = \sqrt{G}\,\mathrm{d}a_2\end{aligned} \qquad \text{. . . . (281a}$$

showing E and G as representing distance change along the a_1 and a_2 curves, respectively.

Angle between parametric axes

From Equation 280a, the angle η between the two parametric axes is defined as

$$\cos\eta = \frac{F}{|\mathbf{a}_1||\mathbf{a}_2|} = \frac{F}{\sqrt{EG}} \qquad \text{. . . . . (281b}$$

showing F as representing angular separation. Adopting the further notation

$$H^2 = EG - F^2 \qquad \text{. . . . . (281c}$$

with H positive, we obtain:

$$\sin\eta = \sqrt{1 - \frac{F^2}{EG}} = \sqrt{\frac{EG - F^2}{EG}} = \frac{H}{\sqrt{EG}} \qquad \text{. . . . . (281d}$$

$$\tan\eta = \frac{H}{F} \quad \text{or} \quad \cot\eta = \frac{F}{H} \qquad \text{. . . . . (281e}$$

with all roots taken as positive. Substitution into Equation 279b gives

$$\mathrm{d}s^2 = E\,\mathrm{d}a_1{}^2 + 2\sqrt{EG}\cos\eta\,\mathrm{d}a_1\,\mathrm{d}a_2 + G\,\mathrm{d}a_2{}^2 \qquad \text{. . . . . (281f}$$

showing a variant of the fundamental form with η replacing F as the third factor.

Areas

The area of the parametric parallelogram is

$$dA = {}_{a_1}ds \cdot {}_{a_2}ds \cdot \sin\eta \quad = \sqrt{E}\,da_1 \sqrt{G}\,da_2 \frac{H}{\sqrt{EG}} = H\,da_1\,da_2 \qquad \ldots\ldots (282a$$

showing H as representing the parallelogram effect. From Equations 281c & 280b, we have

$$H^2 = EF - G^2 = \left\{\left(\frac{\partial x}{\partial a_1}\right)^2 + \left(\frac{\partial y}{\partial a_1}\right)^2\right\}\left\{\left(\frac{\partial x}{\partial a_2}\right)^2 + \left(\frac{\partial y}{\partial a_2}\right)^2\right\} - \left\{\frac{\partial x}{\partial a_1}\frac{\partial x}{\partial a_2} + \frac{\partial y}{\partial a_1}\frac{\partial y}{\partial a_2}\right\}^2$$

$$= \left(\frac{\partial x}{\partial a_1}\frac{\partial y}{\partial a_2}\right)^2 - 2\frac{\partial x}{\partial a_1}\frac{\partial x}{\partial a_2}\frac{\partial y}{\partial a_1}\frac{\partial y}{\partial a_2} + \left(\frac{\partial y}{\partial a_1}\frac{\partial x}{\partial a_2}\right)^2 \quad = \left\{\frac{\partial x}{\partial a_1}\frac{\partial y}{\partial a_2} - \frac{\partial y}{\partial a_1}\frac{\partial x}{\partial a_2}\right\}^2 \qquad \ldots (282b$$

The bearing angles

Considering the right-angled triangle completed by the dotted lines at the right of Exhibit 12–2, for the bearing angle β_1 of the parametric incremental vector relative to the a_1 curve, we have

$$\cos\beta_1 = \frac{\sqrt{E}\,da_1 + \sqrt{G}\,da_2\cos\eta}{ds} \quad = \frac{E\,da_1 + \sqrt{EG}\,da_2\cos\eta}{\sqrt{E}\,ds}$$

$$= \frac{E\,da_1 + F\,da_2}{\sqrt{E}\,ds} \qquad \ldots\ldots\ldots (282c$$

$$= \frac{1}{\sqrt{E}}\left\{E\frac{da_1}{ds} + F\frac{da_2}{ds}\right\}$$

Alternatively, applying the Law of cosines and the Law of sines from Tutorial 11 to the basic triangle, the former gives

$$ds^2 = E\,da_1^{\;2} - G\,da_2^{\;2} + 2\sqrt{E}\,da_1\,ds\cos\beta_1$$

or

$$\cos\beta_1 = \frac{E\,da_1^{\;2} + ds^2 - G\,da_2^{\;2}}{2\sqrt{E}\,da_1\,ds} \qquad \ldots\ldots\ldots (282d$$

then the latter, noting that the angle shown obtuse in the illustration is the supplement, i.e., difference from π, of angle η,

$$\frac{\sin\beta_1}{\sqrt{G}\,da_2} = \frac{\sin(\pi - \eta)}{ds} \quad = \frac{\sin\eta}{ds}$$

or, substituting from Equation 281d,

$$\sin\beta_1 = \frac{\sqrt{G}\,da_2}{ds}\frac{H}{\sqrt{EG}} \quad = \frac{H}{\sqrt{E}}\frac{da_2}{ds} \qquad \ldots\ldots\ldots (282e$$

Dividing the sine expression into the cosine expression gives

$$\cot\beta_1 = \left\{\frac{\sqrt{E}\,\mathrm{d}s}{H\,\mathrm{d}a_2}\right\}\left\{\frac{E\,\mathrm{d}a_1{}^2 + \mathrm{d}s^2 - G\,\mathrm{d}a_2{}^2}{2\sqrt{E}\,\mathrm{d}a_1\,\mathrm{d}s}\right\}$$

$$= \frac{E\,\mathrm{d}a_1{}^2 + \mathrm{d}s^2 - G\,\mathrm{d}a_2{}^2}{2H\,\mathrm{d}a_1\,\mathrm{d}a_2}$$

Substituting from Equation 279b this becomes

$$\cot\beta_1 = \frac{E\,\mathrm{d}a_1{}^2 + E\,\mathrm{d}a_1{}^2 + 2F\,\mathrm{d}a_1\,\mathrm{d}a_2 + G\,\mathrm{d}a_2{}^2 - G\,\mathrm{d}a_2{}^2}{2H\,\mathrm{d}a_1\,\mathrm{d}a_2}$$

$$= \frac{2\left\{E\,\mathrm{d}a_1{}^2 + F\,\mathrm{d}a_1\,\mathrm{d}a_2\right\}}{2H\,\mathrm{d}a_1\,\mathrm{d}a_2}$$

$$= \frac{E\,\mathrm{d}a_1 + F\,\mathrm{d}a_2}{H\,\mathrm{d}a_2} \qquad \ldots\ldots\ldots (283a$$

$$= \frac{E}{H}\frac{\mathrm{d}a_1}{\mathrm{d}a_2} + \frac{F}{H} \qquad = \frac{\sqrt{EG}}{H}\frac{\sqrt{E}\,\mathrm{d}a_1}{\sqrt{G}\,\mathrm{d}a_2} + \frac{F}{H}$$

Reordering then substituting from Equations 281e, 281d & 281a gives

$$\cot\beta_1 = \cot\eta + \frac{\sqrt{EG}}{H}\frac{\sqrt{E}\,\mathrm{d}a_1}{\sqrt{G}\,\mathrm{d}a_2} \qquad = \cot\eta + \csc\eta\frac{{}_{a_1}\mathrm{d}s}{{}_{a_2}\mathrm{d}s} \qquad \ldots\ldots\ldots (283b$$

Similar processes applied to the right-angled triangle completed by the dotted lines in the upper part of Exhibit 12–2 give for bearing angle $\beta_2 = (\eta - \beta_1)$:

$$\cos\beta_2 = \frac{\sqrt{G}\,\mathrm{d}a_2 + \sqrt{E}\,\mathrm{d}a_1\cos\eta}{\mathrm{d}s} \qquad = \frac{F\,\mathrm{d}a_1 + G\,\mathrm{d}a_2}{\sqrt{G}\,\mathrm{d}s} \qquad = \frac{1}{\sqrt{G}}\left\{F\frac{\mathrm{d}a_1}{\mathrm{d}s} + G\frac{\mathrm{d}a_2}{\mathrm{d}s}\right\} \qquad \ldots (283c$$

$$\sin\beta_2 = \frac{\sqrt{E}\,\mathrm{d}a_1\sin\eta}{\mathrm{d}s} \qquad = \frac{H}{\sqrt{G}}\frac{\mathrm{d}a_1}{\mathrm{d}s} \qquad \ldots\ldots (283d$$

$$\cot\beta_2 = \frac{G}{H}\frac{\mathrm{d}a_2}{\mathrm{d}a_1} + \frac{F}{H} \qquad = \cot\eta + \frac{\sqrt{EG}}{H}\frac{\sqrt{G}\,\mathrm{d}a_2}{\sqrt{E}\,\mathrm{d}a_1} \qquad = \cot\eta + \csc\eta\frac{{}_{a_2}\mathrm{d}s}{{}_{a_1}\mathrm{d}s} \qquad \ldots\ldots (283e$$

The angles between parameter axes and main axes

Angles γ_1 and γ_2 in Exhibit 12–2 represent the divergence of the parametric α_1 axis and parametric α_2 axis, respectively, from the $+y$ half-axis. By simple consideration of differential slopes:

$$\gamma_1 = \operatorname{arccot}\frac{\partial y}{\partial a_1}\left(\frac{\partial x}{\partial a_1}\right)^{-1} \qquad = \arctan\frac{\partial x}{\partial a_1}\left(\frac{\partial y}{\partial a_1}\right)^{-1}$$

$$\gamma_2 = \operatorname{arccot}\frac{\partial y}{\partial a_2}\left(\frac{\partial x}{\partial a_2}\right)^{-1} \qquad = \arctan\frac{\partial x}{\partial a_2}\left(\frac{\partial y}{\partial a_2}\right)^{-1} \qquad \ldots\ldots\ldots (283f$$

ORTHOGONAL PARAMETRIC AXES

Simplifications

If the parameter curves are orthogonal then $F = 0$, which removes terms from most of the above formulae. To mark overtly the revised formulae as applicable to the restricted circumstances of orthogonality, an angle capping is applied to the parametric variables. Thus, we have

$$\frac{\partial x}{\partial \hat{a}_1}\frac{\partial y}{\partial \hat{a}_2} + \frac{\partial x}{\partial \hat{a}_2}\frac{\partial y}{\partial \hat{a}_1} = 0 \qquad \text{or} \qquad -\frac{\partial x}{\partial \hat{a}_1}\frac{\partial y}{\partial \hat{a}_2} = \frac{\partial x}{\partial \hat{a}_2}\frac{\partial y}{\partial \hat{a}_1} \qquad \text{...... (284a}$$

which is shown later to be called the *Cauchy-Riemann condition*.

The incremental distances along the axes remain as in Equation 281a, i.e.,

$${}_{\hat{a}_1}\mathrm{d}s = \sqrt{E}\ \mathrm{d}\hat{a}_1, \qquad {}_{\hat{a}_2}\mathrm{d}s = \sqrt{G}\ \mathrm{d}\hat{a}_2 \qquad \text{........ (284b}$$

but the orthogonality changes Equation 279b to

$$(\mathrm{d}s)^2 = E\,\mathrm{d}\hat{a}_1^{\,2} + G\,\mathrm{d}\hat{a}_2^{\,2} \qquad \text{........ (284c}$$

while $\eta = {}^{\pi}\!/_{2}$, $\cos\eta = 0$, and $\sin\eta = 1$.

Equations 283b & 283c for the bearing angles β_1 and β_2 become:

$$\cot\beta_1 = \frac{E}{H}\frac{\mathrm{d}\hat{a}_1}{\mathrm{d}\hat{a}_2} = \frac{E}{\sqrt{EG}}\frac{\mathrm{d}\hat{a}_1}{\mathrm{d}\hat{a}_2} = \sqrt{\frac{E}{G}}\frac{\mathrm{d}\hat{a}_1}{\mathrm{d}\hat{a}_2} \qquad \text{........ (284d}$$

$$\cot\beta_2 = \frac{G}{H}\frac{\mathrm{d}\hat{a}_2}{\mathrm{d}\hat{a}_1} = \frac{G}{\sqrt{EG}}\frac{\mathrm{d}\hat{a}_2}{\mathrm{d}\hat{a}_1} = \sqrt{\frac{G}{E}}\frac{\mathrm{d}\hat{a}_2}{\mathrm{d}\hat{a}_1} \qquad \text{........ (284e}$$

i.e., mutual reciprocals, the two bearing angles now adding to a right angle. With the parameter axes orthogonal, obviously:

$$\cos\beta_1 = \frac{\sqrt{E}\,\mathrm{d}\hat{a}_1}{\mathrm{d}s} = \frac{\sqrt{E}\,\mathrm{d}\hat{a}_1}{\sqrt{E\,\mathrm{d}\hat{a}_1^{\,2} + G\,\mathrm{d}\hat{a}_2^{\,2}}} = \sin\beta_2$$

$$\sin\beta_1 = \frac{\sqrt{G}\,\mathrm{d}\hat{a}_2}{\mathrm{d}s} = \frac{\sqrt{G}\,\mathrm{d}\hat{a}_2}{\sqrt{E\,\mathrm{d}\hat{a}_1^{\,2} + G\,\mathrm{d}\hat{a}_2^{\,2}}} = \cos\beta_2 \qquad \text{........ (284f}$$

and Equation 282a becomes

$$\mathrm{d}A = \sqrt{E\,G}\ \mathrm{d}\hat{a}_1\,\mathrm{d}\hat{a}_2 \qquad \text{........ (284g}$$

Fundamental form in angular terms

Rearranging Equation 284c then using Equation 284d to eliminate $\mathrm{d}\hat{a}_1$ provides expression of the fundamental form in terms of one parameter together with a parametric bearing angle.

$$\begin{aligned} \mathrm{d}s^2 &= \left\{ E\left(\frac{\mathrm{d}\hat{a}_1}{\mathrm{d}\hat{a}_2}\right)^2 + G \right\} \mathrm{d}\hat{a}_2{}^2 \\ &= \left\{ E\left(\cot\beta_1\sqrt{\frac{G}{E}}\right)^2 + G \right\} \mathrm{d}\hat{a}_2{}^2 \\ &= \left\{ G\cot^2\beta_1 + G \right\} \mathrm{d}\hat{a}_2{}^2 \qquad \text{(285a} \\ &= G\left(1+\cot^2\beta_1\right)\mathrm{d}\hat{a}_2{}^2 \\ &= G\csc^2\beta_1\ \mathrm{d}\hat{a}_2{}^2 \qquad = G\sec^2\beta_2\ \mathrm{d}\hat{a}_2{}^2 \end{aligned}$$

since β_1 and β_2 are now complementary. Alternatively, using Equation 284d to eliminate $\mathrm{d}\hat{a}_2$,

$$\mathrm{d}s^2 = E\csc^2\beta_2\ \mathrm{d}\hat{a}_1{}^2 \qquad = E\sec^2\beta_1\ \mathrm{d}\hat{a}_1{}^2 \qquad \text{(285b}$$

With longitude and latitude (when orthogonal)

Longitude and latitude are orthogonal everywhere on the globe and are orthogonal on various maps. Wherever so, they can be substituted in the preceding equations. To maintain orientation, we equate longitude with the first generic parameter, and latitude with the second. Then, assuming the two are orthogonal,

$$\mathrm{d}s^2 = E\,\mathrm{d}\lambda^2 + G\,\mathrm{d}\phi^2$$

$$\text{with } E = \left\{\left(\frac{\partial x}{\partial\lambda}\right)^2 + \left(\frac{\partial y}{\partial\lambda}\right)^2\right\},\quad G = \left\{\left(\frac{\partial x}{\partial\phi}\right)^2 + \left(\frac{\partial y}{\partial\phi}\right)^2\right\},\quad F \text{ being } 0 \qquad \text{(285c}$$

The parametric axes become the parallel and meridian through the point, with:

$$\begin{aligned} &\text{distance along the meridian} = {}_{\mathrm{m}}\mathrm{d}s = \sqrt{G\,\mathrm{d}\phi^2} = \sqrt{G}\ \mathrm{d}\phi \\ &\text{distance transversely thereto} = {}_{\mathrm{t}}\mathrm{d}s = \sqrt{E\,\mathrm{d}\lambda^2} = \sqrt{E}\ \mathrm{d}\lambda \end{aligned} \qquad \text{(285d}$$

Equation 284g becomes

$$\mathrm{d}A = \sqrt{E\,G}\ \mathrm{d}\lambda\,\mathrm{d}\phi \qquad \text{(285e}$$

Bearing angle β_2 is now identically the azimuth, which we have routinely denoted by ζ, yielding, from Equations 285a & 285b, the fundamental form in azimuthal terms as

$$\mathrm{d}s^2 = E\csc^2\zeta\ \mathrm{d}\lambda^2 \quad \text{and} \quad \mathrm{d}s^2 = G\sec^2\zeta\ \mathrm{d}\phi^2$$

We also have, from Equations 284d & 284f:

$$\cot\zeta = \sqrt{\frac{G}{E}}\frac{\mathrm{d}\phi}{\mathrm{d}\lambda} \quad \text{or} \quad \frac{\mathrm{d}\phi}{\mathrm{d}\lambda} = \sqrt{\frac{E}{G}}\cot\zeta \quad \text{or} \quad \mathrm{d}\phi = \sqrt{\frac{E}{G}}\cot\zeta\ \mathrm{d}\lambda \qquad \text{(285f}$$

$$\mathrm{d}s\sin\zeta = \sqrt{E}\ \mathrm{d}\lambda \quad \text{and} \quad \mathrm{d}s\cos\zeta = \sqrt{G}\ \mathrm{d}\phi$$

THE PLANE AS A SURFACE

Parametric representation

The plane, as used for our maps, is represented sometimes by rectilinear Cartesian co-ordinates u and v, and sometimes by polar co-ordinates ρ and θ.[†] We address each, equating in turn each pair with the generic parameters a_1 and a_2 of Equation 279b.

Using Equations 207e for expressing standard 2-dimensional Cartesian co-ordinates to the intermediate parameters u and v we get for a general (rotated) planar surface (Equations 86a):

$$dx = du\cos\chi - dv\sin\chi$$
$$dy = du\sin\chi + dv\cos\chi$$

Adding the squares of each of these two equations gives for this 2-dimensional scene

$$\begin{aligned}(ds)^2 = (dx)^2 + (dy)^2 &= (du)^2\cos^2\chi - 2du\,dv\cos\chi\sin\chi + (dv)^2\sin^2\chi \\ &+ (du)^2\sin^2\chi + 2du\,dv\cos\chi\sin\chi + (dv)^2\cos^2\chi \qquad \ldots\ldots (286a \\ &= (du)^2 + (dv)^2\end{aligned}$$

So, comparing with Equation 279b, and labelling our quantities with a left-subscript p for plane and denoting the parameters involved in their order in a right-subscript, for the plane in terms of Cartesian co-ordinates we have:

$$_{\mathrm{p}}E_{u,v} = 1 \qquad _{\mathrm{p}}F_{u,v} = 0 \qquad _{\mathrm{p}}G_{u,v} = 1 \qquad _{\mathrm{p}}H_{u,v} = 1 \qquad \ldots\ldots\ldots (286b$$

For the plane in terms of 2-dimensional polar co-ordinates, developed by similar means using equations from Tutorial 1, we get:

$$dx = d\rho\cos\theta - d\theta\,\rho\sin\theta$$
$$dy = d\rho\sin\theta + d\theta\,\rho\cos\theta$$

so

$$\begin{aligned}(ds)^2 = (dx)^2 + (dy)^2 &= (d\rho)^2\cos^2\theta - 2d\rho\,d\theta\cos\theta\rho\sin\theta + (d\theta)^2\rho^2\sin^2\theta \\ &+ (d\rho)^2\sin^2\theta + 2d\rho\,d\theta\cos\theta\rho\sin\theta + (d\theta)^2\rho^2\cos^2\theta \qquad \ldots\ldots (286c \\ &= 1(d\rho)^2 + \rho^2(d\theta)^2\end{aligned}$$

Comparing with Equation 279b and labelling our quantities accordingly, for the plane in terms of polar co-ordinates, we have:

$$_{\mathrm{p}}E_{\rho,\theta} = 1 \qquad _{\mathrm{p}}F_{\rho,\theta} = 0 \qquad _{\mathrm{p}}G_{\rho,\theta} = \rho^2 \qquad _{\mathrm{p}}H_{\rho,\theta} = \rho \qquad \ldots\ldots\ldots (286d$$

[†] In these analyses of individual surfaces we omit the overscripts used for distinguishing map from globe.

THE SPHERE AS A SURFACE

Parametric representation

Longitude and latitude are the natural and obvious parameters to use for the sphere. Using the equations of Tutorial 3 expressing 3-dimensional Cartesian co-ordinates in polar terms, we get for the general spherical surface:

$$\begin{aligned} dX &= -R\{\sin\phi\, d\phi \cos\lambda + \cos\phi \sin\lambda\, d\lambda\} \\ dY &= -R\{\sin\phi\, d\phi \sin\lambda - \cos\phi \cos\lambda\, d\lambda\} \\ dZ &= +R\{\cos\phi\, d\phi\} \end{aligned}$$

Summing the squares of each of these three equations, Equation 279a becomes

$$\begin{aligned} (ds)^2 = {} & R^2\{\sin\phi\, d\phi \sin\lambda - \cos\phi \cos\lambda\, d\lambda\}^2 \\ & +R^2\{\sin\phi\, d\phi \cos\lambda + \cos\phi \sin\lambda\, d\lambda\}^2 \\ & +R^2\{\cos\phi\, d\phi\}^2 \\ = {} & R^2\left\{\sin^2\phi\,(d\phi)^2 \cos^2\lambda + 2\sin\phi\, d\phi \cos\lambda \cos\phi \sin\lambda\, d\lambda + \cos^2\phi \sin^2\lambda\,(d\lambda)^2\right\} \\ & +R^2\left\{\sin^2\phi\,(d\phi)^2 \sin^2\lambda - 2\sin\phi\, d\phi \sin\lambda \cos\phi \cos\lambda\, d\lambda + \cos^2\phi \cos^2\lambda\,(d\lambda)^2\right\} \\ & +R^2 \cos^2\phi\,(d\phi)^2 \end{aligned}$$

which resolves to

$$ds^2 = R^2 \cos^2\phi\, d\lambda^2 + R^2 d\phi^2 \qquad \text{. . . . . (287a}$$

Comparing with Equation 285c gives fundamental quantities for the <u>sphere in terms of standard longitude and latitude</u> as:

$${}_sE_{\lambda,\phi} = R^2\cos^2\phi \qquad {}_sF_{\lambda,\phi} = 0 \qquad {}_sG_{\lambda,\phi} = R^2 \qquad {}_sH_{\lambda,\phi} = R^2\cos\phi \qquad \text{. . . . . (287b}$$

using left-subscript s to indicate sphere, and the right-subscript to specify the order λ, ϕ.

Substituting in Equations 285d through 285f we get:

$$\left.\begin{aligned} \text{distance along the meridian} = {}_m ds &= \sqrt{{}_sG_{\lambda,\phi}}\, d\phi &= R\, d\phi \\ \text{distance transversely thereto} = {}_t ds &= \sqrt{{}_sE_{\lambda,\phi}}\, d\lambda &= R\cos\phi\, d\lambda \end{aligned}\right\} \qquad \text{. . . . . (287c}$$

$$\begin{aligned} dA &= \sqrt{{}_sE_{\lambda,\phi} \cdot {}_sG_{\lambda,\phi}}\, d\lambda\, d\phi &&= R^2\cos\phi\, d\lambda\, d\phi \\ ds^2 &= {}_sE_{\lambda,\phi} \csc^2\zeta\, d\lambda^2 &&= R^2\cos^2\phi \csc^2\zeta\, d\lambda^2 \\ ds^2 &= {}_sG_{\lambda,\phi} \sec^2\zeta\, d\phi^2 &&= R^2 \sec^2\zeta\, d\phi^2 \end{aligned} \qquad \text{. . . . . (287d}$$

$$d\phi = \sqrt{\frac{{}_sG_{\lambda,\phi}}{{}_sE_{\lambda,\phi}}} \cot\zeta\, d\lambda = \sqrt{\frac{R^2\cos^2\phi}{R^2}} \cot\zeta\, d\lambda = \cos\phi \cot\zeta\, d\lambda \qquad \text{. . . . . (287e}$$

Isometric latitude

Cartesian co-ordinates give identical values for E and G for the plane — an attribute termed **isometric** (from Greek *equal* + *measure*] — whereas the polar co-ordinates did so for neither plane nor sphere. It is possible, however, to construct a new scheme of co-ordinates for the sphere that does have such result. From Equation 287a, we have

$$(\mathrm{d}s)^2 = R^2\cos^2\phi\,(\mathrm{d}\lambda)^2 + R^2(\mathrm{d}\phi)^2 = R^2\cos^2\phi\left\{(\mathrm{d}\lambda)^2 + \frac{(\mathrm{d}\phi)^2}{\cos^2\phi}\right\}$$

$$= R^2\cos^2\phi\left\{(\mathrm{d}\lambda)^2 + (\mathrm{d}\overline{\varphi})^2\right\} \qquad \text{if we define} \qquad \mathrm{d}\overline{\varphi} = \frac{\mathrm{d}\phi}{\cos\phi}$$

Integrating the definition gives

$$\overline{\varphi} = \int\frac{\mathrm{d}\phi}{\cos\phi} + \text{ constant} = \ln\tan\left(\frac{\pi}{4} + \frac{\phi}{2}\right) + \text{constant}$$

with the constant being zero if we choose $\overline{\varphi} = 0$ for $\phi = 0$. Then

$$\overline{\varphi} = \ln\tan\left(\frac{\pi}{4} + \frac{\phi}{2}\right) \qquad \text{. . . . . (288a}$$

giving

$$(\mathrm{d}s)^2 = R^2\cos^2\phi\left\{(\mathrm{d}\lambda)^2 + (\mathrm{d}\overline{\varphi})^2\right\} \qquad \text{. . . . . (288b}$$

Our new variable is a function purely of latitude, and completely replaces latitude in the equations. Although not truly a latitude, like latitude this new variable is constant for all points a given distance from the Equator. Hence it is routinely called *isometric latitude.* Because it is thus orthogonal to meridians, we can relate Equation 288b to Equation 284c, giving for the sphere in terms of standard longitude and isometric latitude, the fundamental quantities:

$${}_{\mathrm{S}}E_{\lambda,\overline{\varphi}} = R^2\cos^2\phi, \quad {}_{\mathrm{S}}F_{\lambda,\overline{\varphi}} = 0, \quad {}_{\mathrm{S}}G_{\lambda,\overline{\varphi}} = R^2\cos^2\phi, \quad {}_{\mathrm{S}}H_{\lambda,\overline{\varphi}} = R^2\cos^2\phi \qquad \text{. . . . (288c}$$

The integral that occurred above is familiar from Chapter 4, where it was used in establishing conformality. This is not by chance. The equating of quantities E and G corresponds to the need with conformality for identical change of scale along the two parameter lines. The latitudinal plotting equation in the Mercator projection (Equation 81b) becomes

$$\dot{y} = \ln\tan\left(\frac{\pi}{4} + \frac{\phi}{2}\right) = \overline{\varphi}$$

Using more detail from Equation 81b we have

$$\overline{\varphi} = \ln\tan\left(\frac{\pi}{4} + \frac{\phi}{2}\right) = \frac{1}{2}\ln\frac{1+\sin\phi}{1-\sin\phi} = \ln\sqrt{\frac{1+\sin\phi}{1-\sin\phi}}, \quad \text{i.e. } e^{\overline{\varphi}} = \sqrt{\frac{1+\sin\phi}{1-\sin\phi}} \qquad \text{. . . (288d}$$

Hence, for the hyperbolic functions introduced in Tutorial 16 and elaborated in Tutorial 26,

$$\cosh\overline{\varphi} = \frac{1}{2}\left\{e^{\overline{\varphi}} + e^{-\overline{\varphi}}\right\}$$
$$= \frac{1}{2}\left\{\sqrt{\frac{1+\sin\phi}{1-\sin\phi}} + \sqrt{\frac{1-\sin\phi}{1+\sin\phi}}\right\}$$
$$= \frac{1}{2}\left\{\sqrt{\frac{(1+\sin\phi)^2}{(1-\sin\phi)(1+\sin\phi)}} + \sqrt{\frac{(1-\sin\phi)^2}{(1-\sin\phi)(1+\sin\phi)}}\right\} \quad \ldots\ldots (289a$$
$$= \frac{1}{2}\left\{\frac{(1+\sin\phi)+(1-\sin\phi)}{\sqrt{(1-\sin^2\phi)}}\right\}$$
$$= \frac{1}{2}\left\{\frac{2}{\sqrt{\cos^2\phi}}\right\}$$
$$= \sec\phi$$

which provides a more direct interrelationship between the isometric and geodetic latitudes. Further, such relationships between functions of the two are shown later in Tutorial 30, in association with the use of isometric latitude in deriving additional conformal projections.

TUTORIAL 26 HYPERBOLIC FUNCTIONS

The hyperbolic functions

$$\cosh b = 1 + \frac{b^2}{2!} + \frac{b^4}{4!} + .. = \frac{1}{2}\left\{e^b + e^{-b}\right\}$$
$$\sinh b = \frac{b^1}{1!} + \frac{b^3}{3!} + \frac{b^5}{5!} + .. = \frac{1}{2}\left\{e^b - e^{-b}\right\}$$

were defined in Tutorial 16.

Functions tanh, coth, csch and sech are defined from them in the same manner as for the familiar circular trigonometric functions.

The relationships between these new functions vary, however, from the familiar pattern shown in Tutorial 10. Use of the respective series leads to the following formulae:

$$\sinh(-a) = -\sinh a$$
$$\cosh(-a) = \cosh a$$

$$\cosh^2 a - \sinh^2 a = 1$$
$$\coth^2 a - \operatorname{csch}^2 a = 1$$
$$\tanh^2 a + \operatorname{sech}^2 a = 1$$
$$\operatorname{csch}^2 b - \operatorname{sech}^2 b = \operatorname{csch}^2 b \operatorname{sech}^2 b$$

$$\sinh(a+b) = \sinh a\cosh b + \cosh a\sinh b$$
$$\sinh(a-b) = \sinh a\cosh b - \cosh a\sinh b$$
$$\cosh(a+b) = \cosh a\cosh b + \sinh a\sinh b$$
$$\cosh(a-b) = \cosh a\cosh b - \sinh a\sinh b$$

$$\sinh 2a = 2\sinh a\cosh a = \frac{2\tanh a}{1-\tanh^2 a}$$

$$\cosh 2a = \cosh^2 a + \sinh^2 a$$

$$\tanh 2a = \frac{2\tanh a}{1+\tanh^2 a}$$

$$\coth 2a = \frac{\coth^2 a + 1}{2\coth a}$$

SWAPPING BETWEEN PARAMETER PAIRS

Re-expressing the fundamentals

Every map projection is the transformation of one surface into another, essentially of the curved globe into the flat map. We also for some, however, transform one flat map into another, perhaps repeatedly as an iterative process, while adaptation with the ellipsoidal globe can be facilitated by transforming its curved surface into a spherical one. Even within one simple map projection, we have typically two distinct pairs of parameters to consider — the Cartesian else polar intermediate parameters and the essential longitude plus latitude, the former serving our mapping process, the latter being an essential basis for qualitative assessment.

If a_1, a_2 and $\bar{a}_1, \bar{a}_2$ are two pairs of parameters that can each describe a surface, then, assuming appropriate differentiability,

$$\frac{\partial f}{\partial \bar{a}_1} = \frac{\partial f}{\partial a_1}\frac{\partial a_1}{\partial \bar{a}_1} + \frac{\partial f}{\partial a_2}\frac{\partial a_2}{\partial \bar{a}_1}, \qquad \frac{\partial f}{\partial \bar{a}_2} = \frac{\partial f}{\partial a_1}\frac{\partial a_1}{\partial \bar{a}_2} + \frac{\partial f}{\partial a_2}\frac{\partial a_2}{\partial \bar{a}_2}$$

for any f that is a function of those two variables. Denoting the respective fundamental quantities in the established way we get

$$\begin{aligned}
E_{\bar{a}_1,\bar{a}_2} &= \left(\frac{\partial x}{\partial \bar{a}_1}\right)^2 + \left(\frac{\partial y}{\partial \bar{a}_1}\right)^2 = \left(\frac{\partial x}{\partial a_1}\frac{\partial a_1}{\partial \bar{a}_1} + \frac{\partial x}{\partial a_2}\frac{\partial a_2}{\partial \bar{a}_1}\right)^2 + \left(\frac{\partial y}{\partial a_1}\frac{\partial a_1}{\partial \bar{a}_1} + \frac{\partial y}{\partial a_2}\frac{\partial a_2}{\partial \bar{a}_1}\right)^2 \\
&= \left(\frac{\partial x}{\partial a_1}\frac{\partial a_1}{\partial \bar{a}_1}\right)^2 + 2\frac{\partial x}{\partial a_1}\frac{\partial a_1}{\partial \bar{a}_1}\frac{\partial x}{\partial a_2}\frac{\partial a_2}{\partial \bar{a}_1} + \left(\frac{\partial x}{\partial a_2}\frac{\partial a_2}{\partial \bar{a}_1}\right)^2 \\
&\quad + \left(\frac{\partial y}{\partial a_1}\frac{\partial a_1}{\partial \bar{a}_1}\right)^2 + 2\frac{\partial y}{\partial a_1}\frac{\partial a_1}{\partial \bar{a}_1}\frac{\partial y}{\partial a_2}\frac{\partial a_2}{\partial \bar{a}_1} + \left(\frac{\partial y}{\partial a_2}\frac{\partial a_2}{\partial \bar{a}_1}\right)^2 \\
&= \left\{\left(\frac{\partial x}{\partial a_1}\right)^2 + \left(\frac{\partial y}{\partial a_1}\right)^2\right\}\left(\frac{\partial a_1}{\partial \bar{a}_1}\right)^2 \\
&\qquad + 2\left\{\frac{\partial x}{\partial a_1}\frac{\partial x}{\partial a_2} + \frac{\partial y}{\partial a_1}\frac{\partial y}{\partial a_2}\right\}\frac{\partial a_1}{\partial \bar{a}_1}\frac{\partial a_2}{\partial \bar{a}_1} \\
&\qquad\qquad + \left\{\left(\frac{\partial x}{\partial a_2}\right)^2 + \left(\frac{\partial y}{\partial a_2}\right)^2\right\}\left(\frac{\partial a_2}{\partial \bar{a}_1}\right)^2 \\
&= E_{a_1,a_2}\left(\frac{\partial a_1}{\partial \bar{a}_1}\right)^2 + 2F_{a_1,a_2}\frac{\partial a_1}{\partial \bar{a}_1}\frac{\partial a_2}{\partial \bar{a}_1} + G_{a_1,a_2}\left(\frac{\partial a_2}{\partial \bar{a}_1}\right)^2
\end{aligned}$$

Likewise,

$$G_{\bar{a}_1,\bar{a}_2} = \left(\frac{\partial x}{\partial \bar{a}_2}\right)^2 + \left(\frac{\partial y}{\partial \bar{a}_2}\right)^2 = E_{a_1,a_2}\left(\frac{\partial a_1}{\partial \bar{a}_2}\right)^2 + 2F_{a_1,a_2}\frac{\partial a_1}{\partial \bar{a}_2}\frac{\partial a_2}{\partial \bar{a}_2} + G_{a_1,a_2}\left(\frac{\partial a_2}{\partial \bar{a}_2}\right)^2$$

while

$$F_{\bar{a}_1,\bar{a}_2} = \frac{\partial x}{\partial \bar{a}_1}\frac{\partial x}{\partial \bar{a}_2} + \frac{\partial y}{\partial \bar{a}_1}\frac{\partial y}{\partial \bar{a}_2}$$

$$= \left\{\frac{\partial x}{\partial a_1}\frac{\partial a_1}{\partial \bar{a}_1} + \frac{\partial x}{\partial a_2}\frac{\partial a_2}{\partial \bar{a}_1}\right\}\left\{\frac{\partial x}{\partial a_1}\frac{\partial a_1}{\partial \bar{a}_2} + \frac{\partial x}{\partial a_2}\frac{\partial a_2}{\partial \bar{a}_2}\right\} + \left\{\frac{\partial y}{\partial a_1}\frac{\partial a_1}{\partial \bar{a}_1} + \frac{\partial y}{\partial a_2}\frac{\partial a_2}{\partial \bar{a}_1}\right\}\left\{\frac{\partial y}{\partial a_1}\frac{\partial a_1}{\partial \bar{a}_2} + \frac{\partial y}{\partial a_2}\frac{\partial a_2}{\partial \bar{a}_2}\right\}$$

$$= \left(\frac{\partial x}{\partial a_1}\right)^2 \frac{\partial a_1}{\partial \bar{a}_1}\frac{\partial a_1}{\partial \bar{a}_2} + \frac{\partial x}{\partial a_1}\frac{\partial a_1}{\partial \bar{a}_1}\frac{\partial x}{\partial a_2}\frac{\partial a_2}{\partial \bar{a}_2} + \frac{\partial x}{\partial a_2}\frac{\partial a_2}{\partial \bar{a}_1}\frac{\partial x}{\partial a_1}\frac{\partial a_1}{\partial \bar{a}_2} + \left(\frac{\partial x}{\partial a_2}\right)^2 \frac{\partial a_2}{\partial \bar{a}_1}\frac{\partial a_2}{\partial \bar{a}_2}$$

$$+ \left(\frac{\partial y}{\partial a_1}\right)^2 \frac{\partial a_1}{\partial \bar{a}_1}\frac{\partial a_1}{\partial \bar{a}_2} + \frac{\partial y}{\partial a_1}\frac{\partial a_1}{\partial \bar{a}_1}\frac{\partial y}{\partial a_2}\frac{\partial a_2}{\partial \bar{a}_2} + \frac{\partial y}{\partial a_2}\frac{\partial a_2}{\partial \bar{a}_1}\frac{\partial y}{\partial a_1}\frac{\partial a_1}{\partial \bar{a}_2} + \left(\frac{\partial y}{\partial a_2}\right)^2 \frac{\partial a_2}{\partial \bar{a}_1}\frac{\partial a_2}{\partial \bar{a}_2}$$

$$= \left\{\left(\frac{\partial x}{\partial a_1}\right)^2 + \left(\frac{\partial y}{\partial a_1}\right)^2\right\}\frac{\partial a_1}{\partial \bar{a}_1}\frac{\partial a_1}{\partial \bar{a}_2}$$

$$+ \left(\frac{\partial x}{\partial a_1}\frac{\partial x}{\partial a_2} + \frac{\partial y}{\partial a_1}\frac{\partial y}{\partial a_2}\right)\left(\frac{\partial a_1}{\partial \bar{a}_1}\frac{\partial a_2}{\partial \bar{a}_2} + \frac{\partial a_1}{\partial \bar{a}_2}\frac{\partial a_2}{\partial \bar{a}_1}\right)$$

$$+ \left\{\left(\frac{\partial x}{\partial a_2}\right)^2 + \left(\frac{\partial y}{\partial a_2}\right)^2\right\}\frac{\partial a_2}{\partial \bar{a}_1}\frac{\partial a_2}{\partial \bar{a}_2}$$

$$= E_{a_1,a_2}\frac{\partial a_1}{\partial \bar{a}_1}\frac{\partial a_1}{\partial \bar{a}_2} + F_{a_1,a_2}\left(\frac{\partial a_1}{\partial \bar{a}_1}\frac{\partial a_2}{\partial \bar{a}_2} + \frac{\partial a_1}{\partial \bar{a}_2}\frac{\partial a_2}{\partial \bar{a}_1}\right) + G_{a_1,a_2}\frac{\partial a_2}{\partial \bar{a}_1}\frac{\partial a_2}{\partial \bar{a}_2}$$

Adopting matrix notation (Tutorials 20 & 24), these equations can be consolidated as

$$\begin{bmatrix} E_{\bar{a}_1,\bar{a}_2} \\ F_{\bar{a}_1,\bar{a}_2} \\ G_{\bar{a}_1,\bar{a}_2} \end{bmatrix} = \begin{bmatrix} \left(\frac{\partial a_1}{\partial \bar{a}_1}\right)^2 & 2\frac{\partial a_1}{\partial \bar{a}_1}\frac{\partial a_2}{\partial \bar{a}_1} & \left(\frac{\partial a_2}{\partial \bar{a}_1}\right)^2 \\ \frac{\partial a_1}{\partial \bar{a}_1}\frac{\partial a_1}{\partial \bar{a}_2} & \left(\frac{\partial a_1}{\partial \bar{a}_1}\frac{\partial a_2}{\partial \bar{a}_2} + \frac{\partial a_1}{\partial \bar{a}_2}\frac{\partial a_2}{\partial \bar{a}_1}\right) & \frac{\partial \alpha_2}{\partial \bar{a}_1}\frac{\partial a_2}{\partial \bar{a}_2} \\ \left(\frac{\partial a_1}{\partial \bar{a}_2}\right)^2 & 2\frac{\partial a_1}{\partial \bar{a}_2}\frac{\partial a_2}{\partial \bar{a}_2} & \left(\frac{\partial a_2}{\partial \bar{a}_2}\right)^2 \end{bmatrix} \cdot \begin{bmatrix} E_{a_1,a_2} \\ F_{a_1,a_2} \\ G_{a_1,a_2} \end{bmatrix} \qquad \text{. . . . . (291a}$$

the 3 x 3 matrix being called the *fundamental transformation matrix*.

If orthogonality applies to both sets of parameters, this equation reduces to the simpler

$$\begin{bmatrix} E_{\bar{a}_1,\bar{a}_2} \\ G_{\bar{a}_1,\bar{a}_2} \end{bmatrix} = \begin{bmatrix} \left(\frac{\partial a_1}{\partial \bar{a}_1}\right)^2 & \left(\frac{\partial a_2}{\partial \bar{a}_1}\right)^2 \\ \left(\frac{\partial a_1}{\partial \bar{a}_2}\right)^2 & \left(\frac{\partial a_2}{\partial \bar{a}_2}\right)^2 \end{bmatrix} \cdot \begin{bmatrix} E_{a_1,a_2} \\ G_{a_1,a_2} \end{bmatrix} \qquad F_{\bar{a}_1,\bar{a}_2} = F_{a_1,a_2} = 0 \qquad \text{. . . . . (291b}$$

This process is operable in the reverse direction, giving equations differing from the above only by swapping the denotation of the parameter pairs.

The Jacobian

From the definition of H in Equation 281c, we have

$$H_{\bar{a}_1,\bar{a}_2}{}^2 = E_{\bar{a}_1,\bar{a}_2} \cdot G_{\bar{a}_1,\bar{a}_2} - F_{\bar{a}_1,\bar{a}_2}{}^2$$

$$= \left\{ E_{a_1,a_2}\left(\frac{\partial a_1}{\partial \bar{a}_1}\right)^2 + 2F_{a_1,a_2}\frac{\partial a_1}{\partial \bar{a}_1}\frac{\partial a_2}{\partial \bar{a}_1} + G_{a_1,a_2}\left(\frac{\partial a_2}{\partial \bar{a}_1}\right)^2 \right\}$$

$$\cdot \left\{ E_{a_1,a_2}\left(\frac{\partial a_1}{\partial \bar{a}_2}\right)^2 + 2F_{a_1,a_2}\frac{\partial a_1}{\partial \bar{a}_2}\frac{\partial a_2}{\partial \bar{a}_2} + G_{a_1,a_2}\left(\frac{\partial a_2}{\partial \bar{a}_2}\right)^2 \right\}$$

$$- \left\{ E_{a_1,a_2}\frac{\partial a_1}{\partial \bar{a}_1}\frac{\partial a_1}{\partial \bar{a}_2} + F_{a_1,a_2}\left(\frac{\partial a_1}{\partial \bar{a}_1}\frac{\partial a_2}{\partial \bar{a}_2} + \frac{\partial a_1}{\partial \bar{a}_2}\frac{\partial a_2}{\partial \bar{a}_1}\right) + G_{a_1,a_2}\frac{\partial a_2}{\partial \bar{a}_1}\frac{\partial a_2}{\partial \bar{a}_2} \right\}^2$$

Multiplying this out gives nine terms from multiplying the first two curly brackets, and nine from the squaring of the other. Five in each are identical except for the latter line being overall of opposite sign, reducing the cumbersome expression to just eight terms, expressible as

$$H_{\bar{a}_1,\bar{a}_2}{}^2 = \left(E_{a_1,a_2} \cdot G_{a_1,a_2} - F_{a_1,a_2}{}^2\right)\left(\frac{\partial a_1}{\partial \bar{a}_1}\frac{\partial a_2}{\partial \bar{a}_2} - \frac{\partial a_1}{\partial \bar{a}_2}\frac{\partial a_2}{\partial \bar{a}_1}\right)^2$$

$$= H_{a_1,a_2}{}^2\left(\frac{\partial a_1}{\partial \bar{a}_1}\frac{\partial a_2}{\partial \bar{a}_2} - \frac{\partial a_1}{\partial \bar{a}_2}\frac{\partial a_2}{\partial \bar{a}_1}\right)^2$$

The last bracketed term can be written as a determinant (Tutorial 20), i.e., as

$$\left(\frac{\partial a_1}{\partial \bar{a}_1}\frac{\partial a_2}{\partial \bar{a}_2} - \frac{\partial a_1}{\partial \bar{a}_2}\frac{\partial a_2}{\partial \bar{a}_1}\right) = \begin{vmatrix} \dfrac{\partial a_1}{\partial \bar{a}_1} & \dfrac{\partial a_2}{\partial \bar{a}_1} \\ \dfrac{\partial a_1}{\partial \bar{a}_2} & \dfrac{\partial a_2}{\partial \bar{a}_2} \end{vmatrix} = \begin{vmatrix} \dfrac{\partial a_1}{\partial \bar{a}_1} & \dfrac{\partial a_1}{\partial \bar{a}_2} \\ \dfrac{\partial a_2}{\partial \bar{a}_1} & \dfrac{\partial a_2}{\partial \bar{a}_2} \end{vmatrix} = J$$

This determinant is called the *Jacobian*[†] (the array usually written in the latter form but more convenient for us in the preceding). Clearly it too depends on the parameters being used — now the two pairs of parameters. Extending our notation to present the upper parameters in the differentials as a left-superscript we get

$$H_{\bar{a}_1,\bar{a}_2}{}^2 = H_{a_1,a_2}{}^2\left({}^{a_1,a_2}J_{\bar{a}_1,\bar{a}_2}\right)^2 \qquad \text{. . . . . . . (292a}$$

The application of the fundamental form and fundamental quantities to considerations of conformal and equal-area transformations is illustrated in the later sections of Chapter 13, in the context of projections already developed by other means. Conformality is addressed more generally in Chapter 15.

[†] Observed and defined by mathematician C. G. J. Jacobi of Germany (1804 - 1851), who also developed much to do with ellipses and ellipsoids, including the elliptic functions used in the Chapter 15.

CHAPTER 13

Fact and Fiction:
Of scales and distortion

Preamble

Virtually all maps show a scale, either as a graduated bar else as a figure, and often both. The figure is normally written as a ratio; 1:1 000 000 may be given on a regional map in an atlas. On a hiking map 1:50 000 is typical today. The figure 1:63,360 was a familiar sight on the "one inch" (i.e., 1 inch equals one mile) Ordnance Survey maps of Britain and similar elsewhere. However expressed, any scale statement on a map is only precisely true for a very limited part of the map — perhaps as little as one spot, commonly along one meridian and one parallel therein, often along all meridians else all parallels, but then along no more than two other lines. For individual maps covering large portions of Earth, there is no way to keep much rein on the distortion beyond that. The problem is obvious with the cylindrical maps: they have a uniform map-distance between meridians despite the fact that on Earth 1°-spaced meridians converge from over 100 km apart at the Equator to intersection at the Poles. In this chapter the variation of scale is examined further, and the related distortion of areas and of angles is addressed.

The mathematics employed in this chapter is just the differential geometry of the preceding chapter, applied comparatively between the two surfaces of a mapping transformation.

This is another chapter devoid of maps.

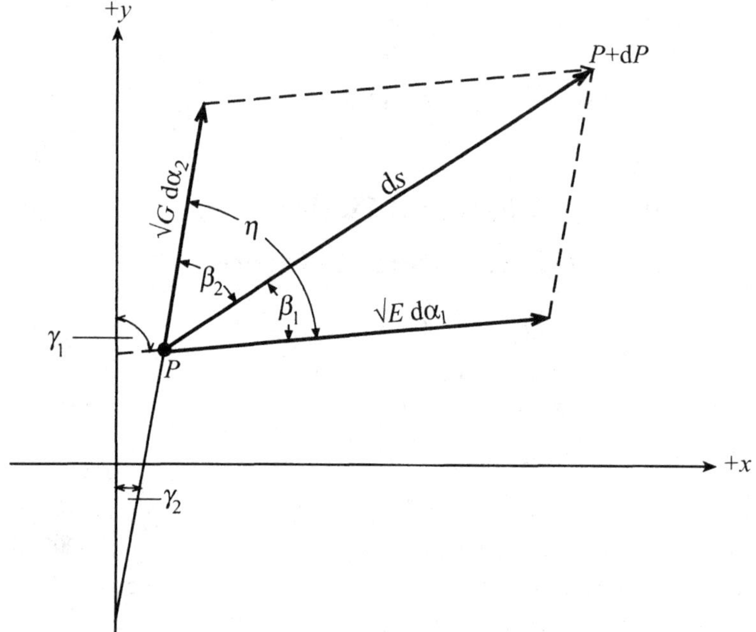

Exhibit 13–1: The general differential parallelogram.

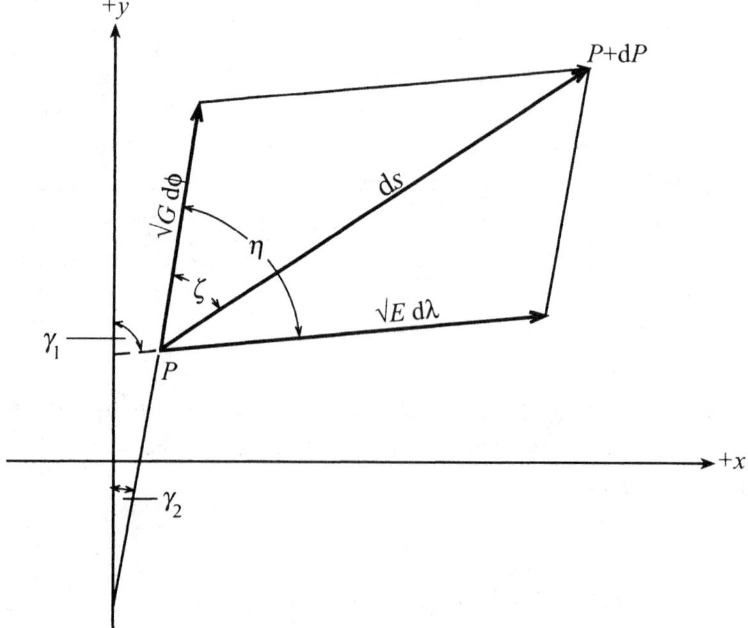

Exhibit 13–2: The differential parallelogram for longitude / latitude.

DIFFERENTIAL GEOMETRY OF DISTORTION

Basis

Distortion relates to the change in lengths, areas, and angles from globe to map and, more generally, between any one surface and a transformation of that surface. The differential geometry developed for general surfaces defined parametrically in Chapter 12 provides a basis for measuring such distortion, with the fundamental quantities and the equations derived there allowing measurement of the changes wrought by a transformation. Exhibit 13–1 is essentially a replica of the key diagram of that previous chapter. Exhibit 13–2 is a graphical replica with longitude and latitude inserted as parameters and the azimuthal angle labelled accordingly.

We again denote entities relating to the derived map with an overscript dot, leaving those of the source, which itself could be a map, unmarked.

Distortion of length (linear distortion)

The distance along the surface between two points an infinitesimal distance apart is given by the first fundamental form, expressed by Equation 279b in general parameters. So, the linear distortion (or relative scale) at a given point, denoted by the positive number $\dot{m}$, is

$$\dot{m} = \frac{\mathrm{d}\dot{s}}{\mathrm{d}s} = \frac{\sqrt{\dot{E}_{\bar{a}_1,\bar{a}_2}\mathrm{d}\bar{a}_1{}^2 + 2\dot{F}_{\bar{a}_1,\bar{a}_2}\mathrm{d}\bar{a}_1\,\mathrm{d}\bar{a}_2 + \dot{G}_{\bar{a}_1,\bar{a}_2}\mathrm{d}\bar{a}_2{}^2}}{\sqrt{E_{a_1,a_2}\mathrm{d}a_1{}^2 + 2F_{a_1,a_2}\,\mathrm{d}a_1\,\mathrm{d}a_2 + G_{a_1,a_2}\mathrm{d}a_2{}^2}}$$

the value generally depending on direction. If parameters a_1,a_2 are common to source and map, this becomes

$$\dot{m} = \frac{\mathrm{d}\dot{s}}{\mathrm{d}s} = \sqrt{\frac{\dot{E}_{a_1,a_2}\,\mathrm{d}a_1{}^2 + 2\dot{F}_{a_1,a_2}\,\mathrm{d}a_1\,\mathrm{d}a_2 + \dot{G}_{a_1,a_2}\mathrm{d}a_2{}^2}{E_{a_1,a_2}\,\mathrm{d}a_1{}^2 + 2F_{a_1,a_2}\,\mathrm{d}a_1\,\mathrm{d}a_2 + G_{a_1,a_2}\mathrm{d}a_2{}^2}} \quad \text{. . . . . . . (295a}$$

where the new fundamental quantities in the numerator can be derived from their predecessors using Equation 291a. Henceforth, we assume that the parameters are common for both surfaces and accordingly omit them as subscripts unless required for clarity.

Appropriately labelling the distortion symbols, Equations 281a show the distortions along the parameter curves to be:

$$\begin{aligned} {}_{a_1}\dot{m} &= \frac{{}_{a_1}\mathrm{d}\dot{s}}{{}_{a_1}\mathrm{d}s} = \frac{\sqrt{\dot{E}}\,\mathrm{d}a_1}{\sqrt{E}\,\mathrm{d}a_1} = \sqrt{\frac{\dot{E}}{E}} \\ {}_{a_2}\dot{m} &= \frac{{}_{a_2}\mathrm{d}\dot{s}}{{}_{a_2}\mathrm{d}\dot{s}} = \frac{\sqrt{\dot{G}}\,\mathrm{d}a_2}{\sqrt{G}\,\mathrm{d}a_2} = \sqrt{\frac{\dot{G}}{G}} \end{aligned} \quad \text{. . . . . . . (295b}$$

Dividing the numerator and denominator of Equation 295a by da_2 gives

$$\dot{m} = \frac{d\dot{s}}{ds} = \sqrt{\frac{\dot{E}\left(\frac{da_1}{da_2}\right)^2 + 2\dot{F}\left(\frac{da_1}{da_2}\right) + \dot{G}}{E\left(\frac{da_1}{da_2}\right)^2 + 2F\left(\frac{da_1}{da_2}\right) + G}} \quad \text{. . . . . . . (296a}$$

showing the dependence of the value on the direction, expressed by the bracketed ratio.

Distortion of area

Using Equation 282a, the distortion of area can be expressed proportionally as

$$\frac{d\dot{A}}{dA} = \frac{\dot{H}\, da_1\, da_2}{H\, da_1\, da_2} = \frac{\dot{H}}{H} = \frac{\sqrt{\dot{E}\cdot\dot{G} - \dot{F}^2}}{\sqrt{E\cdot G - F^2}} \quad \text{. . . . . . . (296b}$$

For an equal-area transformation this ratio must equal 1, hence

$$\sqrt{E\cdot G - F^2} = H = \dot{H} = \sqrt{\dot{E}\cdot\dot{G} - \dot{F}^2}$$

i.e., factor H must be identical for the two surfaces when expressed relative to the same parameters.

Distortion of angles

The proportional distortion of angles can be gauged via the various trigonometric expressions for them on the general surface, for angle η between the parametric axes in Equations 281b through 281e, for angle β_1 in Equations 282c through 283b, for angle β_2 in Equations 283c through 283e. Equation 281b shows

$$\frac{\cos\dot{\eta}}{\cos\eta} = \frac{\dfrac{\dot{F}}{\sqrt{\dot{E}\cdot\dot{G}}}}{\dfrac{F}{\sqrt{E\cdot G}}} = \frac{\dot{F}\sqrt{E\cdot G}}{F\sqrt{\dot{E}\cdot\dot{G}}}$$

while for bearing angle β_2 Equation 283d shows

$$\frac{\sin\dot{\beta}_2}{\sin\beta_2} = \frac{\dfrac{\dot{H}}{\sqrt{\dot{G}}}\dfrac{da_2}{d\dot{s}}}{\dfrac{H}{\sqrt{G}}\dfrac{da_2}{ds}} = \frac{\dot{H}\sqrt{G}\, d\dot{s}}{H\sqrt{\dot{G}}\, ds} = \sqrt{\frac{G\left(\dot{E}\cdot\dot{G} - \dot{F}^2\right)}{\dot{G}\left(E\cdot G - F^2\right)}}\,\frac{d\dot{s}}{ds}$$

For a conformal transformation these ratios must equal 1 regardless of direction. Hence each fundamental quantity for the mapped point must be a common multiple of the corresponding quantity for the point in the source, i.e., for some k:

$$\dot{E} = k^2E, \quad \dot{F} = k^2F, \quad \dot{G} = k^2G \quad \text{. . . . . . . (296c}$$

From Equation 296a, we see that, assuming its positive form is chosen, k equals the linear distortion at the point, which is then identical in all directions (as previously noted for conformality).

Using angle-difference and angle-sum formulae of Tutorial 10 it is also possible to measure the distortion of angles absolutely. For angle β_2 Equations 282d & 282e provide

$$\sin(\dot{\beta}_2 \pm \beta_2) = \sin\dot{\beta}_2 \cos\beta_2 \pm \cos\dot{\beta}_2 \sin\beta_2$$

$$= \frac{\dot{H}}{\sqrt{\dot{E}}}\frac{da_2}{d\dot{s}}\left\{\frac{E\,da_1^2 + ds^2 - G\,da_2^2}{2\sqrt{E}\,da_1\,ds}\right\} \pm \left\{\frac{\dot{E}\,da_1^2 + d\dot{s}^2 - \dot{G}\,da_2^2}{2\sqrt{\dot{E}}\,da_1\,d\dot{s}}\right\}\frac{H}{\sqrt{E}}\frac{da_2}{ds}$$

So

$$\sin(\dot{\beta}_2 - \beta_2) = \left\{\frac{(E\cdot\dot{H} - \dot{E}\cdot H)da_1^2 + ds^2 - d\dot{s}^2 - (G\cdot\dot{H} - \dot{G}\cdot H)da_2^2}{2\sqrt{E\cdot\dot{E}}\;ds\,d\dot{s}}\right\}\frac{da_2}{da_1} \quad \ldots \text{(297a}$$

and

$$\sin(\dot{\beta}_2 + \beta_2) = \left\{\frac{(E\cdot\dot{H} + \dot{E}\cdot H)da_1^2 + ds^2 + d\dot{s}^2 + (G\cdot\dot{H} + \dot{G}\cdot H)da_2^2}{2\sqrt{E\cdot\dot{E}}\;ds\,d\dot{s}}\right\}\frac{da_2}{da_1} \quad \ldots \text{(297b}$$

Though long and clumsy in such general context, these expressions resolve well when the common parametric axes are orthogonal on both surfaces. Using Equations 284f (and denoting the orthogonality by the angled cap on the parameters) we then have:

$$\sin(\dot{\beta}_2 - \beta_2) = \left\{\sqrt{E\cdot\dot{G}} - \sqrt{\dot{E}\cdot G}\right\}\frac{d\hat{a}_1\,d\hat{a}_2}{ds\,d\dot{s}} \quad = \sqrt{E\cdot G}\left\{\sqrt{\frac{\dot{G}}{G}} - \sqrt{\frac{\dot{E}}{E}}\right\}\frac{d\hat{a}_1\,d\hat{a}_2}{ds\,d\dot{s}}$$

and

$$\sin(\dot{\beta}_2 + \beta_2) = \left\{\sqrt{E\cdot\dot{G}} + \sqrt{\dot{E}\cdot G}\right\}\frac{d\hat{a}_1\,d\hat{a}_2}{ds\,d\dot{s}} \quad = \sqrt{E\cdot G}\left\{\sqrt{\frac{\dot{G}}{G}} + \sqrt{\frac{\dot{E}}{E}}\right\}\frac{d\hat{a}_1\,d\hat{a}_2}{ds\,d\dot{s}}$$

Substituting from Equation 295b gives:

$$\sin(\dot{\beta}_2 - \beta_2) = \sqrt{E\cdot G}\left({}_{\hat{a}_2}\dot{m} - {}_{\hat{a}_1}\dot{m}\right)\frac{d\hat{a}_1\,d\hat{a}_2}{ds\,d\dot{s}}$$

$$\sin(\dot{\beta}_2 + \beta_2) = \sqrt{E\cdot G}\left({}_{\hat{a}_2}\dot{m} + {}_{\hat{a}_1}\dot{m}\right)\frac{d\hat{a}_1\,d\hat{a}_2}{ds\,d\dot{s}}$$

Hence, multiplying through each by the right-hand bracketed term of the other, we get

$$\left({}_{\hat{a}_2}\dot{m} + {}_{\hat{a}_1}\dot{m}\right)\sin(\dot{\beta}_2 - \beta_2) = \left({}_{\hat{a}_2}\dot{m} - {}_{\hat{a}_1}\dot{m}\right)\sin(\dot{\beta}_2 + \beta_2) \quad \ldots\ldots\ldots \text{(297c}$$

which shows, for dually orthogonal parametric axes, that distortion is greatest when the right-hand sine term is its maximum of 1 — i.e., when $(\dot{\beta}_2 + \beta_2)$ is an odd multiple of $\pi/2$.

TRANSFORMING THE GLOBE

Basics

Though distortion can occur with a transformation between any pair of surfaces, and relate to various parametric variables, distortion at a point with given longitude and latitude is our prime concern — more particularly the distortion relative to the globe of the finished map (which may be acquired cumulatively over several transformational steps).

With the globe as source and longitude and latitude as common basic parameters (but noting that orthogonality on the globe does not routinely carry over to the map), and using Equations 287b for the fundamental quantities of the sphere, i.e.,

$$ {}_{s}E_{\lambda,\phi} = R^2\cos^2\phi \qquad {}_{s}F_{\lambda,\phi} = 0 \qquad {}_{s}G_{\lambda,\phi} = R^2 \qquad {}_{s}H_{\lambda,\phi} = R^2\cos\phi $$

Equation 295a becomes

$$ \dot{m} = \frac{d\dot{s}}{ds} = \sqrt{\frac{\dot{E}_{\lambda,\phi}\, d\lambda^2 + 2\dot{F}_{\lambda,\phi}\, d\lambda d\phi + \dot{G}_{\lambda,\phi}\, d\phi^2}{E_{\lambda,\phi}\, d\lambda^2 + G_{\lambda,\phi} d\phi^2}} = \sqrt{\frac{\dot{E}_{\lambda,\phi}\, d\lambda^2 + 2\dot{F}_{\lambda,\phi}\, d\lambda d\phi + \dot{G}_{\lambda,\phi}\, d\phi^2}{R^2\cos^2\phi\, d\lambda^2 + R^2 d\phi^2}} \qquad \text{. . . . . (298a} $$

Denoting by ${}_{m}\dot{m}$ the distortion along the meridian (where $d\lambda=0$) and by ${}_{t}\dot{m}$ the distortion transversely thereto (effectively along the parallel, where $d\phi=0$), Equations 295b become:

$$ {}_{t}\dot{m} = \sqrt{\frac{\dot{E}_{\lambda,\phi}}{{}_{s}E_{\lambda,\phi}}} = \sqrt{\frac{\dot{E}_{\lambda,\phi}}{R^2\cos^2\phi}} = \frac{\sqrt{\dot{E}_{\lambda,\phi}}}{R\cos\phi} $$

$$ {}_{m}\dot{m} = \sqrt{\frac{\dot{G}_{\lambda,\phi}}{{}_{s}G_{\lambda,\phi}}} = \sqrt{\frac{\dot{G}_{\lambda,\phi}}{R^2}} = \frac{\sqrt{\dot{G}_{\lambda,\phi}}}{R} \qquad \text{. . . . . . . (298b} $$

while Equation 296b shows the distortion of area to be

$$ {}_{A}\dot{m} = \sqrt{\frac{\dot{E}_{\lambda,\phi}\dot{G}_{\lambda,\phi} - \left(\dot{F}_{\lambda,\phi}\right)^2}{{}_{s}E_{\lambda,\phi}\cdot{}_{s}G_{\lambda,\phi}}} = \sqrt{\frac{\dot{E}_{\lambda,\phi}\dot{G}_{\lambda,\phi} - \left(\dot{F}_{\lambda,\phi}\right)^2}{\left(R^2\cos^2\phi\right)R^2}} = \frac{\sqrt{\dot{E}_{\lambda,\phi}\dot{G}_{\lambda,\phi} - \left(\dot{F}_{\lambda,\phi}\right)^2}}{R^2\cos\phi} $$

Linear distortion in terms of angles

Equations 287d & 283b provide an opportunity to re-express Equation 298a in terms of one parameter plus one angle, with either the azimuth ζ on the globe else bearing $\dot{\beta}_2$ of the map (the latter being, with longitude and latitude as parameters, identically the azimuthal angle $\dot{\zeta}$). Using Equation 287d for the square of the denominator ds and Equation 287e to substitute for dϕ, and again omitting qualification of fundamental quantities with the parameters involved now wholly λ and ϕ, Equation 298a gives

$$\dot{m}^2 = \frac{\left\{\dot{E} + 2\dot{F}\cos\phi\cot\zeta + \dot{G}(\cos\phi\cot\zeta)^2\right\}d\lambda^2}{R^2\cos^2\phi\ \csc^2\zeta\, d\lambda^2}$$

$$= \frac{\dot{E}\sin^2\zeta + 2\dot{F}\cos\phi\ \sin\zeta\ \cos\zeta + \dot{G}\cos^2\phi\ \cos^2\zeta}{R^2\cos^2\phi}$$

$$= \frac{\dot{E}\sin^2\zeta}{R^2\cos^2\phi} + \frac{\dot{F}2\sin\zeta\ \cos\zeta}{R^2\cos\phi} + \frac{\dot{G}\cos^2\zeta}{R^2}$$

$$= \frac{\dot{E}\sin^2\zeta}{R^2\cos^2\phi} + \frac{\dot{F}\sin 2\zeta}{R^2\cos\phi} + \frac{\dot{G}\cos^2\zeta}{R^2}$$

which can be written compactly as

$$\dot{m}^2 = {}_{\zeta}\dot{e}\sin^2\zeta + {}_{\zeta}\dot{f}\sin 2\zeta + {}_{\zeta}\dot{g}\cos^2\zeta \qquad \text{(299a}$$

where: $${}_{\zeta}\dot{e} = \frac{\dot{E}}{R^2\cos^2\phi} \qquad {}_{\zeta}\dot{f} = \frac{\dot{F}}{R^2\cos\phi} \qquad {}_{\zeta}\dot{g} = \frac{\dot{G}}{R^2} \qquad \text{(299b}$$

with all three coefficients being invariant for a given choice of ϕ. Noting the similar ratios for the distortion along meridians and transversely thereto at Equations 298b, and utilizing Equation 281b, these coefficients can also be expressed as:

$${}_{\zeta}\dot{e} = {}_{\mathrm{t}}\dot{m}^2, \quad {}_{\zeta}\dot{f} = \frac{\dot{F}}{R^2\cos\phi} = \frac{\sqrt{\dot{E}\cdot\dot{G}}}{R^2\cos\phi}\frac{\dot{F}}{\sqrt{\dot{E}\cdot\dot{G}}} = {}_{\mathrm{t}}\dot{m}\ {}_{\mathrm{m}}\dot{m}\ \cos\dot{\eta}, \quad {}_{\zeta}\dot{g} = {}_{\mathrm{m}}\dot{m}^2 \qquad \text{(299c}$$

making Equation 299a

$$\dot{m}^2 = {}_{\mathrm{t}}\dot{m}^2\sin^2\zeta + {}_{\mathrm{t}}\dot{m}\cdot{}_{\mathrm{m}}\dot{m}\ \cos\dot{\eta}\sin 2\zeta + {}_{\mathrm{m}}\dot{m}^2\sin^2\zeta$$

A similar equation can be developed expressing the distortion in terms of the azimuthal angle $\dot{\zeta}$ of the map, as detailed in Endnote A to this chapter. The result again has coefficients invariant for a given choice of ϕ, being

$$\dot{m}^{-2} = {}_{\dot{\zeta}}\dot{e}\sin^2\dot{\zeta} + {}_{\dot{\zeta}}\dot{f}\sin 2\dot{\zeta} + {}_{\dot{\zeta}}\dot{g}\cos^2\dot{\zeta} \qquad \text{(299d}$$

where: $${}_{\dot{\zeta}}\dot{e} = \frac{\csc^2\dot{\eta}}{{}_{\mathrm{t}}\dot{m}^2} + \frac{\cot^2\dot{\eta}}{{}_{\mathrm{m}}\dot{m}^2} = \frac{R^2\left(\dot{F}^2 + \dot{G}^2\cos^2\phi\right)}{\dot{G}\cdot\dot{H}^2}$$

$${}_{\dot{\zeta}}\dot{f} = -\frac{\cot\dot{\eta}}{{}_{\mathrm{m}}\dot{m}^2} = -\frac{R^2\dot{F}}{\dot{G}\cdot\dot{H}}$$

$${}_{\dot{\zeta}}\dot{g} = \frac{1}{{}_{\mathrm{m}}\dot{m}^2} = \frac{R^2}{\dot{G}}.$$

The azimuth on the globe and azimuthal angle on the map can be related directly by substituting in Equation 283e successively from Equations 281a, 298b & 285f, giving

$$\cot\dot{\zeta} = \cot\dot{\beta}_2 = \cot\dot{\eta} + \csc\dot{\eta}\frac{\sqrt{\dot{G}}}{\sqrt{\dot{E}}}\cdot\frac{d\phi}{d\lambda} = \cot\dot{\eta} + \csc\dot{\eta}\frac{{}_{\mathrm{m}}\dot{m}}{{}_{\mathrm{t}}\dot{m}}\cot\zeta$$

Hence,

$$\cot\dot{\zeta} = \csc\dot{\eta}\left\{\cos\dot{\eta} + \frac{{}_{\mathrm{m}}\dot{m}}{{}_{\mathrm{t}}\dot{m}}\cot\zeta\right\} \quad \text{or} \quad \cot\zeta = \frac{{}_{\mathrm{t}}\dot{m}}{{}_{\mathrm{m}}\dot{m}}\left\{\frac{\cot\dot{\zeta}}{\csc\dot{\eta}} - \cos\dot{\eta}\right\}$$

The extremes of linear distortion

Identifying the global azimuths occurring with maximal and minimal linear distortion requires differentiation of one of the above equations for linear distortion, then finding the zeros of the resulting derivative. Addressing Equation 299a we get

$$\frac{\mathrm{d}(\dot{m}^2)}{\mathrm{d}\zeta} = 2\,{}_{\zeta}\dot{e}\cos\zeta\sin\zeta + 2\,{}_{\zeta}\dot{f}\cos 2\zeta - 2\,{}_{\zeta}\dot{g}\sin\zeta\cos\zeta$$

$$= 2{}_{\zeta}\dot{f}\cos 2\zeta + \left({}_{\zeta}\dot{e} - {}_{\zeta}\dot{g}\right)\sin 2\zeta$$

The extreme values thus occur at azimuths ζ solving

$$2{}_{\zeta}\dot{f}\cos 2\zeta + \left({}_{\zeta}\dot{e} - {}_{\zeta}\dot{g}\right)\sin 2\zeta = 0$$

i.e., such that

$$\tan 2\zeta = \frac{-2\,{}_{\zeta}\dot{f}}{{}_{\zeta}\dot{e} - {}_{\zeta}\dot{g}} = \frac{2\,{}_{\zeta}\dot{f}}{{}_{\zeta}\dot{g} - {}_{\zeta}\dot{e}}$$

Substituting from Equation 299b gives this in fundamental quantities as

$$\tan 2\zeta = \frac{2\dot{F}}{R^2\cos\phi}\left\{\frac{\dot{G}}{R^2} - \frac{\dot{E}}{R^2\cos^2\phi}\right\}^{-1} = 2\frac{\dot{F}}{R^2\cos\phi}\left\{\frac{R^2\cos^2\phi}{\dot{G}\cos^2\phi - \dot{E}}\right\} = \frac{2\dot{F}\cos^2\phi}{\dot{G}\cos^2\phi - \dot{E}}$$

Alternatively, substituting from Equation 299c,

$$\tan 2\zeta = \frac{2\,{}_{\mathrm{t}}\dot{m}\,{}_{\mathrm{m}}\dot{m}\cos\dot{\eta}}{{}_{\mathrm{m}}\dot{m}^2 - {}_{\mathrm{t}}\dot{m}^2}$$

Hence, extremes of linear distortion occur for

$$\tan(2\zeta + n\pi) = \tan 2\left(\zeta + \frac{n\pi}{2}\right) = \pm\tan 2\zeta = \pm\frac{2\,{}_{\mathrm{t}}\dot{m}\,{}_{\mathrm{m}}\dot{m}\cos\dot{\eta}}{{}_{\mathrm{m}}\dot{m}^2 - {}_{\mathrm{t}}\dot{m}^2}$$

i.e., at successive right-angle increments to any solution, with each pair being the reverse direction of the preceding pair. Thus, extremes of scale distortion at a point occur along directions that are mutually orthogonal on the globe. Interestingly, we find that at that same point on the map the corresponding directions are mutually orthogonal, too.

Addressing Equation 299a, noting Equation 298b, we get a similar conclusion, with extremes occurring at values of azimuthal angle $\dot{\zeta}$ on the map, such that

$$\tan(2\dot{\zeta} + n\pi) = \tan 2\left(\dot{\zeta} + \frac{n\pi}{2}\right) = \pm\tan 2\dot{\zeta} = \pm\frac{{}_{\mathrm{t}}\dot{m}^2\sin 2\dot{\eta}}{{}_{\mathrm{m}}\dot{m}^2 + {}_{\mathrm{t}}\dot{m}^2\cos 2\dot{\eta}}$$

So again, adding a right-angle to any solution gives a second solution, showing that the extremes of linear scale distortion occur along directions that are mutually orthogonal — now on the map.

DUAL ORTHOGONAL PARAMETRIC AXES

Distortion formulae simplified

Equations 284a through 285b provide simplifications of the measurement equations for surfaces that have orthogonal parametric axes, as exemplified in the preceding section with longitude and latitude on the globe for the source surface. The key factor is, of course, that the second fundamental quantity (namely F) is zero for that surface. More potently, those simplifications extend when the axes are orthogonal on both surfaces. Choosing general parameters with an angled overscript to denote their orthogonality and applying Equations 284b & 284c for orthogonal parameter axes to each surface gives:

$$\text{distortion along the } \hat{a}_1 \text{ curve} = {}_{\hat{a}_1}\dot{m} = \frac{{}_{\hat{a}_1}\mathrm{d}\dot{s}}{{}_{\hat{a}_1}\mathrm{d}s} = \sqrt{\frac{\dot{E}_{\hat{a}_1,\hat{a}_2}}{E_{\hat{a}_1,\hat{a}_2}}} \qquad \text{(301a}$$

$$\text{distortion along the } \hat{a}_2 \text{ curve} = {}_{\hat{a}_2}\dot{m} = \frac{{}_{\hat{a}_2}\mathrm{d}\dot{s}}{{}_{\hat{a}_2}\mathrm{d}s} = \sqrt{\frac{\dot{G}_{\hat{a}_1,\hat{a}_2}}{G_{\hat{a}_1,\hat{a}_2}}}$$

$$\text{distortion generally} = \dot{m} = \sqrt{\frac{{}_{\hat{a}_1}\dot{m}^2 \cdot \dot{E}_{\hat{a}_1,\hat{a}_2}\,\mathrm{d}\hat{a}_1^{\,2} + {}_{\hat{a}_2}\dot{m}^2 \cdot \dot{G}_{\hat{a}_1,\hat{a}_2}\,\mathrm{d}\hat{a}_2^{\,2}}{\mathrm{d}s^2}} \qquad \text{(301b}$$

$$= \sqrt{\frac{{}_{\hat{a}_1}\dot{m}^2 \cdot \dot{E}_{\hat{a}_1,\hat{a}_2}\,\mathrm{d}\hat{a}_1^{\,2}}{\mathrm{d}s^2} + \frac{{}_{\hat{a}_2}\dot{m}^2 \cdot G_{\hat{a}_1,\hat{a}_2}\,\mathrm{d}\hat{a}_2^{\,2}}{\mathrm{d}s^2}}$$

Equation 284g shows the distortion of area as

$$\frac{\mathrm{d}\dot{A}}{\mathrm{d}A} = \frac{\sqrt{\dot{E}_{\hat{a}_1,\hat{a}_2} \cdot \dot{G}_{\hat{a}_1,\hat{a}_2}}\,\mathrm{d}\hat{a}_1\mathrm{d}\hat{a}_2}{\sqrt{E_{\hat{a}_1,\hat{a}_2} \cdot G_{\hat{a}_1,\hat{a}_2}}\,\mathrm{d}\hat{a}_1\mathrm{d}\hat{a}_2} = \frac{\sqrt{\dot{E}_{\hat{a}_1,\hat{a}_2} \cdot \dot{G}_{\hat{a}_1,\hat{a}_2}}}{\sqrt{E_{\hat{a}_1,\hat{a}_2} \cdot G_{\hat{a}_1,\hat{a}_2}}} = {}_{\hat{a}_1}\dot{m} \cdot {}_{\hat{a}_2}\dot{m}$$

while Equations 284f show, for β_2 less than the right angle,

$$\frac{\cos\dot{\beta}_2}{\cos\beta_2} = \frac{\dfrac{\dot{E}_{\hat{a}_1,\hat{a}_2}\mathrm{d}\hat{a}_1}{\mathrm{d}\dot{s}}}{\dfrac{E_{\hat{a}_1,\hat{a}_2}\mathrm{d}\hat{a}_1}{\mathrm{d}s}} = \frac{\dot{E}_{\hat{a}_1,\hat{a}_2}}{E_{\hat{a}_1,\hat{a}_2}} \cdot \frac{\mathrm{d}s}{\mathrm{d}\dot{s}}$$

Substituting from Equation 284f reduces Equation 301b to

$$\dot{m}^2 = {}_{\hat{a}_1}\dot{m}^2 \sin^2\beta_2 + {}_{\hat{a}_2}\dot{m}^2 \cos^2\beta_2$$

Differentiating this to find the values of β_2 with extremes of linear distortion gives

$$\frac{\mathrm{d}\left(\dot{m}^2\right)}{\mathrm{d}\beta_2} = 2\ {}_{\hat{a}_1}\dot{m}^2 \cos\beta_2 \sin\beta_2 - 2\ {}_{\hat{a}_2}\dot{m}^2 \sin\beta_2 \cos\beta_2$$

$$= \left({}_{\hat{a}_1}\dot{m}^2 - {}_{\hat{a}_2}\dot{m}^2\right) \sin 2\beta_2$$

This must equal zero for the extremes. Unless the two distortion terms are identical, β_2 must be an integer multiple of the right angle. If the two distortion terms are identical, then, by Equations 301a, the ratios of the two non-zero fundamental forms for the two surfaces are identical. Hence distortion is equal in all directions and the mapping is conformal. Thus proper extremes of linear distortion occur only along the parametric curves, with angle β_2 measured from the $\hat{a}_2$ curve.

Existence of dually orthogonal parametric axes

While orthogonality carries from the globe to the map for meridians and parallels with all 'true' projections, and for any orthogonal axes in the source under conformal transformation, neither holds more generally. Starting with the Sinusoidal of Chapter 6, any retained orthogonality appears minor for most projections. Auguste Tissot [1858a] — one of the earliest essayists into the realm of distortion — showed, however, that any transformation that is continuous in a general sense rather than disjointed in some way, always has a pair of orthogonal parametric lines on the source surface that project into orthogonal lines on the map. He demonstrated this by geometric reasoning. If a pair of intersecting orthogonal lines on the source do not map into orthogonal lines then gyrate those of the source progressively through 90° until those on the map are superimposed on the initial lines there. During the process, those on the map must pass between being less than a right angle and being greater. If it is a continuous function, then somewhere between they must be at precisely a right angle. Later in this chapter, more substantive proof is provided using linear algebra. The parametric lines that stay orthogonal are called the *principal directions*, the axes they provide the *principal parametric axes*. Because gyration of such lines by 90° does not affect the matter, there are technically four choices in a directional sense.

Distortion of azimuth

If longitude and latitude are the parameters, angles β_2 and $\dot{\beta}_2$ become the azimuth ζ on the globe and the azimuthal angle $\dot{\zeta}$ on the map. Extreme values of distortion require, from Equation 297c,

$$f = \left({}_{\mathrm{t}}\dot{m} + {}_{\mathrm{m}}\dot{m}\right)\sin\left(\dot{\zeta} - \zeta\right) - \left({}_{\mathrm{t}}\dot{m} - {}_{\mathrm{m}}\dot{m}\right)\sin\left(\dot{\zeta} + \zeta\right) = 0 \qquad \text{(302a}$$

For conformal projections the two distortion terms are identical, rendering the angles identical. Otherwise, the equation, which can be solved by Newton-Raphson iterative approximation method (Tutorial 18), provides a means for ascertaining azimuthal angle on the map for a given azimuth on the globe. With the global azimuth as the first estimate, this iterative method proves rapidly effective, with

$$\frac{\mathrm{d}f}{\mathrm{d}\dot{\zeta}} = \cos\left(\dot{\zeta} - \zeta\right) - \frac{\left({}_{\mathrm{t}}\dot{m} - {}_{\mathrm{m}}\dot{m}\right)}{\left({}_{\mathrm{t}}\dot{m} + {}_{\mathrm{m}}\dot{m}\right)}\cos\left(\dot{\zeta} + \zeta\right)$$

THE TISSOT INDICATRIX

Concept

The most commonly used visual indicator of distortion is the eponymously named *Tissot indicatrix* [1858a; 1879a; 1881a]. It is essentially a small ellipse whose radii indicate the relative distortion in their respective directions for the point at the ellipse's centre. Hence its shape and size reflect the distortion at the point. The ellipse is defined to be the map's image of a correspondingly small circle on the globe — a circle that must be infinitesimal to provide values for the point itself, but with its mapping enlarged sufficiently to make the ellipse clearly visible. With a conformal projection, all ellipses must be circles. With an equal-area projection, all ellipses must be of identical area. More generally, the ellipses vary in shape and areal size across the map.

Classic derivation

Let $\hat{a}_1, \hat{a}_2$ be principal parametric axes at general point P on the globe such that

$$_{\hat{a}_1}\dot{m} \geq {}_{\hat{a}_2}\dot{m}$$

These values are the extremes of linear distortion. Set x and y axes in the tangent plane at the point coincident with those parametric axes. If $\mathbf{dp} = (\mathrm{d}x, \mathrm{d}y)$ is a small unit-length incremental vector from P in the plane, at angle γ relative to the y axis then, from Equations 284b & 284f, (noting $\mathrm{d}s = |\mathbf{dp}| = 1$)

$$\mathrm{d}x = \mathrm{d}s \sin\gamma = \sin\gamma = \sqrt{E_{\hat{a}_1,\hat{a}_2}}\, \mathrm{d}\hat{a}_1, \qquad \mathrm{d}y = \mathrm{d}s\cos\gamma = \cos\gamma = \sqrt{G_{\hat{a}_1,\hat{a}_2}}\, \mathrm{d}\hat{a}_2 \qquad \ldots (303a$$

If $\mathbf{d\dot{p}} = (\mathrm{d}\dot{x}, \mathrm{d}\dot{y})$ is the mapping of $\mathbf{dp}$ then likewise

$$\mathrm{d}\dot{x} = \mathrm{d}\dot{s}\sin\dot{\gamma} \quad = \sqrt{\dot{E}_{\hat{a}_1,\hat{a}_2}}\, \mathrm{d}\hat{a}_1, \qquad \mathrm{d}\dot{y} = \mathrm{d}\dot{s}\cos\dot{\gamma} \quad = \sqrt{\dot{G}_{\hat{a}_1,\hat{a}_2}}\, \mathrm{d}\hat{a}_2$$

Using these together with Equations 301a gives

$$\mathrm{d}\dot{x} = \sqrt{\frac{\dot{E}_{\hat{a}_1,\hat{a}_2}}{E_{\hat{a}_1,\hat{a}_2}}} \sin\gamma = {}_{\hat{a}_1}\dot{m}\sin\gamma = {}_{\hat{a}_1}\dot{m}\,\mathrm{d}x, \quad \mathrm{d}\dot{y} = \sqrt{\frac{\dot{G}_{\hat{a}_1,\hat{a}_2}}{G_{\hat{a}_1,\hat{a}_2}}} \cos\gamma = {}_{\hat{a}_2}\dot{m}\cos\gamma = {}_{\hat{a}_2}\dot{m}\,\mathrm{d}y \quad \ldots (303b$$

hence,

$$\left(\frac{\mathrm{d}\dot{x}}{_{\hat{a}_1}\dot{m}}\right)^2 + \left(\frac{\mathrm{d}\dot{y}}{_{\hat{a}_2}\dot{m}}\right)^2 = \mathrm{d}x^2 + \mathrm{d}y^2 = \mathrm{d}s^2 = 1 \qquad \ldots\ldots\ldots (303c$$

If we let $\mathbf{dp}$ range around a complete revolution on the tangent plane, then it describes a (small) unit circle, and Equation 303c specifies an ellipse centred on point $\dot{P}$ with semi-axes equal in length to those extremes of linear distortion and lying along axes coincident with the parametric axes. The ellipse represents the distortion brought by the mapping at P. Equations 303a & 303b show that the point $(\sin\gamma, \cos\gamma)$ on the circle has image $({}_{\hat{a}_1}\dot{m}\sin\gamma, {}_{\hat{a}_2}\dot{m}\cos\gamma)$ on the ellipse.

If we superimpose the ellipse on the unit circle, with common origin and axes, the picture is as illustrated below with the dashed circle being the unit circle and the principal directions interpreted as x and y. Point I on the circle becomes point D on the ellipse. Tutorial 27 explains the illustration and derives various formulae.

If the choice of γ puts the radial line OI coincident with the meridian through P, then line $O\hat{I}$ is coincident with the parallel through our point in the tangent plane. Line OD is coincident with the meridian in the image and, since $\hat{D}$ is the image of $\hat{I}$, $O\hat{D}$ lies along the parallel. We therefore immediately have a picture of the mapping of the cardinal lines relative to the principal directions. If we emphasize the reference to those lines by denoting the distortion along the meridian by ${}_{m}\dot{m}$ and that transversely by ${}_{t}\dot{m}$, then, from the tutorial,

$$ {}_{t}\dot{m}^2 = {}_{\hat{a}_1}\dot{m}^2 \cos^2\gamma + {}_{\hat{a}_2}\dot{m}^2 \sin^2\gamma $$

Hence,

$$ {}_{t}\dot{m}^2 = {}_{\hat{a}_1}\dot{m}^2 - {}_{\hat{a}_1}\dot{m}^2 \sin^2\gamma + {}_{\hat{a}_2}\dot{m}^2 \sin^2\gamma \quad \text{or} \quad \sin^2\gamma = \frac{{}_{\hat{a}_1}\dot{m}^2 - {}_{t}\dot{m}^2}{{}_{\hat{a}_1}\dot{m}^2 - {}_{\hat{a}_2}\dot{m}^2} $$

and

$$ {}_{t}\dot{m}^2 = {}_{\hat{a}_1}\dot{m}^2 \cos^2\gamma + {}_{\hat{a}_2}\dot{m}^2 - {}_{\hat{a}_2}\dot{m}^2 \cos^2\gamma \quad \text{or} \quad \cos^2\gamma = \frac{{}_{t}\dot{m}^2 - {}_{\hat{a}_2}\dot{m}^2}{{}_{\hat{a}_1}\dot{m}^2 - {}_{\hat{a}_2}\dot{m}^2} $$

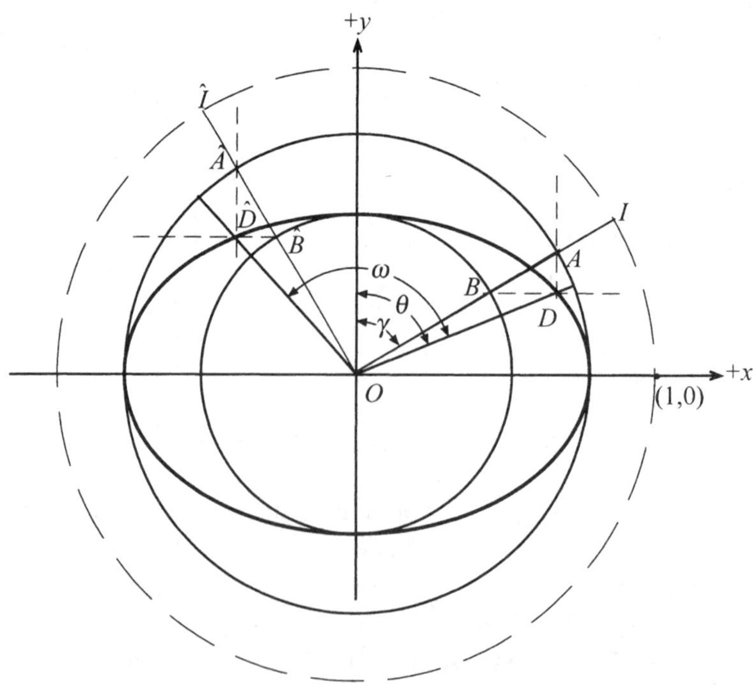

Exhibit 13–3: The ellipse and associated circles.

TUTORIAL 27 THE ELLIPSE AND ASSOCIATED CIRCLES

Given an ellipse

$$\frac{x^2}{a^2}+\frac{y^2}{b^2}=1$$

the circle of radius b touches the ellipse at its crossings of the y axis, the circle of radius a at its crossings of the x axis.

Exhibit 13–3 illustrates this, with $a>b$. We draw the radial line that lies at clockwise angle γ from the y axis, labelling its intersections with the circumscribed and inscribed circles A and B, respectively. Clearly:

$$A=(x_a, y_a) \quad =(a\sin\gamma, a\cos\gamma)$$
$$B=(x_b, y_b) \quad =(b\sin\gamma, b\cos\gamma)$$

We add the two dotted lines, one through A orthogonal to the x axis, and the other through B orthogonal to the y axis. Clearly, if D is their intersection,

$$D=(x_d, y_d)$$
$$=(x_a, y_b) \quad =(a\sin\gamma, b\cos\gamma)$$

Since

$$\frac{(a\sin\gamma)^2}{a^2}+\frac{(b\cos\gamma)^2}{b^2}=\sin^2\gamma+\cos^2\gamma=1$$

point D lies on the ellipse. Thus, we see the geometry corresponding to the parametric form given in Tutorial 8, from which we can note that the angle involved is not that of the direct radial line. If r is the length of line OD and θ is the angle the line makes with the y axis, then:

$$r\sin\theta=a\sin\gamma, \quad r\cos\theta=b\cos\gamma$$
$$\tan\theta=\frac{a\sin\gamma}{b\cos\gamma}=\frac{a}{b}\tan\gamma$$
$$r^2=a^2\sin^2\gamma+b^2\cos^2\gamma$$

If point I on the radial line is unit distance from O (illustrated as outside the ellipse, on dashed unit circle),

$$I=(x_i, y_i)=(\sin\gamma, \cos\gamma)$$

so

$$D=(r\sin\theta, r\cos\theta)$$
$$=(a\sin\gamma, b\cos\gamma) \quad =(ax_i, by_i)$$

If $O\hat{I}$ is the unit-length radial line counter-clockwise by a right angle from OI, then

$$\hat{I}=(-\cos\gamma, \sin\gamma)$$

and its derived intersection point is

$$\hat{D}=(-a\cos\gamma,\ b\sin\gamma)$$

This also lies on the ellipse. If ω is the angle between OD and $O\hat{D}$, then $O\hat{D}$ makes angle $(\omega-\theta)$ with the y axis. If $\hat{r}$ is the length of $O\hat{D}$, then

$$\hat{r}\sin(\omega-\theta)=-a\cos\gamma, \quad \hat{r}\cos(\omega-\theta)=b\sin\gamma$$
$$\hat{r}^2=a^2\cos^2\gamma+b^2\sin^2\gamma$$

Hence, using the equivalent for I,

$$r^2+\hat{r}^2=a^2+b^2$$
$$-\hat{r}\sin(\omega-\theta)\ r\cos\theta=a\cos\gamma\ b\cos\gamma$$
$$\hat{r}\cos(\omega-\theta)\ r\sin\theta=b\sin\gamma\ a\sin\gamma$$

from which adding gives

$$r\hat{r}\sin\omega=ab$$

So, for either choice with the ± sign,

$$(a\pm b)^2=a^2\pm 2ab+b^2$$
$$=r^2\pm 2r\hat{r}\sin\omega+\hat{r}^2$$

Hence, the lengths of the two axes are

$$\sqrt{r^2+2r\hat{r}\sin\omega+\hat{r}^2}\pm\sqrt{r^2-2r\hat{r}\sin\omega+\hat{r}^2}$$

Pairs of radii like r and $\hat{r}$, derived orthogonally, are called *conjoint radii*, their extensions *conjoint diameters*.

The scene is illustrated in Exhibit 13–4, with the origin relabelled as $\dot{P}$ and the various conic sections changed in visual emphasis. The original circle and its cardinal lines are solid, while the ellipse and its cardinal lines are dotted.

Interpreting Equation 281f for the numerator and Equation 284c for the denominator, Equation 295a gives the general distortion as

$$\dot{m}^2 = \frac{d\dot{s}^2}{ds^2} = \frac{\dot{E}_{\lambda,\phi}\,d\lambda^2 + 2\sqrt{\dot{E}_{\lambda,\phi}\cdot\dot{G}_{\lambda,\phi}}\cos\omega\,d\lambda\,d\phi + \dot{G}_{\lambda,\phi}\,d\phi^2}{E_{\lambda,\phi}\,d\lambda^2 + G_{\lambda,\phi}\,d\phi^2}$$

$$= \frac{{}_{t}\dot{m}^2\tan^2\gamma + 2\,{}_{t}\dot{m}\cdot{}_{m}\dot{m}\tan\gamma + {}_{m}\dot{m}^2}{1+\tan^2\gamma}$$

with substitutions from Equations 285f & 298b.

Because the axes are the principal directions, angle γ must correspond to minimum distortion, and its 90° complement to maximum distortion. Differentiating this equation and equating to zero to find the extremes gives

$$\tan^2\gamma - \frac{{}_{t}\dot{m}^2 - {}_{m}\dot{m}^2}{{}_{\hat{a}_1}\dot{m}^2 - {}_{\hat{a}_2}\dot{m}^2}\tan\gamma - 1 = 0$$

the solutions for which are

$$\tan\gamma = \frac{{}_{t}\dot{m}^2 - {}_{m}\dot{m}^2}{2\,{}_{\hat{a}_1}\dot{m}\cdot{}_{\hat{a}_2}\dot{m}\cos\omega} \pm \sqrt{\frac{\left({}_{t}\dot{m}^2 - {}_{m}\dot{m}^2\right)^2}{4\,{}_{t}\dot{m}^2\cdot{}_{m}\dot{m}^2\cos^2\omega} + 1}$$

yielding (for the corresponding angle in the image, labelled θ in the diagram),

$$\tan\theta = \frac{\tan\gamma}{{}_{t}\dot{m}^2 + \tan\gamma\cot\omega}$$

Applying this angle counter-clockwise as with angle χ in Equations 86a allows positioning of the ellipse relative to the cardinal directions. Placing the ellipse on the map requires, of course, the centring of it at $\dot{P}$.

The last formula in the tutorial provides a means for obtaining the extreme values of distortion from those along the cardinal lines:

$${}_{\hat{a}_1}\dot{m} = \frac{1}{2}\left\{\sqrt{{}_{t}\dot{m}^2 + 2\,{}_{t}\dot{m}\cdot{}_{m}\dot{m}\sin\omega + {}_{m}\dot{m}^2} + \sqrt{{}_{t}\dot{m}^2 - 2\,{}_{t}\dot{m}\cdot{}_{m}\dot{m}\sin\omega + {}_{m}\dot{m}^2}\right\}$$

$${}_{\hat{a}_2}\dot{m} = \frac{1}{2}\left\{\sqrt{{}_{t}\dot{m}^2 + 2\,{}_{t}\dot{m}\cdot{}_{m}\dot{m}\sin\omega + {}_{m}\dot{m}^2} - \sqrt{{}_{t}\dot{m}^2 - 2\,{}_{t}\dot{m}\cdot{}_{m}\dot{m}\sin\omega + {}_{m}\dot{m}^2}\right\}$$

Applying linear algebra

Adopting matrix notation and techniques of linear algebra (see Tutorial 20 on page 200 and Tutorial 24 on page 238), Piotr Laskowski [1989f] demonstrated more direct methods for evaluating the characteristics of the distortion at a point, first using *eigenvalues*, then using *Singular Value Decomposition* [1976b]. Using a matrix form, including the transposed form marked by the T right superscript, we have

$$[\mathrm{d}\dot{x}\ \ \mathrm{d}\dot{y}]^{\mathrm{T}} = \mathbf{J}\cdot[\mathrm{d}\lambda\ \ \mathrm{d}\phi]^{\mathrm{T}} \quad \text{where } \mathbf{J} = \begin{bmatrix} \dfrac{\partial\dot{x}}{\partial\lambda} & \dfrac{\partial\dot{x}}{\partial\phi} \\ \dfrac{\partial\dot{y}}{\partial\lambda} & \dfrac{\partial\dot{y}}{\partial\phi} \end{bmatrix}$$

To convert the differentials on the right-hand side to linear rather than angular terms, hence, be consistent with those on the left, Equations 287c gives us

$$\left[{}_{\mathrm{t}}\mathrm{d}s\ \ {}_{\mathrm{m}}\mathrm{d}s\right]^{\mathrm{T}} = \mathbf{K}\cdot[\mathrm{d}\lambda\ \ \mathrm{d}\phi]^{\mathrm{T}} \quad \text{where} \quad \mathbf{K} = \begin{bmatrix} R\cos\phi & 0 \\ 0 & R \end{bmatrix}$$

Hence,

$$[\mathrm{d}\dot{x}\ \ \mathrm{d}\dot{y}]^{\mathrm{T}} = \mathbf{J}\cdot\mathbf{K}^{-1}\cdot\left[{}_{\mathrm{t}}\mathrm{d}s\ \ {}_{\mathrm{m}}\mathrm{d}s\right]^{\mathrm{T}} \quad \text{or} \quad \left[{}_{\mathrm{t}}\mathrm{d}s\ \ {}_{\mathrm{m}}\mathrm{d}s\right]^{\mathrm{T}} = \mathbf{C}\cdot[\mathrm{d}\dot{x}\ \ \mathrm{d}\dot{y}]^{\mathrm{T}} \quad \text{where} \quad \mathbf{C} = \mathbf{K}\cdot\mathbf{J}^{-1}$$

For all points on the infinitesimal unit circle, using some infinitesimal scale,

$$1 = (\mathrm{d}s)^2 = ({}_{\mathrm{t}}\mathrm{d}s)^2 + ({}_{\mathrm{m}}\mathrm{d}s)^2 = \left[{}_{\mathrm{t}}\mathrm{d}s\ \ {}_{\mathrm{m}}\mathrm{d}s\right]\cdot\left[{}_{\mathrm{t}}\mathrm{d}s\ \ {}_{\mathrm{m}}\mathrm{d}s\right]^{\mathrm{T}}$$

so,

$$1 = \left(\mathbf{C}\cdot[\mathrm{d}\dot{x}\ \ \mathrm{d}\dot{y}]^{\mathrm{T}}\right)^{\mathrm{T}}\cdot\left(\mathbf{C}\cdot[\mathrm{d}\dot{x}\ \ \mathrm{d}\dot{y}]^{\mathrm{T}}\right) = [\mathrm{d}\dot{x}\ \ \mathrm{d}\dot{y}]\cdot\mathbf{C}^{\mathrm{T}}\cdot\mathbf{C}\cdot[\mathrm{d}\dot{x}\ \ \mathrm{d}\dot{y}]^{\mathrm{T}} \quad \ldots\ldots \text{(307a}$$

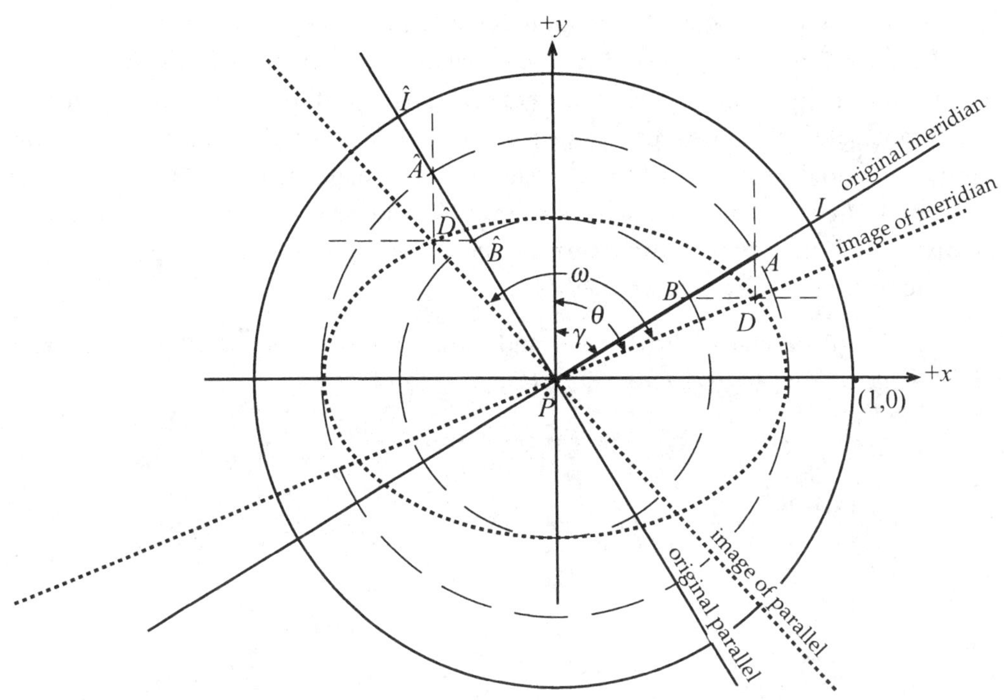

Exhibit 13–4: The Tissot Indicatrix – the ellipse on general background.

We thus have a quadratic form for the infinitesimal entity expressed in terms of the map surface, characterized by the composite matrix $\mathbf{C}^T\cdot\mathbf{C}$. By its progressive definition, **C** must be non-singular so, by Sylvester's Law of Inertia [1997a], the composite will have two positive eigenvalues. This implies that the quadratic form in Equation 307a defines an ellipse, with semi-axial lengths equal to the positive square roots of the reciprocals of those eigenvalues. Because the eigenvalues of any matrix are the reciprocals of those of its reciprocal matrix, we define

$$\mathbf{A} = \mathbf{C}^{-1} = \mathbf{J}\cdot\mathbf{K}^{-1}$$

so that the semi-axial lengths are equal to the square roots of the eigenvalues of $\mathbf{A}\cdot\mathbf{A}^T$. These square roots are known as the *singular values* of **A**.

Laskowski delved yet further into linear algebra, applying Singular Value Decomposition [1997a] — a technique included in standard mathematical software packages — to matrix **A**. Applicable to any non-singular square matrix, the technique derives a pair of orthogonal matrices **U** and **V** and a diagonal matrix **D**, with monotonically decreasing non-zero elements down the diagonal, such that

$$\mathbf{A} = \mathbf{U}\cdot\mathbf{D}\cdot\mathbf{V}^T$$

Our mapping $\mathbf{A} = \mathbf{J}\cdot\mathbf{K}^{-1}$ is thus decomposed into three transformations, successively:

- a rotation $\mathbf{V}^T$ from longitude and latitude to the principal axes,
- a change of axial lengths **D** without rotation of axes,
- a rotation **U** within the map from the principal axes to the co-ordinate axes.

The three new matrices provide measures of each process, and the technique ensures that the intermediate axes are our principle axes with the diagonal elements of **D** corresponding to the extremes of linear distortion, i.e. are the major and minor axes of the Indicatrix If $d_{11} = d_{22}$, the two axes are changed by the same factor and the mapping is conformal. The other two matrices provide the angular positioning of the principal axes relative to the respective reference axes.

The lengths a and b of the major and minor semi-axes of the ellipse of distortion and the angle β between the semi-major axis and the $+y$ half-axis are given by:

$$a = d_{11},\quad b = d_{22},\quad \tan\beta = \frac{u_{21}}{u_{11}}$$

where $\mathbf{U} = \begin{bmatrix} u_{11} & u_{12} \\ u_{21} & u_{22} \end{bmatrix}$, $\mathbf{V} = \begin{bmatrix} v_{11} & v_{12} \\ v_{21} & v_{22} \end{bmatrix}$ and $\mathbf{D} = \begin{bmatrix} d_{11} & 0 \\ 0 & d_{22} \end{bmatrix}$ with $d_{11} \geq d_{22} > 0$.

THE EQUAL-AREA CONDITION

The condition

For a transformation to maintain the equal-area attribute, Equation 296b shows that factor H must be identical for the two surfaces when expressed relative to the same parameters. We look at this formulation of the condition in a practical context, mapping from the globe, and applying it as a test for two projections developed by other means in the early chapters and then as a new construct for two others.

The equal-area condition applied as a test

We look at two simple orthographic projections — the Orthographic Cylindrical, which we know is equal-area, and the Orthographic Azimuthal, which we know is not.

The Orthographic Cylindrical projection (Equations 55d) has intermediate parameters:

$$\dot{u} = R\lambda \qquad \dot{v} = R\sin\phi$$

Hence:

$$\frac{\partial \dot{u}}{\partial \lambda} = R \qquad \frac{\partial \dot{v}}{\partial \lambda} = 0$$
$$\frac{\partial \dot{u}}{\partial \phi} = 0 \qquad \frac{\partial \dot{v}}{\partial \phi} = R\cos\phi \qquad \text{so} \quad {}^{\dot{u},\dot{v}}J_{\lambda,\phi} = R^2\cos\phi$$

From Equation 286b

$$\dot{H}_{\dot{u},\dot{v}} = {}_{\mathrm{p}}H_{\dot{u},\dot{v}} = 1$$

giving, from Equation 292a,

$$\dot{H}_{\lambda,\phi} = \dot{H}_{\dot{u},\dot{v}} \cdot {}^{\dot{u},\dot{v}}J_{\lambda,\phi} = R^2\cos\phi$$

which is identically the quantity for the sphere as shown in Equation 287b, demonstrating the equal-area requirement being met by the Orthographic Cylindrical projection.

The Orthographic (Azimuthal) projection (Equations 47b with Equation 31a) has intermediate parameters of the form:

$$\dot{\theta} = \oplus\lambda, \qquad \dot{\rho} = R\cos\phi$$

Hence:

$$\frac{\partial \dot{\rho}}{\partial \lambda} = 0 \qquad \frac{\partial \dot{\theta}}{\partial \lambda} = \oplus 1$$
$$\frac{\partial \dot{\rho}}{\partial \phi} = -R\sin\phi \qquad \frac{\partial \dot{\theta}}{\partial \phi} = 0 \qquad \text{so} \quad {}^{\dot{\rho},\dot{\theta}}J_{\lambda,\phi} = \oplus R\sin\phi$$

From Equation 286d

$$\dot{H}_{\dot{\rho},\dot{\theta}} = {}_{\mathrm{p}}H_{\dot{\rho},\dot{\theta}} = \dot{\rho}$$

giving, from Equation 292a,

$$\left(\dot{H}_{\lambda,\phi}\right)^2 = \dot{H}_{\dot{\rho},\dot{\theta}} \cdot {}^{\dot{\rho},\dot{\theta}}J_{\lambda,\phi} \quad = \dot{\rho}\left(\oplus R\sin\phi\right) \quad = \oplus\dot{\rho}R\sin\phi$$

This is clearly not identically that for the sphere (see Equation 287b), so the equal-area requirement is not met by the Orthographic Azimuthal projection.

The equal-area condition applied as a construct

Of greater value than simply for testing established projections, differential geometry provides means for developing projections. To illustrate this, we reverse the above process for the Orthographic Cylindrical then apply the process to the Bonne. Starting with the criterion that H for the map equals that for the globe, we apply the essential rules for each mode to reach the established formulae for the intermediate parameters of these equal-area projections.

For any tangential Cylindrical at Simple aspect, the intermediate parameters (Equation 53a) are each functions of one variable, specifically:

$$\dot{u} = R\lambda, \quad \dot{v} \text{ is a function of } \phi, \text{ with the same sign as } \phi$$

so the partial derivatives become ordinary derivatives as:

$$\frac{\partial\dot{u}}{\partial\lambda} = \frac{\mathrm{d}\dot{u}}{\mathrm{d}\lambda} = R \qquad \frac{\partial\dot{v}}{\partial\lambda} = \frac{\mathrm{d}\dot{v}}{\mathrm{d}\lambda} = 0$$

$$\frac{\partial\dot{u}}{\partial\phi} = \frac{\mathrm{d}\dot{u}}{\mathrm{d}\phi} = 0 \qquad \frac{\partial\dot{v}}{\partial\phi} = \frac{\mathrm{d}\dot{v}}{\mathrm{d}\phi} \qquad \text{so} \quad {}^{\dot{u},\dot{v}}J_{\lambda,\phi} = R\frac{\mathrm{d}\dot{v}}{\mathrm{d}\phi}$$

From Equations 286b,

$$\dot{H}_{\dot{u},\dot{v}} =_{\mathrm{P}} H_{\dot{u},\dot{v}} = 1$$

so,

$$\dot{H}_{\lambda,\phi} = \dot{H}_{\dot{u},\dot{v}} \cdot {}^{\dot{u},\dot{v}}J_{\lambda,\phi} \quad = 1 \cdot R\frac{\mathrm{d}\dot{v}}{\mathrm{d}\phi} \quad = R\frac{\mathrm{d}\dot{v}}{\mathrm{d}\phi}$$

All of this applies to any tangential cylindrical projection at Simple aspect. We now apply the specific need for equal-area — i.e., that this function must be identical with that of the sphere as given in Equation 287b. So, we must have

$$R\frac{\mathrm{d}\dot{v}}{\mathrm{d}\phi} =_{\mathrm{S}} H_{\lambda,\phi} = R^2\cos\phi$$

giving

$$\frac{\mathrm{d}\dot{v}}{\mathrm{d}\phi} = \pm R\cos\phi$$

Hence,

$$\dot{v} = \pm R\sin\phi$$

which must have a positive sign to meet our condition of matching sign between intermediate and primary parameters. Thus we have come, through differential geometry, to the formula we know gives equal-area with Cylindricals (Equation 74a).

The above examples of applying differential geometry addressed 'true' projections only, for which necessarily both elements of one diagonal in the Jacobian array are zero, the other two non-zero. With the Pseudo projections one intermediate parameter is a function purely of latitude and the other a function of longitude and latitude. Hence just one Jacobian element is zero. We pursue the Bonne projection, which is more complicated than the Pseudocylindricals and for which treatment by ordinary geometry leaves some question concerning the precise extent of equality of areas.

The Bonne projection was developed as an equal-area Conical in which we maintained the parallels as concentric circular arcs (as with Albers), but demanded equidistance along the central meridian. The consequence was that the taper of the relevant cone varied with latitude — i.e., the constant of the cone — was constant only relative to longitude, varying with latitude, so making the meridians curved. The pertinent equations for Bonne (Equations 162a) are:

$\dot{\theta} = c\,\lambda$ where c is a function solely of ϕ (and necessarily positive)

$$\dot{\rho} = R\{(\phi_0 + \cot\phi_0) - \phi\}$$

The Jacobian array becomes:

$$\frac{\partial\dot{\rho}}{\partial\lambda} = 0 \qquad \frac{\partial\dot{\theta}}{\partial\lambda} = c$$
$$\frac{\partial\dot{\rho}}{\partial\phi} = -R \qquad \frac{\partial\dot{\theta}}{\partial\phi} = \frac{\mathrm{d}c}{\mathrm{d}\phi}\lambda \qquad \text{so} \quad {}^{\dot{\rho},\dot{\theta}}J_{\lambda,\phi} = Rc$$

With the intermediate parameters being polar co-ordinates, Equation 286d gives us:

$$\dot{E}_{\dot{\rho},\dot{\theta}} = {}_{\mathrm{P}}E_{\rho,\theta} = 1, \quad \dot{F}_{\dot{\rho},\dot{\theta}} = {}_{\mathrm{P}}F_{\rho,\theta} = 0, \quad \dot{G}_{\dot{\rho},\dot{\theta}} = {}_{\mathrm{P}}G_{\rho,\theta} = \rho^2, \quad \dot{H}_{\dot{\rho},\dot{\theta}} = {}_{\mathrm{P}}H_{\rho,\theta} = \rho$$

so,

$$\dot{H}_{\lambda,\phi} = \dot{H}_{\dot{\rho},\dot{\theta}} \cdot {}^{\dot{\rho},\dot{\theta}}J_{\lambda,\phi} = \dot{\rho}(Rc) = R\,\dot{\rho}\,c$$

For equal-area this function must be identical with that for the sphere (Equation 287b), giving

$$R\,\dot{\rho}\,c = {}_{\mathrm{S}}H_{\rho,\theta} = R^2\cos\phi$$
$$\dot{\rho}\,c = R\cos\phi$$
$$c = R\,\dot{\rho}^{-1}\cos\phi$$

Hence,

$$\dot{\theta} = c\,\lambda = \frac{R\cos\phi}{\dot{\rho}}\lambda$$

essentially as we had adopted in that earlier development, as shown in Equation 161a.

Distortion values in rectilinear equal-area mappings

If the parameter axes are orthogonal, F is zero and H equals the product $\sqrt{EG}$. Because $\sqrt{E}$ and $\sqrt{G}$ then represent scale change along mutually orthogonal lines, having H unchanged is synonymous with these two fundamental quantities changing reciprocally.

The attribute of equal-area is tautologous with the measure of areal distortion equalling unity. For any rectilinear co-ordinates, Equation 296b thus gives

$$1 = {}_{\mathrm{A}}\dot{m} = \sqrt{\frac{\dot{E}_{\hat{\alpha}_1,\hat{\alpha}_2}\,\dot{G}_{\hat{\alpha}_1,\hat{\alpha}_2}}{E_{\hat{\alpha}_1,\hat{\alpha}_2}\,G_{\hat{\alpha}_1,\hat{\alpha}_2}}} = \sqrt{\frac{\dot{E}_{\hat{\alpha}_1,\hat{\alpha}_2}}{E_{\hat{\alpha}_1,\hat{\alpha}_2}}}\cdot\sqrt{\frac{\dot{G}_{\hat{\alpha}_1,\hat{\alpha}_2}}{G_{\hat{\alpha}_1,\hat{\alpha}_2}}} = {}_{\hat{\alpha}_1}\dot{m}\cdot{}_{\hat{\alpha}_2}\dot{m} \qquad \ldots\ldots (312a$$

So, knowing one linear distortion element on the right side of the equation gives the value of the other. Substituting one for the reciprocal of other in Equation 302a gives

$$f = \sin\left(\dot{\xi} - \xi\right) - \frac{\left({}_{\hat{\alpha}_1}\dot{m} - \dfrac{1}{{}_{\hat{\alpha}_1}\dot{m}}\right)}{\left({}_{\hat{\alpha}_1}\dot{m} + \dfrac{1}{{}_{\hat{\alpha}_1}\dot{m}}\right)}\sin\left(\dot{\xi} + \xi\right) = \sin\left(\dot{\xi} - \xi\right) - \frac{\left(\dfrac{1}{{}_{\hat{\alpha}_2}\dot{m}} - {}_{\hat{\alpha}_2}\dot{m}\right)}{\left(\dfrac{1}{{}_{\hat{\alpha}_2}\dot{m}} + {}_{\hat{\alpha}_2}\dot{m}\right)}\sin\left(\dot{\xi} + \xi\right) = 0$$

Multiplying through the fractions and putting longitude and latitude as parameters gives

$$f = \sin\left(\dot{\xi} - \xi\right) - \frac{\left({}_{\mathrm{m}}\dot{m}^2 - 1\right)}{\left({}_{\mathrm{m}}\dot{m}^2 + 1\right)}\sin\left(\dot{\xi} + \xi\right) \qquad = \sin\left(\dot{\xi} - \xi\right) - \frac{\left(1 - {}_{\mathrm{t}}\dot{m}^2\right)}{\left(1 + {}_{\mathrm{t}}\dot{m}^2\right)}\sin\left(\dot{\xi} + \xi\right) = 0$$

which can be solved for either parameter numerically by the Newton-Raphson technique (Tutorial 18) to give the angular distortion for any point.

All 'true' Cylindricals using Cartesian co-ordinates have one intermediate parameter a function purely of λ, i.e., relative to some chosen central longitude,

$$\dot{u} = R\lambda \qquad \text{hence} \qquad \mathrm{d}\dot{u} = R\,\mathrm{d}\lambda$$

because the other parameter is a function purely of ϕ, Equation 286a gives

$$(\mathrm{d}\dot{s})^2 = (\mathrm{d}\dot{u})^2 + (\mathrm{d}\dot{v})^2 = \left(\frac{\mathrm{d}\dot{u}}{\mathrm{d}\lambda}\mathrm{d}\lambda\right)^2 + \left(\frac{\mathrm{d}\dot{v}}{\mathrm{d}\phi}\mathrm{d}\phi\right)^2 = \dot{E}_{\lambda,\phi}\cdot\mathrm{d}\lambda^2 + \dot{G}_{\lambda,\phi}\cdot\mathrm{d}\phi^2$$

The second of Equations 298b then shows

$$_{\mathrm{t}}\dot{m} = \frac{R}{R\cos\phi} = \frac{1}{\cos\phi} \qquad \ldots\ldots (312b$$

For an equal-area 'true' Cylindrical, Equation 312b with Equation 312a gives

$$_{\mathrm{m}}\dot{m} = \frac{1}{_{\mathrm{t}}\dot{m}} = \cos\phi$$

the relevant tangential projection being that produced by orthographic projecting in the cylindrical mode.

All 'true' Azimuthals and Conicals, using polar co-ordinates, have one intermediate parameter that is a function purely of longitude. Specifically, for some constant c and relative longitude λ, the parameter is of the form $\dot{\theta} = c\,\lambda$. Hence, $\mathrm{d}\dot{\theta} = c\,\mathrm{d}\lambda$.

Since the other parameter is a function purely of ϕ, Equation 286c gives

$$\begin{aligned} \mathrm{d}\dot{s}^2 &= \dot{\rho}^2 \left(\mathrm{d}\dot{\theta}\right)^2 + \left(\mathrm{d}\dot{\rho}\right)^2 \\ &= \dot{\rho}^2 \left(\mathrm{c}\,\mathrm{d}\lambda\right)^2 + \left(\frac{\mathrm{d}\dot{\rho}}{\mathrm{d}\phi}\mathrm{d}\phi\right)^2 \\ &= \dot{E}_{\lambda,\phi} \cdot \mathrm{d}\lambda^2 + \dot{G}_{\lambda,\phi} \cdot \mathrm{d}\phi^2 \end{aligned}$$

The second of Equations 298b then shows

$$_{\mathrm{t}}\dot{m} = \frac{\sqrt{\dot{G}_{\lambda,\phi}}}{R\cos\phi} = \frac{c\,\dot{\rho}}{R\cos\phi} \qquad \text{. . . . . . (313a}$$

For an equal-area 'true' Azimuthal (with $c = 1$), Equation 313a with Equation 312a gives

$$_{\mathrm{m}}\dot{m} = \frac{1}{_{\mathrm{t}}\dot{m}} = \frac{R\cos\phi}{\dot{\rho}}$$

the relevant tangential projection being the (Lambert) Equal-Area Azimuthal.

For any 'true' Conical, c is less than 1. For a single standard parallel at latitude ϕ_0 the value of c is $\sin\phi_0$. Hence, from Equation 313a,

$$_{\mathrm{t}}\dot{m} = \frac{\dot{\rho}\sin\phi_0}{R\cos\phi}$$

The tangential equal-area Conical is the Albers projection with one standard parallel. Using the pertinent formula for $\dot{\rho}$ given in Equation 75b, we get

$$\begin{aligned} _{\mathrm{t}}\dot{m} &= \frac{\sin\phi_0}{R\cos\phi} \cdot \frac{R}{\sin\phi_0}\sqrt{(1 + \sin^2\phi_0 - 2\sin\phi_0\sin\phi)} \\ &= \frac{\sqrt{(1 + \sin^2\phi_0 - 2\sin\phi_0\sin\phi)}}{\cos\phi} \end{aligned}$$

Correspondingly, for this equal-area projection,

$$_{\mathrm{m}}\dot{m} = \frac{1}{_{\mathrm{t}}\dot{m}} = \frac{\cos\phi}{\sqrt{(1 + \sin^2\phi_0 - 2\sin\phi_0\sin\phi)}}$$

For the Albers projection with two standard parallels at ϕ_1 and ϕ_2, we have, from Equation 115a,

$$c = \frac{(\sin\phi_1 + \sin\phi_2)}{2}$$

Substituting the necessarily positive-valued $c\dot{\rho}$ from Equation 115b into Equation 313a gives:

$$
{}_{\mathrm{t}}\dot{m} = \frac{\sqrt{\cos^2\phi_1 + (\sin\phi_1 + \sin\phi_2)(\sin\phi_1 - \sin\phi)}}{\cos\phi}
$$

$$
= \frac{\sqrt{1 + \sin\phi_1 \sin\phi_2 - (\sin\phi_1 + \sin\phi_2)\sin\phi}}{\cos\phi}
$$

$$
{}_{\mathrm{m}}\dot{m} = \frac{1}{{}_{\mathrm{t}}\dot{m}} = \frac{\cos\phi}{\sqrt{\cos^2\phi_1 + (\sin\phi_1 + \sin\phi_2)(\sin\phi_1 - \sin\phi)}}
$$

$$
= \frac{\cos\phi}{\sqrt{1 + \sin\phi_1 \sin\phi_2 - (\sin\phi_1 + \sin\phi_2)\sin\phi}}
$$

Equal-area-maintaining transformations

Maintaining equal-area in a transformation is synonymous with quantity H being unchanged when expressed for both surfaces in the same parameters. If in transforming from equal-area map $\{\dot{x}, \dot{y}\}$ to map $\{\ddot{x}, \ddot{y}\}$ we use the axes of the source as the parameters, then, allowing that H must be positive, Equation 282b translates this to

$$
\frac{\partial\ddot{x}}{\partial\dot{x}}\frac{\partial\ddot{y}}{\partial\dot{y}} - \frac{\partial\ddot{y}}{\partial\dot{x}}\frac{\partial\ddot{x}}{\partial\dot{y}} = \frac{\partial\dot{x}}{\partial\dot{x}}\frac{\partial\dot{y}}{\partial\dot{y}} - \frac{\partial\dot{y}}{\partial\dot{x}}\frac{\partial\dot{x}}{\partial\dot{y}} = 1 - \frac{\partial\dot{y}}{\partial\dot{x}}\frac{\partial\dot{x}}{\partial\dot{y}}
$$

With rectilinear co-ordinates, this shows the necessary and sufficient condition for maintaining equal-area in the transformation is

$$
\frac{\partial\ddot{x}}{\partial\dot{x}}\frac{\partial\ddot{y}}{\partial\dot{y}} - \frac{\partial\ddot{x}}{\partial\dot{y}}\frac{\partial\ddot{y}}{\partial\dot{x}} = 1 \qquad \text{. . . . . (314a}
$$

for which there are innumerable solutions. Besides simple rotation as shown in Equations 86a and inverse rescaling along orthogonal axes as depicted in Equations 113a, equal-area is maintained by any polynomial transformation of the form:

$$
\begin{aligned}
\ddot{x} &= \dot{x}, \\
\ddot{y} &= \dot{y} + q_1\dot{x} + q_2\dot{x}^2 + q_3\dot{x}^3 \ldots + q_n\dot{x}^n
\end{aligned} \qquad \text{. . . . . (314b}
$$

else of the form:

$$
\begin{aligned}
\ddot{x} &= \dot{x} + q_1\dot{y} + q_2\dot{y}^2 + q_3\dot{y}^3 \ldots + q_n\dot{y}^n \\
\ddot{y} &= \dot{y}
\end{aligned} \qquad \text{. . . . . (314c}
$$

where n is a positive integer and the q_i terms are finite constants. With either formulation, of the four quotient terms in Equation 314a, the first two are both equal to 1 while one of the next pair is zero, leaving the condition equation met.

The result of either transformation is a progressive curving of the map, according to the pattern of the extra polynomial — as though the polynomial became an axis. That means with Equations 314b a curvaceous x axis, with Equations 314c a curvaceous y axis. This further algebraic technique for transforming an equal-area map without losing the attribute is exploited in Chapter 14.

THE CONFORMALITY CONDITION

The condition

For a transformation to maintain the conformality attribute, Equation 296c shows that each fundamental quantity in the numerator must be the same multiple of the corresponding one in the denominator, i.e.,

$$\text{for some constant } k > 0 \quad \dot{E}_{a_1,a_2} = k^2 E_{a_1,a_2}, \quad \dot{F}_{a_1,a_2} = k^2 F_{a_1,a_2} \text{ and } \dot{G}_{a_1,a_2} = k^2 G_{a_1,a_2}$$

The factor k is the scale at the point, which is uniform in all directions. We look at the condition in mapping from the globe, applying it as a test for one projection developed by other means in the early chapters, and then as a new construct for another.

If the parameter curves are orthogonal in the source, they must be orthogonal in a conformal map. We then have:

$$\dot{E}_{\hat{a}_1,\hat{a}_2} = k^2 E_{\hat{a}_1,\hat{a}_2}, \quad \dot{F}_{\hat{a}_1,\hat{a}_2} = k^2 F_{\hat{a}_1,\hat{a}_2} = 0, \quad \dot{G}_{\hat{a}_1,\hat{a}_2} = k^2 G_{\hat{a}_1,\hat{a}_2}$$

so:

$$\dot{F}_{\hat{a}_1,\hat{a}_2} = F_{\hat{a}_1,\hat{a}_2} = 0, \quad \frac{\dot{E}_{\hat{a}_1,\hat{a}_2}}{\dot{G}_{\hat{a}_1,\hat{a}_2}} = \frac{E_{\hat{a}_1,\hat{a}_2}}{G_{\hat{a}_1,\hat{a}_2}} = C$$

for another non-zero constant C, which we can think of as a conformality-related ratio. For conformality over a region, it must be identical for source and map at all points within the region (but not identical from point to point).

Mapping from the globe using longitude and latitude as parameters and applying our familiar distinguishing adornments to this new factor, we have:

$$\dot{F}_{\lambda,\phi} = F_{\lambda,\phi} = 0, \quad \frac{\dot{E}_{\lambda,\phi}}{\dot{G}_{\lambda,\phi}} = \frac{E_{\lambda,\phi}}{G_{\lambda,\phi}} = C_{\lambda,\phi}$$

and, from Equations 287b, making the requirement for conformality in mapping from the globe

$$\dot{C}_{\lambda,\phi} = \frac{\dot{E}_{\lambda,\phi}}{\dot{G}_{\lambda,\phi}} = \frac{{}_{\mathrm{S}}E_{\lambda,\phi}}{{}_{\mathrm{S}}G_{\lambda,\phi}} = \cos^2\phi \qquad \text{. . . . . (315a}$$

To relate the primary and intermediate parameters, Equation 291a translates this to

$$\cos^2\phi = \frac{\dot{E}_{\lambda,\phi}}{\dot{G}_{\lambda,\phi}} = \frac{\left(\dfrac{\partial \dot{a}_1}{\partial \lambda}\right)^2 \dot{E}_{\dot{a}_1,\dot{a}_2} + \left(\dfrac{\partial \dot{a}_1}{\partial \lambda}\dfrac{\partial \dot{a}_2}{\partial \lambda}\right)\dot{F}_{\dot{a}_1,\dot{a}_2} + \left(\dfrac{\partial \dot{a}_2}{\partial \lambda}\right)^2 \dot{G}_{\dot{a}_1,\dot{a}_2}}{\left(\dfrac{\partial \dot{a}_1}{\partial \phi}\right)^2 \dot{E}_{\dot{a}_1,\dot{a}_2} + \left(\dfrac{\partial \dot{a}_1}{\partial \phi}\dfrac{\partial \dot{a}_2}{\partial \phi}\right)\dot{F}_{\dot{a}_1,\dot{a}_2} + \left(\dfrac{\partial \dot{a}_2}{\partial \phi}\right)^2 \dot{G}_{\dot{a}_1,\dot{a}_2}}$$

If these more general intermediate parameters are orthogonal, this becomes (again using the angled overscript to indicate orthogonality)

$$\cos^2\phi = \frac{\dot{E}_{\lambda,\phi}}{\dot{G}_{\lambda,\phi}} = \frac{\left(\frac{\partial\hat{a}_1}{\partial\lambda}\right)^2 \dot{E}_{\hat{a}_1,\hat{a}_2} + \left(\frac{\partial\hat{a}_2}{\partial\lambda}\right)^2 \dot{G}_{\hat{a}_1,\hat{a}_2}}{\left(\frac{\partial\hat{a}_1}{\partial\phi}\right)^2 \dot{E}_{\hat{a}_1,\hat{a}_2} + \left(\frac{\partial\hat{a}_2}{\partial\phi}\right)^2 \dot{G}_{\hat{a}_1,\hat{a}_2}} \qquad \text{(316a}$$

The conformality condition applied as a test

To see this at work, we look at the Stereographic projection, which we know is conformal. Equations 46a & 31a give its intermediate parameters as:

$$\dot{\rho} = 2R\tan\left(\frac{\pi}{4} - \frac{\phi}{2}\right), \qquad \dot{\theta} = \oplus\left(\lambda - \frac{\pi}{4}\right)$$

Addressing the first term via its natural logarithm, we have

$$\ln\dot{\rho} = \ln\left\{2R\tan\left(\frac{\pi}{4} - \frac{\phi}{2}\right)\right\} = \ln 2R + \ln\left\{\tan\left(\frac{\pi}{4} - \frac{\phi}{2}\right)\right\} = \ln 2R - \ln\left\{\tan\left(\frac{\pi}{4} - \frac{\phi}{2}\right)\right\}^{-1}$$

$$= \ln 2R - \ln\left\{\tan\left(\frac{\pi}{4} + \frac{\phi}{2}\right)\right\}$$

Differentiating with respect to ϕ gives

$$\dot{\rho}^{-1}\frac{\partial\dot{\rho}}{\partial\phi} = -\cos^{-1}\phi \qquad \text{or} \qquad \frac{\partial\dot{\rho}}{\partial\phi} = -\dot{\rho}\cos^{-1}\phi$$

Hence:

$$\frac{\partial\dot{\rho}}{\partial\lambda} = 0 \qquad \frac{\partial\dot{\theta}}{\partial\lambda} = \oplus 1$$

$$\frac{\partial\dot{\rho}}{\partial\phi} = -\dot{\rho}\cos^{-1}\phi \qquad \frac{\partial\dot{\theta}}{\partial\phi} = 0$$

From Equation 286d:

$$\dot{E}_{\dot{\rho},\dot{\theta}} = {}_{\mathrm{P}}E_{\dot{\rho},\dot{\theta}} = 1, \quad \dot{F}_{\dot{\rho},\dot{\theta}} = {}_{\mathrm{P}}F_{\dot{\rho},\dot{\theta}} = 0, \quad \dot{G}_{\dot{\rho},\dot{\theta}} = {}_{\mathrm{P}}G_{\dot{\rho},\dot{\theta}} = \rho^2$$

Substituting our polar parameters for the generic ones in Equation 316a, we get

$$\dot{C}_{\lambda,\phi} = \frac{\left(\frac{\partial\dot{\rho}}{\partial\lambda}\right)^2 \dot{E}_{\dot{\rho},\dot{\theta}} + \left(\frac{\partial\dot{\theta}}{\partial\lambda}\right)^2 \dot{G}_{\dot{\rho},\dot{\theta}}}{\left(\frac{\partial\dot{\rho}}{\partial\phi}\right)^2 \dot{E}_{\dot{\rho},\dot{\theta}} + \left(\frac{\partial\dot{\theta}}{\partial\phi}\right)^2 \dot{G}_{\dot{\rho},\dot{\theta}}} = \frac{(0)^2 1 + (\oplus 1)^2 \dot{\rho}^2}{(-\dot{\rho}\cos^{-1}\phi)^2 1 + (0)^2 \dot{\rho}^2}$$

$$= \frac{\dot{\rho}^2}{(-\dot{\rho}\cos^{-1}\phi)^2} = \frac{\dot{\rho}^2}{\dot{\rho}^2\cos^{-2}\phi}$$

$$= \cos^2\phi$$

which meets the condition of Equation 315a.

The conformality condition applied as a construct

To demonstrate differential geometry as a construct for conformal projections, we again turn to the Conical mode, starting with the criterion of Equation 315a then following the rules for the mode, to reach the intermediate parameters for the Conformal Conical projection.

For any Conical at Simple aspect the intermediate parameters are each functions of one variable, specifically:

$$\dot{\theta} = c\lambda, \qquad \dot{\rho} \text{ is a function purely of } \phi$$

so:

$$\frac{\partial\dot{\rho}}{\partial\lambda} = \frac{d\dot{\rho}}{d\lambda} = 0 \qquad \frac{\partial\dot{\theta}}{\partial\lambda} = \frac{d\dot{\theta}}{d\lambda} = c$$

$$\frac{\partial\dot{\rho}}{\partial\phi} = \frac{d\dot{\rho}}{d\phi} \qquad \frac{\partial\dot{\theta}}{\partial\phi} = \frac{d\dot{\theta}}{d\phi} = 0$$

From Equation 286d, we have:

$$\dot{E}_{\dot{\rho},\dot{\theta}} =_{\mathrm{P}} E_{\dot{\rho},\dot{\theta}} = 1, \quad \dot{F}_{\dot{\rho},\dot{\theta}} = F_{\dot{\rho},\dot{\theta}} = 0, \quad \dot{G}_{\dot{\rho},\dot{\theta}} =_{\mathrm{P}} G_{\dot{\rho},\dot{\theta}} = \rho^2$$

Substituting our polar parameters for the generic ones in Equation 316a, we get

$$\dot{C}_{\lambda,\phi} = \frac{\left(\frac{\partial\dot{\rho}}{\partial\lambda}\right)^2 \dot{E}_{\dot{\rho},\dot{\theta}} + \left(\frac{\partial\dot{\theta}}{\partial\lambda}\right)^2 \dot{G}_{\dot{\rho},\dot{\theta}}}{\left(\frac{\partial\dot{\rho}}{\partial\phi}\right)^2 \dot{E}_{\dot{\rho},\dot{\theta}} + \left(\frac{\partial\dot{\theta}}{\partial\phi}\right)^2 \dot{G}_{\dot{\rho},\dot{\theta}}} = \frac{(0)^2 1 + (c)^2 \dot{\rho}^2}{\left(\frac{d\dot{\rho}}{d\phi}\right)^2 1 + (0)^2 \dot{\rho}^2}$$

$$= \frac{c^2\dot{\rho}^2}{\left(\frac{d\dot{\rho}}{d\phi}\right)^2}$$

For conformality, this must meet the condition of Equation 315a, giving

$$\frac{c^2\dot{\rho}^2}{\left(\frac{d\dot{\rho}}{d\phi}\right)^2} = \cos^2\phi \qquad \text{or} \qquad \left(\frac{d\dot{\rho}}{d\phi}\right)^2 = \frac{c^2\dot{\rho}^2}{\cos^2\phi}$$

$$\frac{d\dot{\rho}}{d\phi} = -c\,\dot{\rho}\cos^{-1}\phi$$

The minus sign is the appropriate choice, since ρ decreases as ϕ increases. Hence,

$$\dot{\rho}^{-1} d\dot{\rho} = -c\cos^{-1}\phi\, d\phi$$

which is identical to the formula obtained by earlier means in Equation 84a.

Distortion values in rectilinear conformal mappings

Conformality in a mapping requires at each point

$$_{\hat{\alpha}_1}\dot{m} = {}_{\hat{\alpha}_2}\dot{m} = \sqrt{\frac{\dot{E}}{E}} = \sqrt{\frac{\dot{G}}{G}}$$

whatever the parameters used, with distortion in all directions identical, including along the meridian and transversely thereto. Knowing one linear distortion element again gives the value of the other, while multiplying the two gives the measure of areal distortion. As noted earlier, conformality implies no distortion of angles.

For a conformal 'true' Cylindrical, Equation 312b gives

$$_{\mathrm{m}}\dot{m} = {}_{\mathrm{t}}\dot{m} = \frac{1}{\cos\phi}$$

the relevant projection being the Mercator.

For a conformal 'true' Azimuthal, Equation 316a gives, allowing that the constant c equals 1 for all Azimuthals,

$$_{\mathrm{m}}\dot{m} = {}_{\mathrm{t}}\dot{m} = \frac{\dot{\rho}}{R\cos\phi}$$

The relevant projection is the Stereographic. Using the pertinent formula for $\dot{\rho}$ given in Equation 85d, we get

$$\begin{aligned}
_{\mathrm{m}}\dot{m} = {}_{\mathrm{t}}\dot{m} &= \frac{2R}{R\cos\phi}\tan\left(\frac{\pi}{4} - \frac{\phi}{2}\right) \\
&= \frac{2}{\cos\phi}\tan\left(\frac{\pi}{4} - \frac{\phi}{2}\right) \\
&= \frac{2}{\sin\left(\frac{\pi}{2} - \phi\right)}\tan\left(\frac{\pi}{4} - \frac{\phi}{2}\right) \\
&= \frac{2}{2\sin\left(\frac{\pi}{4} - \frac{\phi}{2}\right)\cos\left(\frac{\pi}{4} - \frac{\phi}{2}\right)}\tan\left(\frac{\pi}{4} - \frac{\phi}{2}\right) \\
&= \frac{1}{\cos^2\left(\frac{\pi}{4} - \frac{\phi}{2}\right)}
\end{aligned}$$

For any 'true' Conical with a single standard parallel at latitude ϕ_0 the value of c is $\sin\phi_0$. Hence, from Equation 313a,

$$_{\mathrm{m}}\dot{m} = {}_{\mathrm{t}}\dot{m} = \frac{c\dot{\rho}}{R\cos\phi} = \frac{\dot{\rho}\sin\phi_0}{R\cos\phi}$$

The relevant projection is the (Lambert) Conformal Conical with one standard parallel. Using the pertinent formula for $\dot{\rho}$ given in Equation 85b, we get

$$ {}_{\mathrm{m}}\dot{m} = {}_{\mathrm{t}}\dot{m} = \frac{\sin\phi_0}{R\cos\phi} R\cot\phi_0 \left\{ \frac{\tan\left(\frac{\pi}{4} - \frac{\phi}{2}\right)}{\tan\left(\frac{\pi}{4} - \frac{\phi_0}{2}\right)} \right\}^{\sin\phi_0} = \frac{\cos\phi_0}{\cos\phi} \left\{ \frac{\tan\left(\frac{\pi}{4} - \frac{\phi}{2}\right)}{\tan\left(\frac{\pi}{4} - \frac{\phi_0}{2}\right)} \right\}^{\sin\phi_0} $$

$$ = \frac{\cos\phi_0}{\tan^{\sin\phi_0}\left(\frac{\pi}{4} - \frac{\phi_0}{2}\right)} \left\{ \frac{\tan\left(\frac{\pi}{4} - \frac{\phi}{2}\right)}{\cos\phi} \right\}^{\sin\phi_0} $$

For a 'true' conformal Conical with two standard parallels, at latitudes ϕ_1 and ϕ_2, the constant of the cone is (see Equations 120a & 120b)

$$ c = \frac{\ln(\cos\phi_1 \sec\phi_2)}{\ln\left\{ \tan\left(\frac{\pi}{4} - \frac{\phi_1}{2}\right) \tan\left(\frac{\pi}{4} + \frac{\phi_2}{2}\right) \right\}} = \frac{\ln(\cos\phi_2 \sec\phi_1)}{\ln\left\{ \tan\left(\frac{\pi}{4} - \frac{\phi_2}{2}\right) \tan\left(\frac{\pi}{4} + \frac{\phi_1}{2}\right) \right\}} $$

Using $\dot{\rho}$ from Equation 121d in Equation 313a gives

$$ {}_{\mathrm{m}}\dot{m} = {}_{\mathrm{t}}\dot{m} = \frac{\cos\phi_1}{\cos\phi} \left\{ \frac{\tan\left(\frac{\pi}{4} - \frac{\phi}{2}\right)}{\tan\left(\frac{\pi}{4} - \frac{\phi_1}{2}\right)} \right\}^{c} = \frac{\cos\phi_1}{\tan^{c}\left(\frac{\pi}{4} - \frac{\phi_1}{2}\right)} \left\{ \frac{\tan\left(\frac{\pi}{4} - \frac{\phi}{2}\right)}{\cos\phi} \right\}^{c} $$

and, identically,

$$ {}_{\mathrm{m}}\dot{m} = {}_{\mathrm{t}}\dot{m} = \frac{\cos\phi_2}{\cos\phi} \left\{ \frac{\tan\left(\frac{\pi}{4} - \frac{\phi}{2}\right)}{\tan\left(\frac{\pi}{4} - \frac{\phi_2}{2}\right)} \right\}^{c} = \frac{\cos\phi_2}{\tan^{c}\left(\frac{\pi}{4} - \frac{\phi_2}{2}\right)} \left\{ \frac{\tan\left(\frac{\pi}{4} - \frac{\phi}{2}\right)}{\cos\phi} \right\}^{c} $$

The relevant projection is the (Lambert) Conformal Conical with two standard parallels.

The pattern of distortion

Minimizing the sum of errors over a region inherently produces the best conformal maps. P. L. Chebyshev asserted [1856a] and D. A. Grave proved [1896a] that such conditions result if the bounds of the region have constant scale. The Stereographic, with its concentric circles of a given scale, is a prime example of a conformal projection meeting this test.

Achieving the map with the least overall distortion from nominal scale typically requires adjustment to that scale, as discussed in Chapter 5, so that the scale factor on the periphery is the inverse of that at the centre.

Conformality-maintaining transformations

If a transformation is conformal, then the difference between angles γ_1 and γ_2 of Equations 283f must be a right angle, hence

$$\frac{\partial x}{\partial a_1}\left(\frac{\partial y}{\partial a_1}\right)^{-1} = \tan\gamma_1 \quad = -\cot\gamma_2 = -\frac{\partial y}{\partial a_2}\left(\frac{\partial x}{\partial a_2}\right)^{-1}$$

So,

$$\frac{\partial x}{\partial a_1}\frac{\partial x}{\partial a_2} = -\frac{\partial y}{\partial a_2}\frac{\partial y}{\partial a_1}$$

or

$$\frac{\partial x}{\partial a_1}\frac{\partial x}{\partial a_2} + \frac{\partial y}{\partial a_2}\frac{\partial y}{\partial a_1} = 0$$

which is identical to stating that the fundamental quantity F is zero.

This condition is met if for a function f we have

$$f\frac{\partial x}{\partial a_1} = \frac{\partial y}{\partial a_2} \text{ and } \frac{\partial x}{\partial a_2} = -f\frac{\partial y}{\partial a_1} \qquad \text{. . . . . (320a}$$

Since the lengths of the two parametric vectors must be identical,

$$\left(\frac{\partial x}{\partial a_1}\right)^2 + \left(\frac{\partial y}{\partial a_1}\right)^2 = \left(\frac{\partial x}{\partial a_2}\right)^2 + \left(\frac{\partial y}{\partial a_2}\right)^2$$

Substituting from Equations 320a for elements on the right-hand side gives

$$\left(\frac{\partial x}{\partial a_1}\right)^2 + \left(\frac{\partial y}{\partial a_1}\right)^2 = \left(f\frac{\partial y}{\partial a_1}\right)^2 + \left(f\frac{\partial x}{\partial a_1}\right)^2$$

$$= f^2\left\{\left(\frac{\partial x}{\partial a_1}\right)^2 + \left(\frac{\partial y}{\partial a_1}\right)^2\right\}$$

implying that $f = 1$. Hence,

$$\frac{\partial x}{\partial a_1} = \frac{\partial y}{\partial a_2} \text{ and } \frac{\partial x}{\partial a_2} = -\frac{\partial y}{\partial a_1} \qquad \text{. . . . . (320b}$$

This pair of identities is called the Cauchy-Riemann† condition. Though initially named for its occurrence as a necessary condition for differentiability of functions of a complex variable (see Tutorial 28 on page 322), the condition is a defining — i.e., necessary and sufficient — condition for conformality in a transformation.

---

† Recognized by French mathematician A. L. Cauchy (1789 - 1857) then exploited in his general development of conformal transformations in a mathematical context by G. F. B. Riemann of Germany (1826 – 1866), but actually originated relative to fluid mechanics by J. le R. d'Alembert [1752a].

ENDNOTE A: DERIVING COEFFICIENTS FOR EQUATION 299d

The steps

Inverting the distortion ratio then applying Equations 287a, 283d, 282e & 281d, we get

$$\dot{m}^{-2} = \frac{\mathrm{d}s^2}{\mathrm{d}\dot{s}^2} = \frac{E\,\mathrm{d}\lambda^2 + G\,\mathrm{d}\phi^2}{\mathrm{d}\dot{s}^2}$$
$$= E\left(\frac{\mathrm{d}\lambda}{\mathrm{d}\dot{s}}\right)^2 + G\left(\frac{\mathrm{d}\phi}{\mathrm{d}\dot{s}}\right)^2$$
$$= E\frac{\dot{G}}{\dot{H}^2}\sin^2\beta_2 + G\frac{\dot{E}}{\dot{H}^2}\sin^2\beta_1$$
$$= \frac{\dot{E}\cdot\dot{G}}{\dot{H}^2}\left\{\frac{E}{\dot{E}}\sin^2\beta_2 + \frac{G}{\dot{G}}\sin^2\beta_1\right\}$$
$$= \csc^2\dot{\eta}\left\{\frac{E}{\dot{E}}\sin^2\beta_2 + \frac{G}{\dot{G}}\sin^2\beta_1\right\}$$

Allowing that, with longitude and latitude as parameters, bearing $\dot{\beta}_2$ is identically the azimuthal angle $\dot{\xi}$ of the map, applying basic trigonometric formulae and Equations 298b gives

$$\dot{m}^{-2} = \csc^2\dot{\eta}\left\{\frac{E}{\dot{E}}\sin^2\dot{\xi} + \frac{G}{\dot{G}}\sin^2\left(\dot{\eta} - \dot{\xi}\right)\right\}$$
$$= \csc^2\dot{\eta}\left\{\frac{1}{{}_{\mathrm{t}}\dot{m}^2}\sin^2\dot{\xi} + \frac{1}{{}_{\mathrm{m}}\dot{m}^2}\left(\sin\dot{\eta}\cos\dot{\xi} - \cos\dot{\eta}\sin\dot{\xi}\right)^2\right\}$$
$$= \frac{\csc^2\dot{\eta}}{{}_{\mathrm{t}}\dot{m}^2}\sin^2\dot{\xi} + \frac{\csc^2\dot{\eta}}{{}_{\mathrm{m}}\dot{m}^2}\left(\sin^2\dot{\eta}\cos^2\dot{\xi} - 2\sin\dot{\eta}\cos\dot{\eta}\sin\dot{\xi}\cos\dot{\xi} + \cos^2\dot{\eta}\sin^2\dot{\xi}\right)$$
$$= \left(\frac{\csc^2\dot{\eta}}{{}_{\mathrm{t}}\dot{m}^2} + \frac{\cot^2\dot{\eta}}{{}_{\mathrm{m}}\dot{m}^2}\right)\sin^2\dot{\xi} - \frac{\cot\dot{\eta}}{{}_{\mathrm{m}}\dot{m}^2}\sin 2\dot{\xi} + \frac{1}{{}_{\mathrm{m}}\dot{m}^2}\cos^2\dot{\xi}$$

Reverting to the fundamental quantities using Equations 281d, 298b & 281e, then translating E and G, being those of the sphere, via Equations 287b this becomes

$$\dot{m}^{-2} = \left(\frac{\dot{E}\cdot\dot{G}}{\dot{H}^2}\cdot\frac{E}{\dot{E}} + \frac{\dot{F}^2}{\dot{H}^2}\cdot\frac{G}{\dot{G}}\right)\sin^2\dot{\xi} - \frac{\dot{F}}{\dot{H}}\cdot\frac{G}{\dot{G}}\sin 2\dot{\xi} + \frac{G}{\dot{G}}\cos^2\dot{\xi}$$
$$= \frac{R^2\left(\dot{G}^2\cos^2\phi + \dot{F}^2\right)}{\dot{G}\cdot\dot{H}^2}\sin^2\dot{\xi} - \frac{R^2\dot{F}}{\dot{G}\cdot\dot{H}}\sin 2\dot{\xi} + \frac{R^2}{\dot{G}}\cos^2\dot{\xi}$$

giving us the coefficients in terms of those fundamental quantities.

TUTORIAL 28 FUNCTIONS OF A COMPLEX VARIABLE

As discussed in Tutorial 23, complex numbers have the form

$$z = x + iy = \rho(\cos\theta + i\sin\theta) = \rho \operatorname{cis}\theta$$

where $i = \sqrt{-1}$ and $\rho = |z| = +\sqrt{x^2 + y^2}$ with θ being the one of the two values of $\arctan\frac{y}{x}$ such that $\cos\theta = \frac{x}{\rho}$ and $\sin\theta = \frac{y}{\rho}$.

The bracketed term was cited in Tutorial 16, and shown there to be identically $e^{i\theta}$.

Because the trigonometric functions are themselves periodic, for any integer m the complex number can equally be written

$$z = \rho \operatorname{cis}(2m\pi + \theta) = \rho e^{i(2m\pi+\theta)}$$

With ρ being a simple scalar number, arithmetic operations on complex numbers in this form effectively apply separately to it and to the cis part. Hence, their complexities can be studied assuming $\rho = 1$ (i.e., omitting ρ).

Since

$$\begin{aligned}
&(\cos\theta_1 + i\sin\theta_1)(\cos\theta_2 + i\sin\theta_2) \\
&= \cos\theta_1\cos\theta_2 - \sin\theta_1\sin\theta_2 \\
&\quad + i(\cos\theta_1\sin\theta_2 + \sin\theta_1\cos\theta_2) \\
&= \cos(\theta_1 + \theta_2) + i\sin(\theta_1 + \theta_2)
\end{aligned}$$

it follows that

$$\prod_{j=1,..k} \operatorname{cis}\theta_j = \operatorname{cis} \sum_{j=1,..k} \theta_j$$

Hence,

$$(\operatorname{cis}\theta)^n = \operatorname{cis} n\theta$$

for any integer n and

$$\begin{aligned}
(\operatorname{cis}\theta)^{-1} &= \operatorname{cis}(-\theta) = \cos(-\theta) + i\sin(-\theta) \\
&= \cos\theta - i\sin\theta
\end{aligned}$$

Complex numbers can be used as variables in functions just like real numbers (but with various differences, some covered in Tutorial 23). The real and imaginary parts of any function are separable at any stage to give two independent ordinary functions.

Differentiation and integration apply to functions of a complex variable much as with real numbers, but with an important proviso. If $z = x + iy$ such that x and y are both functions of λ and ϕ, then, for function

$$f(z) = u(\lambda,\phi) + iv(\lambda,\phi)$$

to be differentiable at the point z, the ratio

$$\frac{f(z+\Delta z) - f(z)}{\Delta z}$$

must tend to one definite limit as $\Delta z \to 0$, regardless of approach. Let $\Delta z = \Delta x + i\Delta y$. If $\Delta y \to 0$ and $\Delta x \to 0$ separately, then:

$$\frac{u(x+\Delta x, y) - u(x,y)}{\Delta x} + i\frac{v(x+\Delta x, y) - v(x,y)}{\Delta x} = \frac{\partial u}{\partial x} + i\frac{\partial v}{\partial x}$$

$$\frac{u(x, y+\Delta y) - u(x,y)}{i\Delta y} + i\frac{v(x, y+\Delta y) - v(x,y)}{i\Delta y}$$

$$= \frac{\partial u}{i\partial y} + \frac{\partial v}{\partial y} = -i\frac{\partial u}{\partial y} + \frac{\partial v}{\partial y} = \frac{\partial v}{\partial y} - i\frac{\partial u}{\partial y}$$

Because these two values must be identical, separating their parts gives:

$$\frac{\partial u}{\partial x} = \frac{\partial v}{\partial y}, \quad \frac{\partial v}{\partial x} = -\frac{\partial u}{\partial y}$$

These are known as the *Cauchy-Riemann* equations. On page 320 it was shown that satisfying them is synonymous with conformality in any transformation.

CHAPTER 14

Seeking the Best: Optimizing distortion

Preamble

Going from tangential projections to secantal versions of the same in Chapter 5 was a first step in adjusting a projection to reduce the maximal distortion from nominal over the coverage of a map. Having now, in Chapter 13 acquired sophisticated measures of distortion, we proceed to explore the improvement of the overall scaling of more general projections using both algebraic and statistical techniques. This is of particular value when one of the incompatible attributes conformality and equal-area can be improved for a map that has the other. We look into the pattern of distortion, which leads more deeply into mathematics.

This chapter introduces and uses complex numbers and their functions.

Though their methodologies are widely applicable, the two projections illustrated in this chapter were developed for particular regions, and the maps shown are of those regions.

TUTORIAL 29 DE MOIVRE'S THEOREM

Rules for the addition, multiplication and division of complex numbers $z = x+iy$ were established in Tutorial 23 (page 220). These implied formulae for both positive and negative integral powers.

In polar form every such number is expressed as

$$z = \rho \operatorname{cis}\theta = \rho(\cos\theta + i\sin\theta) = \rho e^{i\theta}$$

where $\rho = |z|$ and θ is a unique angle in the *principal range* $0 \le \theta < 2\pi$. Because the trigonometric functions are themselves periodic, for any integer m the complex number can equally be written

$$z = \rho \operatorname{cis}(2m\pi + \theta) = \rho e^{i(2m\pi+\theta)}$$

Tutorial 28 showed that such numbers could be studied omitting ρ, and that

$$\prod_{j=1,..k} \operatorname{cis}\theta_j = \operatorname{cis} \sum_{j=1,..k} \theta_j$$

hence,

$$(\operatorname{cis}\theta)^n = \operatorname{cis} n\theta$$

for any integer n.

De Moivre's theorem asserts that this is also true for any rational n, but with the note that for fractional n the solution on the right-hand side is not unique. Indeed, if

$$n = \frac{p}{q} \quad \text{for integers } p, q, \ q > 0$$

there are q distinct solutions, namely the complex numbers

$$\operatorname{cis}\left(\frac{p}{q}\theta + 2m\pi\right) \quad m = 0, 1, \ldots q-1$$

Depending on p and q, these are not necessarily within the principal range, but all can be reduced to distinct values within that range. If p and q are relatively prime, the solutions can also be written as

$$\operatorname{cis}\left(\frac{p(\theta + 2m\pi)}{q}\right) \quad m = 0, 1, \ldots q-1$$

Using a formula of Tutorial 16 we get:

$$\begin{aligned} 2\cos n\theta &= e^{in\theta} + e^{-in\theta} \\ &= \operatorname{cis} n\theta + \operatorname{cis}(-n\theta) \\ &= \operatorname{cis} n\theta + (\operatorname{cis} n\theta)^{-1} \\ &= (\operatorname{cis}\theta)^n + (\operatorname{cis}\theta)^{-n} \\ 2i\sin n\theta &= i(e^{in\theta} - e^{-in\theta}) \\ &= \operatorname{cis} n\theta - \operatorname{cis}(-n\theta) \\ &= \operatorname{cis} n\theta - (\operatorname{cis} n\theta)^{-1} \\ &= (\operatorname{cis}\theta)^n - (\operatorname{cis}\theta)^{-n} \end{aligned}$$

Thus, de Moivre's theorem provides expression of all rational powers of $\sin\theta$ and $\cos\theta$ in terms of multiple angles, including the power-2 versions in Tutorial 10. Then:

$$\begin{aligned} (2\sin\theta)^3 &= 2\{3\sin\theta - \sin 3\theta\} \\ (2\cos\theta)^3 &= 2\{3\cos\theta + \cos 3\theta\} \\ (2\sin\theta)^4 &= 2\{3 - 4\cos 2\theta + \cos 4\theta\} \\ (2\cos\theta)^4 &= 2\{3 + 4\cos 2\theta + \cos 4\theta\} \end{aligned}$$

onwards, and the reverse, e.g.,

$$\begin{aligned} \sin 3\theta &= 3\sin\theta - 4\sin^2\theta \\ \cos 3\theta &= -3\cos\theta + 4\cos^3\theta \\ \sin 4\theta &= 4\sin\theta\cos\theta - 8\sin^3\theta\cos\theta \\ \cos 4\theta &= 1 - 8\cos^2\theta + 8\cos^4\theta \end{aligned}$$

ISOCOLS

Purpose

The addition of lines of uniformity of some factor is a common feature of maps, those for elevation — contours — being the most common. Generically called *isolines* or *isopleths* (the prefix being from Greek *equal*), they are typically attuned to the end user of the map. Of note in our context are *isocols* — lines of uniform distortion. This can be linear, areal, else angular distortion. Isocols contrast to the representation of the varying scale about one point provided, for instance, by the Tissot Indicatrix. Typically close to, if not precisely, a conic section of some form, isocols give clear indication of the pertinence of the stated scale when superimposed on any map.

Reshaping

The Chebyshev assertion cited on page 319 prompts the transformation of conformal projections to match the shape of the region being mapped. The use of functions of a complex variable, introduced next, allows unlimited opportunity to effect such transformation without loss of conformality. Such transformations can reshape the circular isolines of the Stereographic to a variety of other shapes. Achieving minimal scale error requires adjustment of the mean overall scale to have a value of 1, using standard least-squares fitting.

A later section describes one such transformation that renders the isolines oval, or oblate. The following section describes projections that render them quite irregular because of the region being covered. J. P. Snyder [1984c] addresses the general question of reshaping the isocols of the Stereographic into polygons, achieving minimum error for areas enclosed by polygons with modestly rounded corners.

The technique involves using the series expansion form of the complex functions, then adjusting the overall scale to have a mean relative value of 1.

Adopting the Equal-area Azimuthal projection as starting point, similar transformations and least-squares fitting can produce equal-area maps of minimal distortion over a given region. Snyder [1988a] provides methods for reshaping its isocols to oval, providing a series of oblate projections for improved equal-area maps of irregular regions. A more elaborate approach is presented in the final section of this chapter.

OPTIMIZING DISTORTION ALGEBRAICALLY

Beginnings

Tissot considered the optimizing of distortion in his major work on the basic subject [1881a]. Working only to third-order approximation, he deduced a projection very close to both the (Lambert) Conformal Conical and the (conformal) Stereographic that became the military map of World War I in France, with its so-called Lambert Grid [1921a].

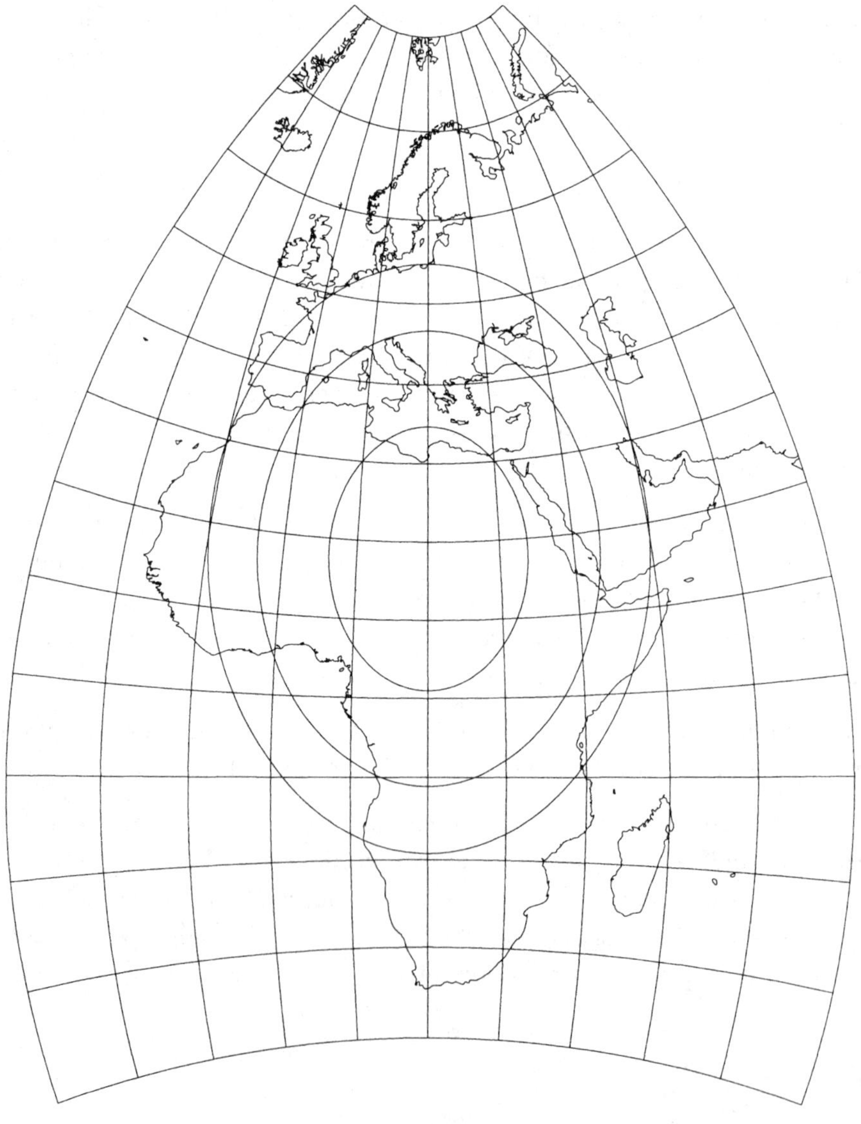

Exhibit 14–1: The Miller Oblated Stereographic projection.
With, reading from centre, isocols for linear scale of 1.06, 1.03 and 1.0 (i.e., true).

The Miller Oblated Stereographic projection

Commissioned to develop a projection covering Africa plus Europe that provided limited distortion of areas while being preferably conformal, O. M. Miller [1953d] used a reshaped Oblique Stereographic, achieving full conformality with a linear scale variation of barely ±8% at the extreme. Noting that the target region is roughly oval, centred about location 20°E, 18°N, Miller established two pairs of symmetric points as outer reference points, then effected a transformation that balanced the scale along the ellipse through them inversely with the scale at the centre. Points on the meridian 20°E near Cape Town and Hammerfest defined the north-south span. The other two points were on a line orthogonal to that meridian, sufficiently wide to bring Madagascar mostly within the ellipse.

If $(\ddot{x}, \ddot{y})$ are rectangular co-ordinates for a second, compounded projection, and $(\dot{x}, \dot{y})$ are rectangular co-ordinates for an existing conformal projection, then Ludovic Driencourt and Jean Laborde [1932c] showed that the transformation:

$$\begin{aligned} \ddot{x} &= q\dot{x}\left\{1 - \frac{k}{12}\left(3\dot{y}^2 - \dot{x}^2\right)\right\} \\ \ddot{y} &= q\dot{y}\left\{1 + \frac{k}{12}\left(3\dot{x}^2 - \dot{y}^2\right)\right\} \end{aligned} \qquad \ldots\ldots (327a$$

maintains conformality for any constants q and k. This is demonstrated by:

$$\begin{aligned} \frac{\partial \ddot{x}}{\partial \dot{x}} &= q\left\{1 - \frac{k}{12}\left(3\dot{y}^2 - \dot{x}^2\right)\right\} + q\dot{x}\left\{\frac{k}{12}(2\dot{x})\right\} \\ &= q\left\{1 - \frac{k}{12}\left(3\dot{y}^2 - 3\dot{x}^2\right)\right\} \\ &= q\left\{1 + \frac{k}{12}\left(3\dot{x}^2 - \dot{y}^2\right)\right\} + q\dot{y}^2\left\{\frac{k}{12}(-2\dot{y})\right\} = \frac{\partial \ddot{y}}{\partial \dot{y}} \\ \frac{\partial \ddot{x}}{\partial \dot{y}} &= q\dot{x}\left\{-\frac{k}{12}(6\dot{y})\right\} = -q\dot{x}\left\{\frac{k\dot{x}\dot{y}}{2}\right\} = -q\dot{y}\left\{+\frac{k}{12}(6\dot{x})\right\} = -\frac{\partial \ddot{y}}{\partial \dot{x}} \end{aligned}$$

showing the transformation meets the Cauchy-Riemann conditions.

If γ is the polar distance of a point relative to the contact point of the Oblique Stereographic, and ζ is its azimuth, then the plotting equations are:

$$\begin{aligned} \dot{x} &= 2\tan\frac{\gamma}{2}\sin\zeta \\ \dot{y} &= 2\tan\frac{\gamma}{2}\cos\zeta \end{aligned} \qquad \ldots\ldots (327b$$

The linear scale distortion in a conformal transformation is

$$\dot{m} = \sqrt{\left(\frac{\partial \dot{x}}{\partial \gamma}\right)^2 + \left(\frac{\partial \dot{y}}{\partial \gamma}\right)^2}$$

Substitution from Equations 327a & 327b, then differentiation, gives

$$\dot{m} = q\left\{\sec^2\frac{\gamma}{2}\right\}\sqrt{1 - 2k\tan^2\frac{\gamma}{2}\cos 2\zeta + k^2\tan^4\frac{\gamma}{2}}$$

whence:

if $\zeta = \pm\frac{\pi}{2}$ as it is for the east and west reference points

$$\dot{m} = q\left\{\sec^2\frac{\gamma}{2}\right\}\sqrt{1 + 2k\tan^2\frac{\gamma}{2} + k^2\tan^4\frac{\gamma}{2}} \quad = q\left\{\sec^2\frac{\gamma}{2}\right\}\left\{1 + k\tan^2\frac{\gamma}{2}\right\}$$

if $\zeta = \frac{\pi}{2} \pm \frac{\pi}{2}$ as it is for the north and south reference points

$$\dot{m} = q\left\{\sec^2\frac{\gamma}{2}\right\}\sqrt{1 - 2k\tan^2\frac{\gamma}{2} + k^2\tan^4\frac{\gamma}{2}} \quad = q\left\{\sec^2\frac{\gamma}{2}\right\}\left\{1 - k\tan^2\frac{\gamma}{2}\right\}$$

If γ_n is the polar distance of the north and south reference points, and γ_e is the polar distance of the east and west points, then demanding equal distortion at all four implies

$$\sec^2\frac{\gamma_n}{2}\left\{1 - k\tan^2\frac{\gamma_n}{2}\right\} = \sec^2\frac{\gamma_e}{2}\left\{1 + k\tan^2\frac{\gamma_e}{2}\right\}$$

giving

$$k = \frac{\sec^2\frac{\gamma_n}{2} - \sec^2\frac{\gamma_e}{2}}{\tan^2\frac{\gamma_n}{2}\sec^2\frac{\gamma_n}{2} + k\tan^2\frac{\gamma_e}{2}\sec^2\frac{\gamma_e}{2}}$$

For such a value of k, the transformation of Equations 327a produces a revised map with uniform scale along the reference ellipse. To minimize maximal distortion within this ellipse requires making the relative scale on the ellipse the reciprocal of that at the centre, which is intrinsically 1. Thus, the multiplier constant q is set to be the reciprocal of the square root of the value initially obtained for the reference points.

Miller's resulting values were $k=0.2522$, $q=0.9245$; his particular projection is illustrated on page 326, with isocols added. The technique could obviously be applied to almost any region — Lee did so effectively for the whole Pacific ocean including its rim [1974d]. Miller's projection has been called both the **Miller Prolated Stereographic** projection and the **Miller Oblated Stereographic** projection, and is also referred to without the creator's name.

Snyder [1988a] applied a similar method to the Equal-area Azimuthal projection to produce a family he called **Oblated Equal-area** projections, with oval isocols of (angular else maximum linear) distortion. The axes of the ovals can be tilted as required, likewise their relative lengths, to provide improved equal-area maps of appropriately shaped regions. Snyder illustrates with the Atlantic ocean, bringing the maximum angular distortion over virtually all of its travelled waters within ±5%. The shaping of the isocols depends on two arbitrary constants, the choice of which can give isocols ranging from circular to rounded rectangles.

USING COMPLEX NUMBERS AND FUNCTIONS

Basics

At this point of our cartographic development we need to employ complex numbers (see Tutorial 23). These creations of mathematics incorporate the fantasy entity $\mathrm{i} = \sqrt{-1}$ and are of the form $\mathbf{z} = x + \mathrm{i}y$ for real numbers x and y. Like 2-dimensional vectors, they allow handling of two independent variables as one item.

Functions of complex variables (see Tutorial 28) can thus represent cartographic projections: first, of a mapping from longitude and latitude into intermediate parameters else final Cartesian co-ordinates; second, of any such mapping into a revised map (and continuing beyond two, when appropriate). As shown in Tutorial 28, any such transformation satisfies the Cauchy-Riemann equations if the function is differentiable (which is essential for our purposes). But on page 320 those conditions, applied to any mapping transformation, were shown as synonymous with the transformation maintaining conformality. Hence any differentiable function of a complex variable used to transform a conformal source produces a conformal map. Besides applying to the mapping from the globe as source, the conformality-maintaining feature is of clear value for re-mapping from the output of such conformal projections as the Stereographic and Mercator. Though they do not routinely maintain any other attribute, and do not produce conformality absent from the source, functions of complex variables have utility for transforming any map.

Like the vectors, complex numbers can be represented in a polar form. Thus using longitude and latitude, we can have

$$\boldsymbol{\mu} = \lambda + \mathrm{i}\phi = \rho(\cos\theta + \mathrm{i}\sin\theta)$$

or, using the special notation introduced in Tutorial 23,

$$\boldsymbol{\mu} = \lambda + \mathrm{i}\phi = \rho\,\mathrm{cis}\,\theta$$

After an initial mapping transformation $\dot{g}$, we get

$$\dot{\boldsymbol{z}} = \dot{x} + \mathrm{i}\dot{y} = \dot{\rho}\,\mathrm{cis}\,\dot{\theta} = \dot{g}(\boldsymbol{\mu})$$

A secondary mapping transformation $\ddot{g}$ then gives

$$\ddot{\boldsymbol{z}} = \ddot{x} + \mathrm{i}\ddot{y} = \ddot{\rho}\,\mathrm{cis}\,\ddot{\theta} = \ddot{g}(\dot{\boldsymbol{z}}) = \ddot{g}(\dot{g}(\boldsymbol{\mu}))$$

with further mappings feasible without limit.

As illustrated, we depict complex numbers routinely in italic bold.

OPTIMIZING DISTORTION STATISTICALLY

Essentials

The Lowry projection in Chapter 3 represented efforts to reduce overall distortion for a perspective Azimuthal by simple averaging — specifically the distance factor for perspective projecting was set by taking the geometric mean of 18 different values. Secantal projecting (Chapter 5) allowed a choice that reduced maximum distortion relative to nominal scale, but only by balancing the two extremes. More sophisticated techniques have been used, particularly with modern computing facilities, resulting in effectively distinct projections that give reduced, if not minimal, error over a pre-set region.

The method of squaring discrepancies to avoid negative signs then adding those squared values over a set of points to provide an overall measure of discrepancy is a common statistical technique. Developed by C. F. Gauss and A. M. Legendre in the early 1800s, the least-squares principle sees the minimum value of this overall measure as a *best* situation. The principle has been used widely in cartography to achieve modified projections that have lower overall distortion. Originally used for projections without regard for the conformality and equal-area, it has been applied more recently, with computer assistance, to modify projections while retaining either such attribute.

G. B. Airy of England was among the pioneers. The **Airy** projection [1861a] is an algebraic azimuthal of the style discussed in Chapter 4, but with the spacing of parallels optimized by the least-squares principle. Little different in appearance from the Equidistant Azimuthal, the optimization algorithm incorporated the distortion along both meridians and parallels (when at Simple aspect) at the selected points. Airy compared his projection (which he referred to as "by Balance of Errors") with that basic projection and with other rivals, finding his beneficial not by an overall reduction in error so much as an avoidance of gross error over the chosen region. A. R. Clarke and Henry James (author of a rival projection) rather surprisingly published [1862b] arithmetical corrections to Airy's calculations that elevated its merits. They also produced a more elegant development of Airy's projection, which we reproduce. That development was effected on a unit globe. We do likewise, then change the result to our globe of radius R by simple multiplication.

Let δ be the (angular) great-circle distance from the centre P_0 of the chosen region to a general point P_i and let β be its bounding value for the desired map. Let the distance between map points $\dot{P}_0$ and $\dot{P}_i$ be $\dot{r}$. Then, if t is the radius of a small circle about the general point, the mapping of that circle will be an ellipse (effectively the Tissot Indicatrix) with semi-axial lengths:

$$t\frac{\mathrm{d}\dot{r}}{\mathrm{d}\delta},\quad \dot{r}\frac{t}{\sin\delta}$$

with the former along the radial line from $\dot{P}_0$ through $\dot{P}_i$ (i.e., along the meridian if at Simple aspect) and the latter transversely thereto. The differences between these quantities and the radius t of the original circle are thus:

$$t\frac{\mathrm{d}\dot{r}}{\mathrm{d}\delta} - t = t\left(\frac{\mathrm{d}\dot{r}}{\mathrm{d}\delta} - 1\right),$$

$$\dot{r}\frac{t}{\sin\delta} - t = t\left(\frac{\dot{r}}{\sin\delta} - 1\right)$$

The strategy for the projection is to minimize the sum over the map of the sum of the squares of these two differences at each point. Allowing for the extending applicability as we radiate out from the centre, and interpreting the least-squares method continuously as an integral rather than discretely to selected points, this means minimizing

$$\int_0^\beta \left\{\left(\frac{\mathrm{d}\dot{r}}{\mathrm{d}\delta} - 1\right)^2 + \left(\frac{\dot{r}}{\sin\delta} - 1\right)^2\right\}\sin\delta\,\mathrm{d}\delta$$

If we put $\dot{r} - \delta = \dot{y}$, then $\dfrac{\mathrm{d}\dot{y}}{\mathrm{d}\delta} = \dfrac{\mathrm{d}\dot{r}}{\mathrm{d}\delta} - \dfrac{\mathrm{d}\delta}{\mathrm{d}\delta} = \dfrac{\mathrm{d}\dot{r}}{\mathrm{d}\delta} - 1$, and the integral becomes

$$\int_0^\beta \left\{\left(\frac{\mathrm{d}\dot{y}}{\mathrm{d}\delta}\right)^2 \sin\delta + \frac{(\dot{y} + \delta - \sin\delta)^2}{\sin\delta}\right\}\mathrm{d}\delta$$

If we then put

$$\dot{L} = 2\frac{\mathrm{d}\dot{y}}{\mathrm{d}\delta}\sin\delta \text{ so } \frac{\mathrm{d}\dot{L}}{\mathrm{d}\delta} = 2\left\{\frac{\mathrm{d}^2\dot{y}}{\mathrm{d}\delta^2}\sin\delta + \frac{\mathrm{d}\dot{y}}{\mathrm{d}\delta}\cos\delta\right\}$$

$$\dot{N} = \frac{2(\dot{y} + \delta - \sin\delta)}{\sin\delta}$$

our requirement becomes

$$\dot{N} - \frac{\mathrm{d}\dot{L}}{\mathrm{d}\delta} = 0 \text{ together with } \dot{L} = 0 \text{ when } \delta = \beta$$

This translates to

$$(\dot{y} + \delta - \sin\delta) - \sin^2\delta\frac{\mathrm{d}^2\dot{y}}{\mathrm{d}\delta^2} - \sin\delta\cos\delta\frac{\mathrm{d}\dot{y}}{\mathrm{d}\delta} = 0$$

or

$$\sin^2\delta\frac{\mathrm{d}^2\dot{y}}{\mathrm{d}\delta^2} + \sin\delta\cos\delta\frac{\mathrm{d}\dot{y}}{\mathrm{d}\delta} - \dot{y} = \delta - \sin\delta \text{ with } \left[\frac{\mathrm{d}\dot{y}}{\mathrm{d}\delta}\right]_P = 0$$

Integration of this differential equation can be effected by the three simplifying steps:

substitution $\delta - \sin\delta = w$

multiplication by $\csc^2\dfrac{\delta}{2}$

multiplication by $\sec^2\dfrac{\delta}{2}$

The result, after back-substituting for *w*, is

$$\dot{y} = -\delta - 2\cot\frac{\delta}{2}\ \ln\cos\frac{\delta}{2} + C\tan\frac{\delta}{2} + D\cot\frac{\delta}{2}$$

$$0 = \csc^2\frac{\beta}{2}\ \ln\cos\frac{\beta}{2} + \frac{1}{2}C\sec^2\frac{\beta}{2} - \frac{1}{2}D\csc^2\frac{\beta}{2}$$

Because $\dot{y}$ must vanish with δ, by the former of these two equations D must be zero, hence from the latter

$$C = -2\sec^{-2}\frac{\beta}{2}\csc^2\frac{\beta}{2}\ \ln\cos\frac{\beta}{2} \quad = 2\left(\cot^2\frac{\beta}{2}\right)\left(-\ln\cos\frac{\beta}{2}\right) \quad = 2\cot^2\frac{\beta}{2}\ln\sec\frac{\beta}{2}$$

$$= \left(\cot^2\frac{\beta}{2}\right)\left(2\ln\sec\frac{\beta}{2}\right) \quad = \cot^2\frac{\beta}{2}\ \ln\sec^2\frac{\beta}{2}$$

This completely defines $\dot{y}$ hence $\dot{r}$, as

$$\dot{r} = y + \delta = -2\cot\frac{\delta}{2}\ \ln\cos\frac{\delta}{2} + 2\cot^2\frac{\beta}{2}\ \ln\sec\frac{\beta}{2}\ \tan\frac{\delta}{2}$$

$$= 2\left\{\cot\frac{\delta}{2}\ \ln\cos^{-1}\frac{\delta}{2} + \cot^2\frac{\beta}{2}\ \ln\sec\frac{\beta}{2}\ \tan\frac{\delta}{2}\right\}$$

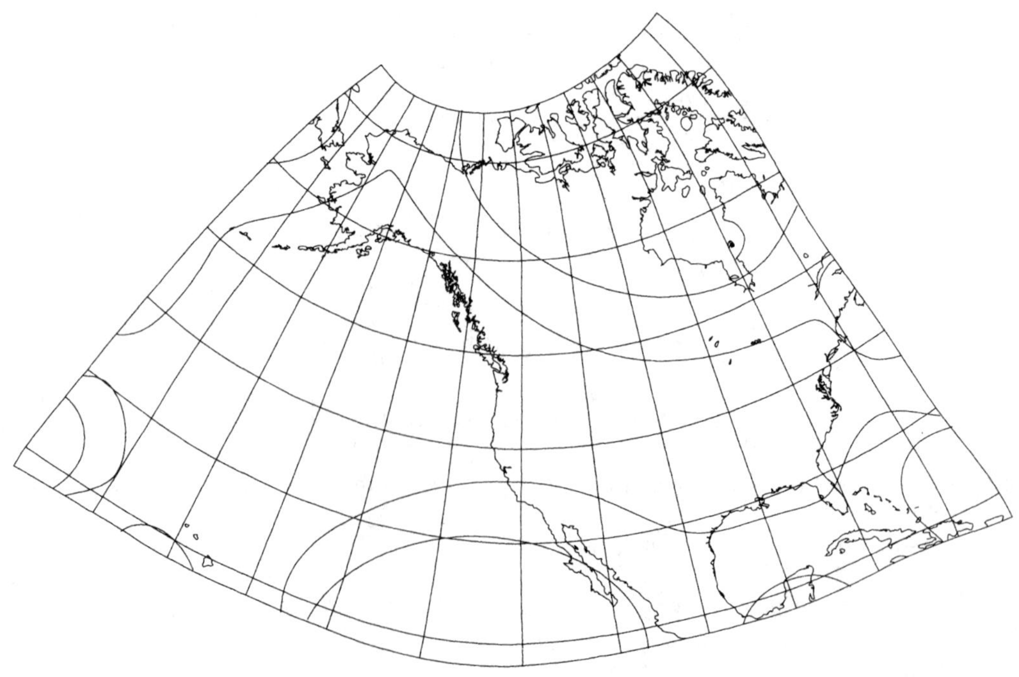

Exhibit 14–2: The GS50 projection.
With isocol for true scale surrounded, in several sections, by that for 1.03 of linear scale.

Turning from the unit globe to our globe of radius R, we get for the radial parameter in Airy's projection

$$\dot{\rho} = 2R\left\{\cot\frac{\delta}{2}\ \ln\sec\frac{\delta}{2} + \cot^2\frac{\beta}{2}\ \ln\sec\frac{\beta}{2}\tan\frac{\delta}{2}\right\} \qquad \text{(333a}$$

At the centre of the map, where $\rho = 0$, we get

$$\frac{\mathrm{d}\dot{r}}{\mathrm{d}\delta} = \frac{1+C}{2}$$

With the unit globe, the radial distance from the centre to the edge of the map is

$$2C\tan\frac{\beta}{2} = 2\cot^2\frac{\delta}{2}\ \ln\sec^2\frac{\beta}{2}\tan\frac{\beta}{2}$$

Interestingly, this function does not increase indefinitely with β but reaches a maximum at β=2.206~ (i.e., at 126.4~°). The projection was thus usable for maps that reached the equivalent of Pole to further Tropic. The projection was characteristically used obliquely (and with ellipsoidal precision), including by the Ordnance Survey as their standard projection for some years.

Clarke, who like Airy created Earth-fitting ellipsoids of note,† in that same publication with James applied Airy's approach to the generalized Perspective Azimuthal.

Maintaining conformality

As shown in Tutorial 28, any transformation using complex variables is conformal hence, if applied to an initial conformal projection, produces a conformal map. Such a process has been applied to various basic conformal projections then subjected to least-squares fitting to gain conformal maps of selected regions with reduced distortion.

W. I. Reilly of New Zealand [1973d] modified an Oblique Mercator map of his country using a sixth-order complex polynomial with optimization over 228 points — the product subsequently being adopted for a new series of national topographic maps. The process alters the shape of meridians and parallels, making them all complicated curves (hence masking the disturbance to appearance inherent with the Oblique aspect). But, as it retains conformality, those cardinal lines intersect orthogonally. With its development being simultaneously aimed at providing a new national grid, the projection is called somewhat misleadingly the **New Zealand Map Grid** (or **NZMG**) [1974f].

† The Clarke 1866 ellipsoid is one piece of English cartography that became entrenched in U.S. practice. It was the standard for American geodesy for a hundred years.

Snyder used the Stereographic at Oblique aspect for his **GS50** projection [1984b], named for the Geological Survey and the 50 states forming his country. The granting of full statehood to Alaska and Hawaii had brought a need for a map of the U.S.A. embracing those outlying regions equally with the rest. Although, like Miller with his oblated projection in using the Stereographic, Snyder minimized distortion for the GS80 using least-squares fitting. After progressive trial and revision, 44 reference points were used. Employing a tenth-order complex polynomial, the resulting linear distortion relative to the spherical globe is a maximum of ±3% for land areas of the country. The map, overlain with isocols, is shown on page 332.

If, using functions of complex variables,

$$\dot{z} = \dot{x} + \mathrm{i}\dot{y} = \dot{\rho}\left(\cos\dot{\theta} + \mathrm{i}\sin\dot{\theta}\right) \qquad \ldots\ldots (334a$$

is a primary mapping of the globe and

$$\ddot{z} = \ddot{x} + \mathrm{i}\ddot{y} = \ddot{\rho}\left(\cos\ddot{\theta} + \mathrm{i}\sin\ddot{\theta}\right)$$

is a secondary mapping transformation thereof, then, since both are conformal, the resulting distortions of linear scale relative to the globe are uniform in all directions at any point. Taking distortion along the meridian as representative of each, we get

$$\dot{m} = {}_{\mathrm{m}}\dot{m} = \sqrt{\left(\frac{\partial \dot{x}}{\partial \phi}\right)^2 + \left(\frac{\partial \dot{y}}{\partial \phi}\right)^2} \quad \text{and} \quad \ddot{m} = {}_{\mathrm{m}}\ddot{m} = \sqrt{\left(\frac{\partial \ddot{x}}{\partial \phi}\right)^2 + \left(\frac{\partial \ddot{y}}{\partial \phi}\right)^2} \qquad \ldots\ldots (334b$$

A polynomial transformation is of the form

$$\ddot{z} = \ddot{x} + \mathrm{i}\ddot{y} = \sum_{j=1}^{n} \boldsymbol{a}_j \cdot \dot{z}^j \quad \text{for complex coefficients } \boldsymbol{a}_j = a_j + \mathrm{i}b_j \quad j = 1,2,\ldots n$$

By de Moivre's theorem (Tutorial 29) we have

$$\begin{aligned} \dot{\mathbf{z}}^j = (\ddot{x} + \mathrm{i}\ddot{y})^j &= \left(\rho\left(\cos\dot{\theta} + \mathrm{i}\sin\dot{\theta}\right)\right)^j \\ &= \dot{\rho}^j\left(\cos\dot{\theta} + \mathrm{i}\sin\dot{\theta}\right)^j \\ &= \dot{\rho}^j\left(\cos j\dot{\theta} + \mathrm{i}\sin j\dot{\theta}\right) \end{aligned}$$

Thus, writing the coefficients in their parts, we get

$$\ddot{z} = \ddot{x} + \mathrm{i}\ddot{y} = \sum_{j=1}^{n} (a_j + \mathrm{i}b_j)\dot{\rho}^j\left(\cos j\dot{\theta} + \mathrm{i}\sin j\dot{\theta}\right) \qquad \ldots\ldots (334c$$

By differentiation of this equation and back substitution using Equation 334a both directly and itself differentiated, it can be shown that

$$\ddot{m} = \dot{m}\sqrt{L_1^{\,2} + L_2^{\,2}} \qquad \ldots\ldots (334d$$

where $\dot{m}$ is the distortion resulting from the initial projection and

$$L_1 = \sum_{j=1}^{n} j\,\dot{\rho}^{j-1}\left\{ a_j \cos\left(\{j-1\}\dot{\theta}\right) - b_j \sin\left(\{j-1\}\dot{\theta}\right)\right\}$$

$$L_2 = \sum_{j=1}^{n} j\,\dot{\rho}^{j-1}\left\{ a_j \sin\left(\{j-1\}\dot{\theta}\right) + b_j \cos\left(\{j-1\}\dot{\theta}\right)\right\} \qquad \text{. . . . . . (335a}$$

The requisite mathematics is included as Endnote A to this chapter.

Applying the least-squares method of fit to k points on the basis of scale distortion at each requires finding a minimum for the function

$$\ddot{M} = \sum_{t=1}^{k} \left(\ddot{m}_t - 1\right)^2$$

where the suffix attached to the distortion term relates it to a particular point.

Differentiating this with respect to representative coefficients a_r and b_r gives

$$\frac{\partial \ddot{M}}{\partial a_r} = \sum_{t=1}^{k} 2\left(\ddot{m}_t - 1\right)\frac{\partial \ddot{m}_t}{\partial a_r} \qquad \text{and} \qquad \frac{\partial \ddot{M}}{\partial b_r} = \sum_{t=1}^{k} 2\left(\ddot{m}_t - 1\right)\frac{\partial \ddot{m}_t}{\partial b_r}$$

both of which must be zero for minimal values of $\ddot{M}$.

With two independent coefficients to choose per algebraic degree, there are $2n$ in total. Fixing $b_1 = 0$, which merely leaves rotation to other b coefficients, changes this to $(2n\text{-}1)$, and the same number of equations. Newton-Raphson approximating techniques (Tutorial 18) must be used to evaluate the coefficients progressively, starting from initial trial values — all zero except for $a_1 = 1$ were adopted.

Using Δ to represent the increments at each cycle of approximation, the $(2n\text{-}1)$ equations are of the form:

$$\sum_{s=1}^{n} \Delta a_s \frac{\partial^2 \ddot{M}}{\partial a_s\,\partial a_r} + \sum_{s=2}^{n} \Delta b_s \frac{\partial^2 \ddot{M}}{\partial b_s\,\partial a_r} = -\frac{\partial \ddot{M}}{\partial a_r} \qquad \text{for } r = 1,\ 2,\ 3,\ ..\ n$$

$$\sum_{s=1}^{n} \Delta a_s \frac{\partial^2 \ddot{M}}{\partial a_s\,\partial b_r} + \sum_{s=2}^{n} \Delta b_s \frac{\partial^2 \ddot{M}}{\partial b_s\,\partial b_r} = -\frac{\partial \ddot{M}}{\partial b_r} \qquad \text{for } r = 2,\ 3,\ 4,\ ..\ n$$

Iteration is continued until these increments are trivial.

The second derivatives can be formulated by differentiating various of the above equations with appropriate substitution. Endnote B to this chapter gives details. The resulting forms for the second derivatives are:

$$\frac{\partial^2 \ddot{M}}{\partial a_s \partial a_r} = \frac{\partial^2 \ddot{M}}{\partial a_r \partial a_s} = 2\sum_{t=1}^{k}\left\{\frac{1}{\ddot{m}_t}\frac{\partial \ddot{m}_t}{\partial a_s}\frac{\partial \ddot{m}_t}{\partial a_r} + S_t \cos\left((q-t)\dot{\theta}\right)\right\}$$

$$\frac{\partial^2 M''}{\partial a_s \partial b_r} = \frac{\partial^2 \ddot{M}}{\partial b_s \partial a_r} = 2\sum_{t=1}^{k}\left\{\frac{1}{\ddot{m}_t}\frac{\partial \ddot{m}_t}{\partial a_s}\frac{\partial \ddot{m}_t}{\partial b_r} + S_t \sin\left((q-t)\dot{\theta}\right)\right\}$$

$$\frac{\partial^2 \ddot{M}}{\partial b_s \partial b_r} = \frac{\partial^2 \ddot{M}}{\partial b_r \partial b_s} = 2\sum_{t=1}^{k}\left\{\frac{1}{\ddot{m}_t}\frac{\partial \ddot{m}_t}{\partial b_s}\frac{\partial \ddot{m}_t}{\partial b_r} + S_t \cos\left((q-t)\dot{\theta}\right)\right\}$$

where $S_t = \left(1 - \frac{1}{\ddot{m}_t}\right)\dot{m}^2 r s \dot{\rho}^{r+s-2}$.

Reference should be made to the original paper [1984b] for detailed results.

Additional usages of functions of a complex variable to create conformal maps with less distortion are discussed in Chapter 15.

Maintaining equal-area

John Dyer and Snyder [1989d] developed an iterative method of using the equal-area-maintaining transformations of Chapter 13 to obtain minimum-error maps of arbitrary regions. Aiming to make the Tissot indicatrices as close to circular as possible, their method was to reduce the sum of the squares of the major and minor axes of each for a chosen number n of selected points $P_i = (\lambda_i, \phi_i)$. The iterative method is fully defined by any one iteration. To simplify notation, we choose the first, applied to the source map, denoted as the set $\{\dot{x}, \dot{y}\}$, to produce the set $\{\ddot{x}, \ddot{y}\}$.

A 3-step approach is used within each iteration (here with hash marks to distinguish the modified variables after each step). The first step rotates the map by angle χ (see Equations 86a) to give:

$$\dot{x}' = \dot{x}\cos\chi - \dot{y}\sin\chi$$
$$\dot{y}' = \dot{x}\sin\chi + y\cos\chi$$

The second step is a polynomial-based transformation of the form shown in Equations 314c, i.e., with polynomial

$$Q(\dot{x}') = q_1(\dot{x}') + q_2(\dot{x}')^2 + q_3(\dot{x}')^3 + \ldots q_n(\dot{x}')^n \quad \text{. . . . . . (336a}$$

and the transformation:

$$\dot{x}'' = \dot{x}'$$
$$\dot{y}'' = \dot{y}' + Q(\dot{x}')$$

The third step is compression/expansion using positive number σ (see Equations 113a):

$$\ddot{x} = \dot{x}''' = \sigma \dot{x}''$$
$$\ddot{y} = \dot{y}''' = \sigma^{-1} \dot{y}''$$

At each iteration, factors $\chi, q_1, q_2, q_3, ..q_n$ and σ are chosen to minimize the least-squares sum.

We now look at the impact of such progressive transformation on a general indicatrix (centred on a transformationally relocated point). If the initial mapping is:

$$\dot{x} = u(\phi, \lambda), \quad \dot{y} = v(\phi, \lambda)$$

then, using a strikethrough cross-bar to distinguish the co-ordinates of the indicatrix, its initial parametric equations are:

$$\dot{\not{x}} = \dot{x} + b_x \cos\theta + b_y \sin\theta$$
$$\dot{\not{y}} = \dot{y} + c_x \cos\theta + c_y \sin\theta$$

where $b_x = \dfrac{\partial u}{\partial \lambda} \sec\phi, \quad b_y = \dfrac{\partial u}{\partial \phi}, \quad c_x = \dfrac{\partial v}{\partial \lambda} \sec\phi, \quad c_y = \dfrac{\partial v}{\partial \phi}.$

The first step, being purely rotational, has no impact on the dimensionality of the indicatrix but changes its equations to:

$$\dot{\not{x}}' = \dot{x}' + b'_x \cos\theta + b'_y \sin\theta$$
$$\dot{\not{y}}' = \dot{y}' + c'_x \cos\theta + c'_y \sin\theta$$

where $b'_x = b_x \cos\chi - c_x \sin\chi, \quad b'_y = b_y \cos\chi - c_y \sin\chi$

$c'_x = b_x \sin\chi + c_x \cos\chi, \quad c'_y = b_y \sin\chi + c_y \cos\chi.$

The polynomial step changes the equations to:

$$\dot{\not{x}}'' = \dot{x}'' + b''_x \cos\theta + b''_y \sin\theta$$
$$\dot{\not{y}}'' = \dot{y}'' + c''_x \cos\theta + c''_y \sin\theta$$

where $b''_x = b'_x, \quad b''_y = b'_y, \quad c''_x = c'_x + b'_x Q^{(1)}(\dot{x}'), \quad c''_y = c'_y + b'_y Q^{(1)}(\dot{x}')$, with Equation 336a giving the derivative as

$$Q^{(1)}(\dot{x}') = q_1 + 2q_2(\dot{x}') + 3q_3(\dot{x}')2 + \ldots nq_n(\dot{x}')^{n-1}$$

The compression/expansion step completes the iteration to give indicatrix equations:

$$\dot{\not{x}}''' = \dot{x}''' + b'''_x \cos\theta + b'''_y \sin\theta$$
$$\dot{\not{y}}''' = \dot{y}''' + c'''_x \cos\theta + c'''_y \sin\theta \qquad \ldots\ldots (337a$$

where $b'''_x = \sigma b''_x, \quad b'''_y = \sigma b''_y, \quad c'''_x = \sigma^{-1} c''_x, \quad c'''_y = \sigma^{-1} c''_y$ ...... (337b

Were the indicatrix to have its major axis parallel with the x axis, its parametric equations would be of the form:

$$\dot{\not{x}}''' = \dot{x}''' + a\cos\alpha$$
$$\dot{\not{y}}''' = \dot{y}''' + a^{-1}\cos\alpha$$

for some angle α. If they were at angle β to the x axis, being effectively rotated by that angle from the above, the equations would be:

$$\dot{\not{x}}''' = \dot{x}''' + a\cos\beta\cos\alpha - a^{-1}\sin\beta\sin\alpha$$

$$\dot{\not{y}}''' = \dot{y}''' + a\sin\beta\cos\alpha + a^{-1}\cos\beta\sin\alpha$$

For equal-area, the length of the minor axis always equals the reciprocal of that of the major axis. Taking a to represent the latter and using A to represent the sum of the squares of the major and minor axes, we get

$$\begin{aligned} A = a^2 + a^{-2} &= a^2\left(\sin^2\beta + \cos^2\beta\right) + a^{-2}\left(\sin^2\beta + \cos^2\beta\right) \\ &= a^2\cos^2\beta + a^{-2}\sin^2\beta + a^2\sin^2\beta + a^{-2}\cos^2\beta \\ &= \left(+a\cos\beta\right)^2 + \left(-a^{-1}\sin\beta\right)^2 + \left(+a\sin\beta\right)^2 + \left(+a^{-1}\cos\beta\right)^2 \end{aligned}$$

i.e., the sum of the squares of the parametric coefficients in the preceding equations. Equating with Equations 337a gives

$$A = \left(b_x'''\right)^2 + \left(b_y'''\right)^2 + \left(c_x'''\right)^2 + \left(c_y'''\right)^2$$

Unless the ellipse is a circle, which implies conformality at the point (which is inherently exceptional with equal-area projections), $A>1$. Substitution from Equations 337b gives

$$A = \left(\sigma b_x''\right)^2 + \left(\sigma b_y''\right)^2 + \left(\sigma^{-1} c_x''\right)^2 + \left(\sigma^{-1} c_y''\right)^2$$

This function of σ is summed over the reference points, then the minimum obtained by differentiation and equating that to zero. The minimum occurs with

$$\sigma^4 = \frac{\sum\left\{\left(c_x''\right)^2 + \left(c_y''\right)^2\right\}}{\sum\left\{\left(b_x''\right)^2 + \left(b_x''\right)^2\right\}}$$

where summation is carried across the set of reference points. The minimum value is

$$\begin{aligned} \left\{\sum A\right\}^2 &= 4\sum\left\{\left(b_x''\right)^2 + \left(b_x''\right)^2\right\}\sum\left\{\left(c_x''\right)^2 + \left(c_y''\right)^2\right\} \\ &= 4\sum\left\{\left(b_x''\right)^2 + \left(b_x''\right)^2\right\}\sum\left\{\left(c_x' + b_x' G^{(1)}\left(\dot{x}'\right)_y\right)^2 + \left(c_x' + b_y' G^{(1)}\left(\dot{x}'\right)\right)^2\right\} \end{aligned}$$

The standard least-squares method is then applied to obtain the minimum for this expression.

These authors applied this method to the entire state of Alaska, inclusive of adjacent waters, using 177 reference points at 2° intervals. Making Q of fourth order (hence having six constants for each pass through the process) and six iterations, they modified an Equal-area Azimuthal centred at 152°W, 64°N to provide an equal-area map with linear distortion less than ±0.5% over almost all their region. Using a like-centred Stereographic, they produced a conformal map with relative scale ranging barely more than ±0.3%.

MISCELLANEOUS MANIPULATIONS†

Azimuthals and Pseudoazimuthals

Innumerable efforts have been made to adjust the radial distance parameter ρ as a means for ameliorating error in azimuthal maps. G. A. Ginsburg, after recognizing the generic formulae for Azimuthals shown at the close of Chapter 3, created two projections that vary the formula of the Equal-area Azimuthal projection to moderate the angular distortion (while necessarily forfeiting the equal-area attribute), the key parametric equations for the **Ginsburg I** projection are

$$\dot{\rho} = R\left(2\sin\frac{\gamma}{2} + 0.00066\gamma^9\right) \qquad \text{. . . . . (339a}$$

where γ is the polar distance For the **Ginsburg II** projection, they are

$$\dot{\rho} = R\left(2\sin\frac{\gamma}{2} + 0.00025\gamma^{10}\right) \qquad \text{. . . . . (339b}$$

Both were used for hemispheric-sized regions, the latter having less areal distortion with such coverage. N. A. Urmayev also proffered two projections labelled I and II that used a similar technique but employed a multiplier factor and integration.

The proportionality of the doubly compounded Stereographic projection of M. D. Solov'vev shown also at the close of Chapter 3, with the equation

$$\dot{\rho} = 4R\tan\frac{\gamma}{4} \quad = 4R\tan\left(\frac{\pi}{8} - \frac{\oplus\phi}{4}\right) \qquad \text{. . . . . (339c}$$

is identical to that proposed earlier by A. E. Young [1920a] and then advocated by F. A. A. Breusing. It is usually called, with the added qualifier, the **Breusing Harmonic** projection. In this latter form, designed for minimum error, the globe radius R is multiplied by a constant depending on the span of the map — much as with the Airy.

The **Ginsburg III** or **TsNIIGAiK with Oval Isolines** projection (those initials representing the government agency involved) is actually a set of Pseudoazimuthals in which at Simple aspect the meridians meet at their true angles but only those of one bimeridian are straight. The other meridians are curved, mirror image about the straight bimeridian; hence the result is close to being a Pseudoconical. While the expression for radial distance is relatively simple, the angle parameter is quite complicated, but each incorporates an arbitrary constant.

† This section is essentially of 20th-Century Russian work, and the remarks are derived from D. H. Maling's extensive review of Russian map projections [1960a]. Reference his work for further details and identification of the original sources.

Cylindricals

The plotting equations for any 'true' cylindrical are of the form:

$$\dot{x} = \sigma\lambda, \quad \dot{y} = f(\phi)$$

where σ is the relative scale along the Equator and f is a function of latitude. Urmayev developed a technique for establishing f to meet preset criteria of distortion at points on three parallels, using polynomial functions. The relative linear distortion along the parallel for latitude ϕ is fixed at $\cos^{-1}\phi$. That along the meridian — hence the distortion of area — can be chosen; alternatively the maximal angular distortion can be chosen. If $n > 1$ is an integer and

$$y = f = a_0 \frac{\phi^1}{1} + a_2 \frac{\phi^3}{3} \ldots . + a_{2n} \frac{\phi^{2n+1}}{2n+1}$$

then the linear distortion along the meridian is

$$\frac{\mathrm{d}y}{\mathrm{d}\phi} = \frac{\mathrm{d}f}{\mathrm{d}\phi} = a_0 + a_2\phi^2 \ldots . + a_{2n}\phi^{2n}$$

Given preset values for this function for three latitudes we have three equations — sufficient to solve for three coefficients (i.e., all coefficients with $n=2$). The **Urmayev III** projection, used for a world map, has only two non-zero coefficients. The plotting equations are:

$$\dot{x} = \lambda$$

$$\dot{y} = 0.016\,199\,\Phi + 0.000\,001\,974\,7\,\Phi^3 \qquad \ldots\ldots\ldots (340a$$

the degree form for latitude being retained along with the constants as expressed by Urmayev. The projection is close to conformal.

Conicals

The opportunity to improve distortion in conical maps having two standard parallels was touched upon with simple rules-of-thumb in Chapter 5. That was specifically for the equidistant form, but similar opportunity occurs with the other forms.

The Russians V. V. Kavraiskiy and V. V. Vitkovskiy have each developed formulae for equidistant, equal-area and conformal forms, while many other authors have addressed the first of these (which is seen as an acceptable compromise between the other two forms for many purposes). Although the products of such development are usually named as distinct projections, most are essentially just the underlying familiar projection for the form. The distinctions are the choice of standard parallels for a given expanse. For convenience, we assume Simple aspect, though the results are applicable correspondingly at any aspect. Unlike with the developments in Chapter 3 through Chapter 5, the standard parallels are now dependent on the formulae for the linear parameter $\dot{\rho}$, to be determined as the latitudes of true scale.

Snyder [1978b] provides a survey of a dozen algorithms for choosing within the equidistant form, beginning with the 18th-Century creations of Patrick Murdoch of Scotland and proceeding to those of the eminent Swiss mathematician Leonhard Euler. The choice is typically dependent on the amount of latitudinal span, which we denote as 2δ.

In most versions the aim is to make linear distortion along the two furthermost parallels identical (*peripheral distortion*) and balance that with the distortion along an inner parallel. That inner parallel can be either the arithmetic mean of the extremes, which we denote by ϕ_m, else that of maximum distortion, which we denote by ϕ_x and determine from the equation

$$\phi_x + \cot\phi_x = \delta\cot\delta\cot\phi_m + \phi_m$$

where ψ is the difference between the extreme latitudes — the latitudinal span of the map. The balancing can be either of the actual scales else of the scale error (the offset from 1 of relative scales), and can be purely arithmetic (i.e., making the peripheral value the negative of the inner value) else the multiplicative reciprocal.

Murdoch developed two forms of the equidistant that provided for equal-area over the latitudinal span of the map between any two meridians. The intermediate parameters for the **Murdoch I** projection are:

$$\dot{\rho} = R\left(\frac{\sin\delta}{\delta}\cot\phi_m + \phi_m - \phi\right) \qquad \dot{\theta} = c\lambda \quad \text{with} \quad c = \sin\phi_m \qquad \text{. . . . . (341a}$$

and those for the **Murdoch III** (or **Everett**) projection are:

$$\dot{\rho} = R\left(\delta\cot\delta\cot\phi_m + \phi_m - \phi\right) \qquad \dot{\theta} = c\lambda \quad \text{with} \quad c = \frac{\sin\delta}{\delta}\frac{\tan\delta}{\delta}\sin\phi_m \qquad \text{. . . . . (341b}$$

Noting that any equidistant conical formula can be written as

$$\dot{\rho} = \dot{\rho}_E - R\phi$$

where the new term represents the value of this parameter at the Equator, the standard parallels for these and all other versions of the equidistant form are the solutions of the geenral eqaution

$$R\cos\phi = c\left(\dot{\rho}_E - R\phi\right) \quad \text{or} \quad R\left(c\phi + \cos\phi\right) = c\dot{\rho}_E$$

The **Euler** projection balances the peripheral scale arithmetically with that for ϕ_m, giving equations:

$$\dot{\rho} = R\left(\frac{\delta}{2}\cot\frac{\delta}{2}\cot\phi_m + \phi_m - \phi\right)$$

$$\dot{\theta} = c\lambda \quad \text{with} \quad c = \frac{\sin\delta}{\delta}\sin\phi_m \qquad \text{. . . . . (342a}$$

Both C. F. Close and A. R. Clarke [1911c] then Vitkovskiy developed versions that balance the peripheral scale error arithmetically with that at latitude ϕ_x, producing equations different from those of Murdoch III only regarding the constant of the cone. This becomes for the **Vitkovskiy I** projection,

$$c = \frac{2}{\delta}\tan\frac{\delta}{2}\cot\phi_m \qquad \text{. . . . . (342b}$$

and, for what can conveniently be called the **Ordnance Survey** projection (because of its use for a time by that British mapping agency),

$$c = 2\left\{\delta\frac{\cot\delta\cot\phi_m + 1}{\cos(\phi_m - \delta)} + \frac{\delta\cot\delta\cot\phi_m + \phi_m - \phi_x}{\cos\delta}\right\}^{-1} \qquad \text{. . . . . (342c}$$

The **Kavraiskiy II** projection balances the peripheral scale multiplicatively with that at latitude ϕ_x, with the equations again differing from those of Murdoch III only regarding the constant of the cone, which is now

$$c = \sqrt{\frac{\sin\delta}{\delta}\tan\phi_x\sin\phi_m} \qquad \text{. . . . . (342d}$$

All of these projections maintain the equidistant attribute. Various authors have introduced versions that slightly impair the attribute while seeking to improve distortion in some way.

The two Russian cartographers developed versions of the equal-area (**Vitkovskiy II** and **Kavraiskiy I** projections) and conformal (**Vitkovskiy III** and **Kavraiskiy III** projections) forms of conical projecting. Reference should be made to Maling [1960a] for details. The **Kavraiskiy IV** projection is another addressing the equidistant form, but it is based on the least-squares method of fitting.

The Polyconic

Ginsburg, Urmayev, and others have similarly modified the basic Polyconic projection to reduce the exaggeration of scale along the outer meridians.

ENDNOTE A: FROM EQUATIONS 334b TO EQUATION 334d

Basics

It was asserted earlier in the chapter that if the polynomial transformation

$$\ddot{z} = \ddot{x} + \mathrm{i}\ddot{y} = \sum_{j=1}^{n} \left(a_j + \mathrm{i}b_j\right)\dot{\rho}^j\left(\cos j\dot{\theta} + \mathrm{i}\sin j\dot{\theta}\right)$$

shown in Equation 334c is applied as a secondary transformation after a primary transformation that produced rectilinear co-ordinates $(\dot{x}, \dot{y})$, then the linear distortion of the compound transformation, expressed relative to the globe, can be related to that of the primary transformation (see Equation 334b) by

$$\ddot{m} = \dot{m}\sqrt{L_1{}^2 + L_2{}^2}$$

where L_1 and L_2 are defined by Equations 335a.

Multiplying within Equation 334c we get

$$\ddot{\mathbf{z}} = \ddot{x} + \mathrm{i}\ddot{y} = \sum_{j=1}^{n} \dot{\rho}^j\left\{\left(a_j \cos j\dot{\theta} - b_j \sin j\dot{\theta}\right) + \mathrm{i}\left(a_j \cos j\dot{\theta} + b_j \sin j\dot{\theta}\right)\right\}$$

hence, by separating real and imaginary parts:

$$\ddot{x} = \sum_{j=1}^{n} \dot{\rho}^j\left(a_j \cos j\dot{\theta} - b_j \sin j\dot{\theta}\right) \qquad \text{...... (343a}$$

$$\ddot{y} = \sum_{j=1}^{n} \dot{\rho}^j\left(a_j \cos j\dot{\theta} + b_j \sin j\dot{\theta}\right) \qquad \text{...... (343b}$$

Differentiating Equation 343a gives

$$\frac{\partial \ddot{x}}{\partial \phi} = \sum_{j=1}^{n} \left\{ j\dot{\rho}^{j-1}\left(a_j \cos j\dot{\theta} - b_j \sin j\dot{\theta}\right)\frac{\partial \dot{\rho}}{\partial \phi} + \dot{\rho}^j\left(-ja_j \sin j\dot{\theta} - jb_j \cos j\dot{\theta}\right)\frac{\partial \dot{\theta}}{\partial \phi} \right\} \qquad \text{... (343c}$$

To remove the differentials of the intermediate polar parameters from this equation, we turn to Equation 334a and differentiate, thus:

$$\dot{\rho}^2 = \left|\dot{\mathbf{z}}\right|^2 = \dot{x}^2 + \dot{y}^2 \quad \text{hence } \dot{\rho}\,\mathrm{d}\dot{\rho} = \dot{x}\,\mathrm{d}\dot{x} + \dot{y}\,\mathrm{d}\dot{y} \quad = \dot{\rho}\cos\dot{\theta}\,\mathrm{d}\dot{x} + \dot{\rho}\sin\dot{\theta}\,\mathrm{d}\dot{y}$$

$$\dot{y} = \dot{x}\tan\dot{\theta} \quad \text{hence } \mathrm{d}\dot{y} = \tan\dot{\theta}\,\mathrm{d}\dot{x} + \dot{x}\sec^2\dot{\theta}\,\mathrm{d}\dot{\theta} \quad = \tan\dot{\theta}\,\mathrm{d}\dot{x} + \dot{\rho}\sec\dot{\theta}\,\mathrm{d}\dot{\theta}$$

so:

$$\mathrm{d}\dot{\rho} = \cos\dot{\theta}\,\mathrm{d}\dot{x} + \sin\dot{\theta}\,\mathrm{d}\dot{y}$$

$$\dot{\rho}\,\mathrm{d}\dot{\theta} = -\sin\dot{\theta}\,\mathrm{d}\dot{x} + \cos\dot{\theta}\,\mathrm{d}\dot{y}$$

Since the new co-ordinates are rectangular these reduce to:

$$\frac{d\dot\rho}{d\dot x} = \cos\dot\theta \qquad \frac{d\dot\rho}{d\dot y} = \sin\dot\theta$$
$$\frac{d\dot\theta}{d\dot x} = \frac{-\sin\dot\theta}{\dot\rho} \qquad \frac{d\dot\theta}{d\dot y} = \frac{\cos\dot\theta}{\dot\rho}$$

allowing the target derivatives in Equation 343c to be expressed as:

$$\frac{\partial\dot\rho}{\partial\phi} = \frac{\partial\dot\rho}{\partial\dot x}\frac{\partial\dot x}{\partial\phi} + \frac{\partial\dot\rho}{\partial\dot y}\frac{\partial\dot y}{\partial\phi} \quad = \cos\dot\theta\frac{\partial\dot x}{\partial\phi} + \sin\dot\theta\frac{\partial\dot y}{\partial\phi}$$
$$\frac{\partial\dot\theta}{\partial\phi} = \frac{\partial\dot\theta}{\partial\dot x}\frac{\partial\dot x}{\partial\phi} + \frac{\partial\dot\theta}{\partial\dot y}\frac{\partial\dot y}{\partial\phi} \quad = \frac{-\sin\dot\theta}{\dot\rho}\frac{\partial\dot x}{\partial\phi} + \frac{\cos\dot\theta}{\dot\rho}\frac{\partial\dot y}{\partial\phi}$$

Substitution in Equation 343c produces

$$\frac{\partial\ddot x}{\partial\phi} = \sum_{j=1}^{n} j\,\dot\rho^{j-1}\left\{\left(a_j\cos j\dot\theta - b_j\sin j\dot\theta\right)\cos\dot\theta\frac{\partial\dot x}{\partial\phi} - \dot\rho\left(a_j\sin j\dot\theta + b_j\cos j\dot\theta\right)\frac{-\sin\dot\theta}{\dot\rho}\frac{\partial\dot x}{\partial\phi}\right\}$$
$$+\sum_{j=1}^{n} j\,\dot\rho^{j-1}\left\{\left(a_j\cos j\dot\theta - b_j\sin j\dot\theta\right)\sin\dot\theta\frac{\partial\dot y}{\partial\phi} - \dot\rho\left(a_j\sin j\dot\theta + b_j\cos j\dot\theta\right)\frac{\cos\dot\theta}{\rho}\frac{\partial\dot y}{\partial\phi}\right\}$$
$$= \sum_{j=1}^{n} j\,\dot\rho^{j-1}\left\{\left(a_j\cos j\dot\theta - b_j\sin j\dot\theta\right)\cos\dot\theta + \left(a_j\sin j\dot\theta + b_j\cos j\dot\theta\right)\sin\dot\theta\right\}\frac{\partial\dot x}{\partial\phi}$$
$$+\sum_{j=1}^{n} j\,\dot\rho^{j-1}\left\{\left(a_j\cos j\dot\theta - b_j\sin j\dot\theta\right)\sin\dot\theta - \left(a_j\sin j\dot\theta + b_j\cos j\dot\theta\right)\cos\dot\theta\right\}\frac{\partial\dot y}{\partial\phi}$$

Rearranging and using angle-difference formulae from Tutorial 10 this simplifies to

$$\frac{\partial\ddot x}{\partial\phi} = \sum_{j=1}^{n} j\,\dot\rho^{j-1}\left\{a_j\left(\cos j\dot\theta\cos\dot\theta + \sin j\dot\theta\sin\dot\theta\right) - b_j\left(\sin j\dot\theta\cos\dot\theta - \cos j\dot\theta\sin\dot\theta\right)\right\}\frac{\partial\dot x}{\partial\phi}$$
$$+\sum_{j=1}^{n} j\,\dot\rho^{j-1}\left\{a_j\left(\cos j\dot\theta\sin\dot\theta - \sin j\dot\theta\cos\dot\theta\right) - b_j\left(\sin j\dot\theta\sin\dot\theta + \cos j\dot\theta\cos\dot\theta\right)\right\}\frac{\partial\dot y}{\partial\phi}$$
$$= \sum_{j=1}^{n} j\,\dot\rho^{j-1}\left\{a_j\cos\left(\{j-1\}\dot\theta\right) - b_j\sin\left(\{j-1\}\dot\theta\right)\right\}\frac{\partial\dot x}{\partial\phi}$$
$$-\sum_{j=1}^{n} j\,\dot\rho^{j-1}\left\{a_j\sin\left(\{j-1\}\dot\theta\right) + b_j\cos\left(\{j-1\}\dot\theta\right)\right\}\frac{\partial\dot y}{\partial\phi}$$
$$= L_1\frac{\partial\dot x}{\partial\phi} - L_2\frac{\partial\dot y}{\partial\phi}$$

$$\text{where: } L_1 = \sum_{j=1}^{n} j\,\dot\rho^{j-1}\left\{a_j\cos\left(\{j-1\}\dot\theta\right) - b_j\sin\left(\{j-1\}\dot\theta\right)\right\}$$
$$L_2 = \sum_{j=1}^{n} j\,\dot\rho^{j-1}\left\{a_j\sin\left(\{j-1\}\dot\theta\right) + b_j\cos\left(\{j-1\}\dot\theta\right)\right\}$$

i.e., identically Equations 335a.

The same process from Equation 343b gives

$$\frac{\partial \dot{y}}{\partial \phi} = \sum_{j=1}^{n} j\,\dot{\rho}^{j-1}\left\{a_j \sin\left(\{j-1\}\dot{\theta}\right) + b_j \cos\left(\{j-1\}\dot{\theta}\right)\right\}\frac{\partial \dot{x}}{\partial \phi}$$

$$+\sum_{j=1}^{n} j\,\dot{\rho}^{j-1}\left\{a_j \cos\left(\{j-1\}\dot{\theta}\right) - b_j \sin\left(\{j-1\}\dot{\theta}\right)\right\}\frac{\partial \dot{y}}{\partial \phi}$$

$$= L_2\frac{\partial \dot{x}}{\partial \phi} + L_1\frac{\partial \dot{y}}{\partial \phi}$$

Substitution in the first of Equations 334b gives

$$\ddot{m} = \sqrt{\left(L_1\frac{\partial \dot{x}}{\partial \phi} - L_2\frac{\partial \dot{y}}{\partial \phi}\right)^2 + \left(L_2\frac{\partial \dot{x}}{\partial \phi} + L_1\frac{\partial \dot{y}}{\partial \phi}\right)^2}$$

$$= \sqrt{\left(L_1\frac{\partial \dot{x}}{\partial \phi}\right)^2 - 2L_1L_2\frac{\partial \dot{x}}{\partial \phi}\frac{\partial \dot{y}}{\partial \phi} + \left(L_2\frac{\partial \dot{y}}{\partial \phi}\right)^2 + \left(L_2\frac{\partial \dot{x}}{\partial \phi}\right)^2 + 2L_1L_2\frac{\partial \dot{x}}{\partial \phi}\frac{\partial \dot{y}}{\partial \phi} + \left(L_1\frac{\partial \dot{y}}{\partial \phi}\right)^2}$$

$$= \sqrt{\left(L_1^2 + L_2^2\right)\left(\left(\frac{\partial \dot{x}}{\partial \phi}\right)^2 + \left(\frac{\partial \dot{y}}{\partial \phi}\right)^2\right)}$$

$$= \sqrt{L_1^2 + L_2^2}\;\dot{m}$$

as asserted with Equation 334d.

Reverting to complex variables, similar but more elaborate expansion shows that

$$\sqrt{L_1^2 + L_2^2} = \left|\sum_{j=1}^{n} j\left(a_j - \mathrm{i}b_j\right)\left(\dot{x} + \mathrm{i}\,\dot{y}\right)^{j-1}\right|$$

ENDNOTE B: DIFFERENTIATING EQUATIONS 334d & 335a

Basics

Differentiating Equation 334d with respect to the representative coefficient a_r gives

$$\frac{\partial \ddot{m}}{\partial a_r} = \dot{m}\frac{1}{2}\left(L_1{}^2 + L_2{}^2\right)^{-\frac{1}{2}}\left(2L_1\frac{\partial L_1}{\partial a_r} + 2L_2\frac{\partial L_2}{\partial a_r}\right)$$

$$= \frac{\dot{m}^2}{\ddot{m}}\left(L_1\frac{\partial L_1}{\partial a_r} + L_2\frac{\partial L_2}{\partial a_r}\right)$$

Differentiating Equations 335a gives

$$\frac{\partial L_1}{\partial a_r} = r\,\dot{\rho}^{r-1}\cos\left((r-1)\dot{\theta}\right) \qquad \text{and} \qquad \frac{\partial L_2}{\partial a_r} = r\,\dot{\rho}^{r-1}\sin\left((r-1)\dot{\theta}\right)$$

Hence,

$$\frac{\partial \ddot{m}}{\partial a_r} = \frac{\dot{m}^2}{\ddot{m}} r\,\dot{\rho}^{r-1}\left\{L_1\cos\left((r-1)\dot{\theta}\right) + L_2\sin\left((r-1)\dot{\theta}\right)\right\}$$

Substituting from Equations 335a, plus similar differentiation with respect to b_r, gives

$$\frac{\partial \ddot{m}}{\partial a_r} = \frac{\dot{m}^2}{\ddot{m}} r\,\dot{\rho}^{r-1}\sum_{j=1}^{n} j\,\dot{\rho}^{j-1}\left\{\,a_j\cos\left((r-j)\dot{\theta}\right) + b_j\sin\left((r-j)\dot{\theta}\right)\right\}$$

$$\frac{\partial \ddot{m}}{\partial a_r} = \frac{\dot{m}^2}{\ddot{m}} r\,\dot{\rho}^{r-1}\sum_{j=1}^{n} j\,\dot{\rho}^{j-1}\left\{\,-a_j\sin\left((r-j)\dot{\theta}\right) + b_j\cos\left((r-j)\dot{\theta}\right)\right\}$$

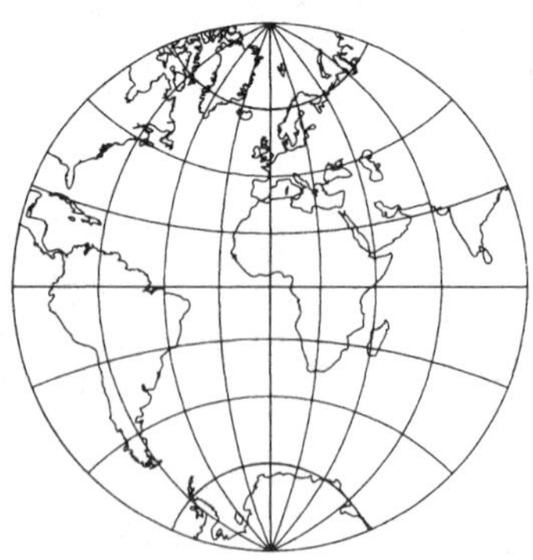

Exhibit 14–3: The Equatorial Stereographic — a hemisphere in a circle.
In miniature to correspond to the inner circle of Lagrange in Exhibit 15–2.

CHAPTER 15

Keeping in Shape: More conformal projections

Preamble

The Mercator projection was created by a drive for a conformal map of 'true' cylindrical mode, and was followed in our development with 'true' conformal maps of azimuthal and conical modes. In Chapter 10 the pursuit of conformality was extended with Lambert's innovation, with a promise of a widening of that to the also conformal Lagrange projection. We now effect that widening, and study the general mathematical requirements for conformality, deriving numerous projections that meet them. This study itself requires considerable elaboration of our mathematical concepts and tools, much of which is carried in the main body of the text rather than as support tutorials.

This chapter extends functions of a complex variable to include elliptic functions.

After development from the Lambert of Chapter 10, the maps become markedly exotic. Again, the general focus of them is on the major land masses with the Greenwich meridian at the centre. The final element in the chapter, reproduced with kind permission of the American Geographical Society, is of an interesting Antarctica-centred usage of a notably exotic projection.

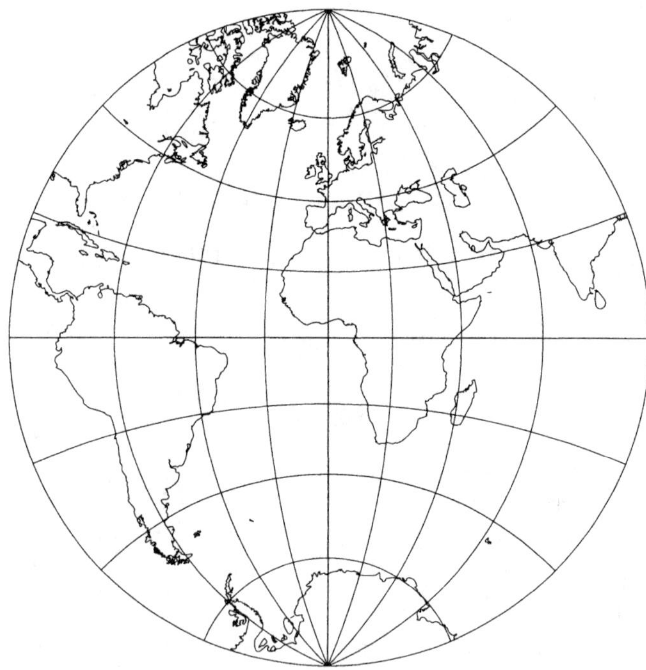

Exhibit 15–1: The Equatorial Stereographic — a hemisphere in a circle.

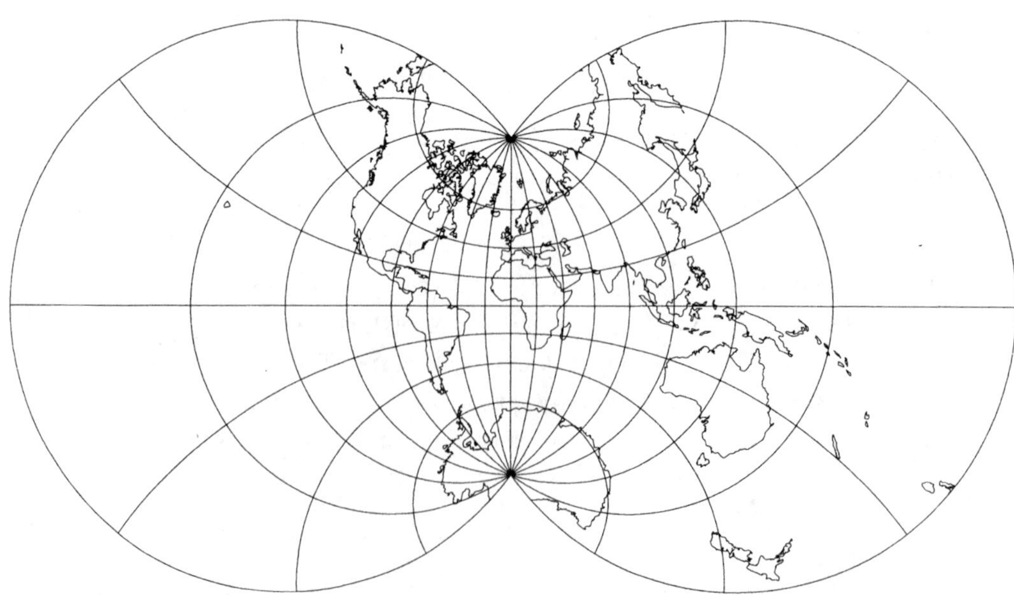

Exhibit 15–2: The Lagrange projection; whole world, with just five eighths in the circle.
[I.e., circle equates with meridians for ±112°30′.]

LAGRANGE PROJECTION

Lambert's beginnings — the straight Equator

Perhaps the most generic of conformal projections is that referred to as the Lagrange. Although promoted as well as somewhat elaborated by J. L. Lagrange[†] [1779a], it is a generalization of a projection, described in Chapter 10, introduced along with several others by Lambert [1772a], who created a conformal projection with the whole world within a circle. As discussed in Chapter 10, Lambert's effort evolved from the Stereographic — the conformal Azimuthal — at Equatorial aspect. Lambert's innovation was to adapt the real values of longitude and latitude to proxy values that fitted a wider world into the circle originally covering only a hemisphere, while maintaining conformality. The basic example of Lambert, illustrated on page 254 and with plotting equations as Equations 259c, fitted the whole world into the circle, and kept the Equator as a straight line bisecting that circle. Both the meridians and the parallels are circular arcs, as is the case for the mapping of any circles with the Stereographic — arcs that are necessarily orthogonal to accord with conformality.

Lagrange's promotions — any one parallel a straight line

While Lambert's espoused form had the Equator a straight line lying along the x axis bisecting the circular map, and had the shrinkage factor such as to put the whole world precisely in a circle, his methodology was significantly more general. It allowed for any one parallel to lie along the x axis and for any degree of shrinkage. In this chapter we address those two options, as the **Lagrange** projection.

Equation 253c gives the general formula derived from applying conformality. If the parallel of polar distance γ_0 is chosen to lie along the x axis, then the constant k in that equation, identically zero for Lambert's choice, is given by

$$0 = \ln \tan^c \frac{\gamma_0}{2} + k$$

so that

$$\ln \tan^2 \frac{\psi}{2} = \ln \tan^c \frac{\gamma}{2} - \ln \tan^c \frac{\gamma_0}{2}$$

Hence, the simple relationship of Equation 253d is elaborated to

$$\tan^2 \frac{\psi}{2} = \tan^{-c} \frac{\gamma_0}{2} \tan^c \frac{\gamma}{2} \qquad \text{(349a}$$

So, the length derived in Equation 254a is similarly elaborated, to

[†] Turin-born French applied mathematician Jean Louis Lagrange (1736 – 1813), who presided over the commission that introduced the metric system, with its geodetic basis.

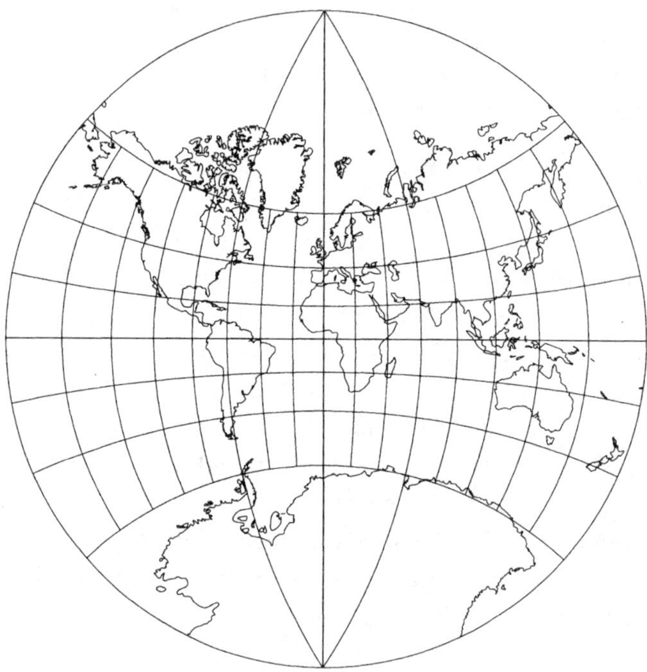

Exhibit 15–3: The Lagrange projection, whole world in a circle; straight line for $\phi = 0°$.

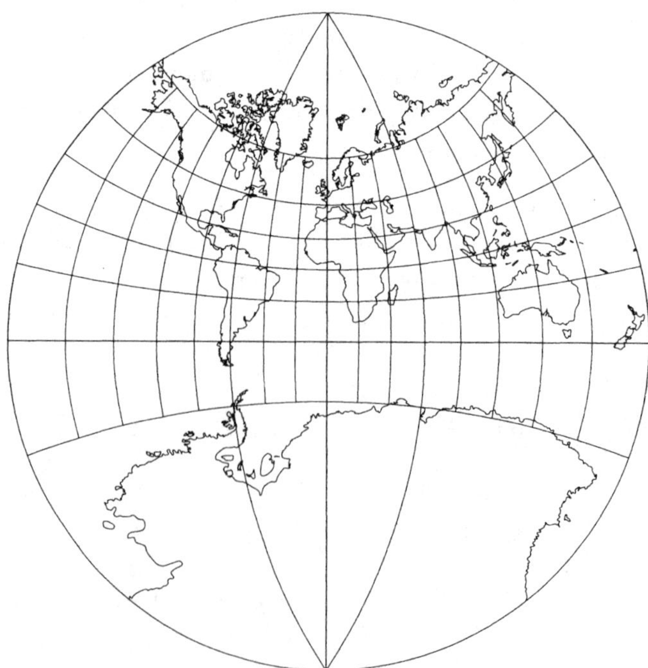

Exhibit 15–4: The Lagrange projection, whole world in a circle; straight line for $\phi = 45°S$.

$$\dot{y}\big]_{\lambda=0} = r\cos\dot{\upsilon} = r\left\{\frac{1-\tan^{-c}\frac{\gamma_0}{2}\tan^{c}\frac{\gamma}{2}}{1+\tan^{-c}\frac{\gamma_0}{2}\tan^{c}\frac{\gamma}{2}}\right\}$$

The development on pages 255 through 257 is independent of this length, so applies identically here. The subsequent development, relating pseudo-latitude β to ordinary latitude ϕ and providing the plotting equations, applies here, too, provided the formulation in Equation 253d is replaced by that in Equation 349a. That means Equation 259a remains valid provided we set

$$V = \cot^{-c}\frac{\gamma_0}{2}\cot^{c}\frac{\gamma}{2} = \left\{\cot^{-2}\frac{\gamma_0}{2}\cot^{2}\frac{\gamma}{2}\right\}^{\frac{c}{2}} = \left\{\frac{1+\cos\gamma_0}{1-\cos\gamma_0}\right\}^{\frac{-c}{2}}\left\{\frac{1+\cos\gamma}{1-\cos\gamma}\right\}^{\frac{c}{2}} \quad \text{. . . . . (351a}$$

$$= \left\{\frac{1+\sin\phi_0}{1-\sin\phi_0}\right\}^{\frac{-c}{2}}\left\{\frac{1+\sin\phi}{1-\sin\phi}\right\}^{\frac{c}{2}}$$

in place of Equation 259b.

Leaving c as an open variable, we thus get plotting equations for the Lagrange projection as:

$$\text{If } \phi = \pm\tfrac{\pi}{2} \quad \dot{x} = 0, \quad \dot{y} = \frac{\phi}{|\phi|}$$

$$\text{else} \quad \dot{x} = 2RW^{-1}\sin\frac{\lambda}{2} \quad \text{. . . . . (351b}$$

$$\dot{y} = 2RW^{-1}\tan\beta$$

where $W = \sec\beta + \cos\frac{\lambda}{2}$,

$$\sec\beta = \frac{V+V^{-1}}{2},$$

$$\tan\beta = \frac{V-V^{-1}}{2},$$

and V is defined in Equation 351a.

Lambert's basic form, with $c = \frac{1}{2}$ and $\gamma_0 = \frac{\pi}{2}$, is shown via a mimic of his original map on page 254 and in more modern form in Exhibit 15–3. Below the latter, in Exhibit 15–4, is a similar map of the world within a circle but with the straight parallel set at 45°S. Exhibit 15–1 is the basic Equatorial Stereographic with just a hemisphere in its circle.

Exhibit 15–2 also has the Equator the straight and central parallel, but has $c = \frac{4}{5}$ bringing only the ±112°30′ meridians down to lie along the bimeridian circle of ±90° of the Equatorial Stereographic. The further-out meridians appear as expanding arcs exceeding semicircles. Were $c = 1$, the outermost meridians would balloon to have infinite radii, the circle would enclose a hemisphere as in the Equatorial Stereographic — shown at the scale of Exhibit 15–2 two pages earlier in Exhibit 14–3.

CONFORMALITY

Isometric latitude and hyperbolic functions

Isometric latitude was defined in Chapter 14 (Equation 288a) as

$$\overline{\varphi} = \ln\tan\left(\frac{\pi}{4} + \frac{\phi}{2}\right)$$

this logarithmic expression occurring routinely in the plotting equations of conformal projections. Using λ to represent relative longitude and adopting for simplicity a unit globe (i.e., $R = 1$) the equations for Mercator at Simple aspect (Equations 119a) become:

$$\dot{x} = \lambda \qquad \dot{y} = \overline{\varphi} \qquad \ldots\ldots\ldots (352a$$

As is shown initially in Equation 289a, and elaborated below in Tutorial 30, this expression for isometric latitude — a function purely of geodetic latitude — relates to the familiar circular trigonometric functions of geodesic latitude in a straightforward manner via hyperbolic functions. Hence, using these less-familiar functions, we can re-express various plotting equations relatively simply in terms of isometric rather than geodetic latitude, including at Oblique aspects. We look particularly at the two most prominent conformal projections — the Stereographic and the Mercator — and re-express their various plotting equations in isometric terms.

TUTORIAL 30 ISOMETRIC-GEODETIC RELATIONSHIPS

From the definition in Chapter 14 of isometric latitude as (Equation 288d)

$$\overline{\varphi} = \ln\tan\left(\frac{\pi}{4} + \frac{\phi}{2}\right) = \ln\sqrt{\frac{1+\sin\phi}{1-\sin\phi}}$$

the use of hyperbolic functions provided (Equation 289a) the relationship

$$\cosh\overline{\varphi} = \sec\phi$$

Using formulae of Tutorials 10 & 26, this leads to the further relationships:

$$\sinh\overline{\varphi} = \tan\phi$$

$$\tanh\overline{\varphi} = \sin\phi$$

$$\tanh\frac{\overline{\varphi}}{2} = \tan\frac{\phi}{2}$$

If γ is the polar distance $\frac{\pi}{2} - |\phi|$, we get:

$$e^{\overline{\varphi}} = \sec\phi + \tan\phi = \cot\frac{\gamma}{2}$$

$$e^{-\overline{\varphi}} = \sec\phi - \tan\phi = \tan\frac{\gamma}{2}$$

$$e^{\frac{\overline{\varphi}}{2}} = \sqrt{\cot\frac{\gamma}{2}}$$

$$e^{-\frac{\overline{\varphi}}{2}} = \sqrt{\tan\frac{\gamma}{2}}$$

$$\cosh\frac{\overline{\varphi}}{2} = \frac{1}{2}\left\{\sqrt{\cot\frac{\gamma}{2}} + \sqrt{\tan\frac{\gamma}{2}}\right\}$$

$$\sinh\frac{\overline{\varphi}}{2} = \frac{1}{2}\left\{\sqrt{\cot\frac{\gamma}{2}} - \sqrt{\tan\frac{\gamma}{2}}\right\}$$

Turning from Simple aspect converts Equation 352a to:

$$\dot{x} = {}_{\mathrm{a}}\lambda \qquad \dot{y} = {}_{\mathrm{a}}\overline{\varphi} \qquad \text{(353a}$$

with the suffix denoting values in the auxiliary system of Chapter 8. Substituting from Equations 207a & 192a, we then get for the plotting equations of the Mercator projection at Oblique aspect the indirect expressions:

$$\tan \dot{x} = \tan {}_{\mathrm{a}}\lambda = \frac{\cos\phi_{\mathrm{N}} \tan\phi - \sin\phi_{\mathrm{N}} \cos\lambda}{\sin\lambda} = \frac{\cos\phi_{\mathrm{N}} \sinh\overline{\varphi} - \sin\phi_{\mathrm{N}} \cos\lambda}{\sin\lambda}$$

$$\tanh \dot{y} = \tanh {}_{\mathrm{a}}\overline{\varphi} = \sin {}_{\mathrm{a}}\phi = \frac{\sin\phi_{\mathrm{N}} \tan\phi + \cos\phi_{\mathrm{N}} \sin\lambda}{\sec\phi} = \frac{\sin\phi_{\mathrm{N}} \sinh\overline{\varphi} + \cot\phi_{\mathrm{N}} \sin\lambda}{\cosh\overline{\varphi}}$$

For the Equatorial aspect, when $\phi_{\mathrm{N}}=0$, $\sin\phi_{\mathrm{N}}=0$, and $\cos\phi_{\mathrm{N}}=1$, these become:

$$\tan \dot{x} = \frac{\tan\phi}{\sin\lambda} = \frac{\sinh\overline{\varphi}}{\sin\lambda}$$

$$\tanh \dot{y} = \frac{\sin\lambda}{\sec\phi} = \frac{\sin\lambda}{\cosh\overline{\varphi}}$$

For the Stereographic at Simple aspect, Equation 118a gives the general form of the linear element $\dot{\rho}$. If we revert from latitude to polar distance γ as in the initial consideration of stereographic projecting [see Equation 46a], and convert using formulae from Tutorial 30 to incorporate isometric latitude, the plotting equations become:

$$\dot{x} = 2\sigma \tan\frac{\gamma}{2} \sin\lambda = +2\sigma e^{-\overline{\varphi}} \sin\lambda$$

$$\dot{y} = 2\sigma \tan\frac{\gamma}{2} \cos\lambda = -2\sigma e^{-\overline{\varphi}} \cos\lambda \qquad \text{(353b}$$

Applying the formulae to Equations 201a gives plotting equations for the Stereographic projection at Oblique aspect as:

$$\dot{x} = \frac{2\sigma R \sec\phi_{\mathrm{N}} \sin\lambda}{\sec\phi_{\mathrm{N}} \sec\phi + \tan\phi_{\mathrm{N}} \tan\phi + \cos\lambda} = \frac{2\sigma R \cosh\overline{\varphi}_{\mathrm{N}} \sin\lambda}{\cosh\overline{\varphi}_{\mathrm{N}} \cosh\overline{\varphi} + \sinh\overline{\varphi}_{\mathrm{N}} \sinh\overline{\varphi} + \cos\lambda}$$

$$\dot{y} = \frac{2\sigma R \{\tan\phi - \tan\phi_{\mathrm{N}} \cos\lambda\}}{\sec\phi_{\mathrm{N}} \sec\phi + \tan\phi_{\mathrm{N}} \tan\phi + \cos\lambda} = \frac{2\sigma R \{\sinh\overline{\varphi} - \sinh\overline{\varphi}_{\mathrm{N}} \cos\lambda\}}{\cosh\overline{\varphi}_{\mathrm{N}} \cosh\overline{\varphi} + \sinh\overline{\varphi}_{\mathrm{N}} \sinh\overline{\varphi} + \cos\lambda} \qquad \text{(353c}$$

For the Equatorial aspect, when $\phi_{\mathrm{N}}=0$, $\sin\phi_{\mathrm{N}}=0$, and $\cos\phi_{\mathrm{N}}=1$, these become:

$$\dot{x} = \frac{2\sigma R \cos\phi \sin\lambda}{1 + \cos\phi \cos\lambda} = \frac{2\sigma R \operatorname{sech}\overline{\varphi} \sin\lambda}{1 + \operatorname{sech}\overline{\varphi} \cos\lambda} = \frac{2\sigma R \sin\lambda}{\cosh\overline{\varphi} + \cos\lambda}$$

$$\dot{y} = \frac{2\sigma R \sin\phi}{1 + \cos\phi \cos\lambda} = \frac{2\sigma R \tanh\overline{\varphi}}{1 + \operatorname{sech}\overline{\varphi} \cos\lambda} = \frac{2\sigma R \sinh\overline{\varphi}}{\cosh\overline{\varphi} + \cos\lambda} \qquad \text{(353d}$$

L. P. Lee provides [1974c] a method for computation of co-ordinates in a conformal mapping using only the difference in longitude and isometric latitude between the general point and the origin.

USING FUNCTIONS OF A COMPLEX VARIABLE

Basics

Complex numbers, covered by Tutorial 23 and introduced in the Chapter 14, allow handling of two independent variables as one item. Further, as shown in Tutorial 28, the transformation produced by any such function is intrinsically conformal if the function is differentiable. Hence, since differentiability is common, functions of a complex variable offer a powerful armoury for producing additional conformal projections. This could be done by directly addressing the globe, but they are more readily used for such purpose as re-transformations of existing conformal projections.

Isometric latitude, being a simplification that accommodates the conformality factor when addressing the globe, is a useful starting point. Using our distinctive notation for isometric latitude, let

$$\mu = \lambda + \mathrm{i}\overline{\varphi}$$

represent a general point on the globe. Then, any transformational function $\dot{g}$ such that

$$\dot{z} = \dot{x} + \mathrm{i}\dot{y} = \dot{\rho}\operatorname{cis}\dot{\theta} = \dot{g}(\mu) = \dot{g}(\lambda + \mathrm{i}\overline{\varphi})$$

produces a conformal map $\{\dot{z}\}$. From Equations 353a the Mercator projection at Simple aspect in these terms is merely the transformation

$$\dot{z} = \dot{x} + \mathrm{i}\dot{y} = \mu = \lambda + \mathrm{i}\overline{\varphi} \qquad \text{(354a}$$

The process of transformation using differentiable functions of a complex variable can be repeated, e.g., with a second transformational function $\ddot{g}$ to give

$$\ddot{z} = \ddot{x} + \mathrm{i}\ddot{y} = \ddot{g}(\dot{z}) = \ddot{g}(\dot{x} + \mathrm{i}\dot{y})$$

which produces a further conformal map $\{\ddot{z}\}$. Such additional transforming of one conformal map into another necessarily conformal map can be carried onward without limit, e.g., to achieve a result progressively closer to some desired criteria.

The conformality-maintaining facet of functions of a complex variable makes them very useful in developing additional conformal transformations and maps. When applied to transform an existing conformal surface, these functions automatically produce a new conformal surface. But, while the obvious candidate source is the surface of the globe, the utility of these functions is more notable when applied to a conformal map to provide a revised conformal map; this is their more usual use. The mapping equations for the Stereographic and Mercator projections are established functions that transform the globe into a conformal map, i.e., examples of function $\dot{g}$. Each of those projections thus provides a convenient starting point for the development of new conformal maps — one favouring a round format, the other a rectangular format.

USING ELLIPTIC FUNCTIONS

Introduction

Expressing the co-ordinates in canonical form (Tutorial 8), the length element along the perimeter of an ellipse (using co-ordinates (x, y)) is given by

$$\mathrm{d}s = \sqrt{\mathrm{d}x^2 + \mathrm{d}y^2} = \int \sqrt{a^2 \sin^2\theta + b^2 \cos^2\theta}\ \mathrm{d}\theta = \int \sqrt{a^2 \sin^2\theta + a^2\left(1-\varepsilon^2\right)\cos^2\theta}\ \mathrm{d}\theta$$

$$= \int a\sqrt{1-\varepsilon^2\cos^2\theta}\ \mathrm{d}\theta$$

Substituting $t = \cos\theta$, this becomes

$$\mathrm{d}s = a\int \sqrt{1-\varepsilon^2 t^2}\ \frac{\mathrm{d}t}{\sqrt{1-t^2}} = a\int \frac{1}{\sqrt{\left(1-t^2\right)\left(1-\varepsilon^2 t^2\right)}}\mathrm{d}t + a\int \frac{\varepsilon^2 t^2}{\sqrt{\left(1-t^2\right)\left(1-\varepsilon^2 t^2\right)}}\mathrm{d}t$$

Integrals of such mathematical form, regardless of context and the particular variables and coefficients involved, are accordingly called *elliptic integrals,* the two in the final line being respectively *of the first kind* and *of the second kind.* Reverse functions, expressing t in terms of s, are called *elliptic functions.* The functions from integrals of the particular style shown here are known as *Jacobian elliptic functions.* The variables in the above equations are real numbers but the terminology applies equally to functions involving complex variables — these of course being of incomparable value for conformal transformations. The versatility of such functions offers opportunity to transform original conformal maps into conformal maps that fit within various geometric constraints.

Cartographic opening

Hermann Schwartz of Germany [1869b] showed that, for any positive integer n, the complex integral

$$\dot{z} = \int_0^z \frac{1}{\sqrt[n]{\left(1-z^n\right)^2}}\mathrm{d}z \qquad \ldots\ldots (355a$$

transformed the interior of the circle $\{\,|z| = 1\}$ into an n-sided polygon. Schwartz went on to produce similar transformations for the interior of an ellipse into a circular disc [1869d] and of the sphere into each of the five regular polyhedra [1872a]. Given a conformal source map — the Stereographic being the notable example — the result is a conformal map of triangular, square else other polygonal outline. C. S. Peirce of the U.S.A. [1879a] applied the work of Schwarz cartographically[‡] with n = 4, to produce a conformal map of

[‡] Bernhard Riemann in 1851 had established conformal mapping as a purely mathematical matter.

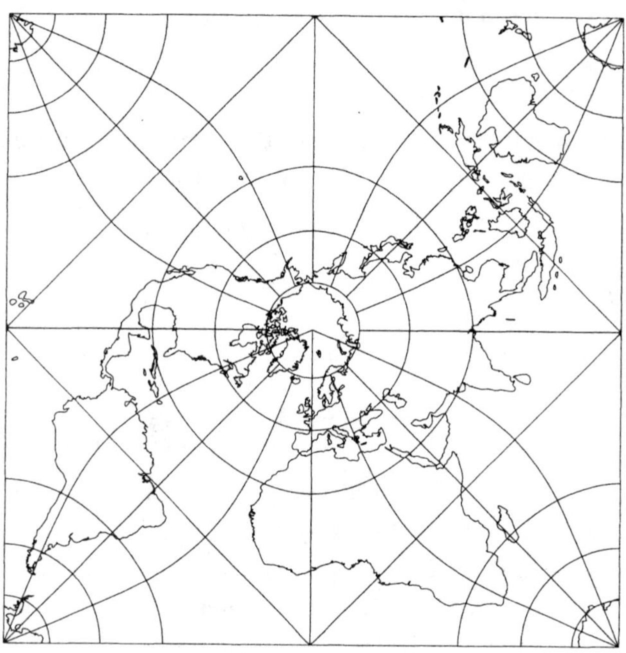

Exhibit 15–5: The Peirce Quincuncial projection — world conformally in a square.

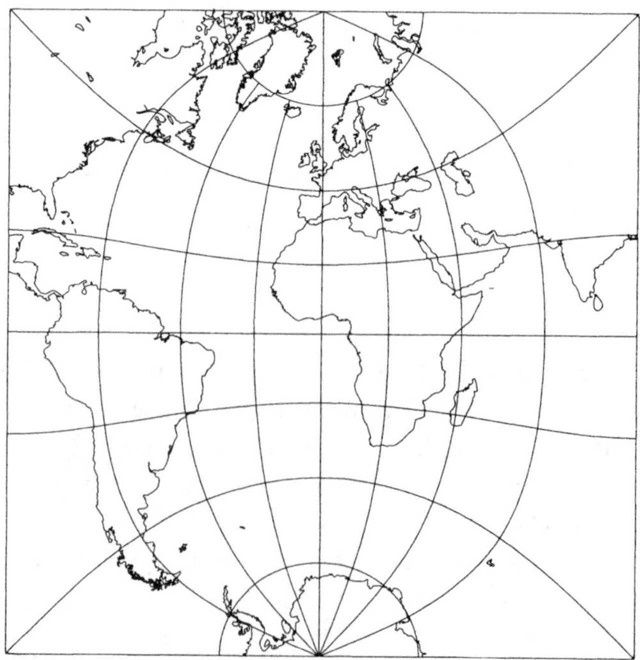

Exhibit 15–6: The Guyou projection — hemisphere conformally in a square.

the world within a square, at azimuthally Simple aspect. He called it the **Quincuncial** projection, because of its ready image as four squares overlain by a fifth in diamond form, the last representing the Equator; Exhibit 15–5 illustrates this pattern. Émile Guyou of France [1886a] elaborated this to produce the transverse aspect, applying it to one hemisphere at a time to cover the world with two squares that can be adjacent on any matched sides; one square of the **Guyou** projection is shown in Exhibit 15–6.

Comprehensive coverage

In a 118-page finely printed monograph, O. S. Adams of the U.S.A. [1925b] provided both a comprehensive discussion of the use of elliptic functions in cartography and several innovative new projections. Half a century later, L. P. Lee of New Zealand produced a similar but more modern exposition of the subject [1976a], replete with further innovations. Lee's monograph also included, aided by the availability of mechanical calculators, many revisions/corrections to the extensive numerical work effected laboriously via logarithms by Adams (personally).

Adams showed that Schwartz's integral (Equation 355a) is, for $n = 3$ (which maps into a triangle), a particular case of the integral

$$\ddot{z} = \int_0^{\dot{z}} \frac{1}{\sqrt[3]{\left(1-\dot{z}^3\right)^2}} d\dot{z}$$

and hence of elliptic functions somewhat different from those above. Labelled *Dixonian elliptic functions*, in contrast to the preceding Jacobian elliptic functions, these are based on the equation $x^3 + y^3 = 1$.

Besides producing a map of a hemisphere in an equilateral triangle, Adams proceeded in his monograph to create maps of a hemisphere in a regular hexagon, of both the whole world and a hemisphere in a rhombus, and of the whole world in a rectangle, a 6-pointed star and an ellipse — all conformally of course (unlike Mollweide, for instance). In reviewing earlier work, he showed that Peirce's projection is a Transverse aspect of the Guyou, and produced an Oblique version that had the hemisphere in a square diamond. He went on [1929a] to show a general method for transforming any part of the world into any polygon, demonstrating maps of the whole world in a square and on the octahedron. The diamond-shaped **Adams World-in-a-Square II** projection is shown on page 359.

Various other authors expanded the repertoire before Lee addressed the subject, beginning with new maps of the world in an ellipse and on a regular tetrahedron [1965b]. The technique, applied to appropriate spherical triangles on the globe, was extended to cover the cube and the regular icosahedron and dodecahedron in later publications.

While conformal generally, these maps have singular points — i.e., points at which conformality fails. In the Peirce map (Exhibit 15–5), the right-angle turns of what should be a straight-line Equator preclude conformality at those four points. Projections of a hemisphere and of the whole world into polygons routinely have singular points at corners. Some projections have singular points within the apparent map, though technically along perimeter lines that coincide. This last occurred with the Adams projection into an ellipse. It was based on a transformation from a circle into a rectangle with the perimeter being totally meridians and the Equator a bisecting horizontal line. A second stage created the ellipse, but with the Equator shrunken to span only between the foci. The outer sections of this axis represented sections of meridians pulled in by the transformation; each one represented, distinctly but in duplicate, the two halves of the sides of the square. Since the parts of the ellipsoidal map on either side of each outer section should not adjoin, the whole of those outer section is singular. Such a plot is sometimes labelled *cut*. Lee [1965b] chose instead to map into a square with the Equator reaching out, bifurcated, along the sides to the four corners. The pieces pulled in by his second stage were also of that main parallel, and the resulting ellipse consistent with expectations, as illustrated in Exhibit 15–7. Conformality still fails at the poles.

Consolidating the work of various predecessors, Lee also produced the **Lee Conformal Projection of the World in a Triangle**, with one pole at an apex the other at

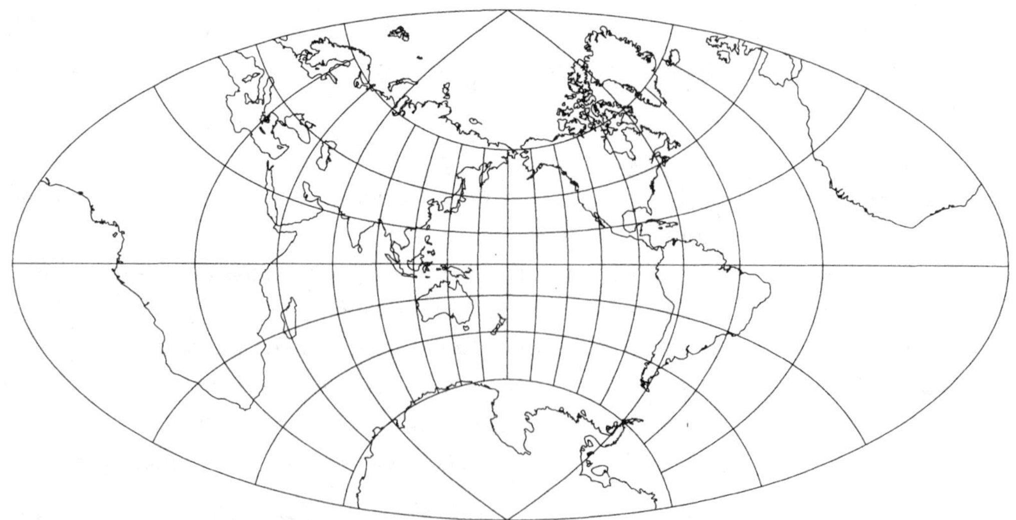

Exhibit 15–7: The Lee Conformal Projection of the World in an Ellipse.

the central point of the opposing side; conformality clearly fails at the three corners of the triangle plus at that other pole.

Evaluating elliptic integrals and elliptic functions occupies a considerable sector of mathematics [1989g], and is much too deep for detailed attention in this work. It involves further functions akin, like the hyperbolic functions, to the circular trigonometric functions, which likewise are periodic in nature. Consequently, when used for maps they produce recurring images. The periodicity applies to both the real part and the imaginary part of the complex values — singly to the former but dually else triply to the imaginary part. For the projections that produce parts of the world in polygons — notably the faces of polyhedra — the result is facility to create a repeating mosaic, allowing any edge to have corresponding pieces of the whole world on its opposite sides with consistency of conformality across (except singular points at corners). Thus, although tetrahedra in their 3-dimensional forms are poor substitutes for globes, their planar surfaces allow flattening into a useful map over whatever area is wanted. Little use has been made of these essentially novelty projections.

The reader interested in delving further into the deep mathematics of elliptic functions should turn to either of the cited monographs [1925b, 1976a].

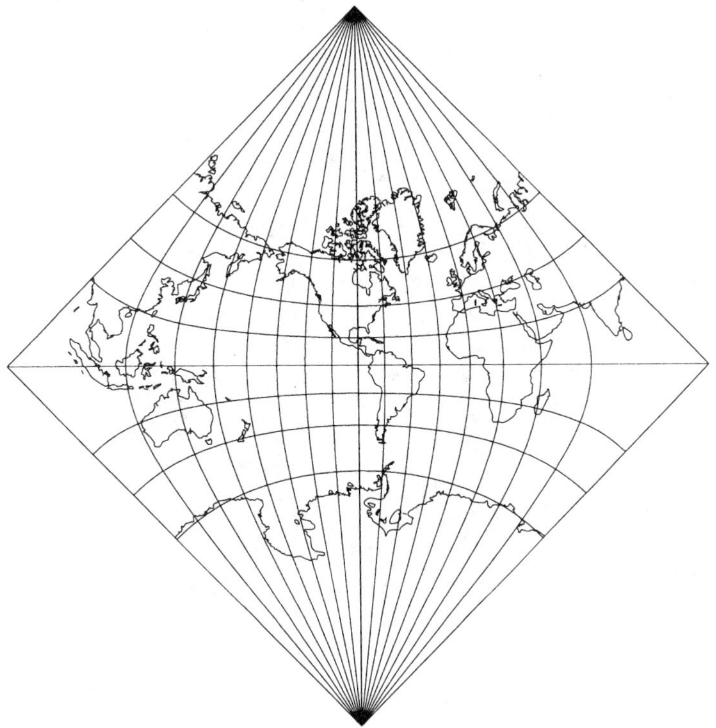

Exhibit 15–8: The Adams World-in-a-Square II projection.

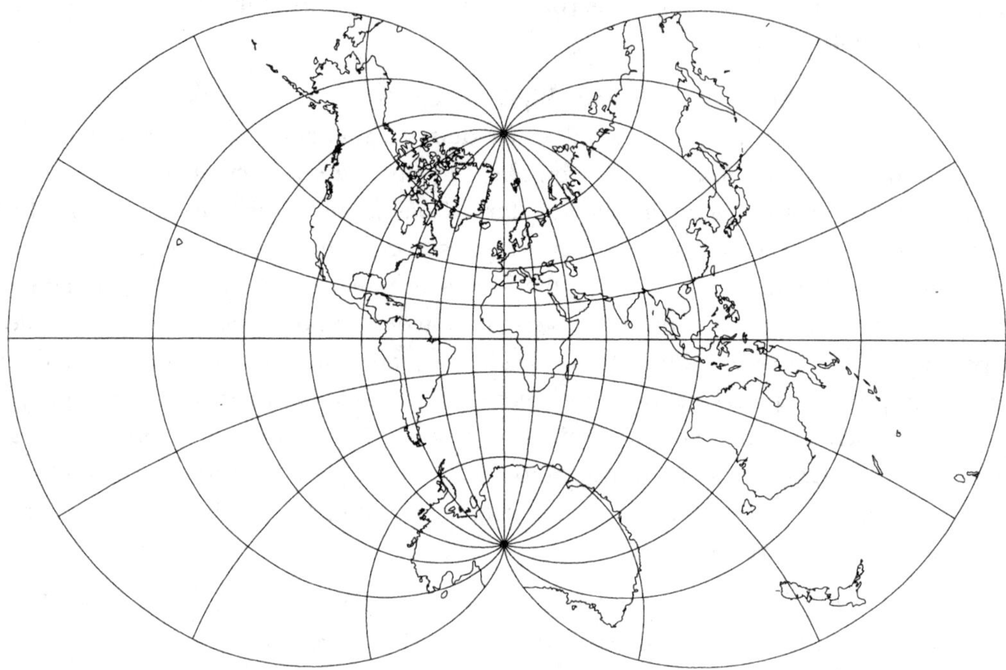

Exhibit 15–9: The Eisenlohr projection.

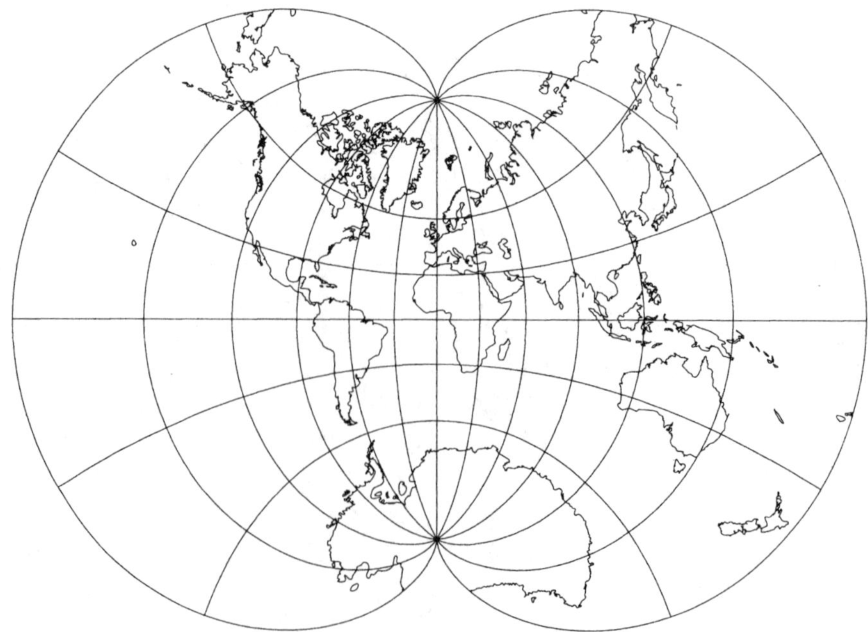

Exhibit 15–10: The August Epicycloidal projection.

TWO-CUSPED CONFORMAL PROJECTIONS

Competitive cartography

Friedrich Eisenlohr and Friedrich August, German contemporaries, pioneered 2-cusped conformal projections. Visually similar to, but decades older than, the so-called Apple-shaped van der Grinten IV [page 237], these two projections are exceptional in being conformal everywhere — even at the poles. The two are extremely similar in appearance, as shown opposite. The two authors and their projections were also archrivals, with published dispute on the relative merits of their projections. Both have very complicated plotting equations which inhibited their use prior to the availability of routine computer plotting, and prompted the publication of plotting tables by the respective authors.

Eisenlohr's projection [1870a} came first; it accords with the Chebyshev principle [1856a] proved by Grave [1896a] that the minimum overall distortion for a map requires uniform scale along its perimeter. The resulting distortion of linear scale relative to the centre point is substantial, being $3+2\sqrt{2}=5.828\,43_{\sim}$ along the perimeter (a figure modest among conformal projections). At the Poles, the recurving meridians meet at their true angles — including the two that form the opposite sides of the perimeter. Snyder [1989a] gives the plotting equations for the **Eisenlohr** projection as:

$$\dot{x}=\left(3+\sqrt{8}\right)R\left\{C\left(V-V^{-1}\right)-2\ln V\right\}$$

$$\dot{y}=\left(3+\sqrt{8}\right)R\left\{C\left(V+V^{-1}\right)-2\arctan T\right\}$$

where $T=\sin\frac{\phi}{2}\left[\cos\frac{\phi}{2}+\sqrt{2\cos\phi}\,\cos\frac{\lambda}{2}\right]^{-1},$ . . . . . . (361a

$$C=\sqrt{\frac{2}{1+T^2}}$$

and $$V=\sqrt{\frac{\cos\frac{\phi}{2}+\left(\cos\frac{\lambda}{2}+\sin\frac{\lambda}{2}\right)\sqrt{\frac{\cos\phi}{2}}}{\cos\frac{\phi}{2}+\left(\cos\frac{\lambda}{2}-\sin\frac{\lambda}{2}\right)\sqrt{\frac{\cos\phi}{2}}}}.$$

Eisenlohr's rival used the epicycloid — the path of a point on a circle that is rolling around the perimeter of a circle twice its diameter — and the projection is often named accordingly. The **August Epicycloidal** projection [1874a] does not accord with the Chebyshev principle. Its perimeter varies in relative scale, up to 8, although the overall distortion is not much greater than that pertaining to the Eisenlohr. Its plotting formulae were marginally simpler, favouring its application (despite the general use of plotting tables). Snyder [1989a] gives the plotting equations for the August Epicycloidal projection as:

$$\dot{x} = 4R\left\{3 + a^2 - 3b^2\right\}$$
$$\dot{y} = 4R\left\{3 + 3a^2 - b^2\right\}$$

where $a = C^{-1}\sin\frac{\lambda}{2}\sqrt{1-\tan^2\frac{\phi}{2}}$, ... (362a

and $b = C^{-1}\tan\frac{\lambda}{2}$

using $C = 1 + \cos\frac{\lambda}{2}\sqrt{1-\tan^2\frac{\phi}{2}}$.

While the shape of the resulting maps gives them novelty appeal, the intrinsic interruption in Polar regions handicaps real use, despite their technical excellence. A. F. Spilhaus, however, found notable and effective use with a map of ocean currents [1942c] shown in Exhibit 15–11. Choosing the August (presumably for relative ease of projecting) and using an oblique version that puts the South Pole at the centre of the map, the map relegates to its outer tracts the land forms of the northern hemisphere while picturing Antarctica as the hub of oceanic circulation.

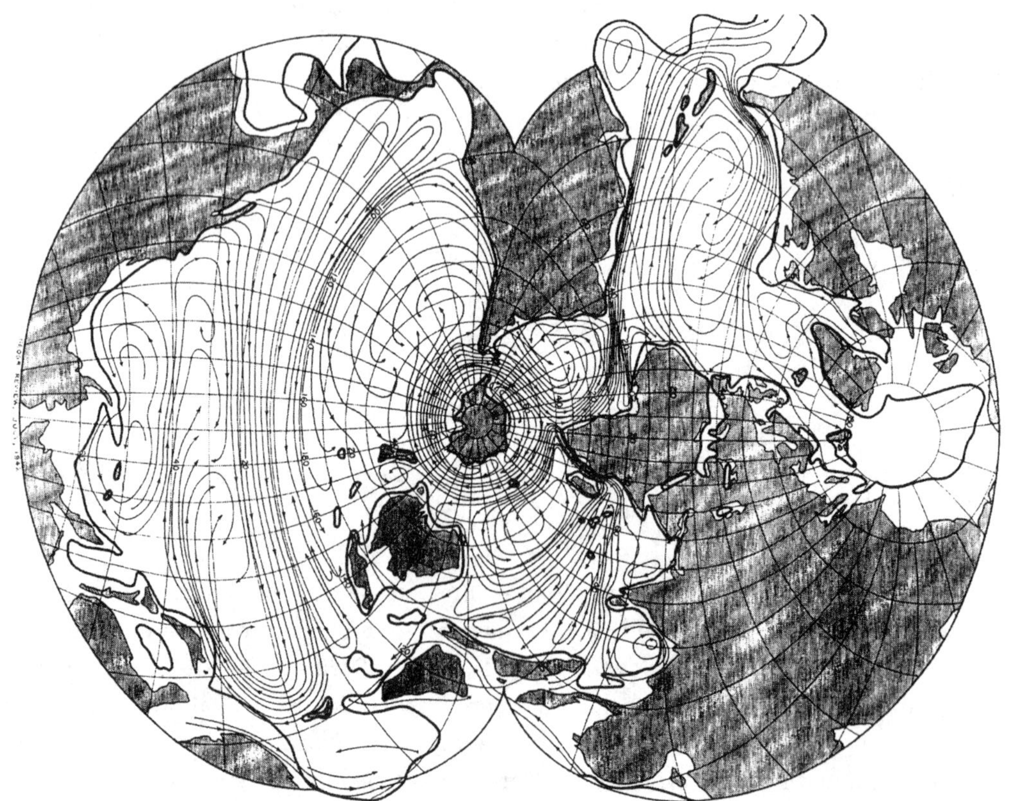

Exhibit 15–11: The August Epicycloidal projection for oceanic circulation.
Reproduced from the Geographical Review [1942c, by A. F. Spilhaus]
with the kind permission of the American Geographical Society.

CHAPTER 16

Having Fun: Novelty projections

Preamble

Various projections developed for serious reasons have, in their appearance, a novelty value — e.g., the cordiform Werner, the apple-shaped van der Grinten IV, the Berghaus Star, the whole-world Polyconic. Others have primarily novelty value, the Wagon-wheel and Collignon being examples already mentioned. These are discussed in this chapter, along with various variants of more serious projections developed in earlier chapters, including some additions to the equal-area repertoire.

Yet again, there is no new mathematics.

The maps in this chapter are visually novel. They are all presented with 30° graticules to minimize clutter while indicating internal shapes.

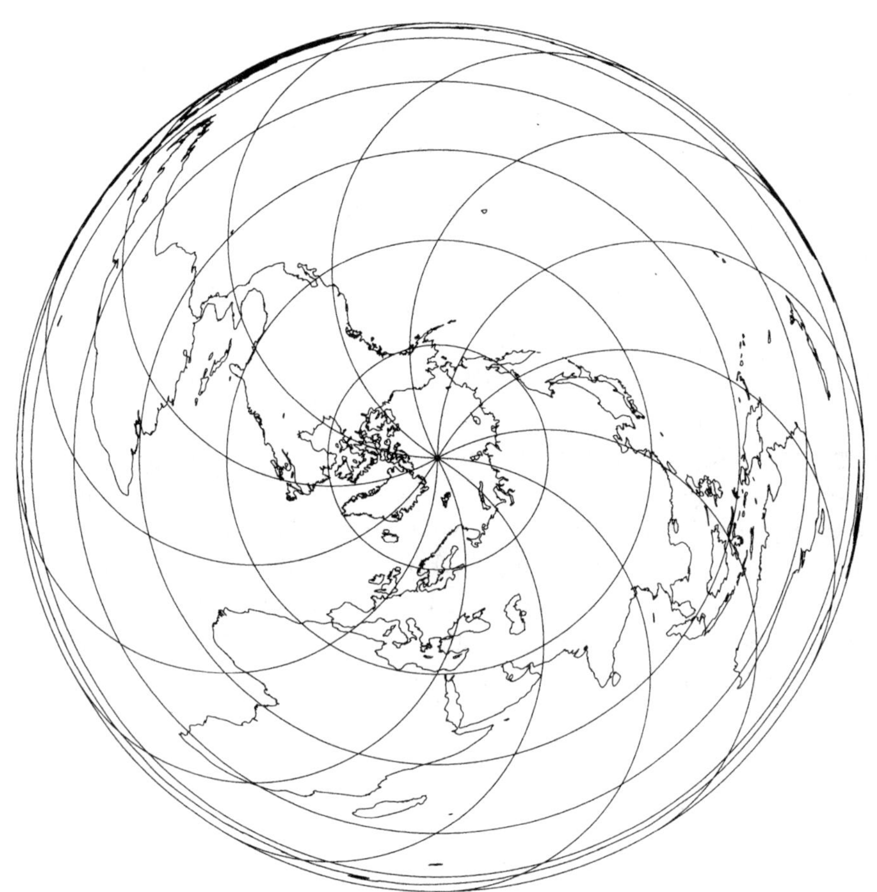

Exhibit 16–1: The Wiechel or Wagonwheel projection.

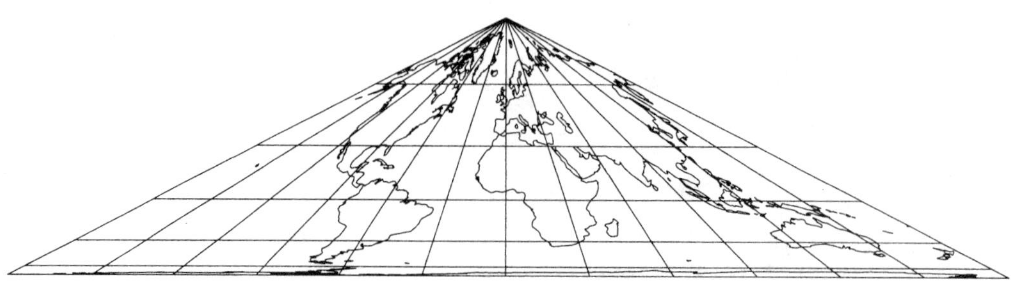

Exhibit 16–2: The Collignon projection.

NOVELTY PSEUDO... PROJECTIONS

Pseudoazimuthal projections

The adoption of curved meridians while retaining concentric circular arcs for parallels is the distinguishing feature of Pseudo projections, as discussed in Chapter 6. Although many Pseudocylindricals were developed in that lengthy chapter, plus one Pseudoconical, there was no similar development relative to the Azimuthal mode. One could develop pseudoazimuthal projections with mirror-image meridians about some selected central meridian, but the only form that has gained attention uses consistently curved meridians that maintain uniform scale along the parallels. As with the other modes, there is limitless choice — of spacing of parallels and of their curvature. The notable example is the **Wiechel** projection, called also, for reasons obvious from the Exhibit 16–1, the **Wagonwheel** projection. H. Wiechel [1879b; see also 1952a] chose the Equal-area Azimuthal projection then curved the meridians to provide true scale along them (assuming true scale at the pole). Clearly, this does not disturb the equal-area attribute. Having true scale along the meridians has obvious merits, though being markedly curved limits usefulness. Intrinsically, but interestingly, this is obtained by making the meridians arcs of circles of radius equal to that of the globe.

If $\lambda_{\text{-y}}$ is to lie along the $-y$ half-axis, Equation 32a shows the general plotting equations for any azimuthal projection, with $\oplus$ being positive for the North Pole and negative for the South Pole, as:

$$\dot{x} = +\,\dot{\rho}\cos\dot{\theta} \quad = -\,\dot{\rho}\sin\left(\lambda - \lambda_{\text{-y}}\right)$$
$$\dot{y} = \oplus\dot{\rho}\sin\dot{\theta} \quad = \oplus\dot{\rho}\cos\left(\lambda - \lambda_{\text{-y}}\right)$$

Having curved meridians means merely a revision of the angle $\dot{\theta}$ to have it vary additionally with latitude. With the curvature chosen by Wiechel, there is a progressively increasing increment/decrement to be applied to that angle — an element more simply seen as dependent on the radial distance of the point from the map's centre than on latitude, i.e., on $\dot{\rho}$. Let it be denoted by $\Delta\dot{\theta}$. Then simple consideration of the triangle between the origin, the variable point and the centre of the circular arc shows that

$$\Delta\dot{\theta} = \arcsin\frac{\dot{\rho}}{2R}$$

Keeping $\dot{\theta}$ intact, the general plotting equations become:

$$\dot{x} = +\,\dot{\rho}\cos\left(\dot{\theta} \pm \Delta\dot{\theta}\right) \quad = -\,\dot{\rho}\sin\left\{\left(\lambda - \lambda_{\text{-y}}\right) \pm \arcsin\frac{\dot{\rho}}{2R}\right\}$$
$$\dot{y} = \oplus\dot{\rho}\sin\left(\dot{\theta} \pm \Delta\dot{\theta}\right) \quad = \oplus\dot{\rho}\cos\left\{\left(\lambda - \lambda_{\text{-y}}\right) \pm \arcsin\frac{\dot{\rho}}{2R}\right\}$$

the ± depending on rotational style, to be negative for the style illustrated.

Using angle-difference formulae, we have

$$\dot{\rho}\sin\left\{\left(\lambda-\lambda_{\text{-y}}\right)\pm\arcsin\frac{\dot{\rho}}{2R}\right\}=\dot{\rho}\left\{\sin\left(\lambda-\lambda_{\text{-y}}\right)\cos\left(\arcsin\frac{\dot{\rho}}{2R}\right)\pm\cos\left(\lambda-\lambda_{\text{-y}}\right)\sin\left(\arcsin\frac{\dot{\rho}}{2R}\right)\right\}$$

The two compounded trigonometrical terms in this equation resolve surprisingly simply. Using the formula for $\dot{\rho}$ for equal-area in Azimuthals (Equation 73b), we get

$$\dot{\rho}\sin\left(\arcsin\frac{\dot{\rho}}{2R}\right)=\dot{\rho}\frac{\dot{\rho}}{2R}=\frac{2R^2(1-\sin\phi)}{2R}=R(1-\sin\phi)$$

Installing the formula for $\dot{\rho}$ in two stages, the second of those compound terms becomes

$$\begin{aligned}\dot{\rho}\cos\left(\arcsin\frac{\dot{\rho}}{2R}\right)&=\dot{\rho}\sqrt{1-\sin^2\left(\arcsin\frac{\dot{\rho}}{2R}\right)}=\dot{\rho}\sqrt{1-\left(\frac{\dot{\rho}}{2R}\right)^2}\\&=\dot{\rho}\sqrt{1-\frac{2R^2(1-\sin\phi)}{4R^2}}=\dot{\rho}\sqrt{1-\frac{(1-\sin\phi)}{2}}=\dot{\rho}\sqrt{\frac{(1+\sin\phi)}{2}}\\&=R\sqrt{2(1-\sin\phi)}\sqrt{\frac{(1+\sin\phi)}{2}}=R\sqrt{(1-\sin\phi)(1+\sin\phi)}\\&=R\sqrt{(1-\sin^2\phi)}=R\sqrt{\cos^2\phi}=R\cos\phi\end{aligned}$$

Hence,

$$\dot{\rho}\sin\left\{\left(\lambda-\lambda_{\text{-y}}\right)\pm\arcsin\frac{\dot{\rho}}{2R}\right\}=R\left\{\cos\phi\sin\left(\lambda-\lambda_{\text{-y}}\right)\pm(1-\sin\phi)\cos\left(\lambda-\lambda_{\text{-y}}\right)\right\}$$

Similar manipulation gives

$$\dot{\rho}\cos\left\{\left(\lambda-\lambda_{\text{-y}}\right)\pm\arcsin\frac{\dot{\rho}}{2R}\right\}=R\left\{-\cos\phi\cos\left(\lambda-\lambda_{\text{-y}}\right)\pm(1-\sin\phi)\sin\left(\lambda-\lambda_{\text{-y}}\right)\right\}$$

Thus the plotting equations for the Wiechel projection as illustrated are:

$$\begin{aligned}\dot{x}&=+R\left\{\cos\phi\sin\left(\lambda-\lambda_{\text{-y}}\right)-(1-\sin\phi)\cos\left(\lambda-\lambda_{\text{-y}}\right)\right\}\\\dot{y}&=-R\left\{\cos\phi\cos\left(\lambda-\lambda_{\text{-y}}\right)+(1-\sin\phi)\sin\left(\lambda-\lambda_{\text{-y}}\right)\right\}\end{aligned}\qquad\text{. . . . . (366a}$$

where $\lambda_{\text{-}y}$ is the longitude that emanates from the Pole along the $-y$ axis. The curvature could just as well be reversed, and the projection could be applied to Antarctica.

The angle subtended at the centre of its circle by the meridional arc is double

$$\begin{aligned}\Delta\dot{\theta}&=\arcsin\frac{\dot{\rho}}{2R}=\arcsin\frac{R\sqrt{2(1-\sin\phi)}}{2R}=\arcsin\sqrt{\tfrac{1}{2}\{1-\sin\phi\}}\\&=\arcsin\sqrt{\tfrac{1}{2}\left\{1-\cos\left(\tfrac{\pi}{2}-\phi\right)\right\}}\\&=\arcsin\sqrt{\sin^2\left(\frac{\pi}{4}-\frac{\phi}{2}\right)}=\arcsin\left\{\sin\left(\frac{\pi}{4}-\frac{\phi}{2}\right)\right\}=\left(\frac{\pi}{4}-\frac{\phi}{2}\right)\end{aligned}$$

which equals half the polar distance — proving that the meridian has true scale.

A simple Pseudocylindrical

In the pseudocylindrical Eckert II projection the curve of the meridians is only a break of line at the Equator, the hemispheric sections being straight. The result is an increasingly abrupt change across that central parallel as longitude ranges outwards. Were the meridians to be complete straight lines, a point-polar Pseudocylindrical would become a single line, while a flat-polar Pseudocylindrical would become a 'true' Cylindrical. Édouard Collignon in a wider context [1865b] chose a composite of point- and flat-polar Pseudocylindricals, and thereby kept meridians straight over their whole length. The result is a triangular map, producing severe narrowing between meridians in regions included near the one pointed pole. Collignon chose to make his projection equal-area, so there is a reciprocal narrowing between parallels toward the flat pole. The projection is routinely shown with the North Pole as the point, so compromising the northern lands while making expansive the southern oceans.

The narrowing between parallels is proportional to the length of the flat pole, that between meridians is inversely so. The area of a triangle with the apex at the point pole and the base parallel towards the flat pole fraction f of the length of the central meridian from the apex, is fraction f^2 of the total area.

If the length of the central meridian on the polar-point side of the Equator is d, then equal area implies its overall length is $d\surd 2$ and the lengths of the parallels for Equator and flat pole are of the same ratio, i.e., $1{:}\surd 2$. Collignon chose to make the flat pole four times the length of the central meridian. To be authalic, the overall area of $4d^2$ must equal $4\pi R^2$, so $d = R\surd\pi$ and the flat pole is of length $4R\surd(2\pi)$. Hence, setting uniform scale along all parallels, assuming the pointed pole is North, and putting the origin on the Equator:

$$\text{if } \phi = +\tfrac{\pi}{2} \qquad \dot{x} = 0, \qquad \dot{y} = R\sqrt{\pi}$$

$$\text{if } \phi = -\tfrac{\pi}{2} \qquad \dot{x} = 2\sqrt{2\pi}\,R\,\lambda, \quad \dot{y} = R\sqrt{\pi}\left(1 - \sqrt{2}\right)$$

If the base of the triangle fraction f down the central meridian is to represent latitude ϕ, then from Equation 73a, we have $2f = 1 - \sin\phi$ (noting that the sine function changes sign across the Equator). This fraction is $d\surd 2 = +R\surd(2\pi)$ and measured downward from $\dot{y} = +d = +R\sqrt{\pi}$; we thereby have an equation for $\dot{y}$. The other co-ordinate is easily obtained by applying the fraction to the length of the flat pole to give plotting equations for the **Collignon** projection, as illustrated in Exhibit 16–2 on page 364, of:

$$\dot{x} = \frac{2R\lambda}{\sqrt{\pi}}\sqrt{1 - \sin\phi}$$
$$\dot{y} = R\sqrt{\pi}\left\{1 - \sqrt{1 - \sin\phi}\right\} \qquad \text{. . . . . (367a}$$

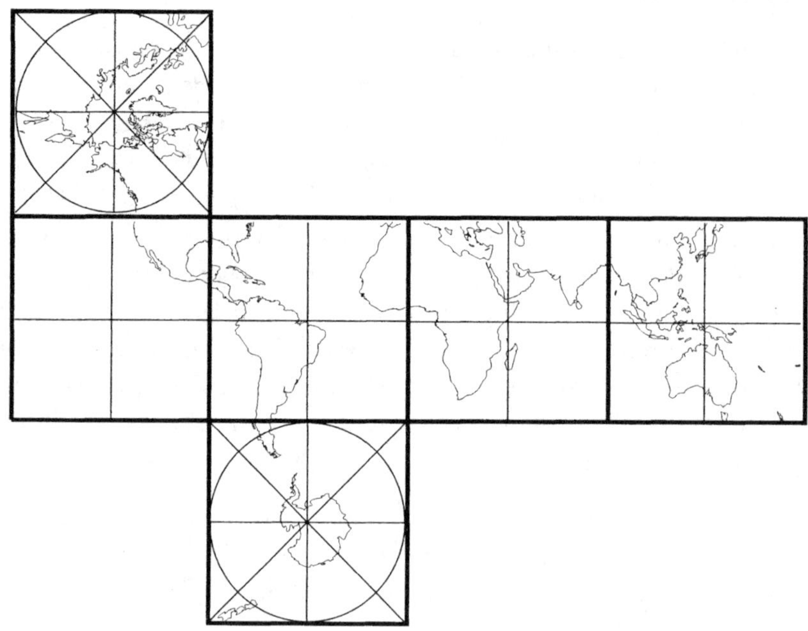

Exhibit 16–3: The six flat regional maps of the Cube projection.

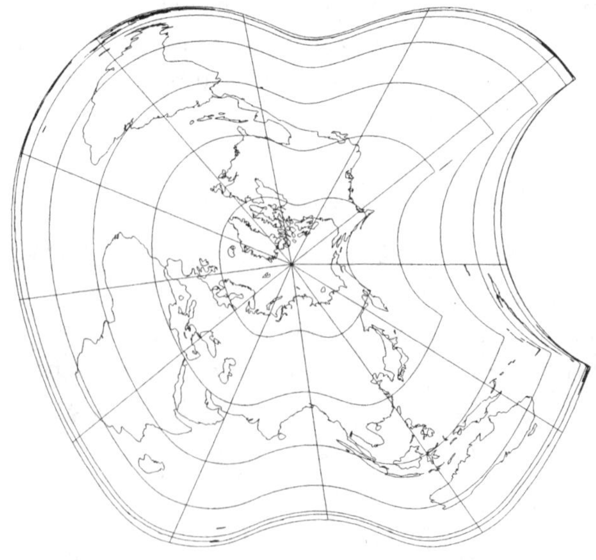

Exhibit 16–4: The Apple projection.

PSEUDOGLOBES

Concept

The globe adopted in Chapter 3 to mimic Earth, and used to this point for developments, is spherical to represent an ostensibly spherical Earth. Later, in Chapter 16, we develop spherical globes that mimic an ostensibly ellipsoidal Earth. Various efforts have been made over centuries, however, to map the round Earth onto 3-dimensional bodies of very different shape than the sphere. Generally, this has been onto polyhedra — closed bodies of multiple flat planes — with the pyramidal tetrahedron and the cube being obvious simple cases, but with limitless opportunity for elaboration. Despite the target of producing flat maps, however, projections have also been developed that involve a curved, markedly non-spherical intermediate body. Both polyhedral and curved bodies can be embraced by the term *pseudoglobes*. We address the two forms in turn below.

Polyhedral pseudoglobes

Beginning with Albrecht Dürer, Leonardo da Vinci and others half a millennium ago, various efforts have been made to map the rounded globe successfully onto a set of flat surfaces forming a polyhedric pseudoglobe and thereby create a flat mosaic map. Primarily these have used the surfaces of the *Platonic solids* in regular form, i.e., with all faces identically shaped and sized. These are the regular:

tetrahedron, with 4 sides, all being equilateral triangles
cube, with 6 sides, all being squares — illustrated opposite in its flat 6-panel form
octahedron, with 8 sides, all being equilateral triangles
dodecahedron, with 12 sides, all being regular pentagons
icosahedron, with 20 sides, all being equilateral triangles.

Additionally, there is the cuboctahedron — a cube with its corners clipped, thus with eight equilateral triangular surfaces added.

Whichever polyhedron is used, it is usually circumscribed to the globe, i.e., with the planar surfaces touching tangentially, including two tangential at the Poles. J. P. Snyder [1993a] mentions the following authors of *polyhedric* projections, several of which achieved recognition with patents:

A. Dürer [1538a], dodecahedron
also proposed using regular tetrahedron, icosahedron and others
Leonardo da Vinci and others (1500s), eight equilateral triangles each covering 90° of longitude within one hemisphere
I.-G Pardies (1673), applying Dürer's work
C. G. Reichard [1803b], regular tetrahedron

J. W. Woolgar (1833), cube — the six panels for which, using the Plate-Carrée, are illustrated flat in Exhibit 16–3
J. N. Adorno (1851), icosahedron, British patent
J. M. Boorman [1877a], non-Platonic solids of 15, 22, 23, 24, 32 and 37 faces, U.S. patent
B. J. S. Cahill [1909a], octants, following Leonardo, U.S. patents [1913a; 1913b]
A. R. Hinks [1921g], cube
I. Fisher [1943a], regular icosahedron, U.S. patent [1948d]
R. B. Fuller [1943c], cuboctahedron, plus custom projection named "Dymaxion", has constant scale along all edges; granted U.S. patent [1946a]
Subsequently applied to the regular icosahedron
J. A. Smith [1939a], dodecahedron, U.S. patent
A. D. Bradley [1946b], icosahedron
F. V. Botley [1949e], tetrahedron
F. V. Botley [1954a], octahedron
L. P. Lee [1965b, 1976a], all five regular Platonic polyhedra
S. Gurba [1970a], dodecahedron.

Because of having all great circles as straight lines, the Gnomonic projection has been the most used, with its rapid distortion away from the centre of any one map favouring this multifaceted approach. Lee, however, produced his polyhedral maps within his larger advanced work with conformal projections mentioned in the Chapter 15.

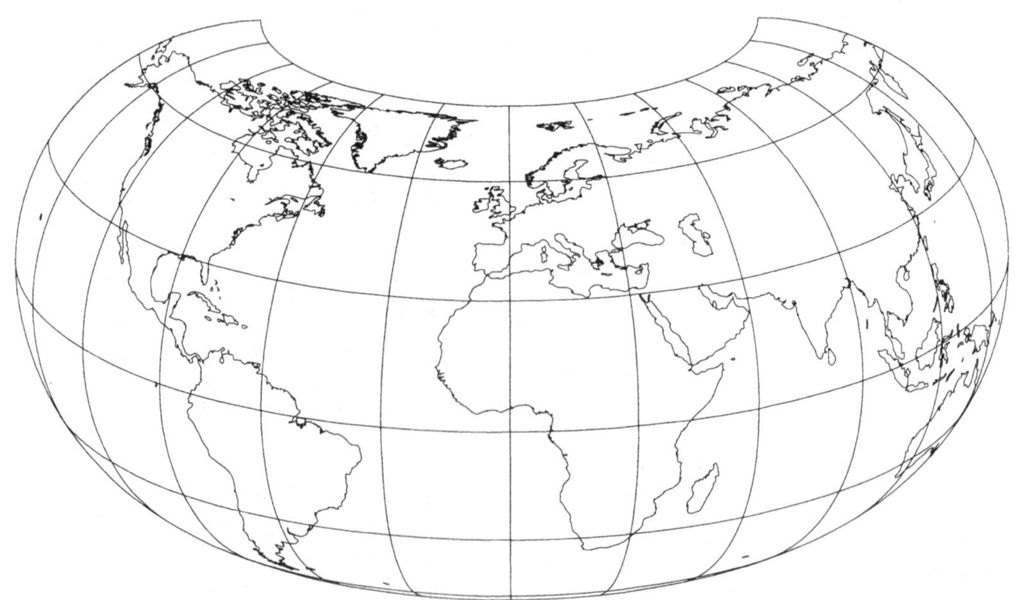

Exhibit 16–5: The Armadillo projection.

Curved pseudoglobes

Erwin Raisz [1943b, reviewed in 1962d] turned to curved pseudoglobes, although only as a transition to flat maps. Drawing a graticule on any suitable curved 3-dimensional body then projecting that orthographically onto flat paper, he produced a variety of innovative projections, for which he coined, invoking the Greek word for *arch*, the name *orthoapsidal*. The most impressive, illustrated in Exhibit 16–5 and was named because of its perceived similarity to a sleeping animal the **Armadillo** projection, elicits a surprising sense of the doubly curved reality of Earth. It uses a torus as the pseudoglobe, and projects that obliquely.

The ellipsoid provides a simpler pseudoglobe — the frontal part of a broad ellipsoid being used for the map; Raisz had the major axis twice the minor. The basic version has the meridians projected at Polar aspect onto the ellipsoid but the parallels projected at Oblique aspect to become elliptical curves. With an emphasis on the major land masses, 20°N was chosen for the obliqueness. An alternative version respaced the parallels to provide an equal-area map, keeping that for 20°N stable. This renders the North Pole into a proper circular arc (whereas it is a point in the basic version). The elliptical meridians make it similar to the Eckert III, but that projection has the Poles as straight lines.

The Armadillo also has the projection of parallels oblique at 20°N, though, like the others, it could be focussed at any latitude. If focussed on a southern latitude, the torus or other pseudoglobe is used correspondingly from the other perspective. Unless the focus is the Equator — i.e., the aspect is actually the Equatorial — the coverage is markedly less than the whole world. If we denote the latitude of focus by ϕ_0, the plotting equations for the Armadillo projection are:

$$\text{for } \phi \geq -\arctan\frac{\cos\frac{\lambda}{2}}{\tan\phi_0} \text{ only}$$

$$\dot{x} = R(1+\cos\phi)\sin\frac{\lambda}{2} \qquad \text{. . . . . (371a}$$

$$\dot{y} = R\left\{\frac{1+\sin\phi_0-\cos\phi_0}{2}\right\} + \sin\phi\cos\phi_0 - (1+\cos\phi)\sin\phi_0\cos\frac{\lambda}{2}$$

Raisz cited beans and scallops as examples from the limitless choices as pseudoglobe, and illustrated the application to an interrupted hyperboloid. B. J. S. Cahill [1914b] produced a somewhat similar interrupted projection centred on the North Pole and ingeniously arranged to have the worst distortions of meridians well out in the oceans. Although offering an attractive multilobed image of Earth, the technique has the peculiar result of duplication of a major sector of the globe.

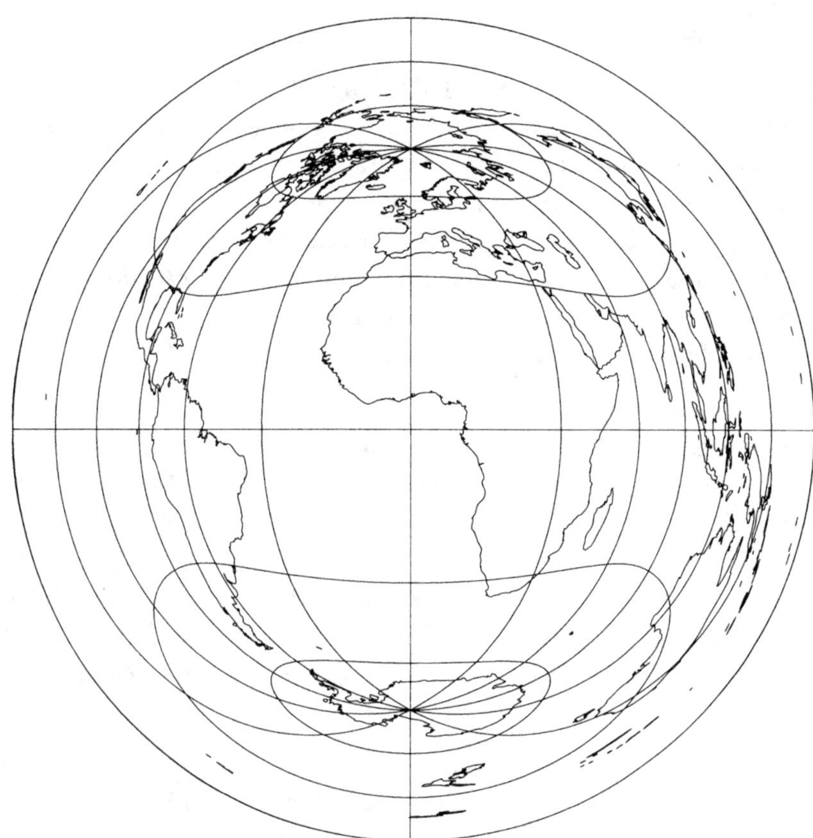

Exhibit 16–6: The Fish Eye projection.

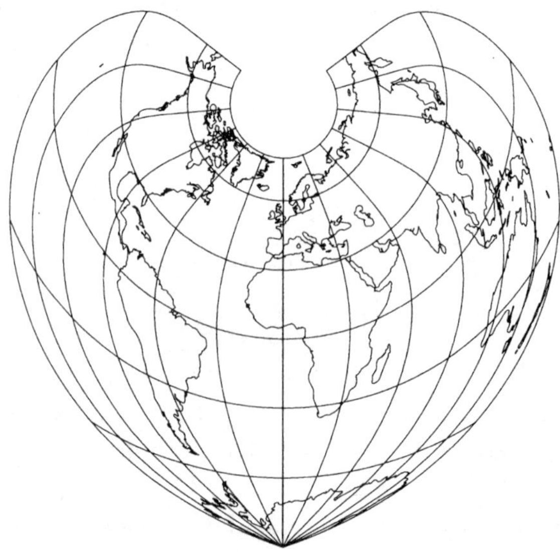

Exhibit 16–7: The Amulet projection.

OTHER RESHAPING

Purpose

Whether of a neighbourhood within a city, else of region within a wider world, there is often the need for a map having more precise detail at the centre with less precision in outer areas. Azimuthal projections inherently serve such purpose, with decline in precision occurring progressively but immediately once the focal point has been departed. Snyder [1987b] created his **Magnifying-glass** projections to maintain precision over an inner zone while having an extended surround of less precision but some consistency. The easiest zone for development is circular, but Snyder also addressed rectangular zones. We address each in turn. Within the zone, the map is true to the projection (although it could be enlarged), but beyond it is shrunken radially.

Magnifying an inner circular zone

With a circular zone and an Azimuthal projection concentrically focused, the question is essentially how the surround is to be scaled. Applying the question to linear scale with the Equidistant and to area scale for the Equal-area Azimuthal, Snyder merely fixed the scale beyond the central zone to a fraction of that within. The fraction chosen depended on the angular span wanted for zone and overall map. We denote by α the great-circle angular radius of the inner zone, and by β the great-circle angular radius of the total area being covered, with $0 < \alpha < \beta \leq \pi$.

For the Equidistant, with its simple formula

$$\dot{\rho} = R\gamma$$

for radial distance at angular distance γ from its centre and linear scale as the fractioned factor, if g_d is the fraction then the effective angular span of the map is

$$C_d = \alpha + g_d(\beta - \alpha) \quad = (1 - g_d)\alpha + g_d\beta$$

If $\dot{r}_\beta$ is the radius of the physical map, then

$$\dot{r}_\beta = RC_d$$

and the radial plotting parameter becomes:

if $\phi \leq \alpha$

$$\dot{\rho} = \dot{r}_\beta \frac{\gamma}{\alpha + g_d(\beta - \alpha)} \quad = \dot{r}_\beta \frac{\gamma}{C_d}$$

if $\alpha < \phi \leq \beta$

$$\dot{\rho} = \dot{r}_\beta \frac{\alpha + g_d(\gamma - \alpha)}{\alpha + g_d(\beta - \alpha)} \quad = \dot{r}_\beta \frac{\alpha + g_d(\gamma - \alpha)}{C_d}$$

For the radius $\dot{r}_\alpha$ of the inner zone on the map, we then have

$$\dot{r}_\alpha = R\alpha = \dot{r}_\beta \frac{\alpha}{\alpha + g_d(\beta - \alpha)}$$

$$= \dot{r}_\beta \frac{\gamma}{(1 - g_d)\alpha + g_d\beta}$$

Hence,

$$\frac{\dot{r}_\alpha}{\dot{r}_\beta} = \frac{R\alpha}{\dot{r}_\beta} = \frac{\alpha}{\alpha + g_d(\beta - \alpha)}$$

$$(\alpha + g_d(\beta - \alpha))r_\alpha = (\alpha + g_d(\beta - \alpha))R\alpha \quad = \alpha\,\dot{r}_\beta$$

$$g_d = \left(\alpha\frac{\dot{r}_\beta}{\dot{r}_\alpha} - \alpha\right)\frac{1}{\beta - \alpha} \quad = \left(\frac{\dot{r}_\beta}{\dot{r}_\alpha} - 1\right)\frac{\alpha}{\beta - \alpha} \quad = \frac{1 - \dot{h}}{\dot{h}} \cdot \frac{\alpha}{\beta - \alpha} \qquad \text{where } \dot{h} = \frac{\dot{r}_\alpha}{\dot{r}_\beta}$$

giving the value of g_d deriving from the choice of angles and radii pertaining to the two circles — those of the inner zone and those of the map.

For the Equal-area Azimuthal the process is complicated first by the basic formula for and then by the need to relate the factoring to the linear element while having it defined by the consequent area scale. Using revised factors g_a for the fraction (of area scale) and

$$C_a = \sqrt{1 - (1 - g_a)\cos\alpha - g_d \cos\beta}$$

then the radial plotting parameter becomes:

if $\phi \le \alpha$

$$\dot{\rho} = r_\beta \frac{\sqrt{1 - \cos\gamma}}{C_a}$$

if $\alpha < \phi \le \beta$

$$\dot{\rho} = r_\beta \frac{1 - (1 - g_a)\cos\alpha - g_a \cos\gamma}{C_a}$$

and the radii in angular terms become:

$$\dot{r}_\beta = \sqrt{2}\,RC_a$$

$$\dot{r}_\alpha = \sqrt{2}\,R\sqrt{1 - \cos\alpha}$$

so

$$g_a = \frac{(1 - \dot{h}^2)(1 - \cos\alpha)}{\dot{h}^2(\cos\alpha - \cos\beta)} \qquad \text{where again } \dot{h} = \frac{\dot{r}_\alpha}{\dot{r}_\beta}$$

giving the value of g_a deriving from the choice of angles and radii for the two circles.

$$C_a = \frac{\dot{r}_\beta}{\dot{r}_\alpha}\sqrt{1 - \cos\alpha}$$

Presuming the choice of factors favours actual relative magnification of the central zone, the resulting map has a relatively spacious eye surrounded by a more densely packed graticule. Since a uniform fraction was applied to scale beyond the central zone, the boundary of that zone shows major breaks of the lines representing parallels and meridians, and their packing is essentially uniform. Snyder, however, also developed versions with a progressively increasing packing, hence less abrupt change of graticule lines across the inner boundary. These he referred to as *tapered*, the rate of such being another parameter of choice for the projecting. Snyder chose to decrease the scale from that of the inner zone at its bounding circle to zero at the map's bounding circle. He applied the methodology to the Stereographic and Gnomonic as well as the two already mentioned, using different tapering curves for each. Tapering destroys the key distinctive features of the projections — conformality and straight great circles for the two additional ones along with equidistance and equal-area for the earlier ones — but keeps them all fully azimuthal. Reference should be made to his paper [1987b] for details. Daniel Strebe's distinct **Fish-eye** projection shown on page 372 illustrates the style of a tapered map.

Magnifying an inner non-circular zone

With elaboration of the algebra to have the distance from centre dependent on azimuth, the erstwhile circle bounding the inner zone and that bounding the map can be changed to almost any other closed shape, while keeping the map fully azimuthal. Thus the magnification effect can be attuned to a corresponding variety of shapes for the zone of interest. Snyder specifically addressed the use for rectangular zones.

Further novelties

Two equal-area projections created by Strebe [private communication] complete the display of novelty projections. The **Amulet** projection shown in Exhibit 16–7 is a composite of the Equal-area Conical and the Bonne projections, welded together where they match in shape and size. The **Apple** projection, shown in Exhibit 16–4 on page 368, has its distinctive outer shell constructed from sections of parabolae then its inside based on the Equal-area Azimuthal with complicated longitude-dependent mathematical adjustment of radial distance. The technique is applicable to any defined outer shell.

TUTORIAL 31 INFINITE SERIES

If the infinite series of real numbers

$$\sum_{n=0}^{\infty} a_n = a_0 + a_1 + a_2 + a_3 + a_4 + \ldots$$

has finite value a, it is said to *converge*, otherwise it is said to *diverge*. It can be shown that it converges if and only if its terms meet the ratio test

$$\lim_{n\to\infty} \frac{|a_{n+1}|}{|a_n|} = \lim_{n\to\infty} \left|\frac{a_{n+1}}{a_n}\right| < 1$$

If n is a non-negative integer, a series of the form

$$\sum_{n=0}^{\infty} c_n x^n = 1 + c_1 x^1 + c_2 x^2 + c_3 x^3 + c_4 x^4 + \ldots$$

is termed a power series. It converges if and only if

$$\lim_{n\to\infty} \left|\frac{c_{n+1}x^{n+1}}{c_n x^n}\right| = \lim_{n\to\infty} \left|\frac{c_{n+1}x}{c_n}\right| = \lim_{n\to\infty} \left\{\left|\frac{c_{n+1}}{c_n}\right| \cdot |x|\right\} < 1$$

Let $f(x)$ be a power series defined at least for a range of values of x that includes a, with

$$f(x) = \sum_{n=0}^{\infty} c_n (x-a)^n = 1 + c_1(x-a)^1 + c_2(x-a)^2 + c_3(x-a)^3 + c_4(x-a)^4 + \ldots$$

Then, differentiating repeatedly, we get:

$$f^{(1)}(x) = 0 \quad + c_1 \quad + 2c_2(x-a)^1 + 3c_3(x-a)^2 \quad + 4c_4(x-a)^3 + \ldots$$

$$f^{(2)}(x) = \quad 0 \cdot c_1 + 1 \cdot 2c_2 \qquad + 2 \cdot 3c_3(x-a)^1 + 3 \cdot 4c_4(x-a)^2 + \ldots$$

etc. Evaluating these at $x = a$ we get:

$$f^{(1)}(a) = f^{(1)}(x)_{x=a} = 1! \cdot c_1$$

$$f^{(2)}(a) = f^{(2)}(x)_{x=a} = 2! \cdot c_2$$

etc. Thus we have

$$f(x) = f(a) + \frac{f^{(1)}(a)}{1!}(x-a)^1 + \frac{f^{(2)}(a)}{2!}(x-a)^2 + \frac{f^{(3)}(a)}{3!}(x-a)^3 + \ldots$$

$$= \sum_{n=0}^{\infty} \frac{f^{(n)}(a)}{n!}(x-a)^n$$

This is called the *Taylor Series* of $f(x)$ at a after its developer Brook Taylor of England (1685 – 1731) but for $a = 0$ it is called the *Maclaurin Series* of $f(x)$, the idea having been first developed in this restricted form by C. Maclaurin of Scotland (1698 – 1746).

PART C

An Ellipsoidal World

That Earth is not a perfect sphere was realized some 300 years ago. In 1669 the young Isaac Newton took his Chair as Professor of Mathematics at Cambridge. In the same year Giovanni Cassini, who had been appointed Professor of Astronomy in Bologna at an even younger age, was persuaded to become the founding Director of the Paris Observatory. Cassini initiated a survey in France that discovered the non-uniformity of the spacing of parallels. Newton, meanwhile, developed his theory of gravitation and asserted that Earth must bulge around its middle. Misled by a lack of appreciation for the phenomenon of geodesic latitude, the French surveyors concluded the opposite. Both eminent men were dead by the time the matter was resolved in the 1730s, using data from the more extreme locations of Finland and Peru. Earth bulges around its middle — which confuses as well as compounds the cartography. Its cross-section along the axis is close to elliptical. We now develop our science on the basis of an ellipsoidal world.

TUTORIAL 32 BINOMIAL SERIES

For any real number p the Taylor Series (Tutorial 31, page 376) gives, for $|x| < 1$,

$$(1+x)^p = 1 + \frac{p}{1!}x^1 + \frac{p(p-1)}{2!}x^2 + \frac{p(p-1)(p-2)}{3!}x^3 + \frac{p(p-1)(p-2)(p-3)}{4!}x^4 + \ldots$$

$$= \sum_{n=0}^{\infty} c_n x^n \qquad \text{where} \quad c_n = \frac{p!}{n!(p-n)!}$$

A generalization of the Binomial Theorem, this is the *Binomial Series* for x and p. We explore its use for a key function with ellipsoids where ellipticity $\varepsilon < 1$, so $|e^2 \sin^2\phi| < 1$.

$$(1-\varepsilon^2\sin^2\phi)^{\frac{-3}{2}} = 1 + \left(-\frac{3}{2}\right)\frac{(-\varepsilon^2\sin^2\phi)}{1!} + \left(-\frac{3}{2}\right)\left(-\frac{5}{2}\right)\frac{(-\varepsilon^2\sin^2\phi)^2}{2!}$$

$$+\left(-\frac{3}{2}\right)\left(-\frac{5}{2}\right)\left(-\frac{7}{2}\right)\frac{(-\varepsilon^2\sin^2\phi)^3}{3!} + \left(-\frac{3}{2}\right)\left(-\frac{5}{2}\right)\left(-\frac{7}{2}\right)\left(-\frac{9}{2}\right)\frac{(-\varepsilon^2\sin^2\phi)^4}{4!} + \ldots$$

$$= 1 + \frac{3}{2}\varepsilon^2\sin^2\phi + \frac{15}{8}\varepsilon^4\sin^4\phi + \frac{35}{16}\varepsilon^6\sin^6\phi + \frac{315}{128}\varepsilon^8\sin^8\phi + \ldots$$

For our geoidal ellipsoids, $\varepsilon^{2n}\sin^{2n}\phi \leq \varepsilon^{2n} < 100^{-n}$. So, using formulae of Tutorial 10,

$$(1-\varepsilon^2\sin^2\phi)^{\frac{-3}{2}} = 1 + \frac{3}{2}\varepsilon^2\left\{\frac{1}{2} - \frac{\cos 2\phi}{2}\right\}$$

$$+\frac{15}{8}\varepsilon^4\left\{\frac{3}{8} - \frac{\cos 2\phi}{2} + \frac{\cos 4\phi}{8}\right\}$$

$$+\frac{35}{16}\varepsilon^6\left\{\frac{5}{16} - \frac{15\cos 2\phi}{32} + \frac{3\cos 4\phi}{16} - \frac{\cos 6\phi}{32}\right\} \qquad \text{. . . (378a}$$

$$+\frac{315}{128}\varepsilon^8\left\{\frac{35}{128} - \frac{7\cos 2\phi}{16} + \frac{7\cos 4\phi}{32} - \frac{\cos 6\phi}{16} + \frac{\cos 8\phi}{128}\right\} + \ldots$$

$$= B_0 - B_2\,\varepsilon^2\cos 2\phi + B_4\,\varepsilon^4\cos 4\phi - B_6\,\varepsilon^6\cos 6\phi + B_8\,\varepsilon^8\cos 8\phi + \ldots$$

where the coefficients, each a series converging to a positive value less than 2, are

$$B_0 = 1 + \frac{3}{4}\varepsilon^2 + \frac{45}{64}\varepsilon^4 + \frac{175}{256}\varepsilon^6 + \frac{11025}{16384}\varepsilon^8 + \ldots$$

$$B_2 = \frac{3}{4} + \frac{15}{16}\varepsilon^2 + \frac{525}{512}\varepsilon^4 + \frac{2205}{2048}\varepsilon^6 + \ldots$$

$$B_4 = \frac{15}{64} + \frac{105}{256}\varepsilon^2 + \frac{2205}{4096}\varepsilon^4 + \ldots \qquad \text{. . . . . (378b}$$

$$B_6 = \frac{35}{512} + \frac{315}{2048}\varepsilon^2 + \ldots$$

$$B_8 = \frac{315}{16384} + \ldots$$

CHAPTER 17

Squeezing the Circle: Ellipses and ellipsoids

Preamble

The circle is merely a special form of ellipse but, as with Simple versus the general Oblique aspect, encompassing the more general case brings considerable complication to all the mathematics. Basing our mapping on an ellipsoidal Earth complicates not only projection geometry, curvature, surface distance, etc., but also how we map our geographic points onto a spherical globe when using the latter for equal-area and conformal purposes. An ellipsoidal Earth also brings more fundamental complication in that it makes the concept of latitude ambiguous, between a surface-based measure and a geocentric measure.

Differentiation of vectors, including of their dot- and cross-products, is an essential part of studying the ellipsoid, as are infinite series and binomial series — all new facets of mathematics introduced in this chapter.

The chapter is devoid of maps.

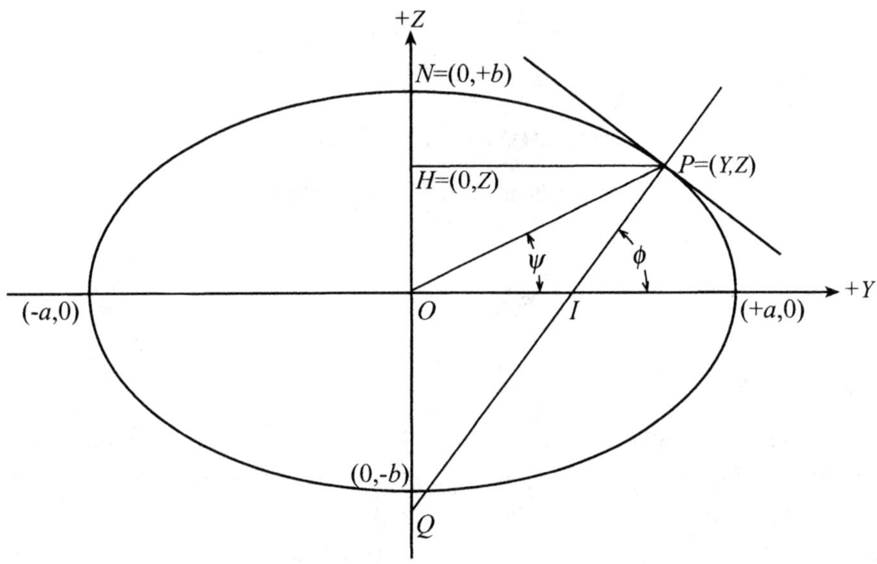

Exhibit 17–1: The tangent to the ellipse.

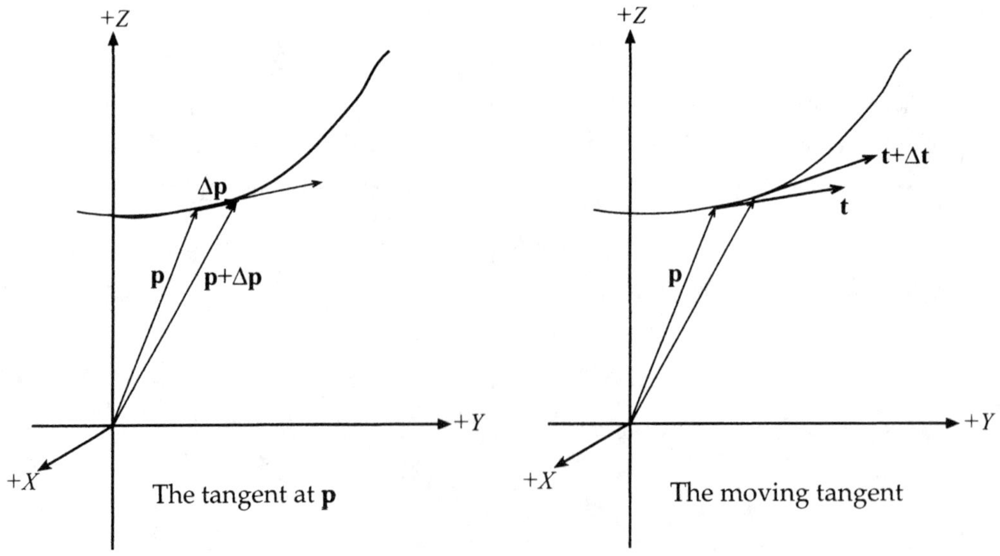

Exhibit 17–2: The space curve.

BASIS

The oblate ellipsoid

The radius of Earth is close to 6378 km everywhere throughout the Equatorial plane, but more than 20 km less to each Pole (varying between the two relative to the Equatorial plane). The cross-section through the Poles is thus roughly elliptical. We adopt a truly elliptical cross-section and revolve it about its minor axis, producing a *surface of revolution* — an ellipsoid[1], more expressly, being revolved about its minor axis, an *oblate ellipsoid*, of formula

$$1 = \frac{X^2}{a^2} + \frac{Y^2}{a^2} + \frac{Z^2}{b^2} \quad = \frac{X^2 + Y^2}{a^2} + \frac{Z^2}{b^2} \quad = \frac{X^2 + Y^2}{a^2} + \frac{Z^2}{\left(1 - \varepsilon^2\right)a^2} \qquad \text{. . . . . . (381a}$$

where ε is the eccentricity of the rotated ellipse. That cross-sectional ellipse, basic to our ellipsoid, we express in terms of the second two co-ordinates, for consistency between the 2- and the 3-dimensional scenes. Hence, its equation is

$$1 = \frac{Y^2}{a^2} + \frac{Z^2}{b^2} \quad = \frac{Y^2}{a^2} + \frac{Z^2}{\left(1 - \varepsilon^2\right)a^2} \quad \text{or} \quad a^2 = Y^2 + \frac{Z^2}{\left(1 - \varepsilon^2\right)} \qquad \text{. . . . . . . (381b}$$

Such an ellipsoid provides a mathematical formulation better approximating the Mean Sea Level surface of Earth, but it is still only an approximation (as discussed in Chapter 19). As a basis for mapping, the minor axis is obviously made coincident with Earth's axis, putting the major axis of any ellipse in the Equatorial plane. While appropriate for whole-world maps, better approximations for regional maps are obtainable by displacing the ellipsoid from this obvious position. For decades, the U.S. authorities, for instance, adopted such displacement for precise mapping of its 48 contiguous states, but modern practice favours using one identical ellipsoid for all regions, as well as the world at large. This reflects the growing globalization of human activity, epitomised by the Global Positioning System which, although created and operated solely by the U.S.A., is a worldwide facility that dictates a matching standard. The pertinent ellipsoid is the WGS84; the dimensions, including the flattening f (see Tutorial 8, page 24, written here, as it usually is, as a ratio) are

$a = 6378.137$ km $b = 6356.752_{\sim}$ km $\varepsilon = 0.081\,819\,79_{\sim}$ $f = 1{:}298.257\,2_{\sim}$

Many ellipses and ellipsoids have been adopted over time. The more important are shown, with indication of usage, on page 406.

[1] The term *spheroid* is frequently used interchangeably with *ellipsoid*. Though the latter is more cumbersome, it has the advantage of being unambiguous, whereas spheroid can be taken to imply an object loosely similar rather than necessarily exactly an ellipsoid. So, ellipsoid is preferred here.

AMBIGUITY OF LATITUDE

Background

Latitude was introduced in Chapter 1 as the angular distance measured at the centre of Earth, and stated as being ascertained by observation of elevation of some celestial object. That elevation can be measured directly relative to a horizontal else as a complementary angle relative to a vertical. While the visible horizon is used for such purpose aboard ship, on land it is preferable to use some gravity-set reference table. Unfortunately, while Earth's gravitational attraction can be regarded as concentrated at its centre for astronomical interactions, within its own vicinity the specific distribution of matter must be considered. On its surface the net result is that, neglecting extremely varied geological circumstances, a plumb line points down orthogonally to the local tangent. Except at the Poles and points on the Equator, such a line does not point to the centre of Earth. Thus, though the difference is never as much as a quarter of a degree, we must recognize two forms of latitude — *geocentric latitude* and *geodesic latitude.* The latter, having the convenience of being the observable form, is the one used in everyday life. The other is clearly the more appropriate for cartographic development. Hence, we need to establish a conversion formula between them.

Exhibit 17–1 on page 380 shows the geometry for a tangent to an ellipse, including the two lines intersecting the x axis defining these two forms of latitude. Angle ϕ is the geodesic latitude and ψ is the geocentric latitude. Differentiating Equation 381b with respect to Y we get

$$\frac{2Y}{a^2}+\frac{2Z}{b^2}\frac{dZ}{dY}=0 \qquad \text{or} \qquad \frac{dZ}{dY}=-\frac{2Y\,b^2}{2Z\,a^2}=-\frac{Y\,b^2}{Z\,a^2}$$

The derivative is the slope of the tangent at point P, which the geometry shows to be identically $\cot\phi$. Similarly it shows $Z = Y\tan\psi$. Hence,

$$\cot\phi=\frac{b^2}{a^2}\cot\psi \qquad \text{or} \qquad \tan\phi=\frac{a^2}{b^2}\tan\psi \qquad \ldots\ldots\ldots \text{(382a}$$

For the WGS84 this becomes

$$\tan\phi = 1.006\,739\,596_{\sim}\ \tan\psi \qquad \text{or} \qquad \tan\psi = 0.993\,305\,522_{\sim}\ \tan\phi$$

While the maximum discrepancy between the two, occurring at 45°, is less than a fifth of a degree, the proportional effect is greatest at the extreme latitudes. Indeed, it has the salient consequence of giving us the *sea mile,* defined as one minute of arc along a meridian — its minimal value at the Equator, despite the radius there being the longest. With geocentric latitude, one minute of arc produces a length exceeding 1855 m at the Equator and less than 1850 m at a Pole. With geodesic latitude, these become less than 1843 m at the Equator and exceeding 1861 m at a Pole. The standardized *nautical mile* is now fixed internationally at 1852 m.

CURVATURE OF A NON-SPHERICAL SURFACE

On the general surface

While the sphere has a universal uniformity of curvature, our oblate ellipsoid has a continually changing curvature from Equator to Pole along any meridian. We now set out to establish the radius of curvature at any point along a meridian. This requires further differential geometry. As in Chapter 12, we develop this first in a general context, ascertaining the formulae for any curve on any surface, then apply the results to the specifics of our circumstances. Exhibit 17–2 on page 380 shows the plain geometry in two diagrams, the left-hand one defining the tangent and the right-hand one expressing the moving tangent.

Although 3-dimensional, any curve on any surface is a function of just one parameter, say v. Hence if $\mathbf{p}$ is the vector to the point $P = (X, Y, Z)$ on the curve, then

$$\mathbf{p} = \mathbf{p}\{v\} = \mathbf{p}(X\{v\}, Y\{v\}, Z\{v\})$$

If $\mathbf{p} + \Delta\mathbf{p}$ is the vector to a nearby point on the curve, and $\Delta s = |\Delta\mathbf{p}|$ (the length of $\Delta\mathbf{p}$) then

$$\breve{\mathbf{t}} = \lim_{\Delta s \to 0} \frac{\Delta\mathbf{p}}{\Delta s} = \frac{d\mathbf{p}}{ds} \qquad \text{. . . . . (383a}$$

is a vector of unit length along the tangent line — the *unit tangent vector* at P. Since the dot-product of any vector with itself is the square of its length, we have

$$1 = \breve{\mathbf{t}} \cdot \breve{\mathbf{t}} = \left|\frac{d\mathbf{p}}{ds}\right|^2 = \left|\frac{d\mathbf{p}}{dv}\frac{dv}{ds}\right|^2 = \left|\frac{d\mathbf{p}}{dv}\right|^2 \left(\frac{dv}{ds}\right)^2 \qquad \text{. . . . . (383b}$$

The variable s represents distance along the curve. Differentiating our unit tangent vector with respect to it provides measure of the change to the tangent vector with distance along the curve. Adopting the up-curved overscript to indicate unit length generally, let

$$\frac{d\breve{\mathbf{t}}}{ds} = -\kappa\breve{\mathbf{n}} \qquad \text{. . . . . (383c}$$

Then, κ represents the rate at which the direction of the tangent vector changes as we move along the curve, referred to as the *curvature.* The new unit vector is called the *principal unit normal vector;* the inclusion of the negative sign ensures that this vector points inward relative to any non-zero curvature.

Because the dot-product of any unit vector with itself is the fixed value 1, hence has value 0 for its derivative, differentiating the product gives

$$0 = \frac{d}{ds}(1) = \frac{d}{ds}\left(\breve{\mathbf{t}} \cdot \breve{\mathbf{t}}\right) = \frac{d\breve{\mathbf{t}}}{ds} \cdot \breve{\mathbf{t}} + \breve{\mathbf{t}} \cdot \frac{d\breve{\mathbf{t}}}{ds} = 2\frac{d\breve{\mathbf{t}}}{ds} \cdot \breve{\mathbf{t}}$$
$$= -2\kappa\breve{\mathbf{n}} \cdot \breve{\mathbf{t}}$$

Hence, the two unit vectors $\breve{\mathbf{n}}$ and $\breve{\mathbf{t}}$, having zero for their dot-product, must be mutually orthogonal.

We introduce a third vector, called the *binormal unit vector,* equal to the cross-product of the two, which must then be orthogonal to both and also of unit length, namely

$$\breve{\mathbf{b}} = \breve{\mathbf{t}} \times \breve{\mathbf{n}} \qquad \text{. . . . . (384a}$$

Each being orthogonal to the others, this trio forms a right-hand triad, with

$$\breve{\mathbf{t}} = \breve{\mathbf{n}} \times \breve{\mathbf{b}} \qquad \text{. . . . . (384b}$$

$$\breve{\mathbf{n}} = \breve{\mathbf{b}} \times \breve{\mathbf{t}} \qquad \text{. . . . . (384c}$$

These three vectors characterize the curve at the point (x, y, z), the last one introduced relating to the twist of the curve out of the instant plane at the point defined by the tangent and normal vectors. Differentiating this third unit vector via Equation 384a gives

$$\begin{aligned} \frac{d\breve{\mathbf{b}}}{ds} = \frac{d}{ds}\left(\breve{\mathbf{t}} \times \breve{\mathbf{n}}\right) &= \frac{d\breve{\mathbf{t}}}{ds} \times \breve{\mathbf{n}} + \breve{\mathbf{t}} \times \frac{d\breve{\mathbf{n}}}{ds} = -\kappa\breve{\mathbf{n}} \times \breve{\mathbf{n}} + \breve{\mathbf{t}} \times \frac{d\breve{\mathbf{n}}}{ds} \\ &= \breve{\mathbf{t}} \times \frac{d\breve{\mathbf{n}}}{ds} \end{aligned} \qquad \text{. . . . . (384d}$$

since the cross-product of any vector with itself is the null vector. Differentiating again the dot-product of a unit vector we get

$$0 = \frac{d}{ds}\left(\breve{\mathbf{n}} \cdot \breve{\mathbf{n}}\right) = \frac{d\breve{\mathbf{n}}}{ds} \cdot \breve{\mathbf{n}} + \breve{\mathbf{n}} \cdot \frac{d\breve{\mathbf{n}}}{ds} = 2\frac{d\breve{\mathbf{n}}}{ds} \cdot \breve{\mathbf{n}}$$

Hence the derivative at the right must be orthogonal to the incoming unit vector, so must lie in the plane of the other two unit vectors, i.e.,

$$\frac{d\breve{\mathbf{n}}}{ds} = \varsigma\,\breve{\mathbf{t}} + \tau\,\breve{\mathbf{b}}$$

for some scalar values ς and τ. Substituting this in Equation 384d gives

$$\begin{aligned} \frac{d\breve{\mathbf{b}}}{ds} = \breve{\mathbf{t}} \times \left(\varsigma\,\breve{\mathbf{t}} + \tau\,\breve{\mathbf{b}}\right) &= \breve{\mathbf{t}} \times \varsigma\,\breve{\mathbf{t}} + \breve{\mathbf{t}} \times \tau\,\breve{\mathbf{b}} = \varsigma\,\breve{\mathbf{t}} \times \breve{\mathbf{t}} + \tau\,\breve{\mathbf{t}} \times \breve{\mathbf{b}} \\ &= -\tau\,\breve{\mathbf{n}} \end{aligned} \qquad \text{. . . . . . . (384e}$$

the variable τ, which is a measure of the twisting, called the *torsion*. Then, from Equation 384c,

$$\begin{aligned} \frac{d\breve{\mathbf{n}}}{ds} = \frac{d}{ds}\left(\breve{\mathbf{b}} \times \breve{\mathbf{t}}\right) = \frac{d\breve{\mathbf{b}}}{ds} \times \breve{\mathbf{t}} + \breve{\mathbf{b}} \times \frac{d\breve{\mathbf{t}}}{ds} &= -\tau\breve{\mathbf{n}} \times \breve{\mathbf{t}} + \breve{\mathbf{b}} \times \left(-\kappa\breve{\mathbf{n}}\right) = -\tau\breve{\mathbf{n}} \times \breve{\mathbf{t}} - \kappa\breve{\mathbf{b}} \times \breve{\mathbf{n}} \\ &= \tau\breve{\mathbf{b}} + \kappa\breve{\mathbf{t}} \end{aligned} \qquad \text{. . . (384f}$$

by Equations 384a & 384b.

Equations 383c, 384f & 384e give the derivatives of the three unit vectors, which are called the *Frenet-Serret* formulae, and can be written together in matrix form as

$$\begin{Bmatrix} d\breve{\mathbf{t}}/ds \\ d\breve{\mathbf{n}}/ds \\ d\breve{\mathbf{b}}/ds \end{Bmatrix} = \begin{bmatrix} 0 & -\kappa & 0 \\ \kappa & 0 & \tau \\ 0 & -\tau & 0 \end{bmatrix} \begin{Bmatrix} \breve{\mathbf{t}} \\ \breve{\mathbf{n}} \\ \breve{\mathbf{b}} \end{Bmatrix}$$

To establish a formula for evaluating the curvature, we begin by combining Equations 383c & 383a, getting

$$-\kappa\breve{\mathbf{n}} = \frac{d\breve{\mathbf{t}}}{ds} = \frac{d}{ds}\left(\frac{d\mathbf{p}}{ds}\right) = \frac{d^2\mathbf{p}}{ds^2} \qquad \text{. . . . . (385a}$$

Using Equation 384a and substituting from Equations 383a & 385a then gives

$$-\kappa\breve{\mathbf{b}} = -\kappa\breve{\mathbf{t}} \times \breve{\mathbf{n}} = \breve{\mathbf{t}} \times (-\kappa\breve{\mathbf{n}}) = \frac{d\mathbf{p}}{ds} \times \frac{d^2\mathbf{p}}{ds^2}$$

Since $\breve{\boldsymbol{b}}$ is a unit vector, this resolves to

$$\kappa = \left|\frac{d\mathbf{p}}{ds} \times \frac{d^2\mathbf{p}}{ds^2}\right| = \left|\frac{d\mathbf{p}}{ds} \times \frac{d}{ds}\left(\frac{d\mathbf{p}}{ds}\right)\right|$$

then, in terms of the basic parameter v,

$$\begin{aligned}
\kappa &= \left|\frac{d\mathbf{p}}{dv}\frac{dv}{ds} \times \frac{d}{ds}\left(\frac{d\mathbf{p}}{dv}\frac{dv}{ds}\right)\right| \\
&= \left|\frac{d\mathbf{p}}{dv}\frac{dv}{ds} \times \left[\frac{d}{ds}\left(\frac{d\mathbf{p}}{dv}\right)\frac{dv}{ds} + \frac{d\mathbf{p}}{dv}\frac{d}{ds}\left(\frac{dv}{ds}\right)\right]\right| \\
&= \left|\frac{d\mathbf{p}}{dv}\frac{dv}{ds} \times \left[\frac{dv}{ds}\frac{d}{dv}\left(\frac{d\mathbf{p}}{dv}\right)\frac{dv}{ds} + \frac{d\mathbf{p}}{dv}\frac{d}{ds}\left(\frac{dv}{ds}\right)\right]\right| \\
&= \left|\frac{d\mathbf{p}}{dv}\frac{dv}{ds} \times \left[\frac{dv}{ds}\left(\frac{d^2\mathbf{p}}{dv^2}\right)\frac{dv}{ds} + \frac{d\mathbf{p}}{dv}\frac{d^2v}{ds^2}\right]\right| \\
&= \left(\frac{dv}{ds}\right)^3 \cdot \left|\frac{d\mathbf{p}}{dv} \times \frac{d^2\mathbf{p}}{dv^2}\right|
\end{aligned}$$

after allowing that the cross-product of a vector with itself is null. Substituting for the cubed term from Equation 383b removes variable s, giving

$$\kappa = \left|\frac{d\mathbf{p}}{dv} \times \frac{d^2\mathbf{p}}{dv^2}\right| \cdot \left|\frac{d\mathbf{p}}{dv}\right|^{-3}$$

If r_c denotes the radius of curvature at the point, which equals the reciprocal of κ, then

$$r_c = \kappa^{-1} = \left|\frac{d\mathbf{p}}{dv}\right|^3 \cdot \left|\frac{d\mathbf{p}}{dv} \times \frac{d^2\mathbf{p}}{dv^2}\right|^{-1} \qquad \text{. . . . . (385b}$$

If our curve is within a plane, then either co-ordinate can act as the sole parameter. Let it be in the X,Y plane. Then $Y = f(X)$ for some function f and, with X the sole parameter,

$$\mathbf{p} = (X, Y\{X\}) = X\,\breve{\mathbf{i}} + Y\{X\}\,\breve{\mathbf{j}}$$

where $\breve{\mathbf{i}}$ and $\breve{\mathbf{j}}$ are unit vectors lying along the X and Y axes, respectively. Differentiating gives

$$\frac{d\mathbf{p}}{dX} = \breve{\mathbf{i}} + \frac{dY}{dX}\breve{\mathbf{j}} \quad \text{then} \quad \frac{d^2\mathbf{p}}{dX^2} = \frac{d^2Y}{dX^2}\breve{\mathbf{j}}$$

Substituting these in Equation 385b, interpreting v as X, gives

$$r_c = \kappa^{-1} = \left|\frac{d\mathbf{p}}{dX}\right|^3 \cdot \left|\frac{d\mathbf{p}}{dX} \times \frac{d^2\mathbf{p}}{dX^2}\right|^{-1}$$

$$= \left|\breve{\mathbf{i}} + \frac{dY}{dX}\breve{\mathbf{j}}\right|^3 \cdot \left|\left(\breve{\mathbf{i}} + \frac{dY}{dX}\breve{\mathbf{j}}\right) \times \frac{d^2Y}{dX^2}\breve{\mathbf{j}}\right|^{-1}$$

$$= \left\{1 + \left(\frac{dY}{dX}\right)^2\right\}^{\frac{3}{2}} \left|\breve{\mathbf{i}} \times \frac{d^2Y}{dX^2}\breve{\mathbf{j}} + \frac{dY}{dX}\breve{\mathbf{j}} \times \frac{d^2Y}{dX^2}\breve{\mathbf{j}}\right|^{-1}$$

Since the two unit vectors are mutually orthogonal, we get the purely scalar equation

$$r_c = \kappa^{-1} = \left\{1 + \left(\frac{dY}{dX}\right)^2\right\}^{\frac{3}{2}} \left|\frac{d^2Y}{dX^2}\right|^{-1} \qquad \text{. . . . . (386a}$$

for the radius of curvature of the plane curve.

On the ellipsoidal surface

Of primary interest are the curvatures along the meridional ellipse and orthogonally transverse thereto, but we first address the radius of the parallel for general latitude ϕ. For the sphere this was $R\cos\phi$, where R was the common radius from the geocentre. For the ellipsoid we find it to be ${}_tR\cos\phi$, where ${}_tR$ is the radius of that transverse curvature — itself a function of ϕ (as is ${}_mR$, the radius of curvature along the meridian).

Referring again to the diagram of the ellipse on page 380 we see that PQ — the normal to the ellipse at P — has (using Equation 381b) a slope of

$$\tan\phi = -\frac{dY}{dZ} = \frac{Za^2}{Yb^2} = \frac{Z}{Y\left(1-\varepsilon^2\right)}$$

giving

$$Z = Y\left(1-\varepsilon^2\right)\tan\phi = -Y\left(1-\varepsilon^2\right)\frac{dY}{dZ} \qquad \text{. . . (386b}$$

Substituting for Z in Equation 381b gives

$$a^2 = Y^2 + \frac{\left\{\left(1-\varepsilon^2\right)Y\tan\phi\right\}^2}{\left(1-\varepsilon^2\right)}$$

$$= Y^2 + \left(1-\varepsilon^2\right)Y^2\tan^2\phi$$

$$= \left(1 + \left(1-\varepsilon^2\right)\tan^2\phi\right)Y^2$$

Multiplying through by $\cos^2\phi$ gives

$$a^2\cos^2\phi = \left(\cos^2\phi + \sin^2\phi - \varepsilon^2\sin^2\phi\right)Y^2 = \left(1 - \varepsilon^2\sin^2\phi\right)Y^2$$

Hence,

$$\text{length of } PH = |Y| = \frac{a\cos\phi}{\sqrt{1-\varepsilon^2\sin^2\phi}} \qquad \text{(387a}$$

It can be shown that the centre of that transverse curvature at any point on the ellipsoid lies on the Z axis. Hence, using triangle PQH, we have for the radius of that curvature

$$_{t}R = \text{length of } PQ = \frac{a}{\sqrt{1-\varepsilon^2\sin^2\phi}} \qquad \text{(387b}$$

The radius of curvature in the YZ plane along the meridian is available from Equation 386a as

$$\begin{aligned} {}_{m}R &= \left|\left\{\sqrt{1+\left(\frac{dZ}{dY}\right)^2}\right\}\right|^3 \cdot \left|\frac{d^2Z}{dY^2}\right|^{-1} \\ &= \left|\left\{\sqrt{1+\cot^2\phi}\right\}\right|^3 \cdot \left|\frac{d^2Z}{dY^2}\right|^{-1} \\ &= \left|\sqrt{\csc^2\phi}\right|^3 \cdot \left|\frac{d^2Z}{dY^2}\right|^{-1} \\ &= \left|\csc^3\phi\right| \cdot \left|\frac{d^2Z}{dY^2}\right|^{-1} \end{aligned} \qquad \text{(387c}$$

Inverting and differentiating Equation 386b gives

$$\frac{dZ}{dY} = -\frac{(1-\varepsilon^2)Y}{Z}$$

$$\frac{d^2Z}{dY^2} = \left(\frac{-(1-\varepsilon^2)}{Z} + \frac{(1-\varepsilon^2)Y}{Z^2}\frac{dZ}{dY}\right) = \frac{1}{Z}\left(-(1-\varepsilon^2) + \frac{(1-\varepsilon^2)Y}{Z}\frac{dZ}{dY}\right)$$

Substituting from Equation 386b gives

$$\begin{aligned} \frac{d^2Z}{dY^2} &= \frac{1}{Z}\left\{-(1-\varepsilon^2) - \left(\frac{dZ}{dY}\right)^2\right\} &&= -\frac{1}{Z}\left\{1+\left(\frac{dZ}{dY}\right)^2 - \varepsilon^2\right\} \\ &= -\frac{1+\cot^2\phi - \varepsilon^2}{Z} &&= -\frac{\csc^2\phi - \varepsilon^2}{Z} \end{aligned} \qquad \text{(387d}$$

Considering triangles POH and PQH, noting that they have side PH in common, we get

$$\frac{\text{length of } OH}{\text{length of } QH} = \frac{\cot\phi}{\cot\psi} = \frac{b^2}{a^2} = (1-\varepsilon^2)$$

using Equation 382a. Clearly, this ratio must apply identically to the division of QP by the Y axis. Hence,

$$\text{length of } IP = (1-\varepsilon^2)\,\text{length of } QP = (1-\varepsilon^2)\,{}_{t}R$$

So, by basic trigonometry with point P and angle ϕ,

$$Z = \left(1-\varepsilon^2\right) {}_{\mathrm{t}}R \sin\phi \quad = \frac{a\left(1-\varepsilon^2\right)\sin\phi}{\sqrt{1-\varepsilon^2 \sin^2\phi}}$$

using Equation 387b. So, from Equation 387d,

$$\frac{\mathrm{d}^2 Z}{\mathrm{d}Y^2} = -\left(\csc^2\phi - \varepsilon^2\right) \frac{\sqrt{1-\varepsilon^2 \sin^2\phi}}{a\left(1-\varepsilon^2\right)\sin\phi}$$

Substituting this into Equation 387c, and ignoring overall signs within the modulus, we get

$${}_{\mathrm{m}}R = \left| \frac{\csc^3\phi}{\left(\csc^2\phi - \varepsilon^2\right) \dfrac{\sqrt{1-\varepsilon^2 \sin^2\phi}}{a\left(1-\varepsilon^2\right)\sin\phi}} \right| = \left| \frac{1}{\sin\phi \left(1-\varepsilon^2 \sin^2\phi\right) \dfrac{\sqrt{1-\varepsilon^2 \sin^2\phi}}{a\left(1-\varepsilon^2\right)\sin\phi}} \right|$$

$$= \left| \frac{a\left(1-\varepsilon^2\right)}{\left(\sqrt{1-\varepsilon^2 \sin^2\phi}\right)^3} \right| \qquad \ldots\ldots (388a$$

$$= \frac{a\left(1-\varepsilon^2\right)}{\left(\sqrt{1-\varepsilon^2 \sin^2\phi}\right)^3}$$

From Equation 387b we then get

$${}_{\mathrm{t}}R = \frac{a}{\sqrt{1-\varepsilon^2 \sin^2\phi}}$$

$$= \left\{ \frac{1-\varepsilon^2 \sin^2\phi}{1-\varepsilon^2} \right\} \frac{a\left(1-\varepsilon^2\right)}{\left(\sqrt{1-\varepsilon^2 \sin^2\phi}\right)^3} \qquad \ldots\ldots (388b$$

$$= \left\{ \frac{1-\varepsilon^2 \sin^2\phi}{1-\varepsilon^2} \right\} {}_{\mathrm{m}}R$$

and, by differentiation,

$$\frac{\mathrm{d}\,{}_{\mathrm{t}}R}{\mathrm{d}\phi} = \frac{a\left(-\frac{1}{2}\right)\left(-2\varepsilon^2 \sin\phi \cos\phi\right)}{\left(\sqrt{1-\varepsilon^2 \sin^2\phi}\right)^3}$$

$$= \frac{a\varepsilon^2 \sin\phi \cos\phi}{\left(\sqrt{1-\varepsilon^2 \sin^2\phi}\right)^3} \qquad \ldots\ldots (388c$$

$$= \left\{ \frac{\varepsilon^2 \sin\phi \cos\phi}{1-\varepsilon^2} \right\} {}_{\mathrm{m}}R$$

Rearranging Equation 388b gives

$$\begin{aligned}\frac{{}_{\mathrm{m}}R}{{}_{\mathrm{t}}R} &= \frac{1-\varepsilon^2}{1-\varepsilon^2\sin^2\phi} \\ &= \frac{1-\varepsilon^2\left(\sin^2\phi+\cos^2\phi\right)}{1-\varepsilon^2\sin^2\phi} \\ &= \frac{1-\varepsilon^2\sin^2\phi-\varepsilon^2\cos^2\phi}{1-\varepsilon^2\sin^2\phi} \\ &= 1-\frac{\varepsilon^2\cos^2\phi}{1-\varepsilon^2\sin^2\phi}\end{aligned} \qquad \text{. . . . . (389a}$$

Factoring the denominator and multiplying it and its numerator by 2 gives

$$\begin{aligned}\frac{{}_{\mathrm{m}}R}{{}_{\mathrm{t}}R} &= 1-\frac{\varepsilon^2\cos^2\phi}{2}\frac{2}{(1-\varepsilon\sin\phi)(1+\varepsilon\sin\phi)} \\ &= 1-\frac{\varepsilon^2\cos^2\phi}{2}\frac{\{(1-\varepsilon\sin\phi)+(1+\varepsilon\sin\phi)\}}{(1-\varepsilon\sin\phi)(1+\varepsilon\sin\phi)} \\ &= 1-\frac{\varepsilon^2\cos^2\phi}{2}\left\{\frac{1}{(1-\varepsilon\sin\phi)}+\frac{1}{(1+\varepsilon\sin\phi)}\right\}\end{aligned} \qquad \text{. . . . . (389b}$$

which we use below in establishing a formula for isometric latitude on the ellipsoid.

Comparing ellipsoidal and spherical surfaces

Division of the two radii given above in Equations 387b & 387c by the radius R of a sphere provides proportional comparison of the incremental distance along the respective surfaces of a unit of longitude and of a unit of latitude. We thus have the lengths for the sphere relative to those for the ellipsoid as:

$$\text{for a unit of longitude} \quad \frac{R}{{}_{\mathrm{t}}R} = \frac{R\sqrt{1-\varepsilon^2\sin^2\phi}}{a}$$

$$\text{for a unit of latitude} \quad \frac{R}{{}_{\mathrm{m}}R} = \frac{R\left(\sqrt{1-\varepsilon^2\sin^2\phi}\right)^3}{a\left(1-\varepsilon^2\right)}$$

Dividing the former by the latter gives

$$\text{local shape factor} \quad \frac{{}_{\mathrm{t}}R}{{}_{\mathrm{m}}R} = \frac{a}{\sqrt{1-\varepsilon^2\sin^2\phi}}\,\frac{\left(\sqrt{1-\varepsilon^2\sin^2\phi}\right)^3}{a\left(1-\varepsilon^2\right)} = \frac{1-\varepsilon^2\sin^2\phi}{\left(1-\varepsilon^2\right)}$$

while multiplying the two gives

$$\text{local area factor} \quad \frac{R\left(\sqrt{1-\varepsilon^2\sin^2\phi}\right)}{a}\cdot\frac{R\left(\sqrt{1-\varepsilon^2\sin^2\phi}\right)^3}{a\left(1-\varepsilon^2\right)} = \frac{R^2\left(1-\varepsilon^2\sin^2\phi\right)^2}{a^2\left(1-\varepsilon^2\right)}$$

THE FIRST FUNDAMENTAL QUANTITIES

Cartesian co-ordinates

From the diagram of the tangent to the ellipse in Exhibit 17–1 on page 380 and the 3-dimensional ellipsoid in Tutorial 3 (page 10), the 3-dimensional co-ordinates for the general point are

$$X = {}_tR\cos\phi\cos\lambda$$

$$Y = {}_tR\cos\phi\cos\lambda$$

$$Z = \left(1-\varepsilon^2\right){}_tR\sin\phi$$

Since ${}_tR$ is independent of λ, differentiating these gives

$$\mathrm{d}X = \left\{\frac{\mathrm{d}_tR}{\mathrm{d}\phi}\cos\phi - {}_tR\sin\phi\right\}\cos\lambda\,\mathrm{d}\phi - {}_tR\cos\phi\sin\lambda\,\mathrm{d}\lambda$$

$$\mathrm{d}Y = \left\{\frac{\mathrm{d}_tR}{\mathrm{d}\phi}\cos\phi - {}_tR\sin\phi\right\}\sin\lambda\,\mathrm{d}\phi + {}_tR\cos\phi\cos\lambda\,\mathrm{d}\lambda$$

$$\mathrm{d}Z = \left(1-\varepsilon^2\right)\left\{\frac{\mathrm{d}_tR}{\mathrm{d}\phi}\sin\phi + {}_tR\cos\phi\right\}\mathrm{d}\phi$$

Noting the matched squares pattern occurring for the first two, we have

$$\mathrm{d}s^2 = \mathrm{d}X^2 + \mathrm{d}Y^2 + \mathrm{d}Z^2$$

$$= \left\{\frac{\mathrm{d}_tR}{\mathrm{d}\phi}\cos\phi - {}_tR\sin\phi\right\}^2(\mathrm{d}\phi)^2 + \left\{{}_tR\cos\phi\right\}^2(\mathrm{d}\lambda)^2 + \left(1-\varepsilon^2\right)^2\left\{\frac{\mathrm{d}_tR}{\mathrm{d}\phi}\sin\phi + {}_tR\cos\phi\right\}^2(\mathrm{d}\phi)^2$$

$$= \left(\left\{\frac{\mathrm{d}_tR}{\mathrm{d}\phi}\cos\phi - {}_tR\sin\phi\right\}^2 + \left(1-\varepsilon^2\right)^2\left\{\frac{\mathrm{d}_tR}{\mathrm{d}\phi}\sin\phi + {}_tR\cos\phi\right\}^2\right)(\mathrm{d}\phi)^2 + \left\{{}_tR\cos\phi\right\}^2(\mathrm{d}\lambda)^2$$

Substituting from Equation 388c for the derivative and from Equation 388b for ${}_tR$ the first of the curly brackets gives

$$\left\{\frac{\mathrm{d}_tR}{\mathrm{d}\phi}\cos\phi - {}_tR\sin\phi\right\}^2 = \left\{\frac{\varepsilon^2\sin\phi\cos\phi}{1-\varepsilon^2}\,{}_mR\cos\phi - \frac{1-\varepsilon^2\sin^2\phi}{1-\varepsilon^2}\,{}_mR\sin\phi\right\}^2$$

$$= \left\{\frac{{}_mR}{1-\varepsilon^2}\sin\phi\right\}^2\left\{\varepsilon^2\cos\phi\cos\phi - \left(1-\varepsilon^2\sin^2\phi\right)\right\}^2$$

$$= \left\{\frac{{}_mR}{1-\varepsilon^2}\sin\phi\right\}^2\left\{\varepsilon^2\cos^2\phi - 1 + \varepsilon^2\sin^2\phi\right\}^2$$

$$= \left\{\frac{{}_mR}{1-\varepsilon^2}\sin\phi\right\}^2\left\{\varepsilon^2 - 1\right\}^2$$

$$= {}_mR^2\sin^2\phi$$

Similar manipulation gives

$$\left(1-\varepsilon^2\right)\left\{\frac{\mathrm{d}_{\mathrm{t}}R}{\mathrm{d}\phi}\sin\phi + {}_{\mathrm{t}}R\cos\phi\right\}^2 = \left\{{}_{\mathrm{m}}R\cos\phi\right\}^2\left\{\varepsilon^2\sin^2\phi + \left(1-\varepsilon^2\sin^2\phi\right)\right\}^2$$

$$= {}_{\mathrm{m}}R^2\cos^2\phi$$

allowing our overall equation to simplify to

$$\begin{aligned}\mathrm{d}s^2 &= \left\{{}_{\mathrm{t}}R\cos\phi\right\}^2(\mathrm{d}\lambda)^2 + \left({}_{\mathrm{m}}R^2\sin^2\phi + {}_{\mathrm{m}}R^2\cos^2\phi\right)(\mathrm{d}\phi)^2 \\ &= \left\{{}_{\mathrm{t}}R\cos\phi\right\}^2(\mathrm{d}\lambda)^2 + {}_{\mathrm{m}}R^2(\mathrm{d}\phi)^2\end{aligned} \quad \text{. . . . . . . (391a}$$

Comparing with Equations 285c, and using Equations 387b & 388a for the radii, gives first fundamental quantities for the ellipsoid in terms of ordinary longitude/latitude, as

$$\begin{aligned}{}_{\mathrm{e}}E_{\lambda,\phi} &= {}_{\mathrm{t}}R^2\cos^2\phi &&= \frac{a^2\cos^2\phi}{\left(1-\varepsilon^2\sin^2\phi\right)} \\ {}_{\mathrm{e}}F_{\lambda,\phi} &= 0 \\ {}_{\mathrm{e}}G_{\lambda,\phi} &= {}_{\mathrm{m}}R^2 &&= \frac{a^2\left(1-\varepsilon^2\right)^2}{\left(1-\varepsilon^2\sin^2\phi\right)^3}\end{aligned} \quad \text{. . . . . (391b}$$

the first being identically the square of the radius of the parallel circle, shown in Equation 387a. Orthogonality of our parameters ensures a zero value for F regardless of eccentricity.

We can also note

$${}_{\mathrm{e}}C_{\lambda,\phi} = \frac{{}_{\mathrm{e}}E_{\lambda,\phi}}{{}_{\mathrm{e}}G_{\lambda,\phi}} = \frac{{}_{\mathrm{t}}R^2\cos^2\phi}{{}_{\mathrm{m}}R^2} = \frac{\left(1-\varepsilon^2\sin^2\phi\right)^2\cos^2\phi}{\left(1-\varepsilon^2\right)^2} \quad \text{. . . . . . . (391c}$$

and

$${}_{\mathrm{e}}H_{\lambda,\phi} = {}_{\mathrm{e}}E_{\lambda,\phi}\cdot{}_{\mathrm{e}}G_{\lambda,\phi} - \left({}_{\mathrm{e}}F_{\lambda,\phi}\right)^2 = \frac{a^4\left(1-\varepsilon^2\right)\cos^2\phi}{\left(1-\varepsilon^2\sin^2\phi\right)^4}$$

Isometric latitude

As we did for the sphere in Chapter 12, we can establish an isometric latitude for the ellipsoid. From Equation 391a, we have

$$\mathrm{d}s^2 = \{{}_{\mathrm{t}}R\cos\phi\}^2(\mathrm{d}\lambda)^2 + {}_{\mathrm{m}}R^2\,(\mathrm{d}\phi)^2$$

$$= \{{}_{\mathrm{t}}R\cos\phi\}^2\left\{(\mathrm{d}\lambda)^2 + \frac{{}_{\mathrm{m}}R^2}{\{{}_{\mathrm{t}}R\cos\phi\}^2}(\mathrm{d}\phi)^2\right\}$$

$$= \{{}_{\mathrm{t}}R\cos\phi\}^2\left\{(\mathrm{d}\lambda)^2 + (\mathrm{d}\overline{\varphi})^2\right\} \qquad \text{if} \qquad \mathrm{d}\overline{\varphi} = \frac{{}_{\mathrm{m}}R}{{}_{\mathrm{t}}R\cos\phi}\mathrm{d}\phi$$

From Equation 389b, we then have

$$\mathrm{d}\overline{\varphi} = \left\{\frac{1}{\cos\phi}\mathrm{d}\phi\right\} + \frac{\varepsilon}{2}\left\{\frac{-\varepsilon\cos\phi}{(1-\varepsilon\sin\phi)}\mathrm{d}\phi\right\} - \frac{\varepsilon}{2}\left\{\frac{\varepsilon\cos\phi}{(1+\varepsilon\sin\phi)}\mathrm{d}\phi\right\}$$

with each of the terms in curly brackets being integrable functions of ϕ shown in Tutorial 15 (page 78), the first by formula (h), the others by formula (b) then (f). The result is

$$\overline{\varphi} = \ln\tan\left(\frac{\pi}{4} + \frac{\phi}{2}\right) + \frac{\varepsilon}{2}\ln(1-\varepsilon\sin\phi) - \frac{\varepsilon}{2}\ln(1+\varepsilon\sin\phi) + \text{constant}$$

The constant is zero if we choose our new variable to be zero for $\phi = 0$. Thus,

$$\overline{\varphi} = \ln\tan\left(\frac{\pi}{4} + \frac{\phi}{2}\right) + \ln\left(\frac{1-\varepsilon\sin\phi}{1+\varepsilon\sin\phi}\right)^{\frac{\varepsilon}{2}}$$

$$= \ln\left\{\left(\frac{1-\varepsilon\sin\phi}{1+\varepsilon\sin\phi}\right)^{\frac{\varepsilon}{2}}\tan\left(\frac{\pi}{4} + \frac{\phi}{2}\right)\right\}$$

Our new variable is the isometric latitude for the ellipsoid of eccentricity ε. For the <u>ellipsoid in terms of ordinary longitude but isometric latitude</u>, and appropriately labelling our first fundamental quantities, we have

$$_{\mathrm{e}}E_{\lambda,\overline{\varphi}} = {}_{\mathrm{e}}G_{\lambda,\overline{\varphi}} = {}_{\mathrm{t}}R^2\cos^2\phi \qquad = \frac{a^2\cos^2\phi}{(1-\varepsilon^2\sin^2\phi)} \qquad \text{(392a}$$

$$_{\mathrm{e}}F_{\lambda,\overline{\varphi}} = 0$$

$$_{\mathrm{e}}C_{\lambda,\overline{\varphi}} = \frac{{}_{\mathrm{e}}E_{\lambda,\overline{\varphi}}}{{}_{\mathrm{e}}G_{\lambda,\overline{\varphi}}} \qquad = 1$$

$$_{\mathrm{e}}H_{\lambda,\overline{\varphi}} = {}_{\mathrm{e}}E_{\lambda,\overline{\varphi}}\cdot{}_{\mathrm{e}}G_{\lambda,\overline{\varphi}} - \left({}_{\mathrm{e}}F_{\lambda,\overline{\varphi}}\right)^2 \qquad = \frac{a^2\cos^4\phi}{(1-\varepsilon^2\sin^2\phi)^2}$$

DISTANCES AND ANGLES ON AN ELLIPSOIDAL SURFACE

Distance along a parallel, along a meridian

Equation 387a provides the radius of and hence the distance ${}_{\mathrm{p}}d$ along a parallel:

$$ {}_{\mathrm{p}}d = \text{length along parallel } \phi \text{ from } \lambda_1 \text{ to } \lambda_2 > \lambda_1 = \frac{a(\lambda_2 - \lambda_1)\cos\phi}{\sqrt{1-\varepsilon^2\sin^2\phi}} $$

Equation 391a gives the formula for incremental distance generally as

$$ \mathrm{d}s = \sqrt{{}_{\mathrm{m}}R^2\,(\mathrm{d}\phi)^2 + \{{}_{\mathrm{t}}R\cos\phi\}^2(\mathrm{d}\lambda)^2} $$

$$ = \sqrt{a^2\left(1-\varepsilon^2\right)^2\left(1-\varepsilon^2\sin^2\phi\right)^{-3}(\mathrm{d}\phi)^2 + \frac{a^2\cos^2\phi}{1-\varepsilon^2\sin^2\phi}(\mathrm{d}\lambda)^2} \qquad \text{. . . . . (393a} $$

Integration of this incremental element provides the means to establish distance, but is beset with difficulty. Even distance along a meridian, where $\mathrm{d}\lambda$ is zero, requires resorting to approximating techniques because of the denominator. We have

$$ {}_{\mathrm{m}}d = \text{length along meridian from } \phi_1 \text{ to } \phi_2 > \phi_1 = \int_{\phi_1}^{\phi_2}\mathrm{d}s = \int_{\phi_1}^{\phi_2}\frac{a\left(1-\varepsilon^2\right)}{\left(1-\varepsilon^2\sin^2\phi\right)^{\frac{3}{2}}}\mathrm{d}\phi $$

Putting ϕ=0 in the binomial series of Equation 378a for the integrand then integrating we have

$$ {}_{\mathrm{m}}d = \left[a\left(1-\varepsilon^2\right)\left(B_0\,\phi - B_2\,\varepsilon^2\frac{\sin 2\phi}{2} + B_4\,\varepsilon^4\frac{\sin 4\phi}{4} - B_6\,\varepsilon^6\frac{\sin 6\phi}{6} + B_8\,\varepsilon^8\frac{\sin 8\phi}{8} + \ldots\right)\right]_{\phi_1}^{\phi_2} $$

with coefficients defined in Equations 378b. Since $\varepsilon^{2n}\sin n\phi \le \varepsilon^{2n} < 100^{-n}$, this converges rapidly. Multiplying the infinite series by the $(1-\varepsilon^2)$ term gives

$$ {}_{\mathrm{m}}d = \left[a\left(\overline{B}_0\,\phi - \overline{B}_2\,\varepsilon^2\sin 2\phi + \overline{B}_4\,\varepsilon^4\sin 4\phi - \overline{B}_6\,\varepsilon^6\sin 6\phi + \overline{B}_8\,\varepsilon^8\sin 8\phi + \ldots\right)\right]_{\phi_1}^{\phi_2} \qquad \text{. . . (393b} $$

where the coefficients, each a series converging to a positive value less than 1, are:

$$ \overline{B}_0 = \left(1-\varepsilon^2\right)B_0 = 1 - \frac{1}{4}\varepsilon^2 - \frac{3}{64}\varepsilon^4 - \frac{5}{256}\varepsilon^6 - \frac{175}{16384}\varepsilon^8 + .. $$

$$ \overline{B}_2 = \left(1-\varepsilon^2\right)B_2 = \frac{3}{8} + \frac{3}{32}\varepsilon^2 + \frac{45}{1024}\varepsilon^4 + \frac{105}{4096}\varepsilon^6 + .. $$

$$ \overline{B}_4 = \left(1-\varepsilon^2\right)B_4 = \frac{15}{256} + \frac{45}{1024}\varepsilon^2 + \frac{525}{16384}\varepsilon^4 + \ldots \qquad \text{. . . (393c} $$

$$ \overline{B}_6 = \left(1-\varepsilon^2\right)B_6 = \frac{35}{3072} + \frac{175}{12288}\varepsilon^2 + \ldots $$

$$ \overline{B}_8 = \left(1-e^2\right)B_8 = \frac{315}{131072} + \ldots $$

Distance in general

On the sphere the shortest route between two points is along the great circle, defined by the intersection of the spherical surface with the plane through the two points and the geocentre. While the expression *great-circle route* is still used relative to an ellipsoidal world (and the geoidal world), the shortest route is generally not a plane curve (i.e., a curve lying within a plane). Called the *geodesic,* it runs very close to the geocentric plane through the two points (exactly within it for routes along the Equator and meridional circles), hence its length is inherently close to the value stated for the sphere in Equation 215c. Getting the precise formula requires integration of ds for both variables λ and ϕ, and then derivation of the minimum to establish the formula applicable to the shortest route.

Rewriting Equation 391a we have

$$\mathrm{d}s^2 = \left\{ {}_{\mathrm{t}}R^2 \cos^2\phi \left(\frac{\mathrm{d}\lambda}{\mathrm{d}\phi}\right)^2 + {}_{\mathrm{m}}R^2 \right\} (\mathrm{d}\phi)^2 = \left\{ {}_{\mathrm{t}}R^2 \cos^2\phi \left(\lambda^{(1)}\right)^2 + {}_{\mathrm{m}}R^2 \right\} (\mathrm{d}\phi)^2 \quad \ldots\ldots \text{(394a}$$

where $\lambda^{(1)}$ is the derivative of λ with respect to ϕ. Denoting the expression within the curly brackets by L, then, noting that it is independent of λ itself, we have

$$\frac{\mathrm{d}}{\mathrm{d}\phi}\left(\frac{\partial L}{\partial \lambda^{(1)}}\right) = \frac{\mathrm{d}}{\mathrm{d}\phi}(0) = 0$$

Hence,

$$2k_0 = \frac{\partial L}{\partial \lambda^{(1)}} = {}_{\mathrm{t}}R^2 \left(\cos^2\phi\right) 2\lambda^{(1)} = {}_{\mathrm{t}}R^2 \left(\cos^2\phi\right) \frac{2\,\mathrm{d}\lambda}{\mathrm{d}\phi} \quad \ldots\ldots\ldots \text{(394b}$$

must be a constant. Rearrangement and integration gives

$$\begin{aligned} \lambda &= k_1 + k_0 \int \frac{1}{{}_{\mathrm{t}}R^2 \cos^2\phi} \mathrm{d}\phi \\ &= k_1 + k_0 \int \frac{\left(1 - \varepsilon^2 \sin^2\phi\right)}{a^2 \cos^2\phi} \mathrm{d}\phi \end{aligned}$$

for some constant k_1, with Equation 387b providing the substitution. Rearranging and then using standard integration formulae of Tutorial 15 gives

$$\begin{aligned} \lambda &= k_1 + k_0 \int \frac{1 - \varepsilon^2\left(1 - \cos^2\phi\right)}{a^2 \cos^2\phi} \mathrm{d}\phi \\ &= k_1 + k_0 \int \left\{ \frac{1 - \varepsilon^2}{a^2 \cos^2\phi} + \frac{\varepsilon^2}{a^2} \right\} \mathrm{d}\phi \\ &= k_1 + k_0 \left[\frac{1 - \varepsilon^2}{a^2} \tan\phi + \frac{\varepsilon^2}{a^2} \phi \right] \end{aligned}$$

Evaluating this for the two end points $P_1 = (\lambda_1, \phi_1)$ and $P_2 = (\lambda_2, \phi_2)$, then differencing to eliminate k_1, resolves to

$$k_0 = \frac{\lambda_2 - \lambda_1}{\dfrac{1-\varepsilon^2}{a^2}(\tan\phi_2 - \tan\phi_1) + \dfrac{\varepsilon^2}{a^2}(\phi_2 - \phi_1)} \qquad \text{(395a}$$

From Equation 393a, substituting successively from Equations 394b, 394a & 395a, we get

$$\mathrm{d}s = \sqrt{{}_{\mathrm{t}}R^2 \cos^2\phi \left(\frac{\mathrm{d}\lambda}{\mathrm{d}\phi}\right)^2 + {}_{\mathrm{m}}R^2}\ \mathrm{d}\phi$$

$$= \sqrt{\frac{k_0^2}{{}_{\mathrm{t}}R^2 \cos^2\phi} + {}_{\mathrm{m}}R^2}\ \mathrm{d}\phi$$

$$= \sqrt{\frac{k_0^2(1-\varepsilon^2\sin^2\phi)}{a^2\cos^2\phi} + \frac{a^2(1-\varepsilon^2)^2}{(1-\varepsilon^2\sin^2\phi)^3}}\ \mathrm{d}\phi \qquad \text{(395b}$$

$$= \sqrt{\frac{(\lambda_2-\lambda_1)(1-\varepsilon^2\sin^2\phi)}{\{(1-\varepsilon^2)(\tan\phi_2 - \tan\phi_1) + \varepsilon^2(\phi_2-\phi_1)\}\cos^2\phi} + \frac{a^2(1-\varepsilon^2)^2}{(1-\varepsilon^2\sin^2\phi)^3}}\ \mathrm{d}\phi$$

This is not analytically integrable. Evaluation of s is achievable only via numerical methods of integration over the range of ϕ defined by the specific values of ϕ_1 and ϕ_2 (with the specific values of λ_1 and λ_2 incorporated).

Angles

Of particular concern are azimuths and loxodromic bearing. Both of these relate the general path to that of a meridian. For any path along a meridian, the longitude is invariant — so $\mathrm{d}\lambda=0$ hence $\mathrm{d}s=\sqrt{G}\,\mathrm{d}\phi$. Using the vector dot-product, with a bar to distinguish the elements of the meridional path from those of the general path, we get the azimuth angle ζ by

$$\cos\zeta = \frac{E\ \mathrm{d}\lambda\ \mathrm{d}\bar{\lambda} + G\ \mathrm{d}\phi\ \mathrm{d}\bar{\phi}}{\mathrm{d}s\ \mathrm{d}\bar{s}} = \frac{G\ \mathrm{d}\phi\ \mathrm{d}\bar{\phi}}{\mathrm{d}s\sqrt{G}\ \mathrm{d}\bar{\phi}} = \sqrt{G}\frac{\mathrm{d}\phi}{\mathrm{d}s}$$

Substitution from Equation 391b gives

$$\cos\zeta = {}_{\mathrm{m}}R\frac{\mathrm{d}\phi}{\mathrm{d}s} = \frac{a(1-\varepsilon^2)}{\left\{\sqrt{1-\varepsilon^2\sin^2\phi}\right\}^3}\frac{\mathrm{d}\phi}{\mathrm{d}s} \qquad \text{(395c}$$

Substitution from Equation 395b for the last term gives us the azimuth for the initial point of the geodesic.

Using the vector cross-product, we get

$$\sin\zeta = \sqrt{EG}\,\frac{\mathrm{d}\phi\, d\bar{\lambda} + \mathrm{d}\bar{\phi}\,\mathrm{d}\lambda}{\mathrm{d}s\,\mathrm{d}\bar{s}}$$

$$= \sqrt{EG}\,\frac{\mathrm{d}\bar{\phi}\,\mathrm{d}\lambda}{\mathrm{d}s\,\sqrt{G}\,\mathrm{d}\bar{\phi}}$$

$$= \sqrt{E}\,\frac{\mathrm{d}\lambda}{\mathrm{d}s}$$

Substitution from Equation 391b gives

$$\sin\zeta = {}_{\mathrm{t}}R\cos\phi\frac{\mathrm{d}\lambda}{\mathrm{d}s} = \frac{a\cos\phi}{\sqrt{1-\varepsilon^2\sin^2\phi}}\,\frac{\mathrm{d}\lambda}{\mathrm{d}s} \qquad \text{. . . . . (396a}$$

Loxodromic distance

Loxodromes maintain constant bearing — i.e., constant azimuth. Combining Equations 395c & 396a, then substituting from Equation 388b, gives

$$\tan\zeta = \frac{\sin\zeta}{\cos\zeta} = \frac{{}_{\mathrm{t}}R\cos\phi}{{}_{\mathrm{m}}R}\,\frac{\mathrm{d}\lambda}{\mathrm{d}s}\,\frac{\mathrm{d}s}{\mathrm{d}\phi} = \left\{\frac{1-\varepsilon^2\sin^2\phi}{1-\varepsilon^2}\right\}\cos\phi\,\frac{\mathrm{d}\lambda}{\mathrm{d}\phi}$$

The length of the loxodrome from $P_1 = (\lambda_1, \phi_1)$ to $P_2 = (\lambda_2, \phi_2)$ is obtained by integrating, over the range of longitudes, the differential

$$\mathrm{d}\lambda = (\tan\zeta)\frac{1-\varepsilon^2}{\left(1-\varepsilon^2\sin^2\phi\right)\cos\phi}\mathrm{d}\phi \qquad \text{. . . . . (396b}$$

for the specific value of $\tan\zeta$. This composite function is analytically integrable, but it must be fragmented to achieve this. The quotient term has already been essentially addressed in Equations 389a & 389b. The result, for some constant k, is

$$\int \mathrm{d}\lambda = (\tan\zeta)\left\{\int\left\{\frac{1}{\cos\phi}\right\}\mathrm{d}\phi + \frac{\varepsilon}{2}\int\left\{\frac{-\varepsilon\cos\phi}{(1-\varepsilon\sin\phi)}\right\}\mathrm{d}\phi - \frac{\varepsilon}{2}\int\left\{\frac{\varepsilon\cos\phi}{(1+\varepsilon\sin\phi)}\right\}\mathrm{d}\phi\right\} + k$$

$$= (\tan\zeta)\left[\ln\tan\left(\frac{\pi}{4}+\frac{\phi}{2}\right) + \frac{\varepsilon}{2}\ln(1-\varepsilon\sin\phi) - \frac{\varepsilon}{2}\ln(1+\varepsilon\sin\phi)\right] + k$$

$$= (\tan\zeta)\left[\ln\tan\left(\frac{\pi}{4}+\frac{\phi}{2}\right) + \ln\left(\frac{1-\varepsilon\sin\phi}{1+\varepsilon\sin\phi}\right)^{\frac{\varepsilon}{2}}\right] + k \qquad \text{. . . (396c}$$

$$= (\tan\zeta)\left[\ln\left\{\left(\frac{1-\varepsilon\sin\phi}{1+\varepsilon\sin\phi}\right)^{\frac{\varepsilon}{2}}\tan\left(\frac{\pi}{4}+\frac{\phi}{2}\right)\right\}\right] + k$$

which must be evaluated for the two extreme latitudes and differenced (eliminating the constant) to obtain the distance along the loxodrome.

IMPACT ON DISTORTION

The differing radii

Having differing radii of curvature along and transverse to a meridian alters the formulae for distortion along meridian, and transversely thereto, to be:

$$_{\mathrm{m}}\dot{m} = \frac{\sqrt{\dot{G}_{\lambda,\phi}}}{_{\mathrm{m}}R} = \frac{\left(\sqrt{1-\varepsilon^2\sin^2\phi}\right)^3}{a\left(1-\varepsilon^2\right)}\sqrt{\left(\frac{\partial x}{\partial\phi}\right)^2+\left(\frac{\partial y}{\partial\phi}\right)^2}$$

$$_{\mathrm{t}}\dot{m} = \frac{\sqrt{\dot{E}_{\lambda,\phi}}}{_{\mathrm{t}}R\cos\phi} = \frac{\sqrt{1-\varepsilon^2\sin^2\phi}}{a\cos\phi}\sqrt{\left(\frac{\partial x}{\partial\lambda}\right)^2+\left(\frac{\partial y}{\partial\lambda}\right)^2}$$

For cylindrical projections, these become:

$$_{\mathrm{m}}\dot{m} = \frac{\mathrm{d}\dot{v}}{_{\mathrm{m}}R\,\mathrm{d}\phi} = \frac{\left(\sqrt{1-\varepsilon^2\sin^2\phi}\right)^3}{a\left(1-\varepsilon^2\right)}\frac{\mathrm{d}\dot{v}}{\mathrm{d}\phi}$$

$$_{\mathrm{t}}\dot{m} = \frac{\mathrm{d}\dot{u}}{_{\mathrm{m}}R\cos\phi\,\mathrm{d}\lambda} = \frac{\sqrt{1-\varepsilon^2\sin^2\phi}}{a\cos\phi}\frac{\mathrm{d}\dot{u}}{\mathrm{d}\lambda}$$

For conical projections (with c the constant of the cone) and polar azimuthal projections (with $c = 1$), these distortion measures become:

$$_{\mathrm{m}}\dot{m} = \frac{-\mathrm{d}\dot{\rho}}{_{\mathrm{m}}R\,\mathrm{d}\phi} = -\frac{\left(\sqrt{1-\varepsilon^2\sin^2\phi}\right)^3}{a\left(1-\varepsilon^2\right)}\frac{\mathrm{d}\dot{\rho}}{\mathrm{d}\phi}$$

$$_{\mathrm{t}}\dot{m} = \frac{c\,\dot{\rho}}{_{\mathrm{m}}R\cos\phi} = +\frac{\sqrt{1-\varepsilon^2\sin^2\phi}}{a\cos\phi}c\,\dot{\rho}$$

The maximum angular distortion is given by

$$\sin\left(\dot{\beta}-\beta\right) = \frac{\left|{}_{\mathrm{m}}\dot{m} - {}_{\mathrm{t}}\dot{m}\right|}{{}_{\mathrm{m}}\dot{m} + {}_{\mathrm{t}}\dot{m}}$$

REPRESENTATIVE PROXY SPHERES

Idea

While plotting equations for various projections could be redeveloped directly from the ellipsoid, there is scope to map the ellipsoid onto the sphere, then apply unchanged the plotting equations already developed as a second transformation. The formula for converting from geodesic to geocentric latitude on the ellipsoid is one possibility that could translate usefully to the sphere, but we look more carefully, seeking spheres that can stand as proxy for the ellipsoid in terms of equal-area, equidistance and conformality — finding that all are feasible without undue complication. The target is to gain revised parameters for the sphere that maintain the chosen attribute, in the form of a proxy latitude and a proxy longitude. We use the differential geometry of Chapter 12, which, being applicable to surfaces in general, accommodates the sphere as the target just as it did the planar map.

Basics

We restrict ourselves to transformations that map Equator into Equator, Poles into Poles, meridians into straight lines connecting Poles, and each parallel into an arc of points equally distant from a Pole. So, on the sphere, proxy latitude is a function of only (geodetic) latitude on the ellipsoid, and proxy longitude is a scalar multiple $\bar{k}$ of relative longitude on the ellipsoid. Then, using a bar overscript to represent proxy parameters:

$$\begin{matrix} \dfrac{\partial\bar{\lambda}}{\partial\lambda} = \bar{k} & \dfrac{\partial\bar{\phi}}{\partial\lambda} = 0 \\ \dfrac{\partial\bar{\lambda}}{\partial\phi} = 0 & \dfrac{\partial\bar{\phi}}{\partial\phi} = \dfrac{\mathrm{d}\bar{\phi}}{\mathrm{d}\phi} \end{matrix}, \qquad \text{so} \quad {}^{\bar{\lambda},\bar{\phi}}J_{\lambda,\phi} = \bar{k}\frac{\mathrm{d}\bar{\phi}}{\mathrm{d}\phi}$$

The values for fundamental quantities on the sphere in Equations 287b apply directly to the proxy parameters. To compare sphere with ellipsoid, however, we need values based on common parameters — i.e., geodetic latitude and true longitude. Using the single-dot overscript for results of the mapping, Equation 291b gives:

$$\begin{aligned} \dot{E}_{\lambda.\phi} &= \left(\frac{\partial\bar{\lambda}}{\partial\lambda}\right)^2 \dot{E}_{\bar{\lambda},\bar{\phi}} + \left(\frac{\partial\bar{\phi}}{\partial\lambda}\right)^2 \dot{G}_{\bar{\lambda},\bar{\phi}} &&= \bar{k}^2 \; {}_{\mathrm{S}}\dot{E}_{\bar{\lambda},\bar{\phi}} &&= \bar{k}^2\,\bar{R}^2 \cos^2\bar{\phi} \\ \dot{G}_{\lambda.\phi} &= \left(\frac{\partial\bar{\lambda}}{\partial\phi}\right)^2 \dot{E}_{\bar{\lambda},\bar{\phi}} + \left(\frac{\partial\bar{\phi}}{\partial\phi}\right)^2 \dot{G}_{\bar{\lambda},\bar{\phi}} &&= \left(\frac{\mathrm{d}\bar{\phi}}{\mathrm{d}\phi}\right)^2 {}_{\mathrm{S}}\dot{G}_{\bar{\lambda},\bar{\phi}} &&= \left(\frac{\mathrm{d}\bar{\phi}}{\mathrm{d}\phi}\right)^2 \bar{R}^2 \end{aligned} \quad \text{. . . . . (398a}$$

We proceed to develop transformations that map the ellipsoid into a sphere for which equal-area applies, then for which equidistance applies, and finally for which conformality applies. Distinct overscripts are used for each of the three.

Authalic sphere, radius and latitude

We turn first to an equal-area transformation, defining a sphere of equal overall area to our ellipsoid, and a longitude/latitude scheme thereon that retains the equal-area attribute locally, too. Simple formulae allow specification of the radius for a sphere of any specified surface area, and express the area between any two parallels thereon. Unfortunately, no such simplicity applies to the ellipsoid. The formula for total surface area A of an oblate ellipsoid is

$$A = 2\pi a^2 + \frac{\pi b^2}{\varepsilon}\ln\left(\frac{1+\varepsilon}{1-\varepsilon}\right) = \pi a^2\left\{2 + \frac{(1-\varepsilon^2)}{\varepsilon}\ln\left(\frac{1+\varepsilon}{1-\varepsilon}\right)\right\} \quad \text{. . . . . (399a}$$

A sphere equal in overall area to an ellipsoid is usually called its *authalic sphere,* its radius the *authalic radius* (see page 456 for discussion regarding this adjective).

Since the surface area of a sphere is 4π times the square of its radius, we have, from Equations 399a,

$$\overline{\overline{R}} = \sqrt{\frac{a^2}{2} + \frac{b^2}{4\varepsilon}\ln\left\{\frac{1+\varepsilon}{1-\varepsilon}\right\}} = \frac{a}{2}\sqrt{1 + \frac{1-\varepsilon^2}{\varepsilon}\ln\left\{\frac{1+\varepsilon}{1-\varepsilon}\right\}} \quad \text{. . . . . (399b}$$

using a double-bar overscript for elements of our equal-area or authalic sphere.

To be equal-area generally, an orthogonal transformation must produce fundamental quantities such that the product EG is unchanged. Hence, using Equation 391b on the left side and Equation 398a (interpreted for our double-bar variables) on the right we have

$$_{t}R^2 \cos^2\phi \cdot {_{m}R^2} = {_{e}E_{\lambda,\phi}} \cdot {_{e}G_{\lambda,\phi}}$$

$$= {_{s}\dot{E}_{\lambda,\phi}} \cdot {_{s}\dot{G}_{\lambda,\phi}}$$

$$= \overline{\overline{k}}^2 \cdot \overline{\overline{R}}^2 \cdot \cos^2\overline{\overline{\phi}} \cdot \left(\frac{\mathrm{d}\overline{\overline{\phi}}}{\mathrm{d}\phi}\right)^2 \cdot \left(\overline{\overline{R}}\right)^2 \quad = \overline{\overline{k}}^2 \cdot \overline{\overline{R}}^4 \cdot \cos^2\overline{\overline{\phi}}\left(\frac{\mathrm{d}\overline{\overline{\phi}}}{\mathrm{d}\phi}\right)^2$$

Clearly, we can achieve an equal-area mapping by merely respacing the parallels, and maintaining the meridians at their correct spacing (i.e., have $\overline{\overline{k}}=1$). Substituting for the ellipsoidal radii from Equations 388a & 388b, this equation then becomes

$$\frac{a^2(1-\varepsilon^2)^2}{\left(1-\varepsilon^2\sin^2\phi\right)^3} \cdot \frac{a^2\cos^2\phi}{\left(1-\varepsilon^2\sin^2\phi\right)} = \overline{\overline{R}}^4 \cdot \cos^2\overline{\overline{\phi}}\left(\frac{\mathrm{d}\overline{\overline{\phi}}}{\mathrm{d}\phi}\right)^2$$

$$\frac{a^2\left(1-\varepsilon^2\right)\cos\phi}{\left(1-\varepsilon^2\sin^2\phi\right)^2}\,\mathrm{d}\phi = \overline{\overline{R}}^2 \cdot \cos\overline{\overline{\phi}}\,\mathrm{d}\overline{\overline{\phi}}$$

The right side is simply integrated using a formula of Tutorial 15; the left side can be made effectively integrable via another binomial expansion (Tutorial 32).

Swapping sides and using

$$\frac{1}{\left(1-\varepsilon^2 \sin^2\phi\right)^2} = 1 + 2\varepsilon^2 \sin^2\phi + 3\varepsilon^4 \sin^4\phi + 4\varepsilon^6 \sin^6\phi + \dots$$

gives via integration

$$\overline{\overline{R}}^2 \sin\overline{\overline{\phi}} = a^2\left(1-\varepsilon^2\right)\left(\sin\phi + \frac{2}{3}\varepsilon^2 \sin^3\phi + \frac{3}{5}\varepsilon^4 \sin^5\phi + \frac{4}{7}\varepsilon^6 \sin^7\phi + ..\right) + \text{constant}$$

Since, on multiplying out, every term other than the constant has a sine multiplier, which is zero for zero latitude, this new constant must be zero. Requiring the Poles to map onto the Poles, where the sine terms are all unity, shows likewise,

$$\overline{\overline{R}}^2 = a^2\left(1-\varepsilon^2\right)\left(1 + \frac{2}{3}\varepsilon^2 + \frac{3}{5}\varepsilon^4 + \frac{4}{7}\varepsilon^6 + \dots\right)$$

giving an alternative expression to Equation 399b for authalic radius. Using either, we have

$$\sin\overline{\overline{\phi}} = \left(\overline{\overline{R}}\right)^{-1}\left(\sin\phi + \frac{2}{3}\varepsilon^2 \sin^3\phi + \frac{3}{5}\varepsilon^4 \sin^5\phi + \frac{4}{7}\varepsilon^6 \sin^7\phi + \dots\right)$$

providing a formula for *authalic latitude* in terms of geodesic latitude which, together with equating longitudes, provides a transformation of the ellipsoid onto a sphere that maintains equal-area for any region.

Except at the Poles and Equator, where the two are identical, the authalic latitude is everywhere greater than the geodesic latitude. This is the inverse of the geodesic-to-geocentric picture. The difference is similar, so that authalic latitude is relatively close to geocentric latitude. At 45° the authalic latitude is about an eighth of a degree less.

Substituting the values for WGS84 (page 407) in Equations 399a & 399b we get:

$$A = 510\,065\,621.7_{\sim}\ \text{km}^2$$

$$\overline{\overline{R}} = 0.998\,882\,1_{\sim}\, a \quad = 6\,371.007_{\sim}\ \text{km}$$

Equidistant sphere, radius, longitude and latitude

Because the length of a meridional ellipse is less than that of the Equator on the ellipsoid, while the two are equal on the sphere, there cannot be a whole-world equidistant sphere that is complete, including point Poles. We can, though, have an equidistant mapping onto a sphere of circumference equalling the length of the meridional ellipse, hence its Equator (and other parallels) minutely short of circumscribing the sphere. We can then establish equidistant longitude and latitude thereon.

Using a down-curved overscript to distinguish this sphere, and equating its circumference with the length of the meridional ellipse given in Equation 393b, we get

$$2\pi\hat{R} = 4\left[a\left(\overline{B}_0\,\phi - \overline{B}_2\,\varepsilon^2 \sin 2\phi + \overline{B}_4\,\varepsilon^4 \sin 4\phi - \overline{B}_6\,\varepsilon^6 \sin 6\phi + \overline{B}_8\,\varepsilon^8 \sin 8\phi + \ldots\right)\right]_0^{\frac{\pi}{2}}$$

with coefficients defined in Equation 393c. With all the angles in the sine terms becoming integer multiples of π — hence the terms themselves equal to zero — we get

$$\hat{R} = a\overline{B}_0 = a\left(1 - \frac{1}{4}\varepsilon^2 - \frac{3}{64}\varepsilon^4 - \frac{5}{256}\varepsilon^6 - \frac{175}{16384}\varepsilon^8 + \ldots\right) \qquad \ldots\ldots (401a$$

Addressing the length along a meridian from the Equator, we get, by similar means,

$$\hat{R}\hat{\phi} = \left[a\left(\overline{B}_0\,\phi - \overline{B}_2\,\varepsilon^2 \sin 2\phi + \overline{B}_4\,\varepsilon^4 \sin 4\phi - \overline{B}_6\,\varepsilon^6 \sin 6\phi + \overline{B}_8\,\varepsilon^8 \sin 8\phi + \ldots\right)\right]_0^{\phi}$$

Hence, the equidistant latitude is

$$\hat{\phi} = \left(\overline{B}_0\right)^{-1}\left\{\overline{B}_0\,\phi - \overline{B}_2\,\varepsilon^2 \sin 2\phi + \overline{B}_4\,\varepsilon^4 \sin 4\phi - \overline{B}_6\,\varepsilon^6 \sin 6\phi + \overline{B}_8\,\varepsilon^8 \sin 8\phi + \ldots\right\}$$

To provide equidistance along the Equator — as is implied in our use of this term — we must multiply longitude to adjust the mapping along parallels. For equidistant longitude we get

$$\hat{\lambda} = \frac{2\pi\hat{R}}{2\pi a}\lambda = \left(1 - \frac{1}{4}\varepsilon^2 - \frac{3}{64}\varepsilon^4 - \frac{5}{256}\varepsilon^6 - \frac{175}{16384}\varepsilon^8 + \ldots\right)\lambda \qquad \ldots\ldots (401b$$

Substituting the values for WGS84 (page 407) in Equations 401a & 401b we get:

$\hat{R} = 0.998\,324\,3_{\sim}\,a \quad = 6\,367.449_{\sim}$ km

$\hat{\lambda} = 0.998\,324\,3_{\sim}\,\lambda$

Conformal sphere, radius, longitude and latitude

With common parameters, the criterion for conformality is

$$\frac{\dot{E}}{E} = \frac{\dot{G}}{G} = \dot{m}^2 \qquad \text{and thereby} \qquad \frac{\dot{E}}{\dot{G}} = \frac{E}{G} \qquad \ldots\ldots (401c$$

where $\dot{m}$ is the variable linear scale factor applicable identically in any direction at any one point – in our circumstances necessarily a function of latitude but not of longitude.

Adopting a tilde as overscript to signify elements of our conformal sphere, and using Equation 391c on the left side and Equation 398a (accordingly interpreted) on the right, we have

$$\frac{{}_{\mathrm{t}}R^2\cos^2\phi}{{}_{\mathrm{m}}R^2} = \frac{{}_{\mathrm{e}}E_{\lambda,\phi}}{{}_{\mathrm{e}}G_{\lambda,\phi}} = \frac{{}_{\mathrm{s}}\dot{E}_{\lambda,\phi}}{{}_{\mathrm{s}}\dot{G}_{\lambda,\phi}} = \frac{\tilde{k}^2\,\tilde{R}^2\cos^2\tilde{\phi}}{\tilde{R}^2}\left(\frac{\mathrm{d}\phi}{\mathrm{d}\tilde{\phi}}\right)^2 = \tilde{k}^2\cos^2\tilde{\phi}\left(\frac{\mathrm{d}\phi}{\mathrm{d}\tilde{\phi}}\right)^2$$

Substituting for the ratio of the ellipsoidal radii from Equation 388b, this becomes

$$\frac{\left(1-\varepsilon^2\sin^2\phi\right)^2\cos^2\phi}{\left(1-\varepsilon^2\right)^2}=\tilde{k}^2\cos^2\tilde{\phi}\left(\frac{\mathrm{d}\phi}{\mathrm{d}\tilde{\phi}}\right)^2$$

Rearranging and taking the square root (attending to necessary signs),

$$\frac{1}{\cos\tilde{\phi}}\,\mathrm{d}\tilde{\phi}=\frac{\tilde{k}\left(1-\varepsilon^2\right)}{\left(1-\varepsilon^2\sin^2\phi\right)\cos\phi}\,\mathrm{d}\phi$$

Integrating these expressions involves the same manipulations and integrable terms as in proceeding from Equation 396b to Equation 396c. The result is

$$\ln\tan\left(\frac{\pi}{4}+\frac{\tilde{\phi}}{2}\right)=\tilde{k}\left[\ln\left\{\left(\frac{1-\varepsilon\sin\phi}{1+\varepsilon\sin\phi}\right)^{\frac{\varepsilon}{2}}\tan\left(\frac{\pi}{4}+\frac{\phi}{2}\right)\right\}\right]+\text{constant}$$

Since the ellipsoidal latitude zero must map into spherical latitude zero, this new constant must be zero. Exponentiating thus gives

$$\tan\left(\frac{\pi}{4}+\frac{\tilde{\phi}}{2}\right)=\left\{\left(\frac{1-\varepsilon\sin\phi}{1+\varepsilon\sin\phi}\right)^{\frac{\varepsilon}{2}}\tan\left(\frac{\pi}{4}+\frac{\phi}{2}\right)\right\}^{\tilde{k}}$$

$$\tilde{\phi}=-\frac{\pi}{2}+2\arctan\left\{\left(\frac{1-\varepsilon\sin\phi}{1+\varepsilon\sin\phi}\right)^{\frac{\varepsilon}{2}}\tan\left(\frac{\pi}{4}+\frac{\phi}{2}\right)\right\}^{\tilde{k}} \qquad \text{(402a)}$$

as transformational equations for latitude. It remains to ascertain constant $\tilde{k}$ for completion of this formula. This constant (the multiplier of longitude) and the radius of the conformal sphere depend on the latitude chosen to have true scale. Before choosing such we develop one further sequence of general equations.

Substituting from Equations 391b & 398a (accordingly interpreted) into the first of Equations 401c, we have:

$$\dot{m}^2=\frac{{}_{\mathrm{s}}\dot{E}_{\lambda,\phi}}{{}_{\mathrm{e}}E_{\lambda,\phi}}=\frac{\tilde{k}^2\,\tilde{R}^2\cos^2\tilde{\phi}}{{}_{\mathrm{t}}R^2\cos^2\phi}=\left\{\tilde{k}\frac{\tilde{R}\cos\tilde{\phi}}{{}_{\mathrm{t}}R\cos\phi}\right\}^2$$

$$\dot{m}^2=\frac{{}_{\mathrm{s}}\dot{G}_{\lambda,\phi}}{{}_{\mathrm{e}}G_{\lambda,\phi}}=\frac{\tilde{R}^2}{{}_{\mathrm{m}}R^2}\left(\frac{\mathrm{d}\tilde{\phi}}{\mathrm{d}\phi}\right)^2=\left\{\frac{\tilde{R}}{{}_{\mathrm{m}}R}\frac{\mathrm{d}\tilde{\phi}}{\mathrm{d}\phi}\right\}^2$$

Since all the terms must be positive, taking square roots we have:

$$\dot{m}=\tilde{k}\frac{\tilde{R}\cos\tilde{\phi}}{{}_{\mathrm{t}}R\cos\phi}\qquad\text{or}\qquad\frac{\dot{m}}{\tilde{k}\tilde{R}}=\frac{\cos\tilde{\phi}}{{}_{\mathrm{t}}R\cos\phi} \qquad \text{(402b)}$$

$$\dot{m}=\frac{\tilde{R}}{{}_{\mathrm{m}}R}\frac{\mathrm{d}\tilde{\phi}}{\mathrm{d}\phi}\qquad\text{or}\qquad\frac{\mathrm{d}\tilde{\phi}}{\mathrm{d}\phi}=\dot{m}\frac{{}_{\mathrm{m}}R}{\tilde{R}} \qquad \text{(402c)}$$

Differentiating the first of these with respect to ϕ gives

$$\frac{\mathrm{d}\dot{m}}{\mathrm{d}\phi} = \frac{-\sin\tilde{\phi}}{{}_{\mathrm{t}}R\cos\phi}\frac{\mathrm{d}\tilde{\phi}}{\mathrm{d}\phi} + \cos\tilde{\phi}\left\{\frac{(-1)(-\sin\phi)}{{}_{\mathrm{t}}R\cos^2\phi} + \frac{(-1)}{{}_{\mathrm{t}}R^2\cos\phi}\frac{\mathrm{d}\,{}_{\mathrm{t}}R}{\mathrm{d}\phi}\right\}$$

$$= \frac{-\sin\tilde{\phi}}{{}_{\mathrm{t}}R\cos\phi}\frac{\mathrm{d}\tilde{\phi}}{\mathrm{d}\phi} + \frac{\cos\tilde{\phi}}{\left({}_{\mathrm{t}}R\cos\phi\right)^2}\left\{{}_{\mathrm{t}}R\sin\phi - \cos\phi\frac{\mathrm{d}\,{}_{\mathrm{t}}R}{\mathrm{d}\phi}\right\}$$

We address first the latter expression in curly brackets, substituting from Equation 388b for the radius in the first term, and from Equation 388c for the derivative.

$$\left\{{}_{\mathrm{t}}R\sin\phi - \cos\phi\frac{\mathrm{d}\,{}_{\mathrm{t}}R}{\mathrm{d}\phi}\right\} = \sin\phi\left(\frac{1-\varepsilon^2\sin^2\phi}{1-\varepsilon^2}\right){}_{\mathrm{m}}R - \cos\phi\left(\frac{\varepsilon^2\sin\phi\cos\phi}{1-\varepsilon^2}\right){}_{\mathrm{m}}R$$

$$= {}_{\mathrm{m}}R\sin\phi\left(\frac{1-\varepsilon^2\sin^2\phi}{1-\varepsilon^2} - \frac{\varepsilon^2\cos^2\phi}{1-\varepsilon^2}\right)$$

$$= {}_{\mathrm{m}}R\sin\phi\left(\frac{1-\varepsilon^2\sin^2\phi-\varepsilon^2\cos^2\phi}{1-\varepsilon^2}\right)$$

$$= {}_{\mathrm{m}}R\sin\phi\left(\frac{1-\varepsilon^2}{1-\varepsilon^2}\right)$$

$$= {}_{\mathrm{m}}R\sin\phi$$

Hence,

$$\frac{\mathrm{d}\dot{m}}{\mathrm{d}\phi} = \frac{-\sin\tilde{\phi}}{{}_{\mathrm{t}}R\cos\phi}\frac{d\tilde{\phi}}{d\phi} + \frac{{}_{\mathrm{m}}R\cos\tilde{\phi}\sin\phi}{\left({}_{\mathrm{t}}R\cos\phi\right)^2} \quad \text{. . . . . (403a}$$

Let ϕ_0 be the latitude chosen to have true scale, implying that $\dot{m}=1$ along that parallel. Denoting with suffix 0 all the associated variables, including the conformal radius resulting from that choice, we thus get from Equation 402b & 402c:

$$\tilde{k}_0 = \frac{{}_{\mathrm{t}}R_0\cos\phi_0}{\tilde{R}_0\cos\tilde{\phi}_0} \quad \text{or} \quad \tilde{k}_0\tilde{R}_0\cos\tilde{\phi}_0 = {}_{\mathrm{t}}R_0\cos\phi_0 \quad \text{. . . . . (403b}$$

$$\left[\frac{\mathrm{d}\tilde{\phi}}{\mathrm{d}\phi}\right]_0 = \frac{{}_{\mathrm{m}}R_0}{\tilde{R}_0}$$

If we choose to have the (first) derivative of $\dot{m}$ with respect to ϕ equal to zero at that latitude, then, from Equation 403b,

$$\frac{\sin\tilde{\phi}_0}{\cos\tilde{\phi}_0}\frac{{}_{\mathrm{m}}R_0}{\tilde{R}_0} = \frac{{}_{\mathrm{m}}R_0\sin\phi_0}{{}_{\mathrm{t}}R_0\cos\phi_0} \quad \text{or} \quad \frac{\tan\tilde{\phi}_0}{\tilde{R}_0} = \frac{\tan\phi_0}{{}_{\mathrm{t}}R_0}$$

Multiplying the respective sides of the second version of this equation with the like from Equation 403b gives

$$\sin\tilde{\phi}_0 = \tilde{k}_0^{-1}\sin\phi_0 \quad \text{or} \quad \tilde{k}_0 = \frac{\sin\phi_0}{\sin\tilde{\phi}_0} \quad \text{. . . . . (403c}$$

for our constant in terms of the chosen latitude, then the derived formulae

$$\cos\tilde{\phi}_0 = \sqrt{1-\sin^2\tilde{\phi}_0} = \sqrt{1-\tilde{k}_0^{-2}\sin^2\phi_0} = \tilde{k}_0^{-1}\sqrt{\tilde{k}_0^{2}-\sin^2\phi_0}$$

$$\tan\tilde{\phi}_0 = \frac{\sin\tilde{\phi}_0}{\cos\tilde{\phi}_0} \quad = \frac{\tilde{k}_0^{-1}\sin\phi_0}{\tilde{k}_0^{-1}\sqrt{\tilde{k}_0^2 - \sin^2\phi_0}} \quad = \frac{\sin\phi_0}{\sqrt{\tilde{k}_0^2 - \sin^2\phi_0}} \quad \text{........ (404a}$$

Combining Equation 402b with Equation 403c gives

$$\frac{{}_tR_0\cos\phi_0}{\tilde{R}_0\cos\tilde{\phi}_0} = \tilde{k}_0 = \frac{\sin\phi_0}{\sin\tilde{\phi}_0}$$

hence, by inverting then multiplying out,

$$\frac{\sin\tilde{\phi}_0}{\cos\tilde{\phi}_0} = \frac{\tilde{R}_0\sin\phi_0}{{}_tR_0\cos\phi_0}$$

$${}_tR_0\tan\tilde{\phi}_0 = \tilde{R}_0\tan\phi_0 \qquad \text{or} \qquad \tan\tilde{\phi}_0 = \frac{\tilde{R}_0}{{}_tR_0}\tan\phi_0$$

If we now choose to have the second derivative of $\dot{m}$ with respect to ϕ equal to zero at latitude ϕ_0, then, by similar development from Equation 403a, we get

$$\tan\tilde{\phi}_0 = \sqrt{\frac{{}_mR_0}{{}_tR_0}}\tan\phi_0$$

Hence,

$$\frac{\sin^2\phi_0}{\cos^2\phi_0} = \tan^2\phi_0 \quad = \frac{{}_tR_0}{{}_mR_0}\tan^2\tilde{\phi}_0 \quad = \frac{{}_tR_0}{{}_mR_0}\frac{\sin^2\tilde{\phi}_0}{\cos^2\tilde{\phi}_0} \quad \text{........ (404b}$$

Substituting from Equation 403c for $\sin\phi_0$ gives

$$\frac{\tilde{k}_0^2\sin^2\tilde{\phi}_0}{\cos^2\phi_0} = \frac{{}_tR_0}{{}_mR_0}\frac{\sin^2\tilde{\phi}_0}{\cos^2\tilde{\phi}_0}$$

$$\tilde{k}_0^2 = \frac{{}_tR_0}{{}_mR_0}\frac{\cos^2\phi_0}{\cos^2\tilde{\phi}_0}$$

Substituting from Equation 403b for $\tilde{k}_0$ then gives

$$\frac{{}_tR_0^2}{\tilde{R}_0^2}\frac{\cos^2\phi_0}{\cos^2\tilde{\phi}_0} = \frac{{}_tR_0}{{}_mR_0}\frac{\cos^2\phi_0}{\cos^2\tilde{\phi}_0}$$

$${}_tR_0\cdot{}_mR_0 = \tilde{R}_0^2$$

yielding a formula for conformal radius for a chosen latitude ϕ_0 of true scale:

$$\tilde{R}_0 = \sqrt{{}_tR_0\cdot{}_mR_0} \quad \text{........ (404c}$$

Returning to Equation 404b and substituting for the ratio of the ellipsoidal radii of curvatures from Equation 388b and for the right-side tangent from Equation 404a gives

$$\frac{\sin^2\phi_0}{\cos^2\phi_0} = \left\{\frac{1-\varepsilon^2\sin^2\phi_0}{1-\varepsilon^2}\right\}\left\{\frac{\sin\phi_0}{\sqrt{\tilde{k}_0^2 - \sin^2\phi_0}}\right\}^2$$

$$= \left\{\frac{1-\varepsilon^2\sin^2\phi_0}{1-\varepsilon^2}\right\}\left\{\frac{\sin^2\phi_0}{\tilde{k}_0^2 - \sin^2\phi_0}\right\}$$

$$\frac{1}{\cos^2\phi_0} = \frac{1-\varepsilon^2\sin^2\phi_0}{\left(1-\varepsilon^2\right)\left(\tilde{k}_0{}^2-\sin^2\phi_0\right)}$$

By inverting and cross-multiplying then rearranging, this becomes

$$\cos^2\phi_0\left(1-\varepsilon^2\sin^2\phi_0\right) = \left(1-\varepsilon^2\right)\left(\tilde{k}_0{}^2-\sin^2\phi_0\right)$$

$$\cos^2\phi_0 - \varepsilon^2\cos^2\phi_0\sin^2\phi_0 = \left(1-\varepsilon^2\right)\tilde{k}_0{}^2 - \sin^2\phi_0 + \varepsilon^2\sin^2\phi_0$$

Regrouping and reversing sides gives

$$\begin{aligned}\left(1-\varepsilon^2\right)\tilde{k}_0{}^2 &= \cos^2\phi_0 + \sin^2\phi_0 - \varepsilon^2\sin^2\phi_0\left(1+\cos^2\phi_0\right)\\ &= 1-\varepsilon^2\left(1-\cos^2\phi_0\right)\left(1+\cos^2\phi_0\right)\\ &= 1-\varepsilon^2\left(1-\cos^4\phi_0\right)\\ &= 1-\varepsilon^2+\varepsilon^2\cos^4\phi_0\end{aligned}$$

This gives us a formula for conformal constant $\tilde{k}_0$ for a chosen latitude ϕ_0 of true scale

$$\tilde{k}_0 = \sqrt{\frac{1-\varepsilon^2+\varepsilon^2\cos^4\phi_0}{1-\varepsilon^2}} = \sqrt{1+\frac{\varepsilon^2\cos^4\phi_0}{1-\varepsilon^2}} \qquad \text{. . . . . (405a}$$

Conformal longitude is then

$$\tilde{\lambda} = \left(\frac{1-\varepsilon^2+\varepsilon^2\cos^2\phi_0}{1-\varepsilon^2}\right)^{\frac{1}{2}}\lambda = \left(1+\frac{\varepsilon^2\cos^2\phi_0}{1-\varepsilon^2}\right)^{\frac{1}{2}}\lambda$$

while, from Equation 402a, the corresponding formula for conformal latitude is

$$\tilde{\phi} = 2\arctan\left\{\left(\frac{1-\varepsilon\sin\phi}{1+\varepsilon\sin\phi}\right)^{\frac{\varepsilon}{2}}\tan\left(\frac{\pi}{4}+\frac{\phi}{2}\right)\right\}^{\tilde{k}_0} - \frac{\pi}{2} \qquad \text{. . . . . (405b}$$

with the value of constant $\tilde{k}_0$ defined in Equation 405a.

With Equation 404c for radius, this completes the formulae for the conformal sphere.

More latitudes

We thus have added authalic ($\bar{\bar{\phi}}$), equidistant ($\hat{\phi}$), and conformal ($\tilde{\phi}$) latitudes for the proxy sphere to the fundamental geodetic latitude (ϕ) of the ellipsoid. C. H. Deetz [1921c] provides many formulae connecting these and geocentric, isometric and parametric latitudes.

SPECIFIC ELLIPSOIDS

Variety

The monumental survey of India by George Everest and the more detached studies of F. W. Bessel, both in the first half of the 18th Century, led to two of the earliest specific ellipsoids used for representing Earth. The two men, both widely commemorated by other means, are also remembered by their ellipsoids. The table below shows the values adopted by them, and for many other ellipsoids used since, culminating with the WGS84 ellipsoid that has become the universal world standard. It was common practice in applying them to displace the minor axis from Earth's axis and the equivalent plane from the Equatorial plane to obtain a best fit for a given regional situation. The Clarke 1866 was the standard for decades in the U.S.A. However, though dimensioned from a worldwide survey, it was used displaced by over 150 m from both axis and Equatorial plane. The last four ellipsoids shown were all developed for true geocentric application, the visible changes reflecting improving measurement of Earth, the last differing from its predecessor only by its detailed definition discussed next.

	- - - semi-axes - - -		*first*	*flattening*	
	major	*minor*	*eccentricity*	f	
	km	km	ε	1:	
Everest	6377.304	6356.103	0.081 473	300.8	S Asia
Bessel	6377.397	6356.082	0.081 690	299.2	Europe, Chile
Airy	6377.563	6356.300	0.081 591	299.93	U.K.
Clarke 1858	6378.294	6356.621	0.082 366	294.3	
Clarke 1866	6378.206	6356.584	0.082 269	294.98	U.S.A., Philippines
Clarke 1880	6378.249	6356.517	0.082 478	293.5	Africa, France
Hayford (1909)	6378.388	6356.912	0.081 992	297	much of world
Krasovski (1940)	6378.245	6356.863	0.081 813	298.3	U.S.S.R.
Hough	6378.270	6356.794	0.081 992	297	
Fischer (1960)	6378.166	6356.784	0.081 813	298.3	
Australian (1965)	6378.160	6356.775	0.082 647	298.25	Australia
Kaula	6378.165	6356.345	0.082 647	292.3	
IUGG.67	6378.160	6356.775	0.081 820	298.25	whole world
Fischer 68	6378.150	6356.330	0.082 647	292.3	
WGS72	6378.135	6356.751	0.081 808	298.34	whole world
IUG75	6378.140	6356.755	0.081 819	298.253	whole world
GRS80	6378.137	6356.752	0.081 819	298.257	U.S.A
WGS84	6378.137	6356.752	0.081 819	298.257	whole world

Basic parameters of various adopted ellipses (with rounded figures).

GRS80 and WGS84

At its General Assembly of 1979 the International Association of Geodesy adopted for comprehensive international use an ellipsoid seen as providing a best fit across the whole world. Named the *Geodetic Reference System 1980* or *GRS80*, this was based necessarily on the geocentre rather than any surface datum point. This reflected the role of orbiting satellites in measuring Earth, and, in turn, the developments that produced the GPS system of positioning — both correspondingly centred. The defining constants were:

a	Equatorial radius of Earth	6 378 137 m
GM	geocentric gravitational constant (including the atmosphere)	$3\ 986\ 005 \times 10^8\ m^3s^{-2}$
J_2	dynamical form factor (excluding permanent tides)	$108\ 263 \times 10^{-8}$
ω	angular velocity of Earth	$7\ 292\ 115 \times 10^{-11}$ rad·s $^{-1}$

While containing the Equatorial radius, these defining factors conspicuously omit any further factor ordinarily used to define an ellipse. The inclusion of the more physical factors relates to the role of GRS80 in accurate surveying and other geodetic activities. The derived geometrical parameters for the ellipse were:

b	semi-minor axis (polar radius)	6 356 752.314 1~ m
ε	first eccentricity	√(0.006 694 380 022 90~)
f	flattening	1 : 298.257 222 101~

These in turn gave:

R_1	mean radius	6 371 008.771 4~ m
R_a	radius of authalic sphere	6 371 007.181 0~ m

Derived physical parameters were normal potential at ellipsoid:

U_0	normal potential	62 636 860.850~ m^2s^{-2}
g_e	normal gravity at Equator	9.780 326 771 5~ $m{\cdot}s^{-2}$
g_p	normal gravity at a Pole	9.832 186 368 5~ $m{\cdot}s^{-2}$

Though promptly implemented, this definition was soon displaced by a revised definition that reverted to a more geometric basis, but such that the practical factors were essentially unchanged [1993b]. The new definition, termed the *World Geodetic System 84* or *WGS84*, was as follows [1997b]:

a	Equatorial radius of Earth	6 378 137 m
f	flattening (derived from the value of the normalized second degree zonal harmonic coefficient of the gravitational field: $-484.16685 \times 10^{-6}$)	1 : 298.257 223 563
GM	Earth's gravitational constant (including the atmosphere)	$3\ 986\ 005 \times 10^8\ m^3s^{-2}$
ω	angular velocity of Earth	7 292 115~ $\times 10^{-11}$ rad·s $^{-1}$

The only numeric change wrought by WGS84 was in the ninth place for flattening, and amounted to less than 0.5 in 10^8 (hence, a like proportional change to eccentricity ε). It also made the figure precise. WGS84 also carefully defines the axial situation as follows:

> "The origin is Earth's centre of mass, the geocentre.
>
> "The z-axis is aligned parallel to the direction of the Conventional Terrestrial Pole (CTP) for polar motion, as originally defined by the Bureau International de l'Heure (BIH), and since 1989 by the International Earth Rotation Service (IERS).
>
> "The x-axis is the intersection of the WGS84 Reference Meridian Plane and the plane of the CTP Equator (the Reference Meridian being parallel to the Zero Meridian defined by BIH/IERS).
>
> "The y-axis completes a right-handed, earth-centred, earth-fixed (ECEF) orthogonal co-ordinate system, measured in the plane of the CTP Equator, 90 east of the x-axis."

The first nine powers of eccentricity ε are given below.

Using these ellipsoids

The adoption of these Earth-centred ellipsoids resulted in widespread revision of detailed positions on maps, especially where they displaced ellipsoids like the Clarke 1866 used in the U.S.A. on a regional-fit rather than geocentric basis. In some regions, places had new positions different from their previous positions by many kilometres — though locally the places were only trivially changed. The new positioning is embraced by the UTM system discussed in the Chapter 18, the impact further discussed in Chapter 19.

ε	0.081 819 191 043~
ε^2	0.006 694 380 023~
e^3	0.000 547 728 758~
ε^4	0.000 044 814 724~
ε^5	0.000 003 666 704~
ε^6	0.000 000 300 007~
ε^7	0.000 000 024 546~
ε^8	0.000 000 002 008~
ε^9	0.000 000 000 164~

Powers of the eccentricity ε for WGS84.

EMPLOYING ELLIPSOIDS

Purpose

As the variation between sphere and ellipsoid is less than 1%, there is little purpose in using an ellipsoidal form of globe for maps whose variation of scale is more than this (and good reason to avoid the complications attending such use). This applies to maps in atlases and general wall maps. Ellipsoids are employed primarily — and necessarily — in large-scale sectional maps used for engineering and similar purposes, where scale variations are kept well below 1%. In the modern world, this relates primarily to the Mercator projection used transversely in narrow sections, along with the secantal Stereographic at Simple aspect for Polar regions. Both are covered in the Chapter 18, but the Polyconic, so long used in North America for highly accurate maps, is an appropriate example to peruse to see an ellipsoid in action.

The Polyconic in ellipsoidal form

Equations 181b give the plotting equations for the (American) Polyconic on the sphere as:

$$\text{if } \phi = 0 \quad \dot{x} = R(\lambda - \lambda_0), \quad \dot{y} = -R\phi_x$$

$$\text{else} \quad \dot{x} = R\cot\phi\,\sin\{(\lambda - \lambda_0)\sin\phi\}$$

$$\dot{y} = R\phi - R\phi_x + R\left(1 - \cos\{(\lambda - \lambda_0)\sin\phi\}\right)\cot\phi$$

Although used so extensively for survey purposes, the Polyconic is neither conformal nor equal-area nor equidistant. To develop an ellipsoidal form, we make direct substitutions in the preceding formulae. The simple factor R becomes elaborated, its product with latitude replaced by meridional distance. The plotting equations for the ellipsoidal Polyconic projection are:

$$\text{if } \phi = 0 \quad \dot{x} = {}_{t}R_0\,(\lambda - \lambda_0), \quad \dot{y} = -{}_{m}d_x$$

$$\text{else} \quad \dot{x} = {}_{t}R\cot\phi\,\sin\{(\lambda - \lambda_0)\sin\phi\} \qquad \text{. . . . . (409a}$$

$$\dot{y} = {}_{m}d - {}_{m}d_x + {}_{t}R\left(1 - \cos\{(\lambda - \lambda_0)\sin\phi\}\right)\cot\phi$$

where ${}_{m}d$ and ${}_{m}d_x$ are the meridional distances from the Equator to general latitude ϕ and to chosen latitude ϕ_x, respectively, and ${}_{t}R$ and ${}_{t}R_0$ are the transverse radii of curvature at general latitude ϕ and at the Equator, respectively. The meridional distances are obtained by appropriate evaluation of Equation 393b. Radii ${}_{t}R$ and ${}_{t}R_0$ are given by appropriate evaluation of Equation 387b, the latter identically a, the length of the semimajor axis.

Further literature concerning the ellipsoidal form

P. D. Thomas in his U.S. Geological Survey special publication on conformal projections [1952b] deals with their use based on the ellipsoid. J. P. Snyder enhances the coverage and addresses the calculation aspect in a paper [1979c] before delving more deeply in his U.S.G.S. Bulletin *Map Projections used by the United States Geological Survey* [1983a]. The last provides the development on an ellipsoidal basis for several key projections, namely the Mercator, Albers, Lambert Conformal Conical, Polyconic, Stereographic, Lambert Equal-area Azimuthal, and Equidistant Azimuthal. Reverse formulae, giving longitude and latitude from map co-ordinates, are also given.

Because of the complications of using the ellipsoid, extensive numeric tables have been prepared over the years for many projections [e.g., 1950b]. Today, computers provide the required calculations, but the involvement of infinite series and the consequential approximating, together with non-integrable functions, still leaves latitude for addressing the numeric aspects. Snyder, advocating the use of the ellipsoid for any larger-scale equal-area maps, discusses [1985c] the use of the mathematical technique of Fourier transforms to eliminate recurring numerical integration and other lengthy computations. He appends Fourier coefficients for the orthographic Equal-area Cylindrical projection at Transverse and general Oblique aspects.

The development in preceding sections of the series representations of the ellipsoid follows the manner pioneered by C F. Gauss [1825a] and effectively made the norm following further work by Louis Krüger [1912c]. Martin Hotine [1946d & 1947c] wrote extensively on the subject in the context of conformal mappings. L. P. Lee [1953e & 1962e], expanding unpublished work by others, developed alternative formulae to those of the Gauss-Krüger school. He went so far as to show that a conformal proxy sphere can be created that contains the whole ellipsoid, with proxy meridians coincident with the natural ones on that sphere. The scale, while obviously uniform along the Equator, varies along all meridians.

CHAPTER 18

Pinpointing the Place: UTM and UPS

Preamble

While longitude and latitude provide an unambiguous means for identifying the positions of points around the globe, and can be reproduced for like purpose on maps, they are inherently awkward on any map other than a true Cylindrical at Simple aspect. Surveyors and many others working with maps need a co-ordinate scheme that is easily applied — i.e., is linear in both key directions and preferably is rectilinear. In our cylindrical maps, we have a rectilinear grid and linear scales with our plotting variables. Conic maps have long been used generally for accurate survey purposes but, since the 1940s, the transverse Mercator has become the accepted base for most survey usage. Applied in strips narrower than 500 km as the Universal Transverse Mercator, and augmented for the Polar regions by the conformal Universal Polar Stereographic projection, it provides a comprehensive grid scheme for the whole Earth that measures to a metre.

This chapter brings no new mathematics.

It also brings no new projections, but there is a key map of the world using the Mercator projection at Transverse aspect.

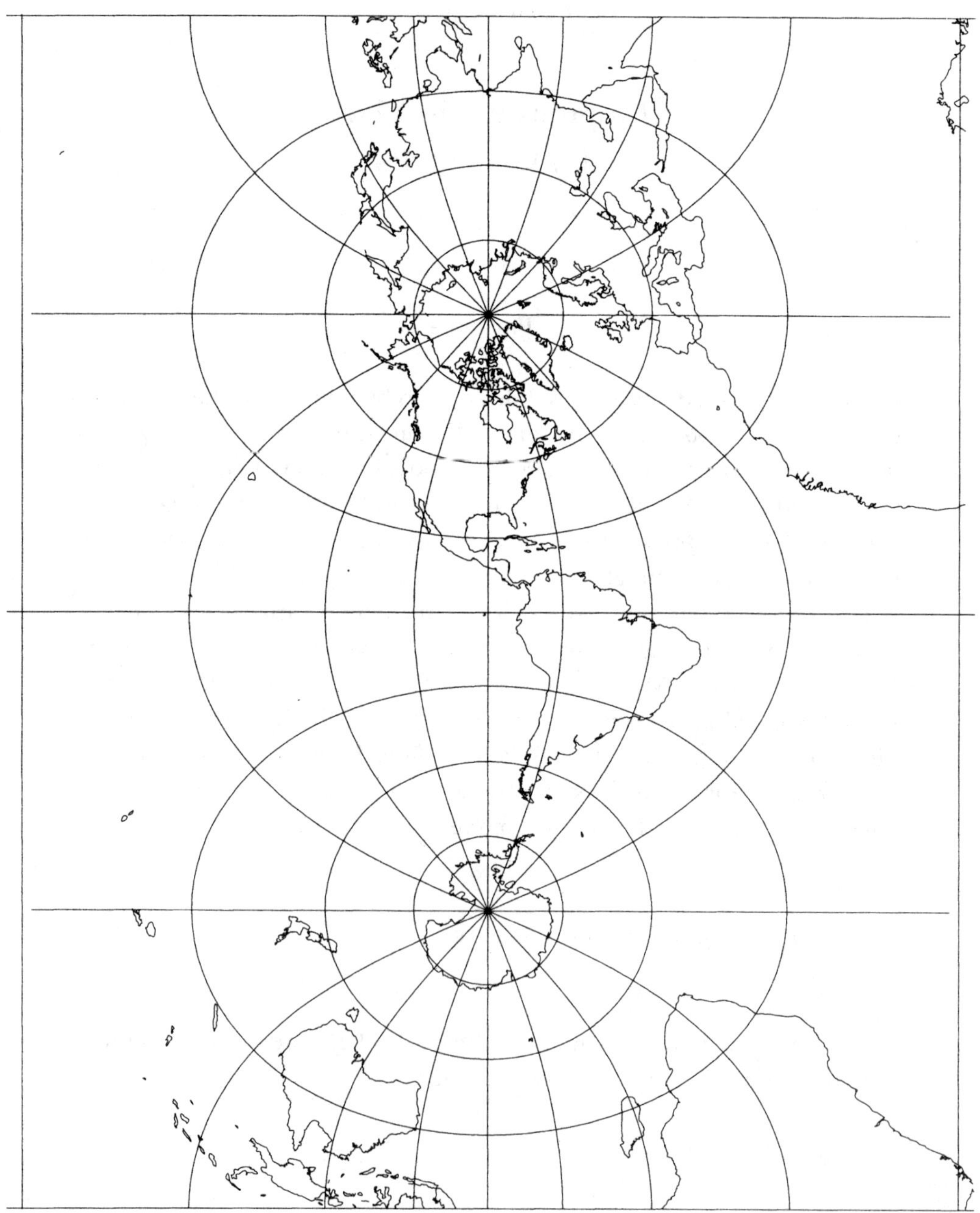

Exhibit 18–1: The Transverse Mercator projection – whole world.

CO-ORDINATE SCHEMES

The need

As stated in the preamble to the chapter, we need a co-ordinate scheme that is linear in both its directions and, preferably, is rectilinear. If it is to serve a reasonable range of purposes — including civil engineering and military activities, we need maximal accuracy of scale, area and angle across the mapped area, without demanding comprehensive and total accuracy in any particular attribute. We do have in our maps an intrinsic rectilinear grid with linear scales, via our plotting co-ordinates x and y. While convenience for the user favours making all co-ordinates positive, this requires only a shift of origin from the position we have routinely adopted. So these co-ordinates could be used for individual maps and, were a common origin adopted, across adjacent maps that are of the same projection, in all details. But if we change any detail — e.g., the central meridian, the standard parallel(s) — then we have difficulties with those co-ordinates and indeed with the adjoining edges of adjacent maps. Ideally we want a scheme that carries unchanged across an extended region and many maps, preferably around the world.

Some history

Various co-ordinate schemes have been adopted. Unitary countries of modest size such as the United Kingdom and France have adopted one scheme across the country while in the larger and federal United States, each individual state has had its own (called overall the State Plane system). Each such state adopted as the basis of its scheme a map projection suitable for its own purposes — usually a (Lambert) Conformal Conical with two standard parallels situated approximately one sixth and five sixths between the extreme parallels for the jurisdiction, but a secantal transverse Mercator with similar spacing otherwise. For the typically larger provinces of Canada and states of Australia, similar solutions have generally been adopted. The two projections cited are both conformal. Within the confines of a specified scale variation, there is generally less distorting of area with a conformal projection — like the two cited above — than there is of angle when using an equal-area projection. Compromises with neither attribute have often been used, but these have typically been inherently local — e.g., the Airy projection in the U.K. (see page 330). Thus, while equal-area might seem to be preferable, conformality is the typical choice.

Numerical constraints and current choices

If we specify that the scale across our map should never be more than one part in a thousand different, we get (from Chapter 5) that the spacing of the standard lines can little exceed 5° angle at the centre of the globe, and that the map can extend barely more than 1° further on each side. Converting to radians and multiplying by the radius of Earth puts the limit on the spacing of our two parallels as about 570 km, the aggregate width of the map at no more than 800 km. Such linear limits apply regardless of aspect.

For a conformal Conical at Simple aspect, this aggregate limit of 7° in latitude range sufficed for most of the states of the U.S.A. but for only a minority of provinces of Canada, for instance. Where necessary, a jurisdiction can be covered by sectional maps with differing details of projection, suitably chosen to minimize mismatching at the edges when maps are juxtaposed. An alternative for jurisdictions that are narrow in longitude terms (and for sectioning larger ones) is the transverse Mercator. While the above limits mean only 7° along the Equator, they embrace progressively more degrees of longitude as one moves north or south there from. At ±60° latitude, 500 km spans 14° of longitude.

The transverse Mercator (illustrated on page 412) has long been used selectively for such purpose, but has now become the ubiquitous worldwide standard. By repeating the projection at intervals of 6° longitude, the world can be covered Pole-to-Pole by 60 strips, with scale variation everywhere within ±0.1%. Fitted to the WGS84 ellipsoid, this is referred to as the *Universal Transverse Mercator* (*UTM*) scheme. Associated with it and often interpreted within those initials is the Stereographic projection for the Polar regions (with maximal scale variation exceeding ±0.1%); this conformal Azimuthal, which is used at Simple (i.e., Polar) aspect, is known then as *UPS* (*Universal Polar Stereographic*). Co-ordinated zone-labelling schemes and an associated metre grid have been adopted across the two projections and their sections.

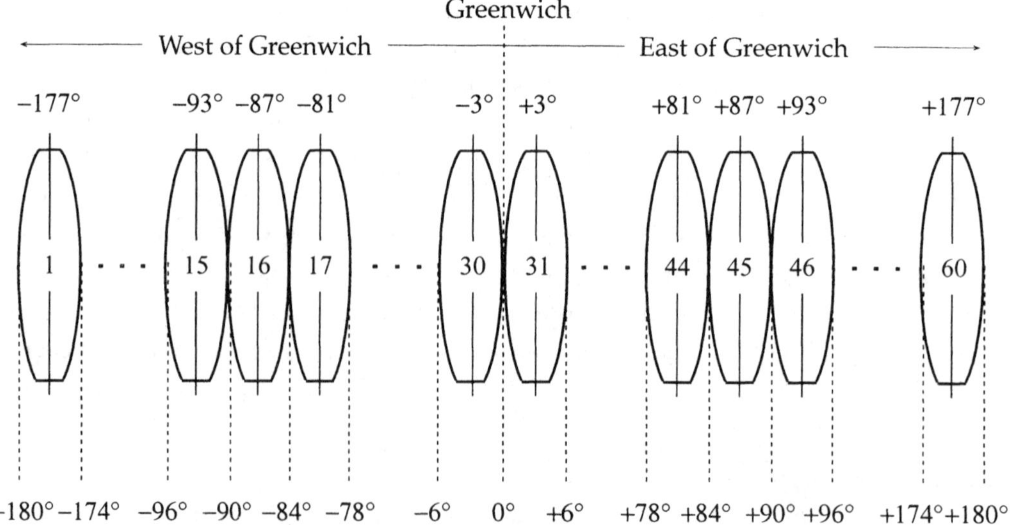

Exhibit 18–2: UTM – primary longitudinal zonation.
[Not of correct proportion/scale.]

UTM — UNIVERSAL TRANSVERSE MERCATOR

Outline

Because any strip defined by two meridians 6° apart — technically a *lune* — comes to a point at each Pole, such strips become impracticably thin for purposes of local map use at higher latitudes. The UTM scheme adopts a limit of 80° in the southern hemisphere. To accommodate all normally settled land, however, it extends the coverage to 84° in the northern hemisphere. The lunes become doubly clipped lunes, as shown in Exhibit 18–2 opposite and in Exhibit 18–3 overleaf.

UTM – the projection

Plotting of the Mercator projection at Transverse aspect is covered by Equations 209f, but those were for the spherical world. The precision intended for the UTM scheme necessitates use of an ellipsoidal model. The relevant plotting equations require protracted mathematical development and are complicated. Standard reference works give formulae applicable within the confines of the lune (plus a little beyond). Retaining σ as the scaling factor (being 1 for a tangential projection but a lesser figure for all points of a secantal projection, as is used for UTM) and ${}_tR$ as the transverse radius of curvature at latitude ϕ established in Equation 387b, then writing the abbreviation t for $\tan\phi$, the ellipsoidal plotting equations are:

$$\frac{\dot{x}}{\sigma_0\,{}_tR} = L + \frac{L^3}{6}\left(1 - t^2 + \eta^2\right) + \frac{L^5}{120}\left(5 - 18t^2 + t^4 + 14\eta^2 - 58t^2\,\eta^2 + 13\eta^4\right) + \ldots$$

$$\frac{\dot{y}}{\sigma_0\,{}_tR} = \frac{{}_md}{{}_tR} + \frac{t\,L^2}{2} + \frac{t\,L^4}{24}\left(5 - t^2 + 9\eta^2 + 4\eta^4\right)$$

$$+ \frac{t\,L^6}{720}\left(61 - 58t^2 + 270\eta^2 - 330\,t^2\eta^2 + 445\eta^4\right) + \ldots$$

where ${}_md$ is the distance along a meridian from Equator to latitude ϕ as formulated in Equation 393b and, denoting the longitude of the central meridian by λ_0:

$$L = (\lambda - \lambda_0)\cos\phi$$

$$\eta = \frac{\varepsilon\cos\phi}{\sqrt{1-\varepsilon^2}} = \frac{\sqrt{a^2 - b^2}\,\cos\phi}{b} \qquad \text{. . . . . (415a}$$

For UTM, the scaling factor adopted is

$$\sigma_{UTM} = 0.9996$$

Hence, the lines of true scale across the Equator are at relative longitudes

$$\pm\arccos(0.9996) = \pm 1.620\,62_{\sim}{}^\circ$$

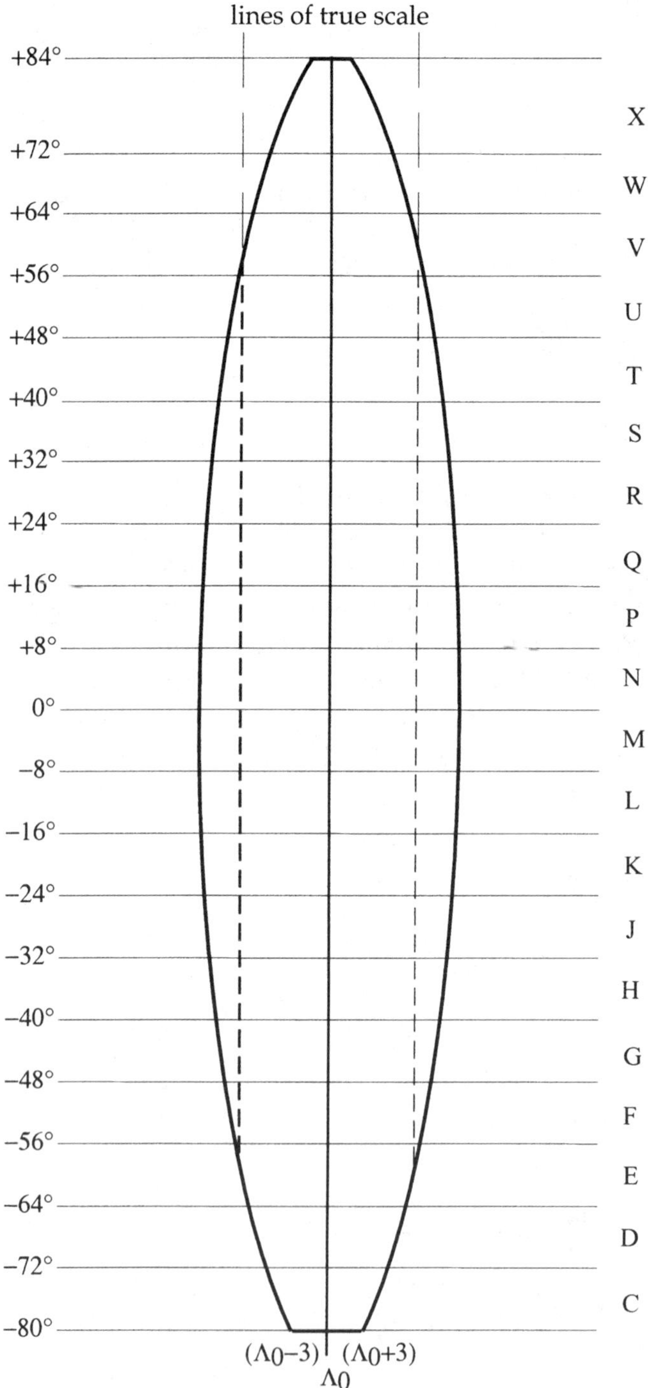

Exhibit 18–3: The Transverse Mercator projection – diagrammatic clipped 6° lune.
[Not of correct proportions.]

With Equatorial radius of 6378.137 km, the corresponding distance along the Equator is $\pm 180._\sim$ km. The same offset distance from the central meridian applies over the whole length of the lune. The scale is true along the lines parallel to and at that distance from the central meridian, on each side. Away from the Equator, the line of true scale is increasingly far, in longitude angular terms, from its central meridian. This can readily be appreciated by reference to Exhibit 18–3. On an ellipsoidal model such lines of true scale are not precisely straight. With Mercator being a conformal projection, the scales are identical and hence true across these lines at any point and in any direction, but only instantaneously. At any point not on the lines the scale is varied, being smaller between the lines (down to the stated 0.9996 of true), progressively larger beyond. Since

$$\pm \arccos(0.9996)^2 = \pm 2.29_\sim{}^\circ$$

the variation from true scale beyond this angular distance along the Equator (i.e., beyond linear distance $255._\sim$ km) exceeds the 0.04% adopted for the central meridian. At the edge of the 6° lune, the relative scale on the Equator is

$$\frac{0.9996}{\cos 3^\circ} = 1.000\,97_\sim,$$

or a percentage variation over twice that at the central meridian. This is the extreme, however, not only in being at the edge of the lune, but occurring only at the widest point of the lune. Even there, however, the variation from true is less than 0.1%, i.e., less than one metre in a kilometre. Beyond ±40° latitude, the lune is less than 256 km wide, hence the scale variation across it is down to at most ±0.04% everywhere. With η and L again as defined in Equations 415a, the relative scale at any point within the lune is

$$\sigma_{\mathrm{UTM}}\left\{1+(1+\eta)\frac{L^2}{2}+\left(5-4t+42\eta+13\eta^2-\frac{28\varepsilon^2}{1-\varepsilon^2}\right)\frac{L^4}{24}+\left(61-148t+16t^2\right)\frac{L^6}{720}+\ldots\right\}$$

UTM – primary zonation

The 60 UTM strips that girdle Earth are centred successively on the meridians 177°W, 171°W, and so on through to 171°E and finally 177°E. They are numbered in that order 1 through 60. Zone 1 thus covers the strip between 180°W and 174°W, zone 2 covers the strip between 174°W and 168°W, and so on — as illustrated in sample form on page 414. While that illustration may suggest individual maps corresponding with each clipped lune, the essence of the scheme is to provide a multitude of large-scale sheets for any region, sheets spanning as little as 25 km. To manage these, systematic labelling has been adopted, with primary zonation independently for longitude and for latitude, plus a rectilinear grid and related reference system. The primary longitudinal zonation, illustrated partially on page 414, is shown completely, as an aside, in the fringes of the diagrams in Exhibit 18–9 on page 426 (counter-clockwise about the North Pole) and in Exhibit 18–10 on page 429 (clockwise about the South Pole).

Zone	Φ	cos Φ	*higher-to-lower ratio*	sec Φ	km *for* 3° *longitude*	km *from Equator*
Z&Y	— but within UPS rather than UTM, and of mixed longitude and latitude					
	+84°	0.104 528 5		9.566 777	35	9 340
X			0.338 261 2			
	+72°	0.309 017 0		3.236 068	103	8 006
W			0.704 920 9			
	+64°	0.438 371 1		2.281 172	146	7 116
V			0.783 935 5			
	+56°	0.559 192 9		1.788 291	187	6 227
U			0.835 700 7			
	+48°	0.669 130 6		1.494 477	223	5 337
T			0.873 488 0			
	+40°	0.766 044 4		1.305 407	256	4 448
S			0.903 303 0			
	+32°	0.848 048 1		1.179 178	283	3 558
R			0.928 304 2			
	+24°	0.913 545 5		1.094 636	305	2 669
Q			0.950 360 8			
	+16°	0.961 261 7		1.040 299	321	1 779
P			0.970 708 6			
	+8°	0.990 268 1		1.009 828	330	890
N			0.990 268 1			
	0°	1		1	334	0
M			0.990 268 1			
	–8°	0.990 268 1		1.009 828	330	890
L			0.970 708 6			
	–16°	0.961 261 7		1.040 299	321	1 779
K			0.950 360 8			
	–24°	0.913 545 5		1.094 636	305	2 669
J			0.928 304 2			
	–32°	0.848 048 1		1.179 178	283	3 558
H			0.903 303 0			
	–40°	0.766 044 4		1.305 407	256	4 448
G			0.873 488 0			
	–48°	0.669 130 6		1.494 477	223	5 337
F			0.835 700			
	–56°	0.559 192 9		1.788 291	187	6 227
E			0.783 935 5			
	–64°	0.438 371 1		2.281 172	146	7 116
D			0.704 920 9			
	–72°	0.309 017 0		3.236 068	103	8 006
C			0.561 937 7			
	80°	0.173 648 2		5.758 770	58	8 896
A&B	— but within UPS rather than UTM, and of mixed longitude and latitude					

Exhibit 18–4: UTM – primary latitudinal zonation (distances approximate).

For latitudinal zonation, each lune is divided by parallels into 8° bands from 80°S to 72°N plus one 12° band from 72°N through to 84°N. These are labelled, starting from 80°S, with the letters C through X excluding the ambiguous I and O. (The remaining two letters at each end of the alphabet are used for the Polar zones, as discussed further below.) The 20-letter sequence of zone letters is shown diagrammatically in Exhibit 18–3 on page 416 then, with various associated dimensional values, tabulated in Exhibit 18–4 opposite. Within a lune, each such band is a tapering entity, narrowing toward its higher latitude; the pertinent amount of narrowing is indicated in the third numeric column of the tabulation, which gives the ratio of the length of the parallel for the higher latitude to that for the lower. The adjacent columns give the cosine and its reciprocal — the trigonometric secant (i.e., sec) — for each boundary latitude, to indicate width relative to that at the Equator. The last two columns show, in rounded kilometres, the approximate half-length of those boundary parallels and their distances from the Equator.

The term *zone* can be used with the number alone to refer to one of the (clipped) lunes, else with the letter alone to refer to a latitudinal zone within a lune. The term can also be used with number then letter to identify uniquely the 60 x 20 = 1200 near-rectangular zones that divide the non-Polar surface of the globe (e.g., 31U for the zone between 0° and 6°E and between 48°N and 56°N, and similarly 48J for that between 102°E and 108°E and between 32°S and 24°S). As can be deduced from the tabulation opposite, these zones span roughly 200 km to 500 km longitudinally, and less than 1000 km latitudinally except for about 1334 km in the X zones.

UTM – the metre grid

A metre-based grid is superimposed on the longitude/latitude graticule to provide a rectilinear rather than angular reference scheme. The grid is defined likewise within each lune but therein uniquely for a given side of the Equator.

The metre grid is built about the Equator and the central meridian, but somewhat differently in the two hemispheres. It is a Cartesian grid, with the first co-ordinate — normally labelled x — being eastward, the second — normally labelled y — northward, in typical mathematical style. However, it is designed to avoid the occurrence of any negative numbers, hence avoid any need for signs. It achieves this by setting the zeros of its co-ordinates (i.e., its 2-dimensional origin) on else beyond the left and lower boundaries of the area to be covered by the grid.

The y co-ordinate represents the distance of a point from the Equator measured along a line parallel to the central meridian (hence, for the general point, not along its own meridian). In the northern hemisphere, y has its zero at the Equator — i.e., on its southern

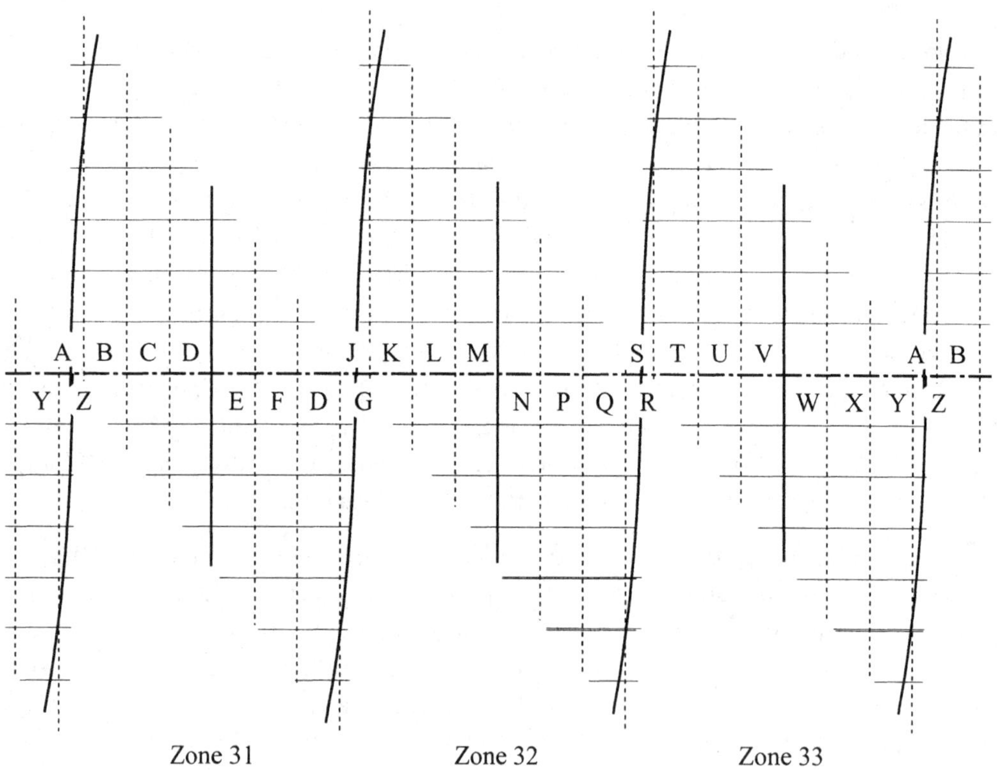

Exhibit 18–5: UTM – MGRS zonation – column lettering pattern.
Illustrated for trio of primary longitudinal zones, shown cut off to accommodate overlap of columns.

0° 3°E 6°E

AB	BB	CB	DB	EB	FB	GB	HB	
								6000000
AA	BA	CA	DA	EA	FA	GA	HA	
								5900000
AV	BV	CV	DV	EV	FV	GV	HV	
								5800000
AU	BU	CU	DU	EU	FU	GU	HU	

200000 300000 400000 500000 600000 700000 800000

Exhibit 18–6: UTM – MGRS zonation – sample from primary zone 31U.

boundary. The value reached at 84°N, therefore, is 9 340 000 approximately. For the southern hemisphere, y has its zero set at exactly 10 000 km south of the Equator. The values begin at latitude 80°S with 1 104 000 approximately then increase progressively to an exact 10 000 000 at the Equator[†]. By convention, the y value is called the *false northing*. It is, of course, truly a northing, the only semblance of falsity being the choice of origin, which is usually regarded as arbitrary anyway.

The x co-ordinate for any latitude, northern or southern, is set for each lune to be 500 km exactly to the west of the central meridian — i.e., every point on the central meridian has x value 500 000. Given the approximate widths of a lune, as shown in the penultimate column of Exhibit 18–4 on page 418, the x value will range approximately from 166 000 to 834 000 across a lune at the Equator but only from 465 000 to 535 000 at 84°N. The x value is called the *false easting*.

Clearly, within one lune, nearly all the (x, y) values that occur in the southern hemisphere occur also in the northern hemisphere, and all values occur repeatedly from lune to lune. Hence, it is routine to begin any reference with the number-then-letter zone label. Calculation of the easting and northing for points known by longitude and latitude is extremely complicated; it is addressed below.

UTM – secondary zonation, the Military Grid Reference System (MGRS)

To facilitate map use, the U.S. military introduced a finer labelling scheme now used routinely on civil maps. It uses a grid with a spacing of the 100 km between gridlines. The resulting squares are identified by pairs of letters, the first letter in each representing the easting, the second letter the northing. For the eastings, three squares and a partial fourth occur either side of the central meridian near the Equator, requiring eight letters (for columns) per lune. To avoid close recurrence of any one letter-pair, the 24 unambiguous letters (i.e., A to Z excluding I and O) are used across successive trios of lunes for eastings. As the distance from Equator to 84°N is about 9350 km, a span of 94 squares is required for northings in the northern hemisphere, but just 90 in the southern. Only the 20 letters A to V excluding I and O are used for northings, where they can be equated with rows. They are assigned in sequence northward, repeatedly, uniformly across all columns within one lune. Hence, a given column-row letter pair repeats within the column, but only at intervals of 2000 km (hence not within any pair of primary latitudinal zones except beyond 80°N).

[†] The metre was originally defined to be one ten-millionth of the distance from Pole to Equator, but early re-surveying showed the initial size slightly too small. Not wanting to hazard the reputation of their recent innovation, the French authorities chose to keep what they had — and to enshrine it in physical-prototype form (therby subverting their aspiration to have it a natural unit) [2004a].

The column-labelling scheme is illustrated in Exhibit 18–5 (with the lunes further clipped off for compaction and, because the fourth column either side of each central meridian overlaps the adjacent lune, with the letters staggered on alternating sides of the Equator). Zones 31, 32, and 33 are cited in the illustration, but any shift of three lunes leaves the picture unchanged.

While row labelling is uniform across one lune, its scheme is also designed to have differing patterns in nearby lunes. Each of the 90 rows in the southern hemisphere, and each the 94 in the northern (the most Polar in each being of partial width) require multiple runs of the 20-letter alphabet. The spread is such as to have an A immediately above the Equator in Zone 1, then to move the multi-alphabet letter sequence up by several letters for Zone 2. The original scheme was elaborate, with the movement being of six letters to that second zone, and then the same again, repeatedly; the result, illustrated on page 430, was a different pattern for ten zones. The revised scheme shifts up only five letters to Zone 2 and then reverts to the starting pattern for the next. It thus applies identically to all odd-numbered zones, and identically to all even-numbered zones; the column-row numbering thus repeats every sixth zone number. The five-letter shift means that across adjacent lunes the row lettering is offset by 500 km.

Though the two-letter identifiers must each occur over a hundred times, the careful assignment of the letters ensures that repetitions are a minimum of 1900 km apart in northing terms and three lunes apart otherwise — i.e., far further than the typical working span in which confusion could occur. A sample of the composite result is shown in the Exhibit 18–6 on page 420 for Zone 31U, with x and y co-ordinates given at the sides. It also shows the line of the meridians that define Zone 31. At about 5900 km, they come within the second-from-centre column (having come within the third by 2900 km). As discussed below, that does not make the externalized ones irrelevant.

UTM – the map

Typical large-scale UTM maps are bounded by parallels and meridians, so they actually taper, although that is usually hard to perceive by the unaided eye. Meridians and parallels are not usually ruled within the map, but the longitude and latitude are shown by dashed lines along the edge with regular partial values plus strategic full values near the corners. The metre-grid values are shown similarly, but the grid itself is usually ruled (typically in blue) — e.g., at 1-km intervals on a 1:50 000 map. Since the grid lines run parallel to and orthogonal to the central meridians, they intersect the boundaries of such a map, at narrow angles. The full grid values are usually expressed with a mixture of print size — e.g., 3645000[†] on a 1:50 000 map with the enlarged numerals being the ones that

[†] False eastings and false northings are written without punctuaion in the MGRS scheme.

vary from line to line on the ruled grid of the particular map, and which are then printed alone except for the full figures shown strategically near corners of the map.

The use of 60 zones is akin to using an interrupted projection, in that interior places can be inconveniently split. As with the interrupted projections [Chapter 7], however, pieces of territory can be relocated across the interrupt. There is nothing in UTM that demands a strict division into the designated lunes, and, clearly, jurisdictions straddling the interrupting meridians must have consolidated maps for their municipal and similar purposes. (For instance, London straddles 0°, New Orleans straddles 90°W, and Calgary straddles 114°W.) The scheme of false eastings allows for this in that the six-digit unsigned numbers used can accommodate points up to 500 km from the central meridian, on either side — i.e., nearly 50% further from the central meridian at the Equator and very much more in further latitudes. Published large-scale maps that adjoin interrupting meridians usually show the UTM grid values relative to the central meridian for the other side of the interrupt, as well as for its own, usually with brown hash marks along the perimeter.

For some regions, the numbered zones have been formally changed to embrace otherwise external points. This has been done for Norway and adjacent areas. Being at high latitudes, the scale criteria can be met there by zones spanning many more than 6°. Though the curvature of parallels becomes markedly greater, the convenience of one integrated zone is overall advantageous. The specific changes are:

Zone 32V (56°N to 64°N) is extended 3° west (to the central meridian of Zone 31V).

Zones 33X and 35X (72°N to 84°N) are doubled in width, thus reaching the central meridian of the adjacent zones; consequently Zone 31X is extended 3° east and

Zone 37X is extended 3° west, then Zones 32X, 34X and 36X are eliminated.

UTM – computing position

For any point on the central meridian for the zone, its easting is precisely 500000 while its northing in the northern hemisphere equals scaling factor $\sigma_{\text{UTM}} = 0.9996$ times its distance in metres from the Equator, with the latter value being obtainable directly from Equation 393b. For southern latitudes, the same product must be subtracted from 10 000 000.

If $P = (\phi, \lambda)$ is not a point on the central meridian the calculation becomes elaborate, involving multiple infinite series that, in practice, are approximated. J. E. Jackson [1978c] gives a full exposition of this calculation; we give an abbreviated one. General point P is the same distance from the real Equator as is point $P_c = (\phi, \lambda_c)$ on the central meridian that has the same latitude, but the grid scheme requires the distance from P to the Equator along a line parallel to the central meridian. Because all parallels of latitude other than the Equator curve away from the Equatorial line on our maps, this distance on the map will be greater than that to point P_c. We relate the point P to the point P_c and establish formulae for the offsetting amounts. For easting, this offset will be subtractive from else additive to the central figure of 500000 depending as to whether P lies west else east of the central meridian. For northing, the offset will be additive to the northing of P_c in the northern hemisphere, but subtractive from it in the southern because the northing numbering there increases towards the Equator.

The applicable local geometry is illustrated below. Besides the complexity of the curve representing the parallel of latitude and the lengths of the radii of curvature of the ellipsoidal surface, the scale factor varies with distance from the central meridian.

The distance from the Equator required for establishing the northing of point P_c can be obtained directly from Equation 393b else by computing the radius of the conformal sphere for the ellipsoid then multiplying by the conformal latitude for ϕ, using Equations 404c & 405b. For GRS80, hence for WGS84, the conformal radius evaluates as

$$\tilde{R} = 6\ 367\ 449.145\ 77_{\sim}\ \text{m}$$

and gives the formula

$$\tilde{\phi} = \phi + \sin\phi\cos\phi\{U_0 + U_2\cos^2\phi + U_4\cos^4\phi + U_6\cos^6\phi + \ldots\}$$

for conformal latitude, where the coefficients shown are defined as:

$$U_0 = -0.005\ 048\ 250\ 776$$
$$U_2 = +0.000\ 021\ 259\ 204$$
$$U_4 = -0.000\ 000\ 111\ 423$$
$$U_6 = +0.000\ 000\ 000\ 626$$

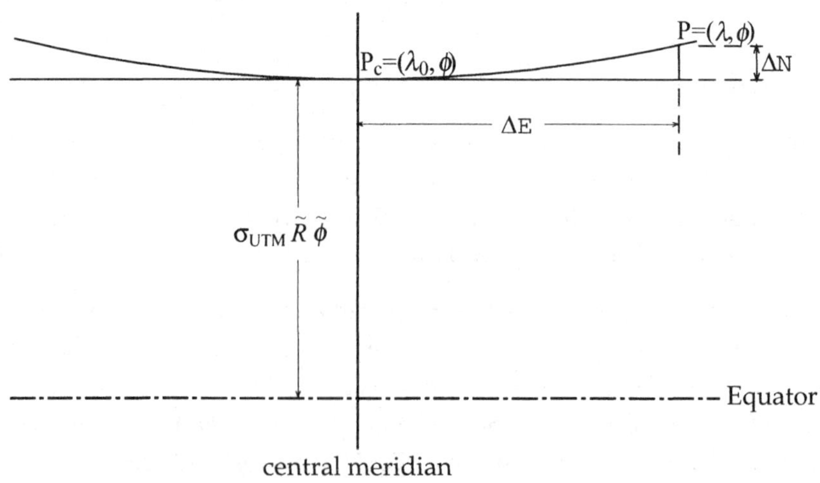

Exhibit 18–7: UTM – positional geometry.
Northern-hemisphere example, for point east of central meridian.

The northing factor for point P_c is then

$$\sigma_{UTM} \cdot \ddot{R}\tilde{\phi}$$

to be subtracted from 10 000 000 to give the northing of P_c if in the southern hemisphere, used directly to give it in the northern hemisphere.

The length factor in the formulae for the offsets is the transverse radius of curvature at latitude ϕ, established in Equation 387b as

$$_tR = \frac{a}{\sqrt{1 - \varepsilon^2 \sin^2\phi}}$$

Adjusting this length for the scaling factor and with η and L again as defined in Equations 415a, the offsets, denoted using a different typeface by the distinctive $\Delta\mathcal{E}$ and $\Delta\mathcal{N}$, are:

$$\Delta\mathcal{E} = \sigma_{UTM} \cdot {_tR}\left\{L + A_3 L^3 + A_5 L^5 + A_7 L^7\right\}$$

$$\Delta\mathcal{N} = \sigma_{UTM} \cdot {_tR}\frac{\tan\phi}{2}\left\{L^2 + A_4 L^4 + A_6 L^6\right\}$$

where the coefficients shown are defined as:

$$A_3 = \frac{1 - \tan^2\phi + \eta^2}{6}$$

$$A_4 = \frac{5 - \tan^2\phi + \eta^2(9 + \eta^2)}{12}$$

$$A_5 = \frac{5 - 18\tan^2\phi + \tan^4\phi + \eta^2(14 - 58\tan^2\phi)}{120}$$

$$A_6 = \frac{61 - 58\tan^2\phi + \tan^4\phi + \eta^2(270 - 330\tan^2\phi)}{360}$$

$$A_7 = \frac{61 - 479\tan^2\phi + 179\tan^4\phi - \tan^6\phi}{5040}$$

Omitting coefficients after A_4, which are influential only when computing for a point outside the 6° lune, the equations become:

$$\Delta\mathcal{E} = \sigma_{UTM} \cdot {_tR}\left(L + \frac{1 - \tan^2\phi + \eta^2}{6}L^3\right)$$

$$\Delta\mathcal{N} = \sigma_{UTM} \cdot {_tR}\frac{\tan\phi}{2}\left(L^2 + \frac{5 - \tan^2\phi + 9\eta^2 + \eta^4}{12}L^4\right)$$

Such calculations are now routinely included in related software, to the fullest precision necessary for engineering work.

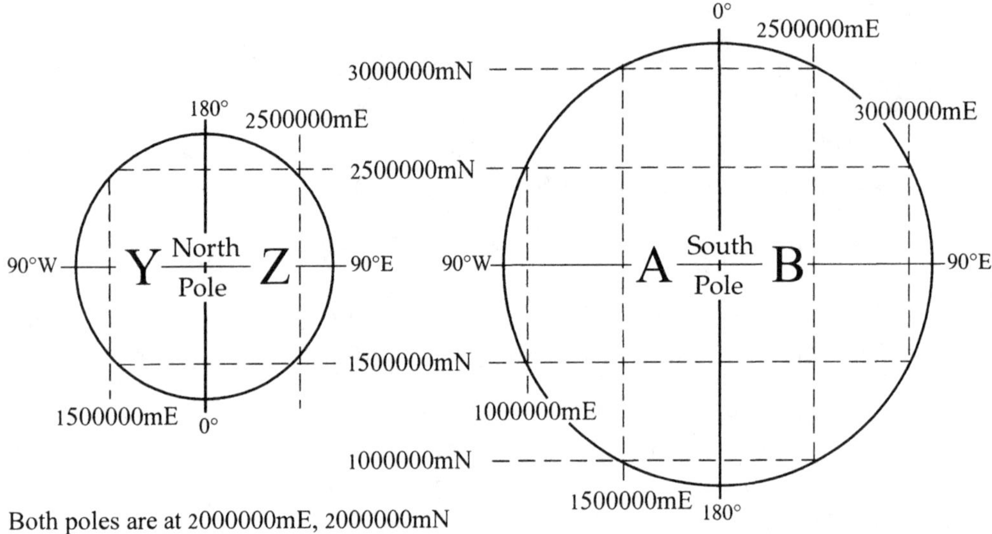

Exhibit 18–8: UPS – primary zonation.

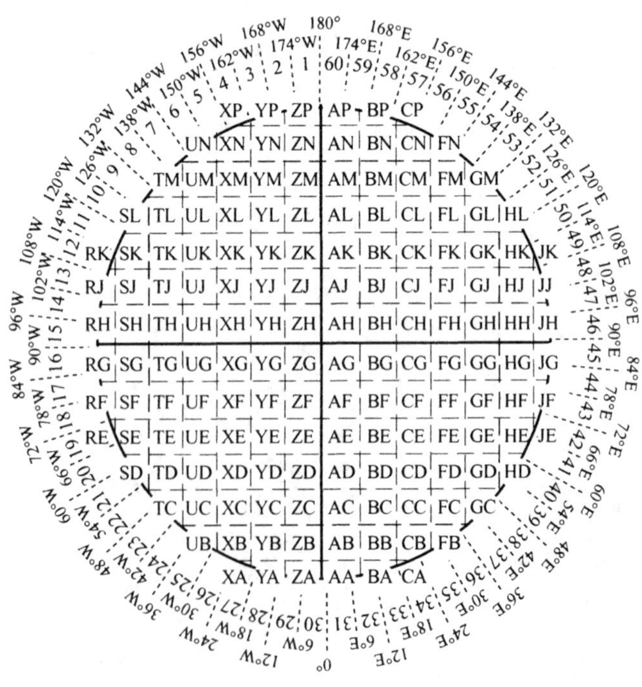

Exhibit 18–9: UPS – MGRS zonation for North Polar region.
With primary UTM longitudinal zones numbered between radiating meridians.

UPS — UNIVERSAL POLAR STEREOGRAPHIC

Outline

To provide for mapping of the Polar regions excluded from UTM, a conformal Azimuthal is used. There is only one such — the Stereographic. It is used singularly for each of the two Polar regions. To complement UTM, it must extend from the Pole to 84° in northern latitudes, from the Pole to 80° in southern latitudes. To avoid difficulties at those joining latitudes, the UPS applies half a degree further, into the area covered by UTM.

UPS – the projection

The Stereographic was shown in Chapter 4 (see Equation 85c and following) to be conformal but, as discussed in Chapter 5, no properly secantal version of it can be conformal. The tangential form, however, can be multiplied in both axial directions by any numeric rescaling factor without compromising conformality. Equation 118b gives the resulting formula for the key variable, for a spherical world. Unfortunately, the scale variation of the Stereographic, which occurs radially in all directions, is considerable over the 10° reach applicable here. This was shown in the table of spacings on page 47 and is elaborated in the table below, for both the basic tangential scene and for UPS, for which the multiplicative scaling factor adopted is $\sigma_{UPS} = 0.994$, putting true scale at 81.11~°. Only within little more than half a degree of this latitude is the scale variation less than 0.1%. Requirements for better accuracy, however, are rare within the Polar zones.

UPS is routinely plotted with latitude 90°E lying along the $+x$ half-axis, so Equations 31c apply. Allowing for the scaling factor, these become:

$$\dot{x} = +\,\sigma_{UPS}\,\rho \sin\lambda$$

$$\dot{y} = \oplus\,\sigma_{UPS}\,\rho \cos\lambda$$

where the ⊕ symbol is positive for the North Pole map, negative for the South.

Latitude	---------- *relative scale* ---------- $\sigma = 1$	$\sigma = 0.994$	*% variation from true* $\sigma = 0.994$
90	1	0.994	–0.6
88	1.000 31~	0.994 30~	–0.570~
86	1.001 22~	0.995 21~	–0.479~
84	1.002 75~	0.996 73~	–0.327~
82	1.004 89~	0.998 86~	–0.113~
80	1.007 65~	1.001 60~	+0.160~

Scale variation in Stereographic projection at Simple aspect.

The distance element $\dot{\rho}$, shown in Equation 118a for the sphere, becomes for an ellipsoid

$$\dot{\rho} = 2a\left(\frac{1+\varepsilon\sin\phi}{1-\varepsilon\sin\phi}\right)^{\frac{\varepsilon}{2}}\left\{(1+\varepsilon)^{1+\varepsilon}(1-\varepsilon)^{1-\varepsilon}\right\}^{\frac{1}{2}}\tan\left(\frac{\pi}{4}-\frac{\phi}{2}\right)$$

UPS – primary zonation

The primary zonation of UPS involves the two pairs of letters at the ends of the alphabet that were omitted from the primary zonation of UTM. They are used as illustrated in the Exhibit 18–8 on page 426.

UPS – the metre grid

For both Poles, the grid for UPS assigns the value 2000000 to the Pole for each plotting axis, with the first figure again being along the x axis (now appropriately called a false easting), which runs along the –90° then +90° meridians. A false northing then runs orthogonally in the familiar direction along the Greenwich bimeridian, specifically along the 0° then 180° meridians for the North Pole, along the 180° then 0° for the South. The pattern of values is shown along with primary zonation in Exhibit 18–8.

UPS – secondary zonation, the Military Grid Reference System (MGRS)

A two-letter column-then-row zonation scheme has been established within MGRS for both Poles. Column labelling has A immediately to the right of the Greenwich bimeridian, then running rightwards and resuming at the extreme left such as to end with Z at that bimeridian. Row labelling starts with A at the bottom and runs continually upwards. As the radius of the North Pole region is close to 700 m, a 14 x 14 array is required there. A radius exceeding 1100 m demands a 24 x 24 array for the South Pole region. Thus the 24-letter unambiguous alphabet suffices. However, an alphabet of just 14 letters, specifically ABC FGHJKL PQR, was adopted for column labelling for the North Pole region and, despite causing duplications (1300 km apart), applied to the South. The results are shown in Exhibit 18–9 on page 426 for the North Pole region, in Exhibit 18–10 on page 429 for the South.

UPS – the map

With radial meridians and concentric circles for parallels, the strategy for sectional mapping is more appropriately attuned to the grid.

UPS – computing position

Except for the incorporation of ellipsoidal distances along the radial meridians, positional computation for this secantal projection at Simple aspect is straightforward.

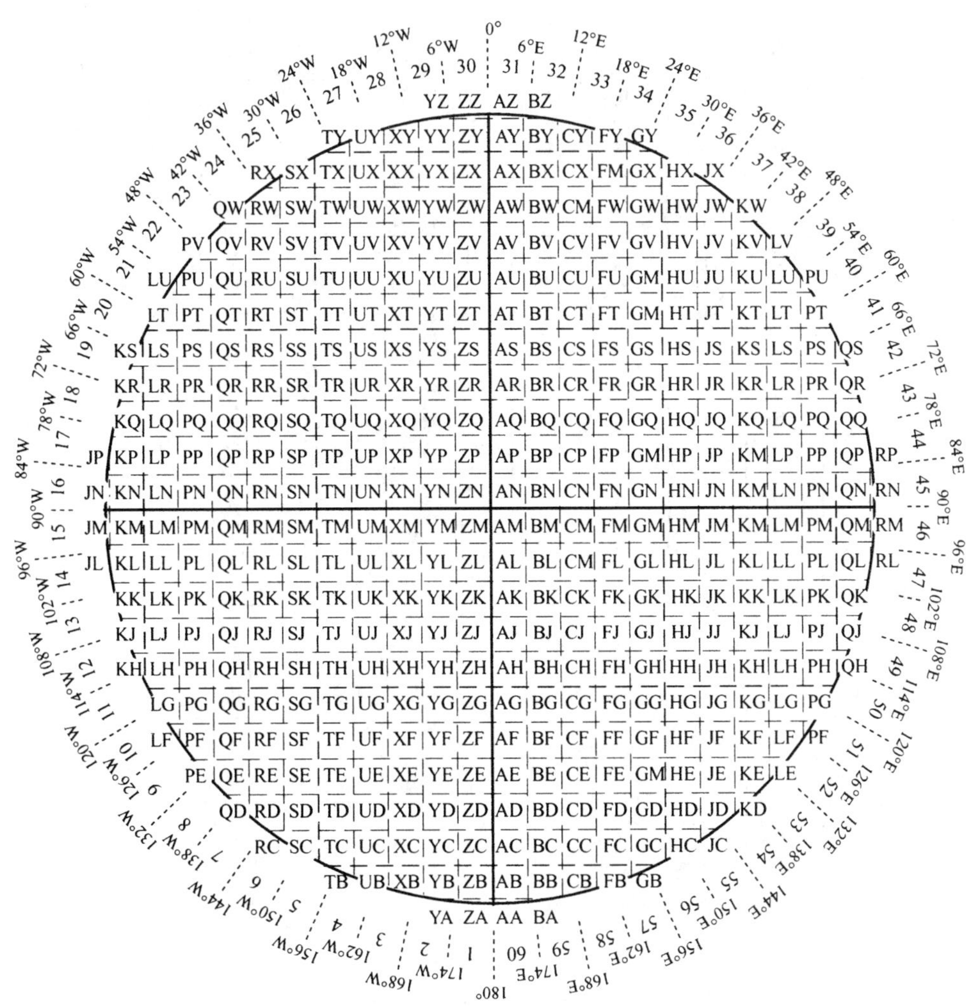

Exhibit 18–10: UPS – MGRS zonation for South Polar region.

Zone:	31	32	33	34	35	36	37	38	39	40
84°N	-P	-H	-B	-R	-K	-D	-T	-F	-F	-V
	-N	-G	-A	-Q	-J	-C	-S	-E	-E	-U
	-M	-F	-V	-P	-H	-B	-R	-D	-D	-T
	-L	-E	-U	-N	-G	-A	-Q	-C	-C	-S
	-K	-D	-T	-M	-F	-V	-P	-B	-B	-R
	-J	-C	-S	-L	-E	-U	-N	-A	-A	-Q
	⋮	⋮	⋮	⋮	⋮	⋮	⋮	⋮	⋮	⋮
	⋮	⋮	⋮	⋮	⋮	⋮	⋮	⋮	⋮	⋮
	⋮	⋮	⋮	⋮	⋮	⋮	⋮	⋮	⋮	⋮
	⋮	⋮	⋮	⋮	⋮	⋮	⋮	⋮	⋮	⋮
	-H	-B	-R	-K	-D	-T	-M	-F	-V	-P
	-G	-A	-Q	-J	-C	-U	-L	-E	-U	-N
	-F	-V	-P	-H	-B	-S	-K	-D	-T	-M
	-E	-U	-N	-G	-A	-Q	-J	-C	-S	-L
	-D	-T	-M	-F	-V	-P	-H	-B	-R	-K
	-C	-S	-L	-E	-U	-N	-G	-A	-Q	-J
	-B	-R	-K	-D	-T	-M	-F	-V	-P	-H
	-A	-Q	-J	-C	-S	-L	-E	-U	-N	-G
Equator										
	-V	-P	-H	-B	-R	-K	-D	-T	-M	-F
	-U	-N	-G	-A	-Q	-J	-C	-S	-L	-E
	-T	-M	-F	-V	-P	-H	-B	-R	-K	-D
	-S	-L	-E	-U	-N	-G	-A	-Q	-J	-C
	-R	-K	-D	-T	-M	-F	-V	-P	-H	-B
	-Q	-J	-C	-S	-L	-E	-U	-N	-G	-A
	-P	-H	-B	-R	-K	-D	-T	-M	-F	-V
	-N	-G	-A	-Q	-J	-C	-S	-L	-E	-U
	-M	-F	-V	-P	-H	-B	-R	-K	-D	-T
	-L	-E	-U	-N	-G	-A	-Q	-J	-C	-S
	-K	-D	-T	-M	-F	-V	-P	-H	-B	-R
	-J	-C	-S	-L	-E	-U	-N	-G	-A	-Q
	-H	-B	-R	-K	-D	-T	-M	-F	-V	-P
	⋮	⋮	⋮	⋮	⋮	⋮	⋮	⋮	⋮	⋮
	⋮	⋮	⋮	⋮	⋮	⋮	⋮	⋮	⋮	⋮
	⋮	⋮	⋮	⋮	⋮	⋮	⋮	⋮	⋮	⋮
	-G	-A	-Q	-J	-C	-U	-L	-E	-U	-N
	-F	-V	-P	-H	-B	-S	-K	-D	-T	-M
	-E	-U	-N	-G	-A	-Q	-J	-C	-S	-L
	-D	-T	-M	-F	-V	-P	-H	-B	-R	-K
	-C	-S	-L	-E	-U	-N	-G	-A	-Q	-J
80°S	-B	-R	-K	-D	-T	-M	-F	-V	-P	-H

Exhibit 18–11: UTM – MGRS zonation – original row-lettering pattern.
Each rectangle represents one full sequence of the 20-letter alphabet.

PART D

The Real World

The real world of Earth is neither spherical nor elliptical. That is true not only of the literal surface but of the entity called Mean Sea Level (MSL). Vagaries of Earth's crust have long been recognized as causing variations in the implied level surface MSL. The 20th Century began with a functional belief that the mean level of the sea was consistent from place to place, but extended transits by surveyors with progressively refined instruments soon showed that not to be the case. This resulted in a common MSL being defined — but generally common only within a country, e.g., the National American Datum of 1927. From their first introduction in 1957, artificial satellites have allowed detached measurement of Earth's surface and of related geophysical phenomena. Their use has necessitated enhancement of the global models used to represent Earth and also the development of new projections to deal with the elaborate relationship between the geocentric path of a satellite and the revolving surface it overflies. The Global Positioning System (GPS) required a uniform representation for the whole world rather than regionally fitted models. In the one chapter in this closing part, we look at factors influencing MSL — the *geoid* as it is also called — and note the cartographic aspects of some of these developments of modern times.

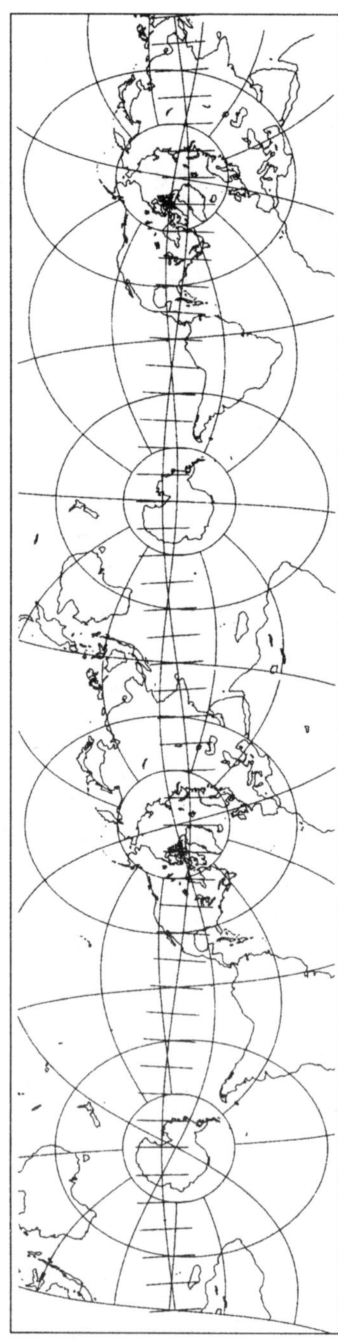

Exhibit 19–0: The Space Oblique projection – two orbits.
Derived, with appreciation, from U.S.G.S. Professional Paper N°1453 [1989a].

CHAPTER 19

Reality Check:
The geoid and geodesy

Preamble

The level of the sea is dictated by gravity, which effectively defines Mean Sea Level — i.e., the geoid. The gravitational attraction of Earth's overall mass is the overwhelming factor in setting the strength and direction of gravity, but both strength and direction are affected by many other factors. In time terms, these range from the gyration of the axis over millennia to the double cycle of tides within a single day. The mixture and distribution of solid materials below the surface provides local variations to gravity, while the movement of the crustal plates actually redistributes that material; the lateral movement exceeds 10 cm per year in places, while the elevational movement can be intermittently far greater. The observational accuracy provided by satellites allows measurement of such movement as well as of the (relatively) static background. With the accompanying adoption of the Earth-centred ellipsoids GRS80 and then WGS84, there has been widespread revision of detailed positions on maps. This applied especially where they displaced ellipsoids like the Clarke 1866 used in the U.S.A. on a regional-fit rather than geocentric basis. In some regions places had to be moved many kilometres in whole-Earth terms — though, locally, the places were trivially changed. Satellites have, in turn, demanded new map projections to accommodate their pattern of overflight.

There is no additional mathematics in this final chapter.

Exhibit 19–1: The Space Oblique projection – partial orbit.
Derived, with appreciation, from U.S.G.S. Professional Paper N°1453 [1989a].

GRAVITY

Competing forces – gravitation and spin

The inward gravitational pull of its nearly 6×10^{24} kg causes Earth to be close to a sphere. The outward thrust of its rotation, giving a linear tangential speed over 1500 km/hr at the Equator, causes it to be ellipsoidal by bulging around the Equator. The centrifugal acceleration of the rotation amounts to barely $\frac{1}{300}$ of the inward gravitational acceleration even at the Equator, and it declines progressively to zero at the Poles. The two together, along with various less-significant factors, produce *gravity* (usually denoted g).

Newton's Universal Law of Gravitation states that two bodies attract each other with a force proportional to the product of their masses and inversely to the square of their distance apart. Because force equals mass times acceleration, we can equivalently say that a body of mass M attracts another body distance d away with a force that causes an acceleration proportional to $M\,d^{-2}$. The value equals $G\,M\,d^{-2}$, where G is the Newtonian constant of gravitation, currently evaluated at 6.674×10^{-11} m^3 kg^{-1} s^{-2} [2002b]. The accelerative force acts along the line between the two masses. If the sizes of the two bodies are small compared with their distance apart, their masses can be regarded as concentrated at their centres. Even if they are not far apart, if they are spherical and their composition spherically uniform they can still be regarded in this way. More generally, the gravitational affects must be ascertained by summation of the affects of the fragmented components of each overall body. If we take Earth as spherical and of uniform composition, a first approximation for the mean value for gravitational acceleration on Earth's surface is 9.820 m s^{-2}. In reality the composition is neither totally uniform nor even radially uniform. And, Earth is not spherical. So, for instance, in mid latitudes the gravitational force is more influenced by the nearby component of the Equatorial bulge than by the opposite component.

The centrifugal force caused by the daily rotation of Earth applies to all Earth-bound objects. For any object spinning at angular speed ω at distance r orthogonally from the spin axis, the resulting centrifugal acceleration equals $r\,\omega^2$. The accelerative force acts outward along the line orthogonal to the axis. At the Equator, this puts it directly opposed to the gravitational acceleration; however, even though it is at the maximal distance from the axis, it amounts there to just 0.034 m s^{-2} or about 0.35% of the gravitational acceleration. Besides r for a point on Earth's surface at latitude ϕ being barely $\cos\phi$ times the value at the Equator (see Equation 387a), the directional line of the centrifugal force diverges progressively from that of the roughly geocentre-based radial line of the gravitational force, hence reducing its counter affect to gravitational force. Compounding the two, the centrifugal force component acting contrary to the gravitational acceleration at latitude ϕ is virtually $\cos^2\phi$ times the value at the Equator.

The latitudinal affect on magnitude is expressed for surveying use by the *International Gravity Formula,* which has been redefined over the years to accommodate revised determinations and adjustments to the reference ellipsoid at which it is applicable. In reverse chronological order, the values that have been set are:

1984 $g_0 = 9.780\,318\,4\,(1 + 0.005\,302\,4 \sin^2\phi + 0.000\,005\,9 \sin^4\phi)$ m s^{-2}

1980 $g_0 = 9.780\,327\,(1 + 0.005\,302\,4 \sin^2\phi + 0.000\,005\,9 \sin^4\phi)$ m s^{-2}

1971 $g_0 = 9.780\,318\,5\,(1 + 0.005\,278\,895 \sin^2\phi + 0.000\,023\,462 \sin^4\phi)$ m s^{-2}

1930 $g_0 = 9.780\,490\,(1 + 0.005\,278\,84 \sin^2\phi + 0.000\,005\,9 \sin^2 2\phi)$ m s^{-2}

Because the gravitational attraction declines with distance, a correction must be applied for altitudes above the reference ellipsoid. For H metres altitude the standard *free air correction* is approximately $3.086 \times 10^{-6} \times H$ m s^{-2}.

Gravity essentially dictates the distribution of the waters of the oceans and seas, and hence their mean level. Because today's solid components of Earth were in geologically far distant time more plastic, if not actually fluid, their distribution too reflects the net affects of Earth's gravitational attraction and spin — which, in turn, affects today's gravity at any place. Tectonic movements and volcanic activity through to tides, prevailing winds and currents, even varying atmospheric pressure, redistribute the mass while various forces cause the position of the axis to move — both rhythmically and randomly. But all are minor compared with the impact on gravitational attraction of the variation in radius of Earth between the Equator and the Poles; it amounts to about 0.7%.

Local gravity

Because of the spin, gravity does not generally point to Earth's centre. Except at the Equator and the Poles, its direction is routinely to a point on the axis beyond the Equatorial plane — i.e., in the other hemisphere. Its direction, as well as magnitude at any location, depends on many factors, including the composition of local materials. Land versus sea provides an obvious difference, with mountains an exaggerator, but there is variation even over the seas. The largest anomaly of value occurs south of India. Measured with the precision available in modern instruments, both value and direction vary continually. (Such variation is a key part of mineral prospecting.) The component of centrifugal force not directly opposed to Earth's gravitational attraction routinely provides a slight tilt to the direction of gravity. When a plumb line is used for observation, there is a further tilt because the bob is not bound to Earth and its spin.

Earth's rate of spin is slowing due primarily to the drag of the tides, but it also fluctuates noticeably in the very short term (prompting the existence within the U.S. government of the *International Earth Rotation Service*).

EQUIPOTENTIAL SURFACES

Gravity potential

Having magnitude and direction, gravity is mathematically a vector. The integral of the gravity force from infinity to a point distance ρ from the attracting body, which represents the work it requires to bring unit mass from infinity to that point, is the *gravity potential* at the point. With unit mass, this is identically the integral of the acceleration g. Writing M for the mass of Earth and V for the potential, we thus have

$$V = \int_{\rho}^{\infty} g \, \mathrm{d}s = \int_{\rho}^{\infty} \rho\left(GMs^{-2} + r\omega^2 \cos\psi\right)\mathrm{d}s = \left[-GMs^{-1} + \tfrac{1}{2} r\omega^2 \cos\psi\right]_{\rho}^{\infty} \mathrm{d}\rho$$

The potential of a gravity force of a given value depends on the proximity to the centre of the attracting mass of its point of action. For Earth, a higher value of gravity at the Poles thus produces the same potential as a lesser value at the Equator.

Points with the same gravity potential

The set of all points with the same value for this potential form a surface enveloping Earth's centre. Each such surface is continuous, enveloping without breaks, and possesses no steps, no sharp edges. Except where mass density changes suddenly, such as at the junction of the ground and the air, the radius of curvature of the surface changes smoothly. Further, those radii always point inwards. The surface is everywhere convex, with no depressions. Such a surface is called an *equipotential surface.*

The direction of gravity at any point defines the local direction of a plumb line; hence what is *vertical* there. Since that direction must be orthogonal to the equipotential surface through the point, that surface must be seen as *horizontal.*

Two equipotential surfaces with differing values do not make contact with each other; that with the smaller value envelops that with the larger value. Their separation varies, however, with the magnitude of gravity, with stronger gravity corresponding to greater proximity (similarly to wind speeds and isobars for atmospheric conditions). Followed inward through surfaces of progressively greater potential, the direction of gravity can change variously. But, after generally pointing across the Equatorial plane at the surface, it must point at the centre ultimately as the surface comes close to that innermost point, giving the mathematical plumb line an overall tendency to be curvaceous, and progressively less angled from the Equatorial plane.

Readers interested in further information in the geodetic matters should consult a specialty geodesy text [e.g., 1981c, 1988c, 1993c].

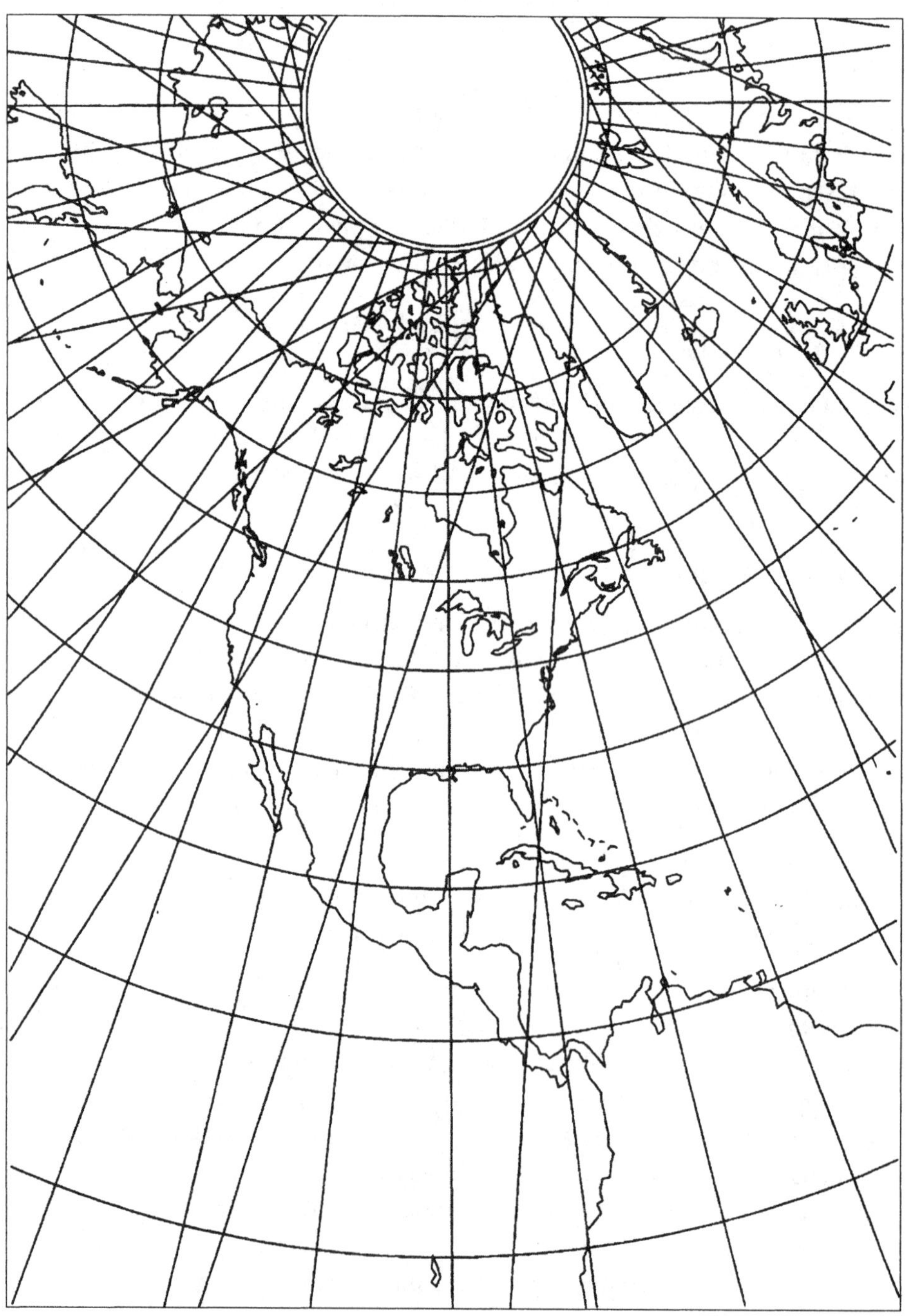

Exhibit 19–2: Satellite tracking.
Derived, with appreciation, from U.S.G.S. Professional Paper N°1453 [1989a].

MEAN SEA LEVEL — THE GEOID

Mean Sea Level

MSL is defined as the equipotential surface enveloping Earth that has a particular value of this potential, not a particular value of gravity. C. F. Gauss first saw Earth as having a mathematical figure. Twenty years later, after extensive study of gravity measurements, G. G. Stokes [1849a] derived formulae for representing Earth's fluidic shape (and thereby his famed integral). Although conceived as a world-wide entity, MSL was for decades essentially a local phenomenon [see 1925c]. While surveyors ran levelling lines connecting sites and rationalizing their readings by averaging, for decades this was only over very limited regions. The results were appropriately called Local Mean Sea Level. Over time, however, the surveying was progressively extended to provide one consistent level across countries and then across continents. [See 1976d for a history of geodetic levelling in the U.S.A. up to the creation of a national datum of 1927.] J. F. Listing in 1873 had coined the term *geoid* for the world-wide MSL, a term that has come into common use with the matching scope for levelling attainable via artificial satellites.

Since the first one was launched in 1957, satellites have provided both an improved picture of Earth and improved reference points for positioning on Earth [1993c]. This has prompted adoption of a succession of geocentred reference ellipsoids. The use of such world-embracing technology as GPS has also demanded a singular world-wide scheme for expressing position. The values adopted for the gravity potential at MSL have been:

in 1967 $U_0 = 62\,636\,860.85\ m^2\,s^{-2}$

in 1980 $U_0 = 62\,636\,860.850\ m^2\,s^{-2}$

The former was incorporated in the IUGG67 ellipsoid, the latter was incorporated in the GRS80 ellipsoid and then in the WGS84 ellipsoid. Of course, these ellipsoids are what were deemed to be best fits to the irregular MSL envelope, each being such that its potential would equal the defining value were Earth rotationally symmetric [see 1981c for discussion of geodetic Earth reference models].

Elevations on maps are relative to the geoid, not to any ellipsoid. But the geoidal surface is in turn expressed relative to such an ellipsoid, with its own contour lines. The extreme values for these elevations amount to about 100 m. The sea south of India is about 110 m below the ellipsoid while MSL across the island of New Guinea is 80 m above the ellipsoid.

Revising benchmarks

The enhanced precision and accuracy available via satellites and contemporary technological innovations [see 1993c] has shown that most of the thousands of enshrined benchmarks are not quite in the longitude/latitude positions and elevations previously established. In the case of North America that establishment was via the North American Datum of 1927 (NAD27), applicable to over a quarter of a million benchmarks across the U.S.A. and southern Canada. Recomputation with satellite-based data showed that some were up to 100 m laterally discrepant [1983c, 1993d], and many were 5 m so. The new figures are embedded in the North American Datum of 1983 (NAD83), but, as so much of accumulated map data was recorded in NAD27 and remains unchanged, that older basis is still widely used in Geographic Information Systems.

Up until 1900, the heights of these benchmarks were covered by relative local traverses referenced to local mean sea level — preferably a mean over several gauges spread along the coast. The North American Vertical Datum of 1929 (NAVD29) established a consolidated continental standard MSL that survived until the development in the 1970s and 80s of North American Vertical Datum of 1988 (NAVD88). NAVD88 is referenced to mean sea level at Father Point, Rimouski on the St. Lawrence estuary, as with the more local International Great Lakes Datum 1985 (IGLD85). The new datum incorporated revisions of up to 25 cm commonly, but some well exceeding 100 cm in volcanic, tectonically active and isostatic rebound areas [1992b].

Similar revisions of longitude/latitude position and of vertical measurements had to be effected over comparable regions outside North America.

SATELLITES AND NEW PROJECTIONS

Sputnik and beyond

Launched by the U.S.S.R. in October 1957, Sputnik I — the first artificial satellite — was an inert metal ball. By circling Earth rhythmically in the uppermost reaches of the atmosphere, it and its immediate successors provided surveyors with benchmarks that all could see. While their inertness made them similar to traditional benchmarks physically, their constant movement made them antitheses of the traditional benchmarks functionally. But their positions at any instant could be ascertained.

Initially this was done with arrays of kinetheodolites designed for tracking aircraft and missiles at an altitude of around 10 km rather than the 150 km of the satellites — hence of poor accuracy. The newly developed Baker-Nunn cameras that photographed the moving object against the star background with tightly timed shutter closures to break the streak images of the stars, however, soon allowed very precise position readings. An extract from this author's computer program of that time, which did the tracking for the hemisphere centred on Australia, is included as the Frontispiece of this book. The program was written for the Department of Supply of the Commonwealth of Australia, to use data from instruments sited at the rocket-testing range at Woomera and an observation station near Perth.

As with planets and natural satellites, artificial satellites follow elliptical orbits with the centre of the controlling body at a focus. Most satellites added to Earth's one natural one (the Moon), however, are put into essentially circular orbits — i.e., ellipses of trivial ellipticity. Once launched, they have lost the rotational motion of Earth, so, while they travel with Earth around its orbit, the plane of their orbits stays virtually fixed relative to the stars. If of small ellipticity, their orbits are essentially symmetric about the Earth's axis and about the Equatorial plane. However, because of Earth not being truly spherical, the axis of every satellite's orbit gyrates in contrary direction to the Earth's orbital travel — it *precesses* (just as the axis of Earth does over a cycle of nearly 26 000 years — one of the minor changes to mass distribution).

Precise observations of the orbit of Sputnik II were a crucial first step into the modern knowledge of Earth's shape [1976b], most notably to refine the measure of its flattening (to the current 1:298.257 used in WGS84), which is discernible from the implied gravity.

To stay aloft, satellites must travel with a speed sufficient to counter the Earth's gravitation at their altitude. Using formulations discussed at the opening of this chapter, with ω now meaning the angular velocity of the satellite and R its distance from the geocentre, this means (noting that the centrifugal force now acts along the same line as the gravitational force) having for a consistent orbit

$$G\,M\,R^{-2} = R\omega^2 \quad \text{or} \quad G\,M = R^3\,\omega^2 \quad \text{or} \quad G\,M\,\omega^{-2} = R^3$$

Using the value adopted for WGS84 (see Chapter 17), we have

$$G\,M = 3.986\,005 \times 10^{14}\ \text{m}^3\text{s}^{-2}$$

allowing direct reciprocal computation of rotational velocity via its square versus radial distance from the geocentre via its cube. Greater altitude corresponds with lower angular speed. If the speed is set to be identical with Earth's and the satellite is deployed in the Equatorial plane with motion of the same direction as Earth's, it is *geostationary*. Using the value set for WGS84 at

$$\omega = 7.292\,115 \times 10^{-5}\ \text{rad s}^{-2}$$

gives for the geostationary satellite a radial distance of

$$R = 4.216\,417_{\sim} \times 10^7\ \text{m} = 6.610\,735_{\sim}a$$

where a is the Equatorial radius adopted for WGS84 of $6.378\,137 \times 10^6$ m. The resulting altitude above the surface of over 35 000 km allows such satellites to have line-of-sight virtually to the Poles, and is perfectly functional for radio transmissions. But it is clearly excessive for optical observation of Earth's surface and for many other purposes, even were the satellite not geostationary. The distance is sufficient to cause noticeable delay in any two-way conversation.

More generally, satellites are deployed to follow paths at a large angle to the Equatorial plane, at an altitude just sufficient to avoid troublesome drag from the

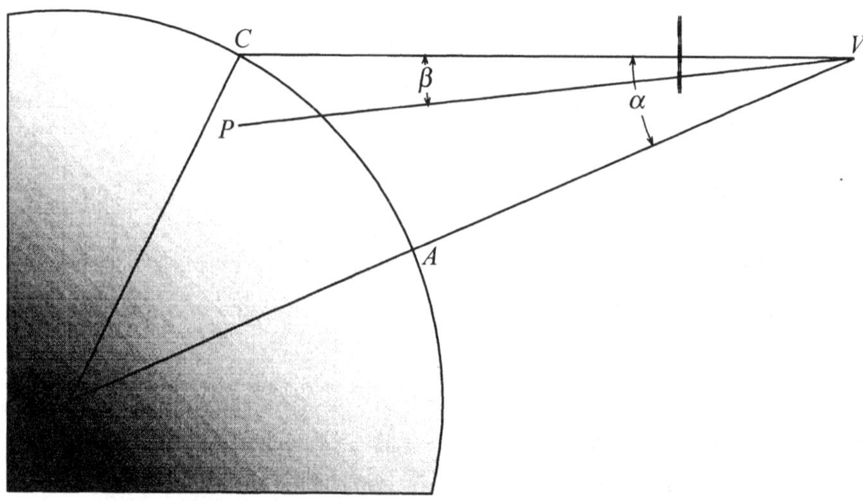

Exhibit 19–3: Geometry of the tilted view of the globe.

atmosphere (at which height their orbital parameters are not so simple). This means being at least 200 km above the surface, i.e., about 1.03 mean radii from the geocentre. (The International Space Station is at a steady altitude of close to 350 km, and has a period barely over 90 minutes.)

With Earth rotating eastwards, the path followed by the typical satellite recurs westwards, to the extent of about 30° each orbit; careful timing can interleave such transits over successive days else make them coincide. Exhibit 2–2, on page 26, shows the restricted region visible from such a satellite, and thereby illustrates the considerable number required for continuous close coverage of any place. (GPS requires multi-coverage, with at least three satellites clearly visible from an observing spot.) Exhibit 2–2 was from the General Perspective projection, presented as the Aerial projection in that early chapter.

A tilted view

Photographing of Earth from satellites and airplanes is not routinely directed vertically downwards. The effective film plate is tilted from the vertical. Because it assumes verticality, the illustration in Exhibit 2–2 is not typical. The **Tilted Perspective** projection was developed to accommodate this deviation, and provides maps simulating aerial views and also positioning markers on real pictures. J. P. Snyder [1981e] provides a detailed (hence lengthy) development, essentially tilting the map of the vertical case. R. E. Deakin [1990b] took a direct approach, using vectors (which Snyder had explicitly abjured), to produce a very compact development. We imitate it.

As with the Aerial projection of Chapter 2, we have V as the viewpoint (the focal point of the camera) and G as the geocentre, but the centre line of the camera points to some other point than G, denoted by C, on the surface of the globe. This is illustrated in Exhibit 19–3. We denote by A the point where the vertical from V to G intersects the surface. The illustration has C and A on the rim of the depicted surface, plus a third, general point P somewhere on the surface. Given the longitude and latitude for each point on the surface, we can compute its Cartesian co-ordinates on the standard frame of reference, using formulae in Tutorial 3. This can be done for a sphere using a single value for R else for an ellipsoid using distinct values based on the latitudes. If h is again the altitude of point V as a proportion of global radius (specifically that at A if using the ellipsoid), then the co-ordinates for F are $(1{+}h)$ times those of A.

We thus have co-ordinates for A, C, P, and V. Denoting them in the usual way, with the respective letters as suffices, we can subtract those of V from those of each of the others in turn to get, in co-ordinate form, the vectors from V to each. If we denote these vectors by the same letters in bold lower-case style, as practiced for vectors, we have:

$$\mathbf{a} = (x_A - x_V, y_A - y_V, z_A - z_V) \quad = |\mathbf{a}|\,\breve{\mathbf{a}}$$
$$\mathbf{c} = (x_C - x_V, y_C - y_V, z_C - z_V) \quad = |\mathbf{c}|\,\breve{\mathbf{c}}$$
$$\mathbf{p} = (x_P - x_V, y_P - y_V, z_P - z_V) \quad = |\mathbf{p}|\,\breve{\mathbf{p}}$$

where the extended expressions merely have the scalar length of the modulus times the corresponding unit vector. Using dot-products, these provide values for the cosines of angles α and β shown in Exhibit 19–3, as

$$\cos\alpha = \breve{\mathbf{a}} \cdot \breve{\mathbf{c}}, \quad \cos\beta = \breve{\mathbf{c}} \cdot \breve{\mathbf{p}}$$

and hence these first-quadrant angles themselves.

We now consider a mapping plane, indicated in the preceding illustration by the neatline, orthogonal to **c** (hence tilted relative to the local vertical represented by **a**). We denote by $\dot{A}$, $\dot{C}$, and $\dot{P}$ the intersections of vectors **a,** **c**, and **p** with that plane. Exhibit 19–4 provides a sketch of this mapping plane, with Cartesian axes having origin at $\dot{C}$ and oriented to have $\dot{A}$ on the $-y$ half-axis — usually clear of the mapping area. [Though sub-optimal with a vector approach, Deakin retained Snyder's arrangement adopted when tilting the Vertical Perspective, with the x axis along the intersection of our mapping plane and a plane orthogonal to **a**.] We have to establish the co-ordinates of $\dot{P}$, identically of the vector $\dot{\mathbf{r}}$ in the diagram. Using polar co-ordinates in the usual style, we have

$$\dot{\mathbf{r}} = (\dot{\rho}\cos\dot{\theta}, \rho\sin\dot{\theta}) = \dot{\rho}(\cos\dot{\theta}, \sin\dot{\theta}) = \dot{\rho}\breve{\dot{\mathbf{r}}}$$

The length element is easily established given the distance d of the plane from V as

$$\dot{\rho} = \text{ length } \dot{C}\dot{P} = d\tan\beta$$

To establish $\dot{\theta}$, let $\breve{\mathbf{i}}$ and $\breve{\mathbf{j}}$ be the unit vectors along the $+x$ and $+y$ half-axes respectively. Noting the mutual orthogonality of these two and **c**, and adding a new unit vector $\dot{\mathbf{w}}$ in our plane counter-clockwise from $\dot{\mathbf{r}}$, vector cross-products give

$$\breve{\mathbf{i}} = \frac{\breve{\mathbf{a}} \times \breve{\mathbf{c}}}{\sin\alpha}, \quad \breve{\mathbf{j}} = \frac{\breve{\mathbf{i}} \times \breve{\mathbf{c}}}{1} \quad \text{and} \quad \breve{\dot{\mathbf{w}}} = \frac{\breve{\mathbf{p}} \times \breve{\mathbf{c}}}{\sin\beta}, \quad \breve{\dot{\mathbf{r}}} = \frac{\breve{\mathbf{c}} \times \breve{\dot{\mathbf{w}}}}{1}$$

Reverting to dot-products, we can then get a measure of the indicated rotational angles by:

$$\cos\dot{\theta}_x = \breve{\mathbf{i}} \cdot \breve{\dot{\mathbf{r}}}, \quad \cos\dot{\theta}_y = \breve{\mathbf{j}} \cdot \breve{\dot{\mathbf{r}}}$$

the two angles together defining the relevant quadrant by their signs. The 360° complement of the chosen value of $\dot{\theta}_x$ equals the needed counter-clockwise angle $\dot{\theta}$.

The track of a satellite

The typical satellite does not follow a Polar orbit; depending on its purpose and its functional aperture it covers any needs at the actual Poles peripherally, with its overhead latitude limited to perhaps ±80°. Its recurrent transits can be represented on a map as on page 438, but the actual path of each across Earth's surface is of an elongated S shape.

This complication prompted the development of probably the most significant projection of modern times — the **Space Oblique Mercator** projection; conceived and named by A. P. Colvocoresses [1974e]. The extended mathematics was developed to completion simultaneously by J. L. Jenkins and J. D. Turner [1977d] and by Snyder [1978d], bringing the accession to cartographic eminence of the latter.

The projection is illustrated for a Pole-crossing satellite on page 432 (for one full orbit) and on page 434 (a quadrant only); its extensive mathematics can be consulted via the references above or in Snyder's later U.S.G.S. Bulletin N°1518 [1981d]; Snyder published a wider survey of map projections for satellite tracking in 1981a. Progressively developed from serving a circular orbit about a spherical Earth to accommodating ellipticity for both Earth and orbit, the Space Oblique projection maintains true scale along the sinusoidally warped ground-track and varies from true by less than 0.01% within the bounds of the usual observational aperture.

The projection is based on the ordinary Mercator projection at Oblique aspect, with the auxiliary equator crossing the Equator at the angle of the satellite's orbit. The satellite's orbital period is that between successive crossings of the Equatorial plane in a northward direction. The rotation of Earth as the satellite travels its orbit does not affect the latitudinal readings but clearly does affect the longitudinal readings. They move proportionally to the satellite's traverse of its orbit, resulting in the sinusoidal path. It also means that the scan lines orthogonal to the path are not quite straight. Along the path the scale is true and conformality only minutely impaired.

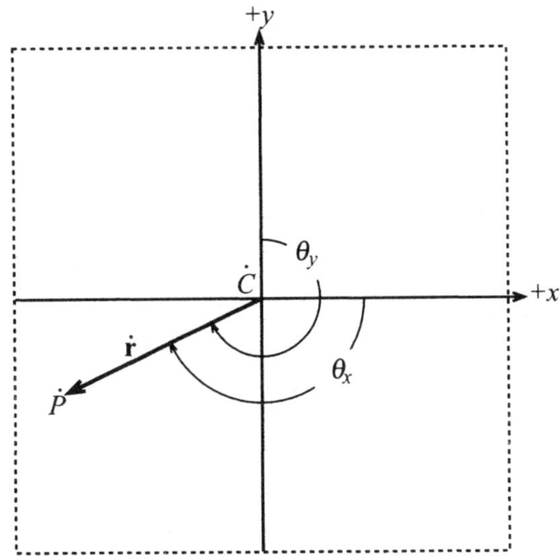

Exhibit 19–4: The tilted mapping plane.

Appendices

There are five appendices before a general index. Those five are:

Appendix A: GREEK LETTERS AND WORDS — Provides a listing of the Greek alphabet then a small selection of words to connect with the terms of cartography.

Appendix B: GLOSSARY OF SYMBOLS — An expansion on the modicum covered in the Preface, and close to comprehensive relative to the vast repertoire used in the text.

Appendix C: GLOSSARY OF TERMS — A modicum covering the less familiar geographic words used in the text, with some included for their particular definition.

Appendix D: INDEX TO PROJECTIONS — An alphabetic listing of every projection cited in this work, with page-based references to the citation, any pertinent plotting equations, and every illustrative map.

Appendix E: HISTORICAL BIBLIOGRAPHY — Year-of-publication-based references to show something of the progress of cartography, including succeeding editions of key works and all known published errata. The bibliography is essentially of published material, but also has some Internet addresses.

The General Index covers a variety of content, but does not cover projections and covers people only when there is content apart from a projection named after them. It defers to Appendix C for most technical words except where there is pertinent discussion within the chapters. It might be seen as an index to general subject material.

APPENDIX A: GREEK LETTERS AND WORDS

Introductory matters

The following list provides the Greek alphabet. Overleaf is a selection of pertinent Greek words. Because they mostly appear so within this book, the lower-case Greek letters are listed in italic form.

The Greek alphabet

graphic		*name*	*English equivalent*
Α	*α*	alpha	a
Β	*β*	beta	b
Γ	*γ*	gamma	g (hard)
Δ	*δ*	delta	d
Ε	*ε*	epsilon	e (short)
Ζ	*ζ*	zeta	z
Η	*η*	eta	long a (mate)
Θ	*θ*	theta	th (as in thing)
Ι	*ι*	iota	I (short I and ee)
Κ	*κ*	kappa	k
Λ	*λ*	lambda	l
Μ	*μ*	mu	m
Ν	*ν*	nu	n
Ξ	*ξ*	xi	ks
Ο	*ο*	omicron	o
Π	*π*	pi	p
Ρ	*ρ*	rho	r
Σ	*σ, ς*	sigma	s
Τ	*τ*	tau	t
Υ	*υ*	upsilon	u, oo
Φ	*ϕ*	phi	f
Χ	*χ*	chi	kh
Ψ	*ψ*	psi	ps
Ω	*ω*	omega	oh
ϑ	φ	phi (alternate form)	
ϖ		omega (alternate form)	

The second graphic for lower-case sigma is used for the final letter of a word.

The usage of this alphabet within this work is shown at the end of Appendix B.

Some Greek words

Greek word	*sound*	*meaning* *derived English words of general usage* *derived English words of cartographic usage*
χαρτος	khartôs	sheet of papyrus card, cartel, chart (as in temperature chart), charter chart (as a map), cartography
γραφη	graphay	drawn line, writing graph, graphics, graphic arts cartographic
ορθος	orthôs	correct, right, straight orthodox, orthogonal orthographic
μορφη	morphay	form, shape metamorphosis orthomorphic
γνωσις	nôsis	knowledge ignorance, know, knowledge see next
γνωμον	gnomon	indicator, carpenter's square gnomon gnomonic
στερεος	stereos	solid, 3-dimensional stereochemistry, stereoscopic vision stereographic
γεο–	geo-	land geography, geometry geodesy

APPENDIX B: GLOSSARY OF SYMBOLS

Introduction

Mathematics is a highly symbolic language, and its extensive and deep involvement in this cartographic work brings demand for a great variety of symbols. Generally, the usage fits with standard practice, but that essentially applies to general mathematical practice, not cartography. The notations in cartography are extremely variable, with little standardization. Given the intent to make the mathematics and cartography as clear and easy to follow as possible, various standards have been adopted for consistent use throughout the work, these being indicated in this appendix. The deepening intricacy of the cartographical mathematics, however, demands much adornment of the letters used to represent variables.

One practice entrenched without exception in this work is to have letters representing angles imply radians when in lower case, and imply degrees when in upper case. All such letters are Greek, α, β, γ, η, ν, ξ, υ, χ, ψ, and ω being examples that occur only in lower case, plus:

θ and Θ as used for general angles in Tutorials 1 & 2, the lower-case θ then becoming the standard for angle in 2-dimensional polar co-ordinates

λ and Λ which always refer to longitude

ϕ and Φ which always refer to latitude

The λ sometimes has an underline to stress when it is not the usual Greenwich-based angle. Because longitude is inherently arbitrarily based, however, and because it appears in so many equations differenced from a central meridian, λ is usually used unadorned to represent longitude relative to that central meridian. Intermittently that fact is overtly stressed in words. The letter ψ is always used for a function dependent solely on latitude, as is the symbol $\overline{\varphi}$, which refers solely and specifically to isometric latitude.

Other Greek letters given standard meaning as variables are:

δ for great-circle arc angle, hence great-circle distance on a sphere of unit radius

ζ for azimuth angle

ρ for radial distance in polar co-ordinates

The letter γ is used mostly for the angle between a central line and a general projection line at the centre of the globe, and the letter ξ is used mostly for the like at any other projection point. The letter γ is also used in later chapters for the angles between cardinal lines and the plotting axes of the map.

Variables are typically written in italics, but upper-case Greek letters are among the exceptions. Excepted also are vectors and matrices, which are always bold Roman letters, the vectors in lower-case, the matrices in upper-case.

The Greek letter π is used in upright form for the constant so familiar with circles and hence radians. The Roman letter e is used likewise in upright form to mean the natural number 2.7182~, leaving the italic version available to be a variable. (The ~ symbol indicates continuing digits.) The Roman letter i is used upright for the invariant $\sqrt{-1}$.

The lower-case duo x, y is used in the standard way for the two co-ordinates of 2-dimensional space. They are used unadorned for mathematics of a general nature but have dotted central superscripts (overscripts) when used for map variables, for which they are the routine final-stage plotting variables. Typically they are then $\dot{x}, \dot{y}$, but if such a map is an intermediate step before a further transformation to a revised map, they are $\ddot{x}, \ddot{y}$ and where necessary $\dddot{x}, \dddot{y}$ for maps successively derived mathematically from the first. Associated variables are likewise adorned. The upper-case trio X, Y, Z is used for 3-dimensional space, oriented in the usual mathematical manner. Where relating to its use in generating the 3-dimensional ellipsoid, the 2-dimensional ellipse is plotted with Y and Z axes as they occur 3-dimensionally.

Consistent with common mathematical practice, the symbol z is used for complex numbers, with real numbers x and y as its constituent parts, as in $\boldsymbol{z} = x + \mathrm{i}y$. The equivalent involving longitude and latitude uses a Greek symbol in the form $\boldsymbol{\mu} = \lambda + \mathrm{i}\phi$. In this work complex numbers are shown in bold italic style.

Derivatives in calculus terms are written either in full form else expressed by a bracketed right-superscript containing an Indo-Arabic numeral — e.g., $f^{(1)}, f^{(2)}, f^{(3)}$,

Besides the rare underlining already mentioned, symbols are adorned with subscripts and superscripts, to the right and left, and even above — the overscripts mentioned in the preceding paragraph. The use of right-superscripts is essentially standard, primarily for exponentiation and, in the bracketed form shown just above, for differentials; some use is also made of prime marks on variables. The practice with right-subscripts is also widely familiar, being an identifier of a choice, e.g.,

λ_0 for a chosen longitude
ϕ_0 for a chosen latitude
$E_{\lambda,\phi}$ where λ,ϕ are the chosen underlying parameters for expressing variable E

also

$P_i = (\lambda_i, \phi_i) = (x_i, y_i)$ for a general to-be-chosen point
$P_{\mathrm{N}} = (\lambda_{\mathrm{N}}, \phi_{\mathrm{N}})$ for the point chosen as auxiliary north pole

Left-subscripts are used to indicate context, e.g.,

${}_{\mathrm{a}}\phi$ to indicate latitude within an auxiliary scheme
${}_{\mathrm{s}}E_{\lambda,\phi}$ for variable $E_{\lambda,\phi}$ in the context of a sphere

The sole use of left-superscripts is

${}^{u,v}J_{\lambda,\phi}$ for expression of parameters u,v in terms of parameters λ,ϕ.

Besides the dots used for map variables throughout the work, and the symbol $\overline{\varphi}$ for isometric latitude already mentioned, overscripts are used extensively in later chapters for identifying competing versions of a symbol, such as:

$\bar{a}_1, \bar{a}_2$ for an alternative parameter pair to a_1, a_2

$\hat{a}_1, \hat{a}_2$ the angled overscript indicating orthogonality of a parameter pair

$\breve{\mathbf{n}}$ the up-curved u-like overscript indicating the vector is of unit length

$\bar{\bar{\phi}}$ the equals symbol as overscript indicating an equal-area variable

$\frown{\phi}$ the down-curved overscript symbol indicating an equidistant variable

$\tilde{\phi}$ the tilde as overscript indicating a conformal variable

The simple, single straight-line overscript is also used in Chapter 16 as generic precursor for the final three. (Note that the word *equivalent* is used here in the ordinary general English sense — not as a synonym for equal-area as it often is in cartography.) The use of each of the last three in this list is essentially constrained to the section in which it is introduced.

Briefly in Chapter 14 the variables applicable to the Tisost indicatrix are distinguished from those of the map by being presented with strikethrough lines.

The Roman alphabet

The following list shows the more particular uses of Roman letters, distinguished according to their being in normal, bold, italic, and bold italic forms. Besides those identified, many of the letters in italic upper-case form are used for points in diagrams and for intermediate functions of a complicated plotting equation.

graphic	*usage*
a	as a left subscript to distinguish auxiliary longitude and latitude [Ch 8]
a	a general line, in a triangle usually that opposite point *A*. Also, the length of the semimajor axis of an ellipse.
A	a general point. Also, the specific point forming the apex of a cone. Also a summation variable on page 338.
b	as $\breve{\mathbf{b}}$, the binormal unit vector [Ch 17].
b	a general line, in a triangle usually that opposite point *B*. Also, the length of the semiminor axis of an ellipse.
c	typically used for the constant of the cone [Ch 3].
C	a general point.
d	the standard mathematical ordinary-differentiation symbol.
d	a variable, usually a distance.
∂	the standard mathematical symbol for partial differentiation.

e	the transcendental number 2.718 28~ [Tutorial 16].
e	a general variable.
E	in the text, East.
E	the first of the three Gaussian fundamental quantities [Ch 12].
$\mathcal{E}$	as $\Delta\mathcal{E}$ the increment of false easting [page 425].
f	a general function.
F	the second of the three Gaussian fundamental quantities [Ch 12]. Also, an intermediate function for a complicated plotting equation.
g	a general variable.
G	typically used for the geocentre or other centre. Also, the third of the three Gaussian fundamental quantities [Ch 12]. Also, an intermediate function for a complicated plotting equation.
h	the hypotenuse of a right-angled triangle. Also, the ratio of the projection and viewing distance to radius R of the globe [Ch 3].
H	the point opposite the hypotenuse in a right-angled triangle.
i	the basic imaginary number √-1 used in complex numbers [Tutorial 23].
i	as $\breve{\mathbf{i}}$, the unit vector along the X axis [Ch 17].
i	a general variable, often an indexing variable running 1, 2, 3, . .
I	a general point. Also in Exhibit 13–3 a point at unit distance from the Origin.
j	as $\breve{\mathbf{j}}$, the unit vector along the Y axis [Ch 17].
j	generally an indexing variable running 1, 2, 3, . .
J	the Jacobian [Ch 12].
k	as $\breve{\mathbf{k}}$, the unit vector along the Z axis [Ch 17].
k	typically used as a constant, notably after integration [Ch 4].
l	an intermediate function of longitude in a complicated plotting equation.
L	an intermediate function of longitude in a complicated plotting equation.
m	typically as $\dot{m}$, the measure of linear distortion [Ch 13].
n	as $\breve{\mathbf{n}}$, the principal unit normal vector [Ch 17].
n	generally an indexing variable running 1, 2, 3, . .
N	in the text, North.
N	typically used in diagrams to denote the North Pole.
$\mathcal{N}$	as $\Delta\mathcal{N}$ the increment of false northing [page 425].
O	the intersection of axes, hence the origin for co-ordinates.
p	the vector to the point P [Ch 17].
r	typically used for a radius.
R	the radius of the globe.
s	distance, along a straight or curving line.

S	in the text, South.
S	typically used for the South Pole.
t	as $\breve{\mathbf{t}}$, the unit tangent vector [Ch 17].
T	an intermediate function for a complicated plotting equation.
u	the first of a pair of Cartesian intermediate parameters (u, v).
v	the second of a pair of Cartesian intermediate parameters (u, v).
V	typically used for the source of projection rays — the focus of viewing rays.
W	in the text, West.
x	the first of a pair of Cartesian 2-dimensional co-ordinates (x, y) [Tutorial 1].
X	the first of a trio of Cartesian 3-dimensional co-ordinates (X, Y, Z) [Tutorial 3].
y	the second of a pair of Cartesian 2-dimensional co-ordinates (x, y) [Tutorial 1].
Y	the second of a trio of Cartesian 3-dimensional co-ordinates (X, Y, Z) [Tutorial 3].
z	a complex number.
Z	the third of a trio of Cartesian 3-dimensional co-ordinates (X, Y, Z) [Tutorial 3].

The Greek alphabet

Excluding the many upper-case letters that are indistinguishable from Roman ones therefore unused, and the many lower-case letters used generally, the following apply

graphic	*name*	*usage*
α	lower-case alpha	a general angle, in radians.
β	lower-case beta	a general angle, in radians. Also, the angle between cardinal lines and vector in a map [Ch 12].
γ	lower-case gamma	a general angle, in radians, often the angle with the axis at the geocentre. Also, the angle between cardinal lines and axes in a map [Ch 12].
Γ	upper-case gamma	the angle γ in degrees.
δ	lower-case delta	angular distance along the surface of the globe.
Δ	upper-case delta	an infinitesimal increment in a variable, e.g., Δx. Also see $\mathcal{E}$ and $\mathcal{N}$ under Roman alphabet.
ε	lower-case epsilon	eccentricity of an ellipse [Tutorial 8 & Ch 16].
ζ	lower-case zeta	azimuth, in radians.
η	lower-case eta	a general angle, in degrees. Also, the angle between cardinal lines in a map [Ch 13].
θ	lower-case theta	a general angle, in radians. Also the second of a pair of 2-dimensional polar co-ordinates (ρ,θ) hence second of a pair of polar intermediate parameters $(\dot{\rho},\dot{\theta})$.
Θ	upper-case theta	the angle θ in degrees.
ι	lower-case iota	not used; too similar to Roman i.

κ lower-case kappa	curvature [Ch 16].
λ lower-case lambda	longitude, in radians.
Λ upper-case lambda	longitude, in degrees.
μ lower-case mu	the complex number $\lambda+i\phi$.
ν lower-case nu	used briefly as a general parameter [Ch 17].
ξ lower-case xi	a general angle, in radians, often the angle with the axis of the projection rays [Ch 3].
Ξ upper-case xi	the angle ξ in degrees.
o lower-case omicron	not used; conflicts with Roman o
π lower-case pi	the transcendental number 3.141 $59_{\sim}$ [Tutorial 16].
Π upper-case pi	standard mathematical product symbol.
ρ lower-case rho	the first of a pair of 2-dimensional polar co-ordinates (ρ,θ) hence first of a pair of polar intermediate parameters $(\dot{\rho},\dot{\theta})$.
σ lower-case sigma	the rescaling factor [Ch 5].
ς terminal sigma	used very briefly on page 384; too similar to lower-case zeta.
Σ upper-case sigma	standard mathematical summation symbol.
τ lower-case tau	torsion [page 383].
υ lower-case upsilon	not used; too similar to Roman *u*.
ϕ lower-case phi	latitude, in radians.
Φ upper-case phi	latitude, in degrees.
χ lower-case chi	rotational angle, in radians [page 86].
ψ lower-case psi	usually an angular-valued function of latitude, in radians, but used relatedly in Chapter 5 for the secant angle.
Ψ upper-case psi	the angle ψ in degrees.
ω lower-case omega	angular velocity — e.g., in the definition of WGS84 [Ch 16].
Ω upper-case omega	not used.

Other glyphs

Besides the various symbols of mathematics used in their normal way, the following distinct usages occur in this work.

$\oplus$ the conditional hemispheric sign symbol used for inverting various factors algebraically. It is negative for the south, positive for the north [page 31]. It applies to the inversion of rotational angles derived by differencing longitude about the South Pole, and to various other measurements, including that of polar distance from latitude.

$_{\sim}$ as a subscript, indicates an incomplete number; the digits displayed represent a rounded value, e.g., $\pi = 3.142_{\sim}$ and $\pi = 3.1416_{\sim}$.

See also the use of tilde and other glyphs as overscripts detailed at the top of page 452.

APPENDIX C: GLOSSARY OF TERMS

Introduction

The following short glossary covers some words used extensively in this work and some that get the barest of mentions.

antapex The circle representing the antipodal point on an azimuthal map.

antipodal Diametrically opposite on the globe.

antipodes Points approximately diametrically opposite on the globe.

aspect The orientation of Earth relative to the projection process [page 41}.

authalic Though in the past a popular synonym for **equal-area**, the term *authalic* is not so used in this work except, in the context of transforming an ellipsoidal surface to a spherical surface in Chapter 17, for the *authalic sphere*. Elsewhere it is used more weakly to mean having equal-area on an overall basis, without implying that the area of any part is in the same proportion. Due to their uniform spacing of meridians, however, many authalic projections produce maps with equal-area for lunes of any width from Pole to Pole; being fully equal-area requires appropriate spacing of parallels.
For **authalic latitude** etc., see page 399. See also **equivalent** on facing page.

auxiliary latitude, **auxiliary longitude** Substitute latitude and longitude used in changing the position of the poles for the oblique mapping; see opening of Chapter 8. Technically also **meta-latitude, meta-longitude**.

azimuth The bearing angle measured clockwise from North.

azimuthal At its Simple aspect, a projection producing parallels that are concentric circles and meridians that are straight lines passing through that centre, intersecting at an angle equal to their longitudinal difference. In its basic literal form, it is the projection onto a plane that is tangential to the globe at a Pole.

bearing (angle) An angle, measured clockwise from some reference line to a general line.

bimeridian A great circle passing through the Poles, and identically two meridians differing in longitude by precisely 180°.

cardinal lines The lines along the cardinal directions (i.e., north, east, south and west), hence the meridians and parallels of the graticule.

cardinal directions The directions north, east, south, and west.

cardinal points North, East, South, and West.

celestial sphere The conceptual sphere upon which lie the stars.

central projecting Projecting from the centre of the globe [Ch 3].

co-latitude For any parallel, its latitude subtracted from that of the indicated Pole; a signed number. Relative to the North Pole the values are 0° to +180°, relative to the South Pole –180° to 0°. *See also* **polar distance**.

conditional sign symbol *See under* Other glyphs on page 455.

conformal (conformality) A conformal mapping of a point is one that does not alter angles of intersection there, which is equivalent to having the scale change at the point identical in every direction. More generally a conformal map meets this condition for all points except possibly some on its boundary.
A conformal mapping is regarded as maintaining shape correctly (hence the sometime more popular *orthomorphic*), but such correctness applies only locally.
For **conformal latitude** etc., see page 401.

conic, conical At its Simple aspect, a projection producing parallels that are concentric circular arcs and meridians that are straight lines passing through that centre. In its basic literal form, it is the projection onto a cone that is co-axial with the globe and tangential thereto.

contact line, contact point The set of points of contact between a developable surface and the globe — a single point for a tangential azimuthal surface otherwise a complete circle [page 41].

conventional Describes a projection that is neither equidistant nor equal-area nor conformal. Not so used in this work.

cylindrical At its Simple aspect, a projection producing parallels and meridians that are mutually orthogonal straight lines. In its basic literal form, it is the projection onto a cylinder that is co-axial with the globe and tangential thereto.

equal-area An equal-area map has its overall area and the area of any part related to that of Earth by the established scale factor.
Synonymous with the never popular **orthembedic**.
See also **authalic**.

equidistant For maps at Simple aspect, having correct scale along all meridians. Correct scale along parallels is simultaneously feasible, but inherently limited to being true for at most two. Applied similarly to maps at any oblique aspect.
For **equidistant latitude** etc., see page 400.

equivalent Though sometimes used elsewhere as a synonym for *equal-area*, the term *equivalent* in this work is used with its normal broad meaning — e.g., a sphere with parallels set to provide true distance along the meridians is just such an equivalent sphere as one with parallels set to provide equal-area.

flexible parameter The latitude-dependent parameter in 'true' projections, being $\dot{v}$ for Cylindricals, and $\dot{\rho}$ for Azimuthals and Conicals.

geocentre (geocentric) The geometric centre of Earth as defined by whatever mathematical approximation is used. Applicable by extension to the model globe.

geocentric latitude The latitude measured relative to the **geocentre** which, for the ellipsoidal globe and for Earth, is not that normally measured; *see* **geodesic latitude**.

geodesic latitude The latitude measured by celestial observation relative to the local vertical and the form of latitude normally implied when that familiar word is not qualified [page 9]. For a spherical world identical with **geocentric latitude** but for the ellipsoidal world generally smaller, hence also **reduced latitude** [page 382].

graticule The mesh of meridians and parallels drawn on a map.

grid, gridlines A mesh of usually orthogonal lines drawn on a map for purpose of easy location of points within it.

homalographic, homolographic *See* Mollweide projection in Appendix C.

isocol A line drawn on a map connecting points for which distortion (e.g., general linear, maximal linear, areal, angular) is uniform [page 325].

isoline, isopleth A line drawn on a map connecting points for which some variable is uniform. For attention to distortion, *see* **isocol.**

isometric latitude A mathematical latitude that provides symmetry to the fundamental forms of differential geometry; *see* pages 288, 352 & 392.

lune The section of a spherical surface between two circular arcs, here only between meridians. A **clipped lune** in this work is short of the North Pole by 6° and of the South by 10°.

meridians Semicircular arcs extending from one Pole to the other, each uniquely identified by its longitude.

meridional adjective from meridian

meta-latitude, meta-longitude *See* **auxiliary latitude, auxiliary longitude**.

mode Used in this work to cover the three basic forms of perspective projecting, according with the three developable surfaces — i.e., azimuthal, cylindrical and conical [Ch 3]. Those three words are capitalized when used as the name of a mode.

orthogonal Intersecting at right angles.

orthographic Projecting from infinity — i.e., with parallel lines [Ch 3].

orthomorphic Of correct shape, *see* **conformal**.

overscript A term used in this work to mean the superscript-level adornment centred above a basic character.

parallel Of two planes and of two lines in a plane, such that they do not intersect at a finite point. Hence, and much more frequently in this work, the closed loop on the world's surface of points in a plane parallel with the Equatorial plane, each uniquely identified by its latitude, or the line on a map representing that loop.

parametric latitude See Tutorial 8 [page 24].

polar distance For any parallel, the angular distance from the indicated Pole; a non-negative number. *See also* **co-latitude**.

proper A proper fraction is one in which the denominator exceeds the numerator, a proper subset is one that is not identically the whole.

pseudoglobes Used in this work for cubes, dodecahedra and other polyhedra, plus other 3-dimensional bodies different from the sphere whose surface is made into a map of the world [page 369].

reduced latitude = **geodesic latitude**.

retro-azimuthal A projection that provides, at a focal point and for every point included in the map, a readable angle that represents the azimuth to the focal point from the other.

secantal Of developable surfaces and their derived projections, intersecting the surface being mapped. Applied also to algebraic non-geometric projections that are notionally so [Ch 5]. *See also* **tangential**.

sidereal day The day measured relative to the stars. Its length is almost 4 minutes less than that of the mean solar day. *See also* **solar day**.

sinusoidal Having the same curvature as the plot of the sine function.

solar day The day between recurrent ascension angles of the sun (e.g., between high noons). Its mean value is the familiar day of 24 hours. *See also* **sidereal day**.

standard framework The 3-dimensional framework with origin at the geocentre, the X and Y axes in the Equatorial plane with the first cutting the Greenwich meridian, the second at 90°E, then the Z axis northward along Earth's axis.

standard parallels Of a secantal projection, the contact lines (just one for azimuthal cases, otherwise two), hence lines of correct scale.

stereographic Projecting from the opposite point of the globe [Ch 3].

tangential Of developable surfaces and their derived projections, making contact with but not crossing the surface being mapped. Applied also to algebraic non-geometric projections that are notionally so [Ch's 3 & 4]. *See also* **secantal**.

transpose Of a matrix, the same values mirrored about the main diagonal [page 200].

UPS = Universal Polar Stereographic [Ch 18].

UTM = Universal Transverse Mercator [Ch 18].

zenithal An old synonym for azimuthal.

zone In UTM, a quadrilateral section of Earth's surface delineated longitudinally by a 6° gore (identified with a number) and latitudinally by the established banding (typically consecutive parallels on an 8° graticule, identified with a letter), but also used to mean a complete (80°S to 84°N) gore, identified by just the number, and to mean the latitudinal belt extended laterally, identified by just the letter; *see* page 419.

APPENDIX D: INDEX TO PROJECTIONS

Introduction

The following table provides an alphabetical index to the projections cited in the main text. Every name used has an entry in bold; additional lines in regular type are included to cover additional equations and illustrations. Data is generally provided for only one preferred name, other names providing direction thereto. Entries for such use the equals symbol to relate it to the preferred name.

The columns provide information as follows:

Projection the name of the projection augmented where pertinent by the number of standard parallels.

Azi. = Azimuthal
Con. = Conic(al)
Cyl. = Cylindrical
E-a. = Equal-area
Interntnl = International
proj = projection

Ch= Chapter number

Aspect

Obliq = Oblique
Equat = Equatorial
Transv = Transverse else as stated
— indicates case that is intrinsically oblique

Page = page number, being that number plus a letter when referring to equations — when the discussion may begin on some preceding page

Map the page of a pertinent exhibit

Notes include cross reference to other name and discriminatory factors re exhibits.

ww=whole world
Further abbreviations as for first column

C= coverage of the projection at the place identified within the line

c = cited only
d =described
e = equations
f = fully developed mathematically
v =variant
x = extra information
y =map only

Projection	*Ch*	*Aspect*	*Page*	*Map*	*Notes*	C
Collignon	16	Simple	367a	364		f
Conformal Azi.	4	Simple	=		Stereographic	
Conformal Azi.	13	Simple	318		*re distortion*	x
Conformal Con. [1	4	Simple	85b		*also* Lambert Conformal Con.	f
Conformal Con. [1	13	Simple	318		*re distortion*	x
Conformal Con. [2	5	Simple	121d	120		f
Conformal Con. [2	13	Simple	317		*from differential geometry*	x
Conformal Con. [2	13	Simple	319		*re distortion*	
Conformal Con. [2	14	Simple	340		*various variants*	f
Conformal Cyl.	4		=		Mercator	
Conformal Proj of the World in an Ellipse	15			358		c
Conic			=		Equidistant Con. [2	
Craig Retro-Azi.	11	—	272a	272		e
Craster Parabolic	6	Simple	138b		= Putnins P4	f
Craster Parabolic	7	Simple	176		*part of* McBryde P3	
Cube	16	—	369	368		f
de la Hire	3	Simple	50		*in* General Perspective Azi. [1	fv
Denoyer Semi-elliptical	6		151			d
Dent Poly-centred	7		185			d
Donald Elliptical	11		265		= Two-point Equidistant	c
Donis	6		143		= Trapezoidal	
Doubly . . .	11				*see* Two-point . . .	
Dürer polyhedric projs	16		369			c
Dymaxion or **"Dymaxion"**	16		370			d
Eckert I	6	Simple	144a			e
Eckert II	6	Simple	144b	142		e
Eckert III	6	Simple	144c			e
Eckert IV	6	Simple	145b	142		e
Eckert V	6	Simple	145c			e
Eckert VI	6	Simple	146a			e
Eckert VI	7	Simple	176		*part of* McBryde S2	
Eckert-Greifendorff	10	Simple	243a	242		f
Edwards	5	Simple	114		= Equal-area Cyl. [2 at ±37.4°	fv
Eisenlohr	15	Simple	361a	360		e
Elliptical (**Equal-area**)	6		=		Mollweide	
Elliptical Equal-area	8	Transv	214		*version of* Mercator	
Epicycloidal	15		=		August Epicycloidal	e
Equal-area Azi.	3	Simple	63		*re double projecting only*	
Equal-area Azi.	4	Simple	73c	72		f
Equal-area Azi.	8	Obliq	197c	198		v
Equal-area Azi.	8	Equat	203d	198		v
Equal-area Azi.	10	Equat		242	*re Hammer proj*	v
Equal-area Azi.	10	Obliq		244	45°N,10°E *re Briesemeister*	v
Equal-area Azi.	13	Simple	313		*re distortion*	x
Equal-area Azi.	14	Obliq	338		*optimized for Alaska*	v
Equal-area Azi.	16	Obliq	374		*in* Magnifying-glass *form*	d

Projection	*Ch*	*Aspect*	*Page*	*Map*	*Notes*	C
Gall			=		Gall Stereographic	
Gall Isographic	5	Simple	109		= Equidistant Cyl. [2 at ±45°	
Gall Orthographic	5	Simple	114	112	= Equal-area Cyl. [2 at ±45°	fv
Gall Stereographic	5	Simple	108	102	= Stereographic Cyl. [2 at ±45°	fv
General Perspective [1	2	Simple	29		*as* Aerial	f
General Perspective [1	2	Simple	32c		*as* Aerial	f
General Perspective [1	2	Obliq	32d	26	*as* Aerial, from 12 km	v
General Perspective [1	2	Obliq		26	*as* Aerial, from 500 km	y
General Perspective [1	2	Obliq		33	*as* Aerial, from 3000 km	y
General Perspective [1	2	Obliq		35	*as* Aerial, from 350 000 km	y
General Perspective [1	3	Simple	50c			f
General Perspective [1	8	Obliq	199b			v
General Perspective [1	8	Equat	201b			v
General Perspective [2	5	Simple	107a			f
General Perspective Azi.			=		General Perspective	
General Perspective Con. [1	3	Simple	61e			f
General Perspective Con. [2	5	Simple	108c			f
General Perspective Cyl. [1	3	Simple	56b			f
General Perspective Cyl. [1	8	Obliq	209a			v
General Perspective Cyl. [2	5	Simple	107b			f
General Vertical Perspective					= General Perspective	
Ginsburg I	14		339a			e
Ginsburg II	14		339b			e
Ginsburg III	14		339		*or* TsNIIGAiK	d
Glareanus	9		=		Bacon Globular	
Gnomonic [1	3	Simple	45b	42		f
Gnomonic [1	8	Obliq	199a	188		v
Gnomonic [1	8	Transv	201b			v
Goode's Homolosine	7		=		Homolosine	
Gretschel	3	Simple	50		*in* General Perspective Azi. [1	fv
GS50	14	Obliq	354	332		f
Gurba polyhedric projs	16		370			c
Guyou	15		357	356		d
Hammer (-Aitoff)	10	Simple	243a	242		f
Hammer blended composites	7		178			c
Hammer Elliptical	10		=		Hammer	
Hammer Retro-Azi.	11	—	273a	274		f
Hatano Asymmetrical	7	Simple	176a	172		f
Hinks polyhedric projs	16		370			c
Homalographic			=		Mollweide	
Homolographic			=		Mollweide	
Homolosine	7	Simple	175	172	*interrupted: 6 as 1*	f
Immler	11		269		Two-point Azi.	
Isospherical Stenoteric			=		Equal-area Con.	
Jäger Star	7	Simple	171			d
James	5		108		*in* General Perpsective Con. [2	fv

Projection	*Ch*	*Aspect*	*Page*	*Map*	*Notes*	C
Kavraiskiy I	14		342		*version of* Equidistant Con. [2	c
Kavraiskiy II	14		342d		*version of* Equidistant Con. [2	e
Kavraiskiy III	14		342		*version of* Equidistant Con. [2	c
Kavraiskiy IV	14		342		*version of* Equidistant Con. [2	c
Kavraiskiy V	6	Simple	159a			□
Kavraiskiy VI	6	Simple	157a		= Wagner I	□
Kite	7		179			c
La Carte Parallelogrammatique	5	Simple	=		Equidistant Cyl. [2	
Lagrange	15	Simple	351b	348	ww, five eighths in circle	f
Lagrange	15	Simple		350	ww in circle, 0° straight	v
Lagrange	15	Simple		350	ww in circle, 30°N straight	v
Lagrange					*see also* Lambert	
Lambert (Lagrange)	10	Simple	259c	254	*see also* Lagrange	f
Lambert Conformal Azi.			=		Stereographic	
Lambert Conformal Con.			=		Conformal Con.	
Lambert Equal-area Azi.			=		Equal-area Azi.	
Lambert Equal-area Con.	5	Simple	117a	117	*version of* Albers [2	v
Lambert Equal-area Cyl.			=		Equal-area Cyl.	
Lee Conformal Proj of the World in a Triangle	15		358			d
Lee Conformal Proj of the World in an Ellipse	15			358		d
Lee polyhedric projs	16		370			c
Leonardo da Vinci	16		369			c
Littrow	11	–	273b	275		f
Lotus	7		179			c
Lowry	3	Simple	50		*in* General Perspective [1	fv
Loximuthal	6	—	155b	154		e
Magnifying-glass projs	16	—	373			d
Marinus of Tyre			=		Equidistant Cyl. [2	
McBryde P3	7	Simple	176			f
McBryde Q3	7	Simple	176			f
McBryde S2	7	Simple	176			f
McBryde S3	7	Simple	176	174	*Interrupted; 6 as 1 oceans*	f
McBryde-Thomas I	6	Simple	159			f
McBryde-Thomas III	6	Simple	151a	150	[Flat-Polar Sinusoidal	f
McBryde-Thomas III	7	Simple	176		*part of* McBryde S3	
McBryde-Thomas IV	6	Simple	151b	150	[Flat-Polar Quartic	f
McBryde-Thomas IV	7	Simple	176		*part of* McBryde Q3	
McBryde-Thomas V	6	Simple	153a	150	[Flat-Polar Parabolic	f
McBryde-Thomas V	7	Simple	176		*part of* McBryde P3	
McGaw (polyconic variant)	7		184			c
McGaw (two-point azimuthal)	11		269		= Two-point Azi.	
Mecca	11		272		= Craig Retro-Azi.	

Projection	*Ch*	*Aspect*	*Page*	*Map*	*Notes*	C
Mercator [1	4	Simple	81b	83	*also* Conformal Cyl.	f
Mercator [1	8	Obliq	209e	206		v
Mercator [1	8	Transv	209f	208		v
Mercator [1	13	Simple	318		*re distortion*	x
Mercator [1	14	Obliq	333		*for* New Zealand Map Grid *[q.v.*	v
Mercator [1	15	Simple	354a		*isometric, hyperbolics*	x
Mercator [2	5	Simple	119a			f
Mercator [2	18	Transv		412	*as* Universal Transverse Merc . . .	y
Mercator Equal-area	6		=		Sinusoidal	
Mercator-Sanson	6		=		Sinusoidal	
Miller Cylindrical	10	Simple	247a	246		f
Miller Oblated Stereographic	14	Obliq	327b	326		f
Miller Prolated Stereographic	14		=		Miller Oblated . . .	
Modified Polyconic for the Interntnl Map of the World	7		184			d
Modified van der Grinten Equatorial Arbitrary Oval	9		237			d
Mollweide	6	Simple	133b	126		f
Mollweide	6	Simple	141		*part of* Boggs Eumorphic	x
Mollweide	7	Simple		170	*interrupted; 6 as 1, oceans*	
Mollweide	7	Simple	175		*part of* Homolosine	
Mollweide	7	Simple	175		*part of* Sinu-Mollweide	
Murdoch I	14		341a		*version of* Equidistant Con. [2	e
Murdoch III	14		341b		*version of* Equidistant Con. [2	e
Neil blended composites	7		178			c
New Zealand Map Grid (proj)	14		333		= NZMG	
Ney's	5		121		*version of* Conformal Con. [2	v
Nicolosi Globular	9	Simple	225d	222		f
Nordic	10		245		= Hammer Obliq	
Nowicki	3	Simple	50		*in* General Perspective Azi. [1	fv
NZMG			=		New Zealand Map Grid (proj)	d
Oblated Equal-area	14		328			c
Oblique . . .					*see . . .*	
Ordinary Polyconic	7				= Polyconic	
Ordnance Survey	14		342		*version of* Equidistant Con. [2	e
Ortelius Oval	10	Simple	249a	249		f
Orthoapsidal	16		370			
Orthodromic	11		269		= Two-point Azi.	
Orthographic (**Azi.**) [1	3	Simple	47c	42		f
Orthographic (Azi.) [1	8	Obliq	199b			v
Orthographic (Azi.) [1	8	Transv	201b			v
Orthographic (Azi.) [1	13	Simple	309		*testing with differential geometry*	x
Orthographic Con. [1	3	Simple	61d			f
Orthographic Cyl. [1	3	Simple	55d	82	= Equal-area Cyl. (q.v.)	f
Orthomorphic . . .			=		Conformal . . .	

Projection	*Ch*	*Aspect*	*Page*	*Map*	*Notes*	C
Sinu-Mollweide	7	Simple	175		*also* Philbrick	f
Sinusoidal	6	Simple	130a	126		f
Sinusoidal	6	Simple	141		*part of* Boggs Eumorphic	
Sinusoidal	7	Simple		168	*interrupted: 3 as 1*	d
Sinusoidal	7	Simple		168	*interrupted: 6 as 1*	y
Sinusoidal	7	Simple	175		*options for composites*	
Sinusoidal	7	Simple	175		*part of* Homolosine	
Sinusoidal	7	Simple	175		*part of* Sinu-Mollweide	
Sinusoidal	7	Simple	176		*part of* McBryde S2	
Sinusoidal	7	Simple	176		*part of* McBryde S3	
Sinusoidal	7	Simple	178		*in* blended composites	
Sinusoidal	8	Obliq		212		v
Sinusoidal	8	Transv		212		v
Sinusoidal	8	Transv		213	*rotated*	y
Smith polyhedric projs	16		370			c
Space Oblique Mercator	19	Obliq	445	432		d
Space Oblique Mercator	19	Obliq		434		x
Stereographic (**Azi.**) [1	3	Simple	46b			f
Stereographic (Azi.) [1	3	Simple	63		*re double projecting*	c
Stereographic (Azi.) [1	4	Simple	85c	86		v
Stereographic (Azi.) [1	4	Simple	85d		*with true scale at centre*	v
Stereographic (Azi.) [1	8	Obliq	201a	196		v
Stereographic (Azi.) [1	8	Equat	201b	196		v
Stereographic (Azi.) [1	13	Simple	316		*testing with differential geometry*	x
Stereographic (Azi.) [1	13	Simple	318		*re distortion*	x
Stereographic (Azi.) [1	14	Equat		346	hemi in circle, *re* Lagrange	f
Stereographic (Azi.) [1	15	Equat		348	hemi in circle, *with* Lagrange	f
Stereographic (Azi.) [1	15	Simple	353b		*isometric*	v
Stereographic (Azi.) [1	15	Obliq	353c		*isometric, hyperbolics*	v
Stereographic (Azi.) [1	15	Equat	353d		*isometric, hyperbolics*	v
Stereographic (Azi.) [2	5	Simple	107a		*in literal secantal form*	f
Stereographic (Azi.) [2	5	Simple	118b		*in continuing conformal form*	f
Stereographic (Azi.) [2 *ellipsoidal*	18	Simple			*as* Universal Polar Stereographic	x
Stereographic Conic (1	3	Simple	61c			f
Stereographic Cyl. [1	3	Simple	55b			f
Stereographic Cyl. [2	5	Simple	108a			f
Stereographic Cyl. [2	5	Simple	108	102	as Gall Stereographic	f
Tilted Perspective			443		*form of* General Perspective	
Tobler blended composites	7		177			f
Transverse . . .					*see . . .*	
Trapezoidal	6		143			
TsNIIGAiK	14		339		= Ginsburg III	d
Two-point Azi.	11	Obliq	269c			f
Two-point Equidistant	11	Obliq	265a	262		f

Projection	*Ch*	*Aspect*	*Page*	*Map*	*Notes*	C
Universal Polar Stereographic	18	Simple	427		Stereographic (Azi.) [2 *ellipsoidal*	
Universal Transverse Mercator	18	Transv	415		Mercator [2 *ellipsoidal*	
UPS	18	Simple	=		Universal Polar Stereographic	
Urmayev I & II	14		339			e
Urmayev III	14		340a			e
Urmayev IV	6	Simple	158a			f
UTM	18		=		Universal Transverse Mercator	
van der Grinten	9		=		van der Grinten I	
van der Grinten Globular	9		=		van der Grinten IV	
van der Grinten I	9	Simple	235a	234		f
van der Grinten II	9	Simple	235b	236		f
van der Grinten III	9	Simple	237a	236		f
van der Grinten IV	9	Simple	232a	230	*hemisphere in circle*	f
van der Grinten IV	9	Simple		237	*whole world in circle*	x
Vertical Perspective	2		32		*form of* General Perspective	f
Vitkovskiy I	14		342b		*version of* Equidistant Con. [2	e
Vitkovskiy II	14		342		*version of* Equidistant Con. [2	c
Vitkovskiy III	14		342		*version of* Equidistant Con. [2	c
Wagner I	6	Simple	157a		= Kavraiskiy VI	□
Wagner II	6	Simple	157d			f
Wagner III	6	Simple	157e			f
Wagner IV	6	Simple	147b	162	= Putnins P2′	
Wagner IV	6		157			x
Wagner V	6	Simple	157			f
Wagner VI	6				*see* Putnins P1′	f
Wagner VII	10	Simple	248a	246		e
Wagonwheel	16		=		Wiechel	
War Office	7				= Rectangular Polyconic	
Watts (side-by-side composites)	7		185			d
Weir Azimuth Diagram	11		275			d
Werenskiold II	6	Simple	157c			f
Werenskiold III	6	Simple	147			f
Werner	6	Simple		160	*version of* Bonne	fv
Werner IV	8	Obliq	201		= Stereographic	
Wiechel	16	Polar	366a	364		f
Winkel I	6	Simple	146b	146		e
Winkel Tripel	10	Simple	260a	258		f
Woolgar polyhedric projs	16		370			c
World in an Ellipse			=		Conformal Proj of . . .	
Zenithal			=		Azimuthal	

APPENDIX E: HISTORICAL BIBLIOGRAPHY

Introduction

The following listing of referenced publications is in order of their dates, the year of every one being qualified by a lower-case letter to separate coincidences. A direct reading of the bibliography thus provides a synopsis of the development of cartography — though references to the very earliest works are conspicuously absent.

Besides the books and papers referred to in the chapters of this book, this bibliography includes several texts of a more general nature that could prove useful further reading. They are distinguished by bold labels. Item 1988d is itself a bibliography should a reader want to search further.

To assist readers in locating copies, each author's name is given as fully as possible, but such that the version appearing on the specific publication is clear.

Where appropriate, notes are appended to any bibliographic entry.

In addition to abbreviations used in the main text, identifications of publishing sites include the following standard North American abbreviations for U.S. states:

CA	California
FL	Florida
MA	Massachusetts
NJ	New Jersey
VA	Virginia

and the following:

Assoc.	Association
ed.	editor
edn.	edition
HMSO	His/Her Majesty's Stationery Office
J.	Journal
N°	Number
Trans.	Transactions
USC&GS	U.S. Coast and Geodetic Survey
USGS	U.S. Geological Survey

Use of the ampersand is extended from lists of numbers to sequences of authors. European quotation symbols are used to enclose translated titles, i.e., as « ».

Chronological listing of monographs and journal papers, plus Internet sites.

1266a *The Opus Majus* Roger Bacon [London: date approximate] 2 vols. See 1962b for translation.

1524a *Cosmographicus Liber* Peter Apian [Landshut]. See 1551a for reprint.

1538a *Unterweysung des Messung mit dem Zirckel und Richtschey, in Linien, Ebnen und ganzen Corporan* Albrecht Dürer [Nuremberg].
• See vol 4.

1551a *Cosmographicus Liber* Peter Apian [Paris]. A reprint of 1524a.

1570a *Theatrus Orbis Terrarum* Abraham Ortelius.
• See also 1993e.

1643a *Hydrographe, contenent la théorie et la pratique de toutes les parties de la navigation* Georges Fournier [Paris].

1660a *Hercole Siciliano, Studio Geografico* Giovanni Battista Nicolosi [Rome].

1752a *Essai d'une nouvelle théorie de la résistance des fluids* Jean le Rond d'Alembert [Paris].

1772a "Anmerkungen und Zusätze zur Entwerfung der Land- und Himmelscharten" Johann Heinrich Lambert *Beiträge zum Gebrauche der Mathematik und deren Anwendung* Part 3, Section 6. See 1972b for English translation.

1779a "Sur la construction des cartes géographiques" Joseph Louis de Lagrange *Nouveaux memoires de l'Acadamie Royale des Sciences et Belles-Lettres* 161-210. See 1869a for reprint, 1894a for German translation.

1803a *Handbuch der Naturlehre* George Gottlieb Schmidt [Germany, Giessen].
• See "Projection der Halbkugelfläche".

1803b *Atlas des ganzen Erdkreises, nach den besten astronomischen Bestimmungen, neuesten Entdeckungen in eigenen Untersuchungen in der Central-Projection auf 6 Tafeln entworfen* Christian Gottlieb Reichard [Weimar: Lander Industrie Comptoirs].

1805a "Beschreibung einer neuen Kegelprojection" Heinrich Christian Albers *Zach's Monatliche Correspondenz* 12:97-114, 240-50, 450-9.

1805b "Über die vom Prof. Schmidt in Giessen in der zweyten Abtheilung seines Handbuchs der Naturlehre S.595 angegebene Projection der Halbkugelfläche" Karl Brandan Mollweide *Zach's Monatliche Correspondenz zur beförderung der Erd- und Himmels-Kunde* 12:152-63.

1825a "Allgemeine Auflössung der Aufgabe: Die Theile einer gegebnen Fläche auf einer andern gegebnen Fläche so abzubilden dass die Abbildung dem Abgebildeten in den kleinsten Theilen ähnlich wird" Carl Friedrich Gauss *Schumachers Astronomische Abhandlungen* [Altona] 3:5-30. See 1894b for later reprint.

1825b various papers including "On the mechanical organisation of a large survey, and the particular application to the Survey of the Coast" Ferdinand Rudolph Hassler *American Philosophical Society Trans.* New Series 2:232-408.

1833a *Chorographie; oder Anleitung, alle Arten von Land- See- und Himmelskarten zu verfertigen* Joseph Johann von Littrow [Vienna: F. Beck].

1845a *Physikalischer Atlas oder Sammlung von Karten* Heinrich Berghaus [Gotha: Justus Perthes].

1849a "On the variation of gravity at the surface of the Earth" G(eorge) G(abriel) Stokes *Trans. of the Cambridge Philosophical Society* 8:672.

1854a "Tables for projecting maps, with notes on map projections" Edward Bissell Hunt & Charles A. Schott *Report of the Superintendent of the Coast Survey: 1953* [Washington: Robert Armstrong, Public Printer] Appendix 39:96-163.

1855a "On improved monographic projections of the world" James Gall *Report of the Twenty-fifth meeting of the British Association for the Advancement of Science.*
- See also 1871a, 1885a.

1856a "Sur la construction des cartes géographiques" Pafnutiy L'vovich Chebyshev *Bulletin de l'Académie Impériale des Sciences* «St. Petersburg» *classe physico-mathématique* 14:257-61.

1859a "Sur les cartes géographique" Nicolas Auguste Tissot *Comptes Rendus des Séances de l'Académie des Sciences* 49:673-6.

1860a "Description of the projection used in the Topographical department of the War Office for maps embracing large portions of Earth's surface" Henry James *J. Royal Geographical Society* 30:106-11.

1861a "Explanation of a projection by balance of errors for maps applying to a very large extent of the Earth's surface" George Biddell Airy *London, Edinburgh, and Dublin Philosophical Magazine* 4th Series 22:409-21.
- See 1862b for corrections.

1862a *Notice sur la construction de nouvelles mappemondes et de nouveaux atlas de geeographie* H.-C.-A. de Prépetit Foucaut [France, Arras].

1862b "On projections for maps applying to a very large extent of the Earth's surface; and a comparison of this projection with other projections" Henry James & Alexander Ross Clarke *London, Edinburgh, and Dublin Philosophical Magazine* 4th Series 23:306-12 (with corrections to 1861a).

1865a "Der Nordpol. Ein thiergeographisches Centrum" G. Jäger *Das Ausland* 37:865-7.

1865b "Recherches sur la représentationplane de la surface du globe terrestre" Édouard Collignon *J. del l'École Polytechnique* 24:125-32.

1869a "Sur la construction des cartes géographiques" Joseph Louis de Lagrange *Oeuvres de Lagrange* [Paris: Gauthier-Villars] 4:635-92. A reprint of 1779a. See 1894a for German translation.

1869b "Über einige Abbildungaufgaben" Hermann Amandus Schwarz *Crelle's J. für reine und angewandte Mathematik* 70:105-20. Also in *Gesammelte Mathematische Abhandlungen* ii:65-83.

1869c "Conforme Abbildung der Oberfläche eines Tetraeders auf die Oberfläche einer Kugel" Hermann Amandus Schwarz *Crelle's J. für reine und angewandte Mathematik* 70:121-136. Also in *Gesammelte Mathematische Abhandlungen* ii:84-101.

1869d "Notizia sulla representazione conforme di un'area ellittica sopra un'area circolare" Hermann Amandus Schwarz *Annali di Matematica pura ed applicata* ii 3:166-70 also *Gesammelte Mathematische Abhandlungen* ii:102-7.

1870a "Über Flächenabbildung" Friedrich Eisenlohr *Crelle's J. für reine und angewandte Mathematik* 72:143-51.

1871a "On a new projection for a map of the world" James Gall *Proceedings of the Royal Geographic Society* 15:159.

1872a "Über diesjenigen Fälle, in welchen die Gaussische hypergeometrische Reihe eine algebraischebFunction ihres vierten elements darstellt" Hermann Amandus Schwarz *Crelle's J. für reine und angewandte Mathematik* 75:292-325 also *Gesammelte Mathematische Abhandlungen* ii:211-59.

1874a "Über eine conforme Abbildung der Erde nach der epicycloidischen Projection" Friedrich Wilhelm Oscar August *Zeitschrift der Gesellschaft für Erdkunde zu Berlin* 9:1-22.

1877a *Improvement in Geometrical Blocks for Mapping* J. Marcus Boorman [U.S. patent 185,889 dated 1877-01-02].

1878a "Sur la représentation des surfaces et les projections des cartes géographique" Nicolas Auguste Tissot *Nouvelle Annales de Mathématiques* 2nd series 17:49-55, 145-63, 351-66.

1879a "A quincuncial projection of the sphere" C(harles) S(anders) Peirce *American J. of Mathematics, Pure and Applied* 2:394-6.
- See also 1880a.

1879b "Rationelle Gradnetzprojectionen" H. Wiechel *Civilingenieur* New series 25:401-22.

1879c "Geography: mathematical geography" Alexander Ross Clarke *Encyclopædia Brittanica* 9th edn. 10:197-210.

1880a "A quincuncial projection of the sphere" C(harles) S(anders) Peirce *U.S. Coast Survey Report for Year Ending with June 1877* [Washington: U.S. Government Printing Office] Appendix 15:191-2.
- See also 1879a.

1881a *Memoire sur la représentation des surfaces et les projections des cartes géographique* Nicolas Auguste Tissot [Paris: Gauthier Villars].

1885a "Use of cylindrical projections for geographical, astronomical, and scientific purposes" James Gall *Scottish Geographical Magazine* 1:119-23.

1886a "Sur un nouveau système de projection de la sphère" Émile Guyou *Comptes Rendus de l'Académie des Sciences* 102:308-10.

1887a "Nouveau système de projection de la sphere: généralisation de la projection de Mercator" Émile Guyou *Annales Hydrographiques* 2nd Series, 9:16-35. Also in *Revue Maritime et Coloniale* 94:228-47.

1889a "Projections des cartes géographiques" David Aitoff *Atlas de géographie moderne* [Paris: Hachette].

1890a "Äquivalante Kartenprojektionen" A(dam) M(aximilian) Neil *Petermanns Geographische Mitteilungen* 36:93-8.

1892a "Über die Planisphere von Aitow und verwandte Entwürfe, insbesondere neue flächentreue änliche Art" Ernst (Herman) (Heinrich) Hammer *Petermanns Geographische Mitteilungen* 38:85-7.

1894a "Über die Construction geographischer Karten" Jean-Louis de Lagrange *Ostwald's Klassiker der exakten Wissenschaften* 55:1-56 with notes by Albert Wengerin [ed.] at 86-97. A translation from the French of 1779a.

1894b "Allgemeine Auflösung der Aufgabe: Die Theile einer gegebnen Fläche auf einer andern gegebnen Fläche so abzubilden, dass die Abbildung dem Abgebildeten in den kleinsten Theilen Ähnlich wird" Karl Friedrich Gauss *Ostwald's Klassiker der exakten Wissenschaften* 55:57-85 with notes by Albert Wengerin [ed.] at 97-101. A reprint of 1825a.

1896a «*Basic problems of the mathematical theory of geographical map construction*» Dmitry Aleksandrovich Grave [St. Petersburg: Academy of Science].

1900a "Unechtcylindrische und unechtkonische flächentreue Abbildungen. Mittel und Auftragengegebener Bogenlängen auf gezeichneten Kreisbögen von bekannten Halbmessern" E(rnst) (Herman) (Heinrich) Hammer *Petermanns Geographische Mitteilungen* 46:42-6.

1904a "Darstellung der ganzen Erdoberfläche auf einer Kreisförmigen Projectionsebene" Alphons J. van der Grinten *Petermanns Geographische Mitteilungen* 50:42-6.
• See 1904c & 1905b for corrections.

1904b *Map* Alphons J. van der Grinten [U.S. patent 751,226 dated 1904-02-02] covering one map of 1904a plus the map of 1905a.

1904c Corrections to 1904a *Petermanns Geographische Mitteilungen* 50:250.

1905a "New circular projection of the whole Earth's surface" Alphons J. van der Grinten *American J. of Science* 2nd Series 19:357-66.

1905b Further corrections to 1904a *Petermanns Geographische Mitteilungen* 51:48.

1906a "Neue Entwürfe für Erdkarten" Max Eckert *Petermanns Geographische Mitteilungen* 52:97-109.

1908a *The Thirteen Books of Euclid's Elements* Thomas L Heath [Cambridge, U.K.: Cambridge University Press]. Re-published as 1956a.

1909a "An account of a new land map of the world" Bernard J. S. Cahill *Scottish Geographical Magazine* 25:449-69.

1910a "Die beste bekannte fläschentreue Projektion der ganze Erde" W[alther] Behrmann *Petermanns Geographische Mitteilungen* 56:141-4.

1910b *Map-Projections (Technical lecture, 1909). The Theory of Map-Projections, with Special Reference to the Projections used in the Survey Department* James Ireland Craig [Cairo: Ministry of Finance] Survey Department Paper N°13.
• See 1911b for review.

1910c "Gegenazimutale Projektionen" Ernst Hermann (Heinrich) Hammer *Petermanns Geographische Mitteilungen* 56:153-5.

1911a "Sur les deformations resultant du mode de construction de la carte internationale du monde au millionème" Charles Lallemande *Comptus Rendus des Séances del'Académie des Sciences* 153:559-67.

1911b Review of 1910b Charles F(rederick) (Arden-)Close *Geographical J.* 37:208-9.

1911c "Map Projections" C(harles) F(rederick) (Arden-)Close & A(lexander) R(oss) Clarke *Encyclopaedia Britannica* [11th edn.] 17:653-63.

1912a *Leitfarden der Kartenentwurfslehre* Karl J. Zöppritz & Alois Bludau. An expanded 3rd edn. of earlier work.

1912b *Map Projections* Arthur Robert Hinks [Cambridge, U.K.: Cambridge University Press]. See 1921d for 2nd edn.

1912c *Konforme Abbildung des Erdellipsoids in der Ebene. Veröffentlichung* Louis Krüger [Potsdam: Köngilich Preussiches Geodätisches Institut].

1913a *Map of the World* [U.S. patent 1,054,276 dated 1913-02-25].

1913b *Geographical Globe* [U.S. patent 1,081,207 dated 1913-12-09].

1914a "Die Definition in der Kartenentwurfeslehre im Anschluss an die Begriffe zenital, azimutal und gegenazimutal" Hans Maurer *Petermanns Geographische Mitteilungen* 60:61-7 & 116-21.

1914b "A new projection" (B. J. S. Cahill) *Geographical J.* 43:86-7.

1919a *A Study of Map Projections in General* Oscar S(herman) Adams [Washington: U.S. Government Printing Office] [USC&GS Special Publication N°60].

1919b *General Theory of Polyconic Map Projections* Oscar S(herman) Adams [Washington: U.S. Government Printing Office] [USC&GS Special Publication N°57]. See 1934c for reprint.

1919c *The Lambert Conformal Conic Projection with Two Standard Parallels, including a Comparison of the Lambert Projection with the Bonne and Polyconic Projections* Charles H(enry) Deetz [Washington: U.S. Government Printing Office] [USC&GS Special Publication N°47].

1919d "Das winkeltreue gegenazimutale Kartennetz nach Littrow (Weirs Azimutdiagramm)" Hans Maurer *Annalen der Hydrographie und Maritimen Meteorologie* 47:14-22.

1919e "Ein doppelazimutaler gnomonischer Kartenentwurf und seine Anwendung auf kreuzpeilungen für grosse Entwurfnungen" Von W. Immler *Annalen der Hydrographie und Maritimen Meteorologie* 47:22-36.

1919f "'Doppelbuschelstrahlige, ortodromische' statt 'doppelazimutale, gnomonische' kartenentwürfe . Doppel mittabstrandstreue Kartogramme" Hans Maurer *Annalen der Hydrographie und Maritimen Meteorologie* 47:75-8.

1920a *Some Investigations in the Theory of Map Projections* Alfred Ernest Young [London: Royal Geographical Society] Royal Geographical Society Technical Series N°1.

1921a "On the projection adopted for the Allied maps on the Western Front" Arthur R(obert) Hinks *Geographical J.* 57:448-54.

1921b "Neue Gradnetzkombinationen" Oswald Winkel *Petermanns Geographische Mitteilungen* 67:248-52.

1921c *Latitude Developments Connected with Geodesy and Cartography* Oscar S(herman) Adams [Washington: U.S. Government Printing Office] [USC&GS Special Publication N°67].

1921d "Note on a doubly-equidistant projection" Charles F(rederick) (Arden-)Close *Geographical J.* 57:446-8.
- See also 1922a.

1921e *Map Projections* Arthur Robert Hinks [Cambridge: CUP] The 2nd edn. of 1912b.

1921f *Elements of Map Projection with Applications to Map and Chart Construction* Charles H(enry) Deetz & Oscar S(herman) Adams [Washington: U.S. Government Printing Office] [USC&GS Special Publication N°68]. See 1934e for 4th edn., 1944b for 5th edn.

1921g "The projection of the sphere on the circumscribed cube" Arthur R(obert) Hinks *Geographical J.* 57:454-7.

1921h "A modified rectangular polyconic projection" George Tyrell McGaw *Geographical J.* 57:451-4.

1922a *Note on Two Double, or Two-Point, Map Projections* Charles (Arden-)Close [London: Ordnance Survey] Ordnance Survey Professional Paper (New series) Nª5].
- See 1922b for comments.

1922b "Two new map projections" Alfred Ernest Young *Geographical J.* 60:297-9. Relates to 1922a.

1922c "Über Bezeichnungen von Kartenentwürfen, insbesondere über doppelsymmetrische uneht-abstandstreue Weltkarten mit Pol-Punkten" Hans Maurer *Zeitschrift der Gesellschaft für Erdkunde zu Berlin* (3-4):115-26.

1925a "The Homolosine projection: a new device for portraying the Earth's surface entire" J(ohn) Paul Goode *Annals of the Assoc. of American Geographers* 15:119-25.

1925b *Elliptic Functions applied to World Maps* Oscar S(herman) Adams [Washington: U.S. Government Printing Office]. [USC&G Special Publication N°112].

1925c "Mean Sea Level and its variations" H. A. Marmer *Annals of the Assoc. of American Geographers* 15:106-118.

1927a *An Introduction to the Study of Map Projections* J(ames) A(lfred) Steers [London: University of London Press]. See 1962f for 13th edn., 1972d for 15th edn.

1927b *The Transverse Elliptical Equal-Area Projection of the Sphere; Otherwise Transverse Mollweide* Charles F(rederick) (Arden-)Close [London: Ordnance Survey] [Ordnance Survey Professional Paper (New Series) N°11].

1927c "Some new map projection" Charles (Arden-)Close *Geography* 14:101-10.

1928a "A new projection" J. Fairgrieve *Geography* 14:525-6.

1928b *The Opus Majus of Roger Bacon* (2 vols) translated by Robert Belle Burke from the Latin of 1266a. See 1962c for reprint.

1928c *Ortelii Catalogus Cartographorum* (A biography of the cartography covered by Ortelius) Leo A. Bagrow (1928-30).
- See also 1993e

1929a *Conformal Projection of the Sphere within a Square* Oscar S(herman) Adams [Washington: U.S. Government Printing Office] [USC&G Special Publication N°153].

1929b "Some equal-area projections of the sphere" J(ohn) E(velyn) E(dmund) Craster *Geographical J.* 74:471-4.

1929c "A new equal-area projection for world maps" S(amuel) Whittemore Boggs *Geographical J.* 73:241-5.

1929d "An oblique Mollweide projection of the sphere" Charles F(rederick) (Arden-)Close *Geographical J.* 73:251-3.

1929e "A chart showing the true bearing of Rugby from all parts of the world" E(dward) A. Reeves *Geographical J.* 73:247-8.

1929f "A retro-azimuthal equidistant projection of the whole sphere" Arthur R(obert) Hinks *Geographical J.* 73:244-7.

1931a *A Concise Etymological Dictionary of Latin* T(homas) G(eorge) Tucker [Halle, Germany: M. Niemeyer].

1931b "Seasonal variations in daylight, twilight, and darkness" S(amuel) W(hittemore) Boggs *Geographical Review* 21:656-9.

1931c *Elements of Map Projection, with Application to Map and Chart Construction* Charles H(enry) Deetz & Oscar S(herman) Adams [Washington DC: U.S. Government Printing Office] [USC&GS Special Publication N°68] 3rd edn. of 1921f. See 1934e for 4th edn.

1931d *An Introduction to the Mathematics of Map Projections* R(obert) K(eith) Melluish [Cambridge, U.K.: University of Cambridge Press].

1932a *Geography of Ptolemy* [New York: N.Y. Public Library] translated by Edward Luther Stevenson. Limited edition of 250 copies. See 1991a for republication.

1932b "Die unechten Zylinderprojektion" Karl-Heinrich (or Karlheinz) Wagner *Aus dem Archiv der Deutschen Seewarte* 51(4):68.

1932c *Traité des projections des cartes géographiques á l'usage des cartographs et des geodesiens* Ludovic Driencourt & Jean Laborde [Paris: Hermann et Cie] 4 vols. See particularly Vol 4, Chap 3 & 4.
• Review in 1934f.

1934a "Jaunas projkcijas pasaules kartem" Reinholds V. Putnins *Geögrafiski Raski Folio Geographica Riga* III & IV:180-209. (Latvian p180-200, plus French résumé p201-9.)

1934b «*Mathematical cartography*» Vladimir Vladimirovich Kavraiskiy [Moscow].
• See English review in 1962a.

1934c *General Theory of Polyconic Map Projections* Oscar S(herman) Adams [Washington: U.S. Government Printing Office] [USC&GS Special Publication N[a]57]. A reprint of 1919b.

1934d "A doubly equidistant projection of the sphere" Charles F(rederick) (Arden-)Close *Geographical J.* 83:144-5.

1934e *Elements of Map Projection with Applications to Map and Chart Construction* Charles H(enry) Deetz & Oscar S(herman) Adams [Washington: U.S. Government Printing Office] [USC&GS Special Publication N°68] The 4th edn. of 1921f. See 1944b for 5th edn.

1934f "A new treatise on map projections" (a review of 1932c) A(rthur) R(obert) H(inks), G(eorge) T(yrell) M(cGaw) & R(obert) K(eith) M(elluish) *Geographical J.* 83:145-50.

1935a "Ebene Kugelbilder: Ein linisches System der Kartenwürfe Mitteilungen" Hans Maurer *Petermanns Geographische Mitteilungen* Supp 221:141-4. For translation into English see 1968a.

1935b "Eine neue flächentreue (azimutaloide) Erdkarte" Max Eckert-Greifendorff *Petermanns Geographische Mitteilungen* 81:190-2.

1935c "Wegtreue Ortskurskarten" Karl Siemon *Mitteilungen des Reichsamts für Landesaufnahme* 11:88-95.

1935d "Two-point azimuthal-equidistant projection" Charles F(rederick) (Arden-)Close *Geographical J.* 86:445-6.

1937a "Flächenproportinales Umgraden von Kartenentwürfen" Karl Siemon *Mitteilungen des Reichsamts für Landesaufnahme* 13:88-102.

1938a "Oblique conical orthomorphic projection for New Zealand" John Evelyn Edmund Craster *Geographical J.* 92:537-8.

1938b *General Cartography* Erwin Josephus Raisz [New York: McGraw-Hill] See 1948b for 2nd edn., 1962d for 3rd edn.

1939a *Globe* James R. Smith [U.S. Patent 2,153,053 dated 1939-04-04].

1941a "More world maps on oblique Mercator projections" Arthur R(obert) Hinks *Geographical J.* 97:353-6.

1941b "Murdoch's third projection" Arthur R(obert) Hinks *Geographical J.* 97:353-6.

1941c "Dr. J. R. Baker's duration of sunlight diagram" Arthur R(obert) Hinks *Geographical J.* 97:358-63.

1941d "A conformal map projection for the Americas" O(sborn) M(aitland) Miller *Geographical Review* 31:100-4.

1942a *The Citizen's Atlas of the World* John Bartholomew [Edinburgh: John Bartholomew & Son].

1942b "Notes on cylindrical world map projections" O(sborn) M(aitland) Miller *Geographical Review* 32:424-30.

1942c "Maps of the whole world ocean" Athelstan F. Spilhaus *Geographical Review* 32:431-5.

1943a "A world map on a regular icosahedron by gnomonic projection" I(rving) Fisher *Geographical Review* 33:605-19.

1943b "Orthoapsidal world maps" Erwin Josephus Raisz *Geographical Review* 33:132-4.

1943c "Dymaxion World" R. Buckminster Fuller *Life* 1943-03-01:41.

1943d "Additional notes on cylindrical map projections and the plotting of great-circle courses" O(sborn) M(aitland) Miller *Geographical Review* 33:328-9.

1944a "Class of equal-area projections" W(erner) Werenskiold *Det Norske Videnskaps-Akademi i Oslo – matematisk-naturvidenskapselig* Klasse 11.

1944b *Elements of Map Projection with Applications to Map and Chart Construction* Charles H(enry) Deetz & Oscar S(herman) Adams [Washington: U.S. Government Printing Office] [USC&GS Special Publication N°68] The 5th edn. of 1921f.

1944c "Grid navigation" Samuel Herrick *Geographical Review* 34:436-56.

1945a *Theory of Equivalent Projections* Oscar S(herman) Adams [Washington: U.S. Government Printing Office] [USC&GS Special Publication Nª236].

1946a *Dymaxion World* R. Buckminster Fuller [U.S. patent 2,393,676 dated 1948-01-29].

1946b "Equal-area projection on the icosahedron" A. D. Bradley *Geographical Review* 36:101-4.

1946c *Map Projections* George P. Kellaway [London: Methuen]. See 1949d for 2nd edn.

1946d "The orthomorphic projection of the spheroid" M(artin) Hotine *Empire Survey Review* 8(62):300-11. See 1947c for continuation.

1947a "The manipulation of projections for world maps" Edward J. Baar *Geographical Review* 37:112-20.

1947b *The Round Earth on Flat Paper* Wellman Chamberlin [Washington: National Geographic Society]. See 1950a for revised edn.

1947c "The orthomorphic projection of the spheroid" M(artin) Hotine *Empire Survey Review* 9(63):25-35, (64):52-70, (65)112-23 & (66)157-66 (being Parts II – V continuing from 1946d).

1948a *The Regional Atlas of the World* John Bartholomew [Edinburgh: Geographical Institute].

1948b *General Cartography* Erwin Josephus Raisz [New York: McGraw-Hill] The 2nd edn. of 1938b. See 1962d for successor text.

1948c *Global Map* Irving Fisher [U.S. patent 2,436,860 dated 1948-03-02].

1949a *Maps and Mapmakers* R. V. Tooley [London: B. T. Batsford].

1949b *Equal-Area Projections for World Statistical Maps* F. Webster McBryde & Paul D. Thomas [Washington: U.S. Government Printing Office] [USC&GS Special Publication Nª245].

1949c *Kartographische Netzentwürfe* Karlheinz (or Karl-Heinrich) Wagner [Leipzig: Bibliographische Institut] See 1962b for 2nd edn.

1949d *Map Projections* George P. Kellaway [London: Methuen] The 2nd edn. of 1946c.

1949e "A tetrahedral gnomonic projection" F. V. Botley *Geography* 34:131-6.

1950a *The Round Earth on Flat Paper – map projections used by cartographers* Wellman Chamberlin [Washington: National Geographic Society] A revised edition of 1947b.

1950b *Constants, Formulae and Methods used by the Ordnance Survey for computing in the Transverse Mercator Projection* [London: HMSO] Ordnance Survey.

1951a "A new use for the Plate Carrée projection" F. V. Botley *Geographical Review* 41:640-4.

1952a "A forgotten pseudo-zenithal projection" Charles (Frederick) Arden-Close & F(rank) George *Geographical J.* 118:237.

1952b *Conformal Projections in Geodesy and Cartography* Paul D. Thomas [Washington: U.S. Government Printing Office] [USC&GS Special Publication N°251].

1953a *A New Map of the World: The Trystan Edwards Homolographic Projection* Trystan Edwards [London: B. T. Basford].

1953b "An oblique equal-area map for world distributions" Allen K. Philbrick *Annals of the American Assoc. of Geographers* 43:201-15.

1953c "A new oblique equal-area projection" W(illiam) Briesemeister *Geographical Review* 43:260-1.

1953d "A new conformal projection for Europe and Asia" (Note: actually applies to Africa, not Asia) O(sborn) M(aitland) Miller *Geographical Review* 43:405-9.

1953e "A transverse Mercator projection of the spheroid alternative to the Gauss-Krüger form" L(aurence) P(atrick) Lee *Empire Survey Review* 12(87):12-7.

1953f *Elements of Cartography* Arthur H(oward) Robinson [New York: John Wiley & Sons]. See 1995e for 6th edn.

1954a "An octahedral gnomonic projection" F. V. Botley *Empire Survey Review* 12(94):379-81.

1955a "The history of geographical map projections until 1600" Johannes Keuning *Imago Mundi* 12:1-25.

1956a *Euclid: The Thirteen Books of the Elements* Thomas L Heath [New York: Dover] A republication of 1908a.

1957a *Donald Elliptical Projection* Jay K. Donald [New York: American Telephone & Telegraph Co.,] unpublished manuscript.

1958a "Use of artificial satellites to explore the Earth's gravitational field: results from Sputnik 2 (1957β)" R. H. Merson & D(esmond) G. King-Hele *Nature* 182:640-1.

1958b *The Times Atlas of the World* John Bartholomew [ed.] [London: Times Publishing].

1960a "A review of some Russian map projections" D(erek) H(ylton) Maling *Empire Survey Review* 15(115):203-15, (116):255-66 & (117):294-303.

1962a "Evolution of interrupted map projections" R(ichard) E. Dahlberg *International Yearbook of Cartography* 2:36-54.

1962b *Kartographische Netzentwürfe* Karlheinz (or Karl-Heinrich) Wagner [Mannheim, Germany: Bibliographische Institut] The 2nd edn. of 1949c.

1962c *The Opus Majus* Roger Bacon (year approximate) [New York: Russell and Russell] 2 vols. A reprint of 1266a.

1962d *Principles of Cartography* Erwin Josephus Raisz [New York: McGraw-Hill] The successor to 1938b & 1948b.

1962e "The transverse Mercator projection of the entire spheroid" L(aurence) P(atrick) Lee *Empire Survey Review* 16(123):208-17.

1962f *An Introduction to the Study of Map Projections* J(ames) A(lfred) Steers [London: University of London Press] The 13th edn of 1927a. See 1972d for 15th edn.

1965a "Mollweide modified" Robert H. Bromley *Professional Geographer* 17:24.

1965b "Some conformal projections based on elliptic functions" L(aurence) P(atrick) Lee *Geographical Review* 55:563-80.

1966a "Notes on two projections" W(aldo) R(udolph) Tobler *The Cartographic J.* 3:87-9.

1967a "On retro-azimuthal projections" J(ohn) E(ric) Jackson *Survey Review* 19(149):319-28.

1967b *Computer Solution of Linear Algebraic Systems* George E. Forsythe & Cleve B. Moler [Englewood Cliffs, NJ: Prentice-Hall].

1968a *Plane Globe Projection: A Linnean System of Map Projection* Peter Ludwig [Cambridge, MA: Harvard University Press] [N°23 in Harvard Papers on Theoretical Geography, Geography and the Properties of Surfaces, William Warntz (ed.)]. A translation of 1935a.

1968b «Combined equal-area projection for world maps» György Érdi-Krauss *Hungarian Cartographical Studies* 1956-1958:44-9.

1968c «The problems of the representation of the globe on a plane with special reference to the preservation of the forms of continents» János Baranyi *Hungarian Cartographical Studies* :19-43.

1970a «The orthodrome on a dodecahedron globe» Stefania Gurba *Polski Przeglad Kartograficzny* 2:160-8.

1970b "Some new map projections of the world" D. G. Watts *The Cartographic J.* 7:41-6.

1970c "Singular Value Decomposition and Least Squares solutions" G. Golub & C. Rensch *Numerical Mathematics* 14:403-20.

1972a *Maps and Man: An Examination of Cartography in Relation to Culture and Civilization* N(orman) J(oseph) W(illiam) Thrower [New York: Prentice Hall]. Re-issued as 1996a.

1972b *Notes and Comments On The Composition of Terrestrial and Celestial Maps by J. H. Lambert* Translated and introduced by Waldo R(udolph) Tobler [Ann Arbor, MI: Dept. of Geography, University of Michigan] [Michigan Geographical Publication N°8]. A translation of 1772a.

1972c "Consideration of the Projection Suitable for Asia-Pacific" Masataka Hatano *Geographical Review of Japan* 45:637-47.

1972d *An Introduction to the Study of Map Projections* J(ames) A(lfred) Steers [London: University of London Press]. The 15th edn. of 1927a.

1972e *Map Projections: for geodesists, cartographers and geographers* Peter Richardus & Ron K. Adler [Amsterdam: North-Holland + New York: American Elsevier].

1973a *Coordinate Systems and Map Projections* D(erek) H(ylton) Maling [London: George Philip & Son]. See 1992a for 2nd edn.

1973b *Facsimile Atlas* A. E. Nordenskjöld [New York: Dover].

1973c "The hyperelliptical and other new pseudo-cylindrical equal-area map projections" W(aldo) R(udolph) Tobler *J. of Geophysical Research* 78:1753-9.

1973d "A conformal mapping projection with minimum scale error" W. I. Reilly *Survey Review* 22(168):57-71.

1974a "A new map projection: its development and characteristics" Arthur H(oward) Robinson *International Yearbook of Cartography* 14:145-55.

1974b *The Seven Aspects of General Map Projection* Thomas Wray [Toronto: B V Gutsell, Dept of Geography, York Univ.] Cartographica Monograph N°11.

1974c "The computation of conformal projections" L(aurence) P(atrick) Lee *Survey Review* 22(172):245-56.

1974d "A conformal projection for a map of the Pacific" L(aurence) P(atrick) Lee *New Zealand Geographer* 30:75-7.

1974e "Space oblique Mercator: a new map projection of the Earth" Alden P. Calvocoresses *Photogrammetric Engineering and Remote Sensing* 40:921-6.

1974f "The new map projection" I. F. Stirling *New Zealand Cartographic J.* 4:3-9.

1975a "On map projections (with special reference to some inspired ones)" Martin Gardner *Scientific American* Nov1975:120-5.

1976a *Conformal projections based on elliptic functions* L(aurence) P(atrick) Lee [Toronto: University of Toronto Press] Cartographica Monograph N°16. Published as a supplement to *Canadian Cartographer* v. 16.

1976b "The Shape of the Earth" Desmond (G.) King-Hele *Science* 192(4246):1293-9.

1976c "An analysis of a modified van der Grinten equatorial arbitrary oval projection" William D. Brooks & Charles E. Roberts Jr. *American Cartographer* 3:143-50.

1976d "History of geodetic leveling in the United States" R. M. Berry *Surveying and Mapping* 36:137-53.

1977a "A comparison of pseudocylindrical map projections" John P(arr) Snyder *American Cartographer* 4:59-81.

- See 1979 for corrections.

1977b *Map Projection Equations* Frederick Pearson II [Dahlgren, VA: Naval Surface Weapons Center]. See 1977b for revised version, 1990a for typeset edn.

1977c "A note on the van der Grinten projection of the whole Earth on a disk" John A. O'Keefe & Allen Greenberg *The American Cartographer* 4:127-32.
- See 1980c for corrections.

1977d *Formulation of a Space Oblique Mercator Map Projection* John L. Jenkins & James D. Turner Final Report N° UVA/525023/ESS77/105 of Nov to Computer Science Laboratory, U.S. Army Engineer Topographic Laboratories .

1978a "A new series of composite equal-area world map projections" Felix Webster McBryde presented at the 9th International Cartographic Conference and shown in *Abstracts* :76-7.

1978b "Equidistant conic map projections" John P(arr) Snyder *Annals of the Assoc. of American Geographers* 68:373-83.

1978c "Transverse Mercator projection" J. E. Jackson *Survey Review* 24(188):278-85.

1978d "The space oblique Mercator projection" John P(arr) Snyder *Photogrammetric Engineering and Remote Sensing* 44:585-96.

1979a Corrections to 1977a John P(arr) Snyder *The American Cartographer* 6:81.

1979b *Introduction to Map Projections* Peter W. McDonnell Jr. [New York: Marcel Dekker].

1979c "Calculating map projections for the ellipsoid" John P(arr) Snyder *The American Cartographer* 6:67-76.
- See footnote in 1981e for corrections.

1980a "The Peters phenomenon" John Loxton *The Cartographic J.* 22:106-8.

1980b "The so-called Peters projection" [Board of the German Geographical Society.] *The Cartographic J.* 22:108-10.

1980c "Van der Grinten formulae" David Parry Rubincam *The American Cartographer* 7:176-7 being corrections to 1977c.

1981a "Map projections for satellite tracking" John P(arr) Snyder *Photogrammetric Engineering and Remote Sensing* 47:205-13.

1981b "Latitude and longitude from van der Grinten grid coordinates" David Parry Rubincam *The American Cartographer* 8:177-80.

1981c "Modern geodetic earth reference models" B. H. Chovitz *American Geophysical Union Trans.* 62:165-7.

1981d *Space Oblique Mercator Projection: Mathematical Development* John P(arr) Snyder [Washington: U.S. Government Printing Office] [USGS Bulletin N°1518].
- See 1982b for review and corrections.

1981e "The perspective map projection of the Earth" John P(arr) Snyder *The American Cartographer* 8:149-60 Includes corrections to 1979c.
- See 1982c for corrections to itself.

1982a *Map Projections used by the United States Geological Survey* John P(arr) Snyder [Washington: U.S. Government Printing Office] [USGS Bulletin N°1532]. See 1983a for 2nd edn.

1982b A review with corrections of 1981d Joseph C. Loon *Photogrammetric Engineering and Remote Sensing* 48:1581.

1982c Corrections to 1981e John P(arr) Snyder *The American Cartographer* 9:84.

1983a *Map Projections used by the United States Geological Survey* John P(arr) Snyder [Washington: U.S. Government Printing Office] [USGS Bulletin N°1532] The 2nd edn. of 1982a.

1983b "Toward an understanding of scale and its relevance to cartographic communication" Mervin D. Henning *Cartographic J.* 20:119-20.

1983c "Latitude, longitude, and ellipsoidal height changes NAD-27 to Predicted NAD-83" B. K. Meade *Surveying & Mapping* 43:65-71.

1984a *Map Projection Methods* Frederick Pearson II [Blacksburg, VA: Sigma Scientific Inc] A revised version of 1977b. See 1990a for typeset edn.

1984b "A low-error conformal map projection for the 50 States" John P(arr) Snyder *The American Cartographer* 11:27-39.

1984c "Minimum-error map projections bounded by polygons" John P(arr) Snyder *Cartographic J.* 22:112-20.

- See 1985b p73 for corrections.

1984d *Elements of Cartography* A(rthur) H(oward) Robinson, R. Sale, J. L. Morrison & P. C. Muehrke [Toronto: Wiley] 5th edn.

1985a *The Story of Maps* Lloyd A. Brown [New York: Bonanza Books].

1985b Note by John P(arr) Snyder *Cartographic J.* 22:73 being corrections to 1984c.

1985c "The transverse and oblique cylindrical equal-area projection of the ellipsoid" John P(arr) Snyder *Annals of the Association of American Geographers* 75:431-42.

1986a "Polycylindric map projections" Waldo R(udolph) Tobler *American Cartographer* 13:43-50.

1987a *Map Projections – a working manual* John P(arr) Snyder [Washington: U.S. Government Printing Office] [USGS Professional Paper N°1395]. See 1994B for reprint with corrections.

1987b "'Magnifying Glass' azimuthal map projections" John P(arr) Snyder *The American Cartographer* 14:61-8.

1987c "Continental shapes on world projections: the design of a poly-centred oblique orthographic world projection" Borden D. Dent *Cartographic J.* 24:117-24.

1987d *The History of Cartography* J(ohn) B(rian) Harley [Chicago: University of Chicago Press].

1988a "New equal-area map projections for noncircular regions" John P(arr) Snyder *The American Cartographer* 15:341-55.

1988b *Choosing a World Map: attributes, distortions, classes, aspects* [Falls Church, VA: American Congress on Surveying and Mapping] [a report of American Cartographic Association Committee on Map Projections].

1988c *Basic Geodesy* J. R. Smith [Rancho Cordova, CA: Landmark Enterprises].

1988d *Bibliography of Map Projections* John P(arr) Snyder & Harry Steward [eds.] [Washington: U.S. Government Printing Office] [USGS Bulletin N°1856].

1989a *An Album of Map Projections* John P(arr) Snyder & Philip M. Voxland [Amsterdam: North Holland] [USGS Professional Paper Nª1453]. For reprint with corrections see 1994a.

1989b "World map on an interrupted Transverse Mercator" Victor Schrader *The American Cartographer* 16:167.

1989c *The World in Perspective – A Directory of World Map Projections* Frank Canters & Hugo Decloir [Chichester, U.K.: Wiley].

1989d "Minimum-error equal-area map projections" John A. Dyer & John P(arr) Snyder *American Cartographer* 16:39-43.

1989e "Area deformation of the Robinson projection" Robert T. Richardson *The American Cartographer* 16:164-6.

1989f "The traditional and modern look at Tissot's Indicatrix" Piotr H. Laskowski *American Cartographer* 16:122-33.

1989g *Elliptic Functions and Applications* Derek F. Lawden [New York: Springer] [Volume 80 in Applied Mathematical Sciences].

1990a *Map Projections: theory and applications* Frederick Pearson II [Boca Raton, FL: CRC Press] A printed edition of the typewritten original [1984a].

1990b "The 'Tilted camera' perspective projection of the Earth" R. E. Deakin *Cartographic J.* 27:7-14.

1991a *The Geography* Cladius Ptolemy [New York: Dover] A re-publication of 1932a.
- See 2000a for explanation of the theoretical chapters.

1992a *Coordinate Systems and Map Projections* D(erek) H(ylton) Maling [Oxford: Pergamon] The 2nd edn. of 1973a.

1992b "Results of the General Adjustment of the North American Vertical Datum of 1988" David B. Zilkoski, John H. Richards & Gary M. Young *Surveying & Land Information Systems* 52:133-49.

1993a *Flattening the Earth: Two Thousand Years of Map Projections* John P(arr) Snyder [Chicago: University of Chicago Press].
- Re Fahey see page 213.

1993b "World Geodetic System 1984: a reference frame for global mapping, charting and geodetic applications" M. Kumar *Surveying and Land Information Systems* 53:53-6.

1993c *Satellite Geodesy: foundations, methods and applications* G. Seeber [Berlin: Walter de Gruyter].

1993d "Update your database – the North American continent has moved" N. T. Olsen *GIS World* 6(9):40-1.

1993e *Mapmakers of the Sixteenth Century and Their Maps* Robert J. Karrow Jr. [Chicago: Speculum Orbis Press].

1994a *An Album of Map Projections* John P(arr) Snyder & Philip M. Voxland [Amsterdam: North Holland] [USGS Professional Paper Nª1453]. A revised printing with corrections of 1989a.

1994b *Map Projections – a working manual* John P(arr) Snyder [Washington: U.S. Government Printing Office] [USGS Professional Paper Nº1395]. A corrected reprint of 1987a.

1995a *Map Projections: A Reference Manual* Lev Bugayevskiy & John P(arr) Snyder [London: Taylor & Francis].

1995b *Longitude: the True Story of a Lone Genius Who Solved the Greatest Scientific Problem of His Time* David Sobel [London: Penguin].

1995c *Elements of Cartography* Arthur H(oward) Robinson, Randall D. Sale, Joel L. Morrison & Phillip C. Muehrcke [New York: John Wiley & Sons] The 6th edn of 1953f.

1996a *Maps and Civilization: Cartography in Culture and Society* N[orman] J(oseph) W(illiam) Thrower [Chicago: Uinversity of Chicago Press] A re-issue of 1972a.

1997a *Linear Algebra with Applications* John T. Scheick [New York: McGraw-Hill].

1997b *Department of Defense World Geodetic System 1984, Its Definition and Relationships With Local Geodetic Systems* [Reston, VA: USGS] 3rd edn. USGS National Imagery and Mapping Agecy Technical Report TR8350.2

1998a *Elsevier's Encyclopedic Dictionary of Measures* Donald Fenna [Amsterdam: Elsevier Science NV].

- See entry for 'mile'.

2000a *Ptolemy's geography; An Annotated translation of the Theoretical Chapters* J. Lennart Berggren & Alexander Jones [Princeton, NJ: Princeton University Press].

2000b *Map Projection Transformation: principles and applications* Qihe H. Yang & John P(arr) Snyder [London and New York: Taylor & Francis].

2001a *Geographic Information Systems and Science* Paul A. Longley, Michael F. Goodchild, David J. Maguire & David W. Rhind [Chichester, U.K.: Wiley].

2002a *Oxford Dictionary of Weights, Measures and Units* Donald Fenna [Oxford: Oxford University Press].

- For a discussion of the metre, see entries for 'metric system' and 'metre'.
- For a discussion of UT, see entry for 'Universal Time'.

2002b http://physics.nist.gov/cuu/Constants/index.html.

2002c "Qibla, and related map projections" Waldo (Rudolph) Tobler *Cartography and Geographic Information Science* 29:17-23.

General Index

- See Appendix A for notes on Greek-based terminology.
- See Appendix B for a glossary of terms.
- See Appendix C for a glossary of symbols.
- See Appendix D for an index to projections by name.
- See list of tutorials following table of contents for mathematical subjects
- Individual cartographers are included in this index in regard to citations separate from projections named for them.

OTHER RELATED TITLES OF INTEREST INCLUDE:

Algorithmic Foundation Multi-Scale Spatial Representation
by Zhilin Li, Hong Kong Polytechnic University, Kowloon
ISBN: 0849390729

Large-scale 3D Data Integration: Challenges and Opportunities
by Sisi Zlatanova, Delft University of Technology, Netherlands and David Prosperi, Florida Atlantic University, Fort Lauderdale, USA
ISBN: 0849398983

Quantitative Methods and Applications in GIS
by Fahui Wang, Northern Illinois University, Dekalb, USA
ISBN: 0849327954